ADVANCES IN MINERAL RESOURCES, GEOTECHNOLOGY AND GEOLOGICAL EXPLORATION

Advances in Mineral Resources, Geotechnology and Geological Exploration focuses on the research of mineral resources, geotechnology and geological exploration. The proceedings features the most cutting-edge research directions and achievements related to geology. Subjects in this proceedings include:

- Materials of geography
- Resource exploration
- Geotechnical engineering
- Rock mechanics and rock engineering

The works of this proceedings can promote development of geology, resource sharing, flexibility and high efficiency. Thereby, promote scientific information interchange between scholars from top universities, research centers and high-tech enterprises working all around the world.

PROCEEDINGS OF THE 7TH INTERNATIONAL CONFERENCE ON MINERAL RESOURCES, GEOTECHNOLOGY AND GEOLOGICAL EXPLORATION (MRGGE 2022), XINING, CHINA, 18–20 MARCH, 2022

Advances in Mineral Resources, Geotechnology and Geological Exploration

Edited by

Ahmad Safuan Bin A Rashid

Department of Geotechnics and Transportation, Faculty of Civil Engineering, Universiti Teknologi Malaysia, Malaysia

Junwen Zhang

School of Energy and Mining Engineering, China University of Mining and Technology, China

CRC Press
Taylor & Francis Group
Boca Raton London New York Leiden

CRC Press is an imprint of the Taylor & Francis Group, an **informa** business

A BALKEMA BOOK

CRC Press/Balkema is an imprint of the Taylor & Francis Group, an informa business

Typeset in Times New Roman by MPS Limited, Chennai, India

Library of Congress Cataloging-in-Publication Data

A catalog record has been requested for this book

First published 2023
Published by: CRC Press / Taylor & Francis Group
4 Park Square, Milton Park / Abingdon, Oxon OX14 4RN / UK
e-mail: enquiries@taylorandfrancis.com
www.routledge.com – www.taylorandfrancis.com

ISBN: 978-1-032-33377-9 (hbk)
ISBN: 978-1-032-33380-9 (pbk)
ISBN: 978-1-003-31941-2 (ebk)
DOI: 10.1201/9781003319412

Advances in Mineral Resources, Geotechnology and Geological Exploration – Rashid & Zhang (eds)
© 2023 Copyright the Editor(s), ISBN 978-1-032-33377-9

Table of contents

Remote sensing modeling measurement technology and numerical analysis

Advances in Mineral Resources, Geotechnology and Geological Exploration – Rashid & Zhang (eds)

Preface

Nowadays, mass gatherings are not permitted by the government. It is uncertain when the COVID-19 would end, so it remains unclear for postponement time, while many scholars and researchers wanted to attend this long-waited conference and have academic exchanges with their peers. Therefore, in order to actively respond the call of the government, and meet author's request, the MRGGE 2022, which was planned to be held in Xining, China from March 18-20, 2022, was changed to be held online through Zoom software. This approach not only reduces people gathering, but also meets their communication needs.

MRGGE 2022 is to bring together innovative academics and industrial experts in the field of Mineral Resources, Geotechnology and Geological Exploration to a common conference. The conference brings together leading researchers, engineers and scientists in the domain of interest from around the world. We warmly welcome previous and prospected authors submit your new research papers to MRGGE 2022, and share the valuable experiences with the scientist and scholars around the world.

During the conference, the conference model was divided into three sessions, including oral presentations, keynote speeches, and online Q&A discussion. In the first part, some scholars, whose submissions were selected as the excellent papers, were given about 5-10 minutes to perform their oral presentations one by one. Then in the second part, keynote speakers were each allocated 30-45 minutes to hold their speeches. There were over 110 participants attended the meeting.

In the second part, we invited four professors as our keynote speakers. The first keynote speaker, Prof. Hong Li, Dalian University of Technology, China. His research interests including Seepage Flows in Engineering Fractured Rocks; Hot Dry Rock Geothermal Extraction (EGS based on Tunnelling); CO2- Enhanced Coalbed Methane Recovery as well as Its Geological Storage. And then we had Assoc. Prof. Jiwei Wen, Shijiazhuang Tiedao University, China. He performed a speech: *Research on Borehole Hydraulic Mining Technology and Matched Key Tools of Underground Solid Resources.* Assoc. Prof. Ahmad Safuan Bin A Rashid, our third keynote speaker, from Universiti Teknologi Malaysia, Malaysia. Lastly, we invited Assoc. Prof. Yu Chen, Sun yat-sen university, China as our finale keynote speaker. Their insightful speeches had triggered heated discussion in the third session of the conference. Every participant praised this conference for disseminating useful and insightful knowledge.

The proceedings of MRGGE 2022 span over 4 topical tracks, that include: Resource Exploration, Geotechnical Engineering, Geography and Geology, Structural engineering and other related fields. All the papers have been through rigorous review and process to meet the requirements of International publication standard.

We would like to acknowledge all of those who have supported MRGGE 2022. Each individual and institutional help were very important for the success of this conference. Especially we would like to thank the organizing committee for their valuable advices in the organization and helpful peer review of the papers. We hope that MRGGE 2022 will be a forum for excellent discussions that will put forward new ideas and promote collaborative researches. We are sure that the proceedings will serve as an important research source of references and the knowledge, which will lead to not only scientific and engineering progress but also other new products and processes.

The Committee of MRGGE 2022

Advances in Mineral Resources, Geotechnology and Geological Exploration – Rashid & Zhang (eds)
© 2023 Copyright the Editor(s), ISBN 978-1-032-33377-9

Committee member

Mineral resources exploration and mining restoration technology

Advances in Mineral Resources, Geotechnology and Geological Exploration – Rashid & Zhang (eds)

Semi-industrial test research on beneficiation of a lead-zinc ore with high oxidation rate in southern Shaanxi

Junjie Wu
Faculty of Land Resource Engineering, Kunming University of Science and Technology, Kunming, Yunnan, China
Shaanxi Institute of Geology and Mineral Resources Experiment Co, Ltd., Xi'an, China

Qi Nie*
Faculty of Mining Engineering, Kunming Metallurgy College, Kunming, Yunnan, China

Huixin Dai*
Faculty of Land Resource Engineering, Kunming University of Science and Technology, Yunnan, Kunming, China

Ping Lu, Lei Ma & Wanhong Li
Faculty of Mining Engineering, Kunming Metallurgy College, Yunnan, Kunming, China

ABSTRACT: In this paper, a semi-industrial experimental study on a lead-zinc ore with a high oxidation rate in southern Shaanxi is carried out. Based on the lead-zinc flotation process flow in which the lead oxide ore and sulfide ore are mixed together, and the zinc oxide ore and sulfide ore are mixed together to float the lead-zinc flotation process, applying the new beneficiation agent, and finally, the obtained lead grade in the lead concentrate is 22.33% under the premise that the raw ore grades of zinc and lead are 2.39% and 0.46%, and the oxidation rates of lead and zinc are 14.77% and 78.54%, respectively. The recovery rate is 48.76%, the zinc grade in the zinc concentrate is 43.61%, and the recovery rate of zinc is 80.44%. It provides reference and guiding significance for the flotation of the same type of high oxidation rate lead-zinc ore.

1 INTRODUCTION

The resource reserve report of a lead-zinc oxide mine in southern Shaanxi shows that the amount of lead-zinc ore is 5,842,232 tons, and the amount of lead metal is 30,768 tons, and the amount of zinc metal is 212,903 tons. The average grade of lead is 2.60%, and the average grade of zinc is 3.67%. In November 2011, the former mineral processing enterprise built and put into operation of a lead-zinc mineral processing plant with a scale of 2,000 tons/day. Since March 2012, the mine has exploited 5 open pits. During the production process, there are objective phenomena such as low ore grade and poor ore body continuity, and the beneficiation recovery rate and concentrate grade of oxidized ore are not ideal. Therefore, the processing plant has not been able to produce at full capacity, and can only recover mixed ore with an oxidation rate of about 20%, and the oxidized ore has not been used for the time being.

The ore dressing process adopts the ordinary flowsheet (Shi et al. 2012; Xu 1995; Zhang & Sun 2008), that is, three-stage one-closed circuit crushing (crushing particle size-12mm)—one-stage closed grinding (grinding fineness-200 mesh accounts for 65%)—priority flotation process (lead selection first—sulfurization Zinc second —Zinc oxide is the last choice). The final grade of zinc concentrate obtained was 30.79%, and the zinc recovery rate was 69.97%.

*Corresponding Authors: 158961326@qq.com, dhxk_must@163.com

DOI 10.1201/9781003319412-1

In order to solve the technical bottleneck of the development and utilization of lead-zinc ore with a high oxidation rate (Jin & Li 2006; Xue 1983; Zhou 2010), this paper aims at the shortcomings of the common technological process of lead-zinc oxide ore, on the basis of adopting new collectors and optimizing the flotation process, the oxidation lead-zinc ore in this mining area is carried out. The semi-industrial test has been carried out, and the ideal mineral processing index has been obtained, so as to realize the efficient development and utilization of mineral resources.

2 PROCESS MINERALOGY RESEARCH

In the semi-industrial test, ore samples for process mineralogy research and semi-industrial test samples were collected from the PD970m, PD1030m, and PD1300m mining development flats. After the raw ore is crushed, screened, mixed, and divided, the properties of the ore are analyzed. The multi-element chemical analysis results of the raw ore are shown in Table 1. The original ore X-ray diffraction analysis results are shown in Table 2. The phase analysis results of the raw ore zinc and lead are shown in Tables 3 and 4.

Table 1. Multi-element chemical analysis results of raw ore.

Element	Pb	Zn	Cu	S	C	TFe	Co	Ni	As
Content	0.41	2.37	0.02	2.15	1.56	1.68	0.0006	0.001	0.036
Element	Sb	K2O	Na2O	CaO	MgO	SiO2	Al2O3	Au*	Ag*
Content	0.03	0.22	0.68	50.71	20.40	10.62	0.72	0.01	1.56

*: measured in g/t

Table 2. Raw ore X-ray diffraction analysis results.

Element	Dolomite	Quartz	Smithsonite	Goethite	Sphalerite
Content	88	8	3	/	1

Table 3. Phase analysis of zinc (mass%).

Phase	Zinc oxide	Zinc sulfide	Bound zinc	Total
Content	1.543	0.544	0.452	2.54
Distribution	60.75	21.46	17.79	100

Table 4. Phase analysis of lead (mass%).

Phase	Lead oxide	Lead sulfide	Bound lead	Total
Content	0.065	0.351	0.024	0.44
Distribution	14.77	79.77	5.46	100.00

The ore type is mainly breccia dolomite, and the weathering phenomenon is obvious. The components in the ore for beneficiation and recovery are lead and zinc, whose grades are 0.41% and 2.37% respectively. The occurrence state of zinc in the ore is relatively simple, mainly in the form of smithsonite and sphalerite, and the oxidation rate of zinc is 78.54%; the occurrence state of lead in the ore is relatively simple, mainly in the form of galena. The oxidation of lead is relatively simple, and the rate is 14.77%. Gangue minerals are mainly dolomite and quartz, with a content of 88% and 8%, respectively. Black pitch generally appears in the ore, and there is the phenomenon of reconstructing other minerals. The pitch is the dense block, fine-grained aggregate, close to cryptocrystalline. Because most of them exist in the cracks, the boundary is straight and there are many shrinkage cracks.

It can be seen from the above data that the low grade of lead, the high oxidation rate of zinc, the severe weathering of the ore, and the black asphalt contained in the gangue minerals are the main factors causing the difficulty in beneficiation.

3 LABORATORY VALIDATION TEST STUDY

The laboratory verification test is carried out with the ore samples produced on-site, in order to obtain stable and good test indicators. The principle flow of the laboratory verification test process is shown in Figure 1, and the closed-circuit process test results of the laboratory verification test are shown in Table 5.

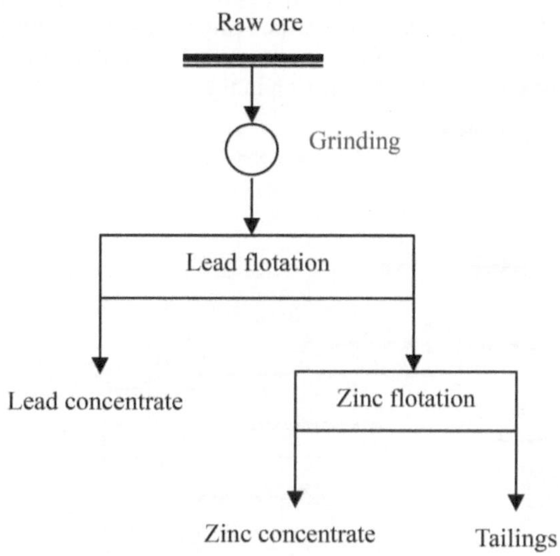

Figure 1. The principle flow of the laboratory verification test process.

Table 5. Laboratory verification test closed-circuit process test results.

		Grade (%)		Recovery (%)	
Product	Yield (%)	Pb	Zn	Pb	Zn
Lead concentrate	1.02	25.13	4.26	58.89	1.79
Zinc concentrate	4.56	1.44	44.32	15.08	83.40
Tailings	94.42	0.12	0.38	26.03	14.81
Total	100.00	0.44	2.42	100.00	100.00

The closed-circuit process test results of the laboratory verification test show that the grade and recovery rate of lead and zinc concentrates are 25.13% and 44.32%, respectively, and the recovery rates of lead and zinc concentrates are 58.89% and 83.40%, respectively. The lead-zinc oxide ore can be fully recovered under this process flow, and it is reasonable and feasible to carry out semi-industrial tests based on this test process.

The reasons for the low grade of lead concentrate are the low grade of lead (only 0.44%) and the presence of organic asphalt in the ore, and the organic asphalt is easily ground during the ore crushing and grinding process. In the process of beneficiation, it is very easy to adsorb with air bubbles and enter the products. Although experimental research on the factors affecting the elimination of asphalt has been carried out, it is difficult to eliminate them (Yao et al. 2015, Beijing: Metallurgical Industry Press 2007). Therefore, lead concentrates that meet the quality standards cannot be obtained.

4 SEMI-INDUSTRIAL TEST RESEARCH

4.1 *The process and results of the semi-industrial test*

The semi-industrial test is based on the chemical system and process flow determined by the laboratory verification test. The biggest innovation of the mineral processing chemical is the application of the lead collector and foaming agent SDT-1 with the authorization of the invention patent (the patent number is ZL201510219679.8), Zinc oxide collector SDP-6 (patent number ZL201510218669.2) and foaming agent SDQ-4. Lead oxide ore and sulfide ore mixed floating, zinc oxide ore and sulfide ore mixed floating lead- Zinc are used, in turn, which prioritizes the flotation process. The process flow of the semi-industrial beneficiation test is shown in Figure 2, and the cumulative weighted average beneficiation index after stable production in the semi-industrial test is shown in Table 6.

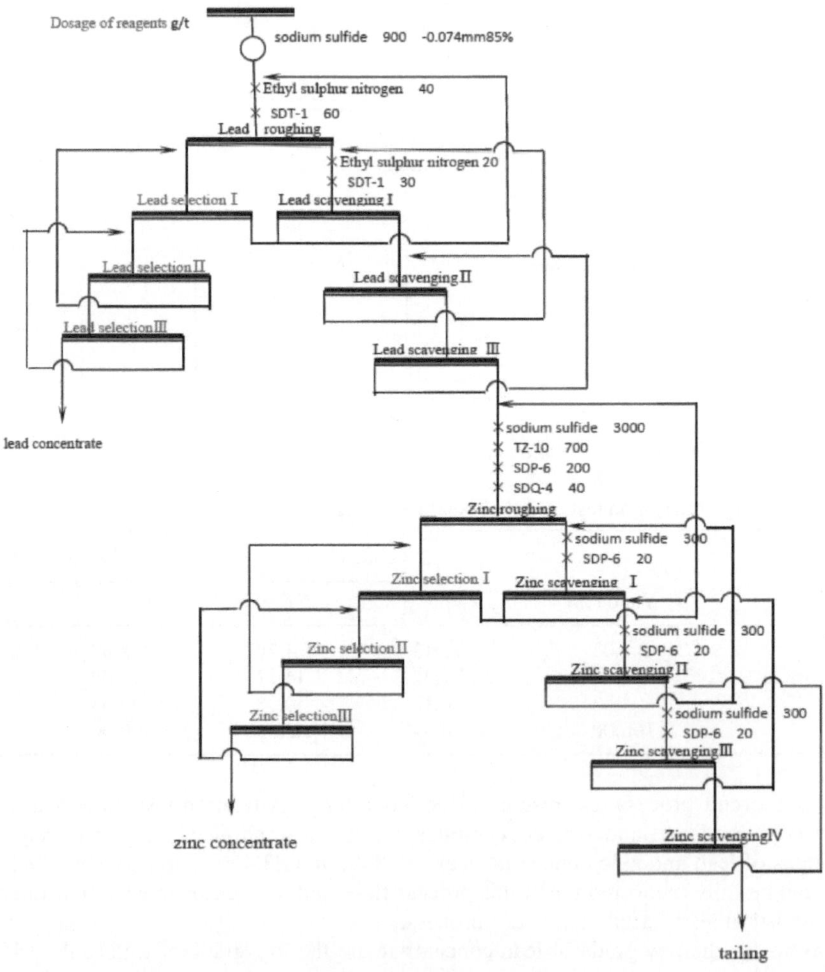

Figure 2. The process flow of the semi-industrial beneficiation test.

Table 6. Cumulative weighted average beneficiation index after stable production in semi-industrial trials.

		Grade (%)		Recovery (%)	
Product	Yield (%)	Pb	Zn	Pb	Zn
Lead concentrate	0.99	22.33	1.36	48.76	0.57
Zinc concentrate	4.41	1.43	43.61	13.85	80.44
Tailings	94.60	0.18	0.48	37.39	18.99
Total	100.00	0.46	2.39	100.00	100.00

4.2 *Evaluation of semi-industrial test results*

The semi-industrial beneficiation process is that the test sample enters the mixing tank after crushing, grinding, and grading. The priority is given to floating lead. The lead selection process is one rough selection, three selections, and three sweeps. The lead flotation tailings enter the zinc series mixing tank and are sorted according to the process flow of one coarse, three refining, and four sweeping. Finally, on the premise that the raw ore grades of zinc and lead are 2.39% and 0.46%, and the oxidation rates of lead and zinc are 14.77% and 78.54%, respectively, the lead grade in the lead concentrate is 22.33%, and the zinc content is 1.36%. The recovery rate is 48.76%, the zinc grade in the zinc concentrate is 43.61%, the lead content is 1.43%, and the zinc recovery rate is 80.44% of the beneficiation index.

4.3 *Economic benefit estimate*

During the semi-industrial test and commissioning period, the processing capacity of a lead-zinc oxide mine in southern Shaanxi was 100t/d, the average lead grade of the raw ore was 0.46%, and the zinc grade was 2.39%. Compared with the actual production in 2012, the lead concentrate grade and recovery rate were similar. In this case, the zinc grade is increased by nearly 13 percentage points, and the zinc recovery rate is increased by more than 10 percentage points. If other factors such as costs are not taken into account, only the economic benefits increased by estimating the improvement of the beneficiation index can be seen: the zinc concentrate is calculated at 16,900 yuan/ton and the discount coefficient is calculated at 70%: 100t/d*12.82%*10.47%*70%* 16,900 yuan/ton * 300d $\approx$ 476,365,900 yuan/year.

5 CONCLUSIONS

A semi-industrial beneficiation test was carried out for a lead-zinc ore with a high oxidation rate in southern Shaanxi. Based on the flotation process of lead-zinc mixed flotation of lead oxide ore and sulfide ore, and lead-zinc flotation of zinc oxide ore and sulfide ore, the lead collector and foaming agent SDT-1 (patent number ZL201510219679.8), zinc oxide collector SDP-6 (patent number ZL201510218669.2) and foaming agent SDQ- 4 are applied. The use of the new agent not only greatly simplifies the beneficiation process, but also greatly improves the beneficiation index. The sluggish lead-zinc oxide ore resources can be fully utilized, and the service life of the mine can be extended. The research results can promote the industrialization of lead-zinc oxide ore and bring good economic and social benefits to the development of the enterprise.

ACKNOWLEDGMENTS

1. This work was financially supported by Shaanxi Natural Resources Fund Project No. 2019jm-395.

2. This work was financially supported by Yunnan Provincial Department of Education Scientific Research Fund Project No. 2021J0945.
3. This work was financially supported by Kunming Metallurgical College Research Fund Project No. 2020XJZK06

REFERENCES

Wang Zi. Flotation mineral processing technology [M]. Beijing: Metallurgical Industry Press, 2007, 22–24.

Jin Mingshui, Li Fushou. Research on beneficiation technology of a lead-zinc mine in Inner Mongolia. Non-ferrous mining and metallurgy [J]. 2006.12, Vol. 22, No. 6, 19–21.

Shi Defeng, Zhang Shugen, Han Shili, et al. Metallogenic system analysis of Dingjiashan lead-zinc deposit in Meixian, Fujian [J]. Chinese Journal of Nonferrous Metals, 2012, 22(3):10.

Xu Shi. Research on ore selectivity [M]. Beijing: Metallurgical Industry Press, 1995, 169–175.

Xue Wenya. Flotation [M]. Northeast Institute of Technology, 1983, 23–27.

Yao Yingqiang, Wang Zhigang, Xu Wei, et al. Industrial test research on beneficiation of high-sulfur lead-zinc ore [J]. Nonferrous Metals (Beneficiation Part), 2015(3): 24–26.

Zhang Jinghui, Sun Lijun. Research on beneficiation technology of refractory lead-zinc oxide ore. Mining Engineering [J]. August 2008, Vol. 6, No. 4, 38–42.

Zhou Jing. Experimental research on non-toxic and high-efficiency beneficiation agents for refractory lead-zinc-sulfur ore [J]. Nonferrous Metals (Beneficiation Part), 2010(4): 43–48.

Advances in Mineral Resources, Geotechnology and Geological Exploration – Rashid & Zhang (eds)

Application progress of reversed-phase emulsion polymerization system in oil and gas field development

Ke Xu*, Yang Shi, Jin Chang & Yang Li
PetroChina Research Institute of Petroleum Exploration & Development, Beijing, China

ABSTRACT: Reviewing the development status of the reversed-phase emulsion polymerization system used in the field of oil and gas field development, this paper discusses the polymerization mechanism of reversed-phase emulsion in depth, analyzes five important factors affecting the polymerization process of reversed-phase emulsion, and elaborates on the application of reversed-phase emulsion polymerization system in hydraulic fracturing, acidification acid pressure, water regulation and blockage, tertiary oil recovery, drilling, and oilfield wastewater sewage treatment process. In addition, the future development direction of the reversed-phase emulsion polymerization system is predicted according to the complex process of the reversed-phase emulsion polymerization system. There are many influencing factors, etc., which point out the direction of the next step of development.

1 INTRODUCTION

Reversed-phase emulsion polymerization technology was first proposed by Vanderhoff in 1962 (Vanderhoff 1962), the traditional emulsion is made of lipophilic monomer as the internal phase, water as the outer phase, the addition of an emulsifier to make the two phases mixed to form a water-in-oil (O/W) type emulsion, and the structure of the monomer droplets and micelles is water-in-oil. The composition structure of the reversed-phase emulsion is opposite, with the monomeric aqueous solution as the internal phase and the oil phase as the outer phase, and the emulsifier is added to form an emulsion, which is a water-in-oil (W/O) type structural system, and the monomer droplets and micelles are also in the form of oil-in-water (Liang et al. 2019). The polymerization rate of reversed-phase emulsion is fast, the heat dissipation is fast, and the molecular weight of the product is high.

Reversed-phase emulsion polymerization can prepare high molecular weight hydrophilic polymers, and this viscoelastic system has been widely used in the field of oilfield gas development. Reversed-phase emulsion polymerization of large-scale industrial applications of the product is mainly polyacrylamide and its derivatives, the molecular weight can reach more than 20 million, its products have powder form, there is also high solid content of sol form, the advantage of the latter is easy to disperse and rapid dissolution. The ionic forms of the product are anionic, cationic, nonionic, and zwitterionic.

2 POLYMERIZATION MECHANISM OF REVERSED-PHASE EMULSION

Reversed-phase emulsion polymerization is opposite to conventional emulsions in the composition of the continuous phase and dispersed phase, but in terms of reaction mechanism, both are free

*Corresponding Author: xuke69@petrochina.com.cn

DOI 10.1201/9781003319412-2

radical reactions, triggered by free radicals generated by initiators, chain radicals through new monomer radicals to prolong growth and form polymer chains, in this case, it will be accompanied by chain transfer to form branched chains and reduce the relative molecular mass (Figure 1). The droplet and micellar nucleation mechanism is the polymerization mechanism of reversed-phase emulsion with a relatively high degree of recognition at this stage, and the polymerization process has four main stages (Du et al. 2016; 2017; Liang et al. 2017).

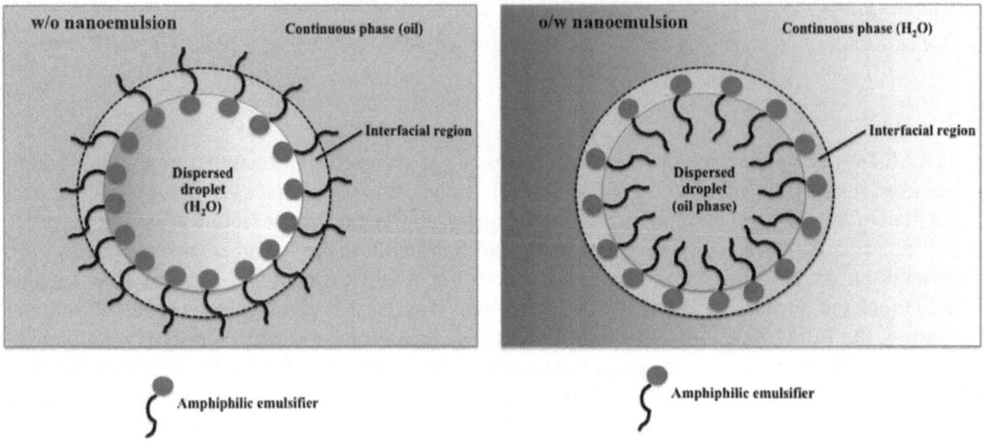

Figure 1. Diagram of a reversed-phase emulsion polymerization system.

2.1 *Dispersion stage*

In the emulsification system, the emulsifier is dissolved in water in a single-molecule form, and after exceeding the critical micellar concentration (CMC), the amount of emulsifier continues to increase, which is manifested in the form of micelles in solution. Micelles can increase the solubility of monomers, and solubilized micelles are formed when monomers dissolve in micelles (Zhai 2012). The stirring effect can break up the monomers in the solution to form bead droplets, and the surface of the bead droplets forms a molecular layer covered with emulsifiers. The hydrophilic end of the emulsifier in the molecular layer is the water phase, so as to maintain the presence of the bead droplets in the aqueous phase.

2.2 *Latex granule generation stage*

The number of monomer beads in the system is small, and the number of micelles is large, so the free radicals will preferentially enter the micelles for a reaction after decomposition. Free radicals trigger reactions in the micelle to form polymer chains, causing the micelles to swell into emulsion colloidal particles, that is, latex particles at the reaction site (Carman Paul et al. 2014; Liu et al. 2018; Yang et al. 2018).

In the process of nucleation, a large number of latex particles are formed, the surface is attached to the emulsifier, a large number of latex particles adsorb the emulsifier of the aqueous phase on the surface, and the surface of the micelle loses the emulsifiers and ruptures. When the nucleation is consumed, the number of micelles begins to decrease and finally disappears.

2.3 *Latex granules growth stage*

At this stage, the initiator continues to release free radicals, the number of latex particles is much more than the number of monomer beads, free radicals first enter the latex particles to consume the monomer for reaction, the latex particle volume increases and the monomer decreases, and the monomer in the monomer bead drops is also consumed and gradually disappears. This process is the growth stage of latex particles, including the disappearance of micelles and the disappearance of monomer bead droplets.

2.4 *Aggregation completion stage*

Entering the polymerization completion stage, the system no longer contains micelles and monomer beads. The conversion rate continues to increase, and when a certain value is reached, the conversion rate no longer increases, and the rate becomes zero, which is called the vitrification effect (Wang 2004). The polymer concentration continues to increase, increasing the glassing temperature of the system. When this temperature reaches the reaction temperature, the latex particles are glassed, the internal polymer chain and the remaining monomers no longer react, the chain growth rate decreases and eventually becomes zero, and the reaction process is terminated.

3 APPLICATION OF REVERSED-PHASE EMULSION POLYMERIZATION SYSTEM IN THE FIELD OF OIL AND GAS DEVELOPMENT

Polymer products prepared by a reversed-phase emulsion polymerization system have the characteristics of high molecular weight and rapid dissolution, and reversed-phase emulsion-type polyacrylamide has the effects of increasing viscosity and thickening, producing flocculation, etc., and has been widely used in the fields of petroleum fracturing fluid, regulation, and dissection, drilling fluid, acid pressure acidification, oilfield wastewater treatment, and secondary and tertiary oil recovery.

Tan et al. (Tan 2020) researched three salt-resistant polymer resistive inhibitors, AM/AA/AMPS (PHWAM-1), PHWAM-2, and PHWAM-3, all of which contain special hydrophobic association groups (C12AM and DiC12AM, respectively), and they were synthesized by reversed-phase emulsion polymerization. PHWAM-1 can achieve high drag reduction performance in freshwater and in brine. The stronger the hydrophobic association effect, the more obvious the drag reduction effect. When the inhibitor is used at a concentration of 0.15wt%, the resistance reduction rates in 1.0×105mg/L NaCl solution are 61.24%, 65.38%, and 68.44%, respectively. Ca2+ of 15000 mg/L was added to the 1.0×105 mg/L NaCl solution, and the resistance reduction rates were 41.12%, 53.41%, and 57.35%, respectively, and mg2+ of 3000 mg/L was added to the 1.0×105 mg/L NaCl solution, and the resistance reduction rate was maintained at 45.32%, 49.58%, and 58.24%, respectively, which had excellent resistance reduction and salt resistance.

Gao et al. (Gao 2017) synthesized nano-scale polyacrylamide microspheres by reversed-phase microemulsion polymerization method, and the particle size of the microspheres ranged from 50–250 nm, and the decomposition began when the temperature resistance reached more than 300°C, which could meet the use environment of the reservoir. When the mineralization degree of the solution is 100,000 mg/L, the equilibrium state can be reached in 10 days, and the microspheres expand by about 2.66 times. This nano-microsphere solution is mainly elastic and good, and the modulus of elasticity and viscosity can reach 10Pa, which meets the conditions of reservoir profiling.

Chang et al. (Chang 2003) used reversed-phase emulsion polymerization to obtain a diameter of $0.05 \sim 1\ \mu m$ of the double crosslinked microspheres, a crosslinking agent using polyethylene glycol 200 bis acrylate (PEG-20ODA), and N, N′-methylene bisacrylamide. Microspheres can achieve a better delayed swelling effect, and can slowly distend for more than 98 days.

Sun et al. (Sun 2018) prepared a double cross-linked polyacrylamide microsphere by reversed-phase emulsion polymerization method, and the crosslinking agent used polyethylene glycol diacrylate (PEGDA) and N, N-methylene bisacrylamide, for which the particle size was uniform, distributed between 10 and 20 μm, and the dispersibility was good. The double crosslinked polyacrylamide microspheres have good swelling performance, and the effect of delaying swelling can be achieved in high mineralization environments; the three-dimensional mesh structure of the microspheres is obtained by crosslinking, the shear resistance is better, and the formation with high moisture content can be effectively blocked after injection.

Li et al. (Li 2019) used the reversed-phase emulsion polymerization method. The self-made long-chain hydrophobic monomer (acrylic long-chain ester, WLHM) synthesized the emulsion type drag reducer WDRA-M, using 0.7g/L of emulsion drag reducer, when the flow rate is 32 m/s, WDRA-M drag reduction rate can exceed 60%, which is higher than the foreign similar products 57.3% drag reduction rate. At a higher flow rate, the resistance reduction rate of foreign products declines rapidly, and WDRA-M still maintains a relatively stable drag reduction rate at high flow rates.

Ma et al. (Ma 2019) synthesized the new drag reducer TF through microemulsion polymerization, and the resistance reduction rate reached 61.7%. The salt resistance is good, and the resistance of TF slip water with an amount of 0.7 g/L is maintained at more than 55% under high mineralization conditions.

He et al. (He 2013) prepared an emulsion-type drag reducing agent FA30, which has good instant performance, and has been tested and applied in shale gas wells in southwest China. The results show that the resistance reduction rate is high when the displacement is 16 m3/min, which can exceed 80%.

Wang et al. (Wang 2013) prepared a tetrasocial hydropolytic polyacrylamide AAMS-1 by reversed-phase emulsion polymerization. The main monomers are AM, AMPS, dimethylhexadecylaminoethanol methacrylate chloride, and sodium 4-acrylamide benzene sulfonate. The critical binding concentration of the polymer is 0.15%, formulated into a 0.6% aqueous polymer solution, and the viscosity can be maintained above 50 mPa·s after shearing at 150°C.

4 CONCLUSION

The stability of the reversed-phase emulsion determines the stability of the reaction and storage, affected by some factors, such as oil-water ratio, emulsifier type and content, stirring conditions, preemulsification temperature, etc. These factors will affect the stability time of product emulsion storage, thus a lot of experiments and evaluations are still needed to obtain more stable emulsification conditions so that the polymerization reaction process remains stable and the time of stable storage of products is prolonged.

There are many influencing factors in the polymerization process of reversed-phase emulsion, and the process is more complex, such as reaction temperature and pH value, and monomer concentration. The proportion of different monomers, the type and amount of compound initiator, the amount of use, slight differences, etc., will affect the effect of polymerization and product molecular weight, etc. There is still in need of a lot of research on the preparation of high solids content, high conversion rate, and high molecular weight of the reversed-phase emulsion, so that the product performance and efficiency are significantly improved.

The amount of oil phase and emulsifier used in reversed-phase emulsion polymerization is larger, the polymerization cost is larger, and the non-ionic emulsifier is mainly dispersed through the space thrombotic resistance system to achieve the emulsification effect so that a smaller and more efficient emulsifier can be found. How to recover the oil phase to achieve the goal of recyclability and find an efficient new emulsifier is also an urgent problem for researchers to solve.

ACKNOWLEDGMENTS

This work was supported by the Scientific research and technology development projects (CNPC): Study of Ultra-high temperature cleaning fracturing fluid and Variable viscosity of slickwater (No. 2020B-4120).

REFERENCES

Carman Paul S. & Gupta D. V. Satyanarayanan. Method of fracturing with phenothiazine stabilizer [P], US8691734.2014-04-08

Chang K T, Frampton H, Morgan J C, et al. Method of recovering hydrocarbon fluids from a subterrabean reservoir: US, 20030155122[P]. 2003-04-15.

Du Tao, Yao Yiming, Jiang Tingxue, et al. The latest research and application progress of synthetic polymer fracturing fluids[J] Advances in Fine Petrochemicals. 2016,17(01):1–5.

Du Tao, Yao Yiming, Jiang Tingxue, et al. A reversible physical crosslinking agent and its preparation method [P]. CN106318366A, 2017-01-11

Gao Y S & Li Y P. Research and application of profile control and water shutoff technology in high water cut oil wells[J]. China Petrochem, 2017(8): 94–95.

He Youhao. Preparation of polymer thickener and research on rheological properties of ultra-high temperature fracturing fluid system [D]. East China University of Science and Technology, 2013

Liang Feng, Reddy B. Raghava, Li Leiming, et al. High-Temperature Crosslinked Fracturing Fluids[P], US2019093005.2019-03-28

Liang Feng, Reddy B. Raghava, Li Leiming, et al. High-Temperature Crosslinked Fracturing Fluids[P], WO2017096055. 2017-06-08

Liu Ping, Guan Baoshan, Xu Minjie, et al. Performance of 220°C ultra-high temperature polymer fracturing fluid [J]. Petrochemical Industry Application, 2018, 37(08): 12–15.

Li Leiming & Liang Feng. Enhanced high temperature crosslinked fracturings [P], WO2019046161.2019-03-07

Ma Yingxian, Ma Leyao, Guo Jianchun, et al. A temperature-resistance and salt-resistance fracturing fluid thickener with a double network structure and its preparation method [P]. Sichuan Province: CN109438618A, 2019-03-08

Sun L, Zhang G Q, Xia Y, et al. Research on double-crosslinked polymer microspheres profile control agent[J]. Tianjin Science and Technology, 2018, 45(9): 45–47.

Tan H Z, Mao J C, Zhang W L, et al. Drag Reduction Performance and Mechanism of Hydrophobic Polymers in Fresh Water and Brine[J]. Polymers, 2020, 12(4): 955.

Vanderhoff J W. Inverse emulsion polymerization[J]. Advances in Chemistry Sciences, 1962, 34:32-51.

Wang Zhonghua. Overview of the research and application of new oilfield chemicals in my country [C]. National Industrial Surfactant Center. The 15th National Industrial Surfactant Development Seminar and the 13th National Industrial Surfactant Development Symposium Proceedings. 2004:203-217

Wang Yongji. Development of a high-temperature resistant fracturing fluid thickener[D]. Southwest Petroleum University, 2017

Yang Zhenzhou, Liu Fuchen, Song Lulu, et al. A new ultra-high temperature fracturing fluid system with a temperature resistance of 230°C [J]. Drilling Fluids and Completion Fluids, 2018, 35(01): 101-104

Zhai Wen. A polymer fracturing fluid system resistant to 200°C ultra-high temperature [C]. Chinese Chemical Society·Chinese Society of Mechanics Rheology Professional Committee. Rheology Progress (2012)-Eleventh Proceedings of the National Conference on Rheology. Chinese Chemical Society·Chinese Society of Mechanics Rheology Professional Committee: Chinese Society of Mechanics, 2012: 286–290.

Advances in Mineral Resources, Geotechnology and Geological Exploration – Rashid & Zhang (eds)

Review of simultaneous hydraulic fracturing technology and field application

Xi Wu*, Lingyu Mu, Kun Ning, Yan Qiao, Guohai Yuan, Litao Shang & Shiling Zhang
CNPC Engineering and Technology R&D Company Limited, Beijing, China
National Engineering Laboratory of Oil and Gas Drilling Technology, Beijing, China

ABSTRACT: Over the past decade, several technologies and completion designs have been developed to make hydraulic fracturing more efficient and economical, from multistage systems for single horizontal wells to complex "zipper fracture" designs. As the staggered zipper frac became more and more prevalent, new challenges caused by the oil price drop required the operators to step up the game again. To increase the efficiency of fracturing operations and improve the financial and environmental impact of hydraulic fracturing on the participating companies furthermore, simultaneous fracturing completion was introduced to the game. With the advances achieved in equipment and auto-control systems, nowadays simul-frac technique has surprised the industry with its impressive efficiencies and impact on reducing greenhouse gas emissions. This paper is aiming to illustrate the details of simultaneous fracturing technology. Case studies in North America are also provided to illustrate the benefits of this new technology in unconventional reservoir development. The results are presented to demonstrate the efficiency improvements in a fracturing operation, which possibly turns a new page of fossil oil development undergoing with Global Energy Transition, and hopefully inspires other operators to make further innovations.

1 INTRODUCTION

The simultaneous fracturing technique has recently become one of the hottest topics among the shale gas operators in North America, because of its impressive benefit in budget. In the past decade, along with the blooming of unconventional development, hydraulic fracturing plays the main role as it critically impacts the production capacity of unconventional wells. Hence, to achieve better returns, operators tried to pump as much fluid and proppant down to the formation as possible, and recently, as fast as possible facing the oil price drop. Given this background, the "simul-frac" concept was born with the expectation of taking operational efficiency and budget-saving to another level. The objective of this project is to systematically review this novel technology and identify the challenges and benefits brought by the simul-frac technique. Analyzing the field implementations and results could further help engineers to decide if the technique is technically applicable.

2 SIMULTANEOUS FRACTURING TECHNIQUE

Simultaneous fracturing was firstly proposed by Continental and experimented with in the Woodford shale block as early as 2008 years at the beginning of the first shale revolution (Mutalik et al. 2008; Waters et al. 2009), with simultaneous modification of two adjacent wells using two individual sets of equipment with a pump rate at 80 bpm (12.7 m^3/min). Each well had 5 to 7 stages, for a total of four wells, with each stage injecting 1600 m^3 of slickwater, 21.5 m^3 of one hundred

*Corresponding Author: wuxidri@cnpc.com.cn

DOI 10.1201/9781003319412-3

mesh sand, and 57.5 m^3 of 30/50 mesh sand per stage. The overall stimulation performance of this experimental platform was fairly good compared to the non-simultaneously fractured wells (i.e., single well fracturing wells), but possibly due to the higher cost of early simultaneous fracturing when compared to single well fracturing, the use of this technology was aimed at creating more complex fractures, and therefore the technology was not widely promoted.

Until the oil price dramatically falls in 2015 and 2018, the major oil companies, under the economic pressure of the oil price's harsh decline, improved their operational efficiency by using large platform zipper operations which minimize the operation cost. However, with the outbreak of the COVID-19 epidemic in 2020, the upstream operators are once again facing the even more crucial challenges of cost control. In such a context, simultaneous fracturing technology, which can further improve efficiency, has become a new solution.

2.1 *Nowadays simul-frac technique*

Unlike earlier simultaneous fracturing techniques, the new simultaneous fracturing technique uses one set of equipment to simultaneously treat two wells. According to QEC, simultaneous fracturing projects are completed 45% faster than zipper fracturing projects (Russell et al. 2021), which allows for achieving higher yields without significantly compromising fracturing performance. Although the simultaneous fracturing technique requires more hydraulic horsepower, it can significantly reduce diesel consumption compared to traditional zipper fracturing, thereby reducing the cost of diesel delivery and realizing environmental benefits through reduced greenhouse gas emissions.

To maximize the efficiency of simultaneous fracturing, the technique is generally implemented on a pad with an even number of wells and above four. In this case, simultaneous fracturing continuously moves forward on all the wells (mostly four of them), which eliminates the idle time of the remaining wells caused by conventional zipper operations that can only work two wells at a time. In a simultaneous fracturing operation, two of the wells are fractured and pumped while the other two wells are simultaneously working on the perforation and setting plugs. The simultaneous fracturing technique enables more horizontal wells to be completed in the same amount of time as the current zipper fracturing operation.

To accomplish such a heavy-duty job, there are some prerequisites for job preparation.

In terms of equipment, as shown in Figure 1, simultaneous fracturing requires a significant amount of hydraulic horsepower provided by a fracturing pump (each pump truck has a separate flow meter to ensure equal pump rates of both wells), specialized manifolds, and a surface control system (pressure and flow sensors for each well and corresponding software). Because pump groups are physically separated from each other, each well can be treated at different rates and pressures (as Figure 2 shown).

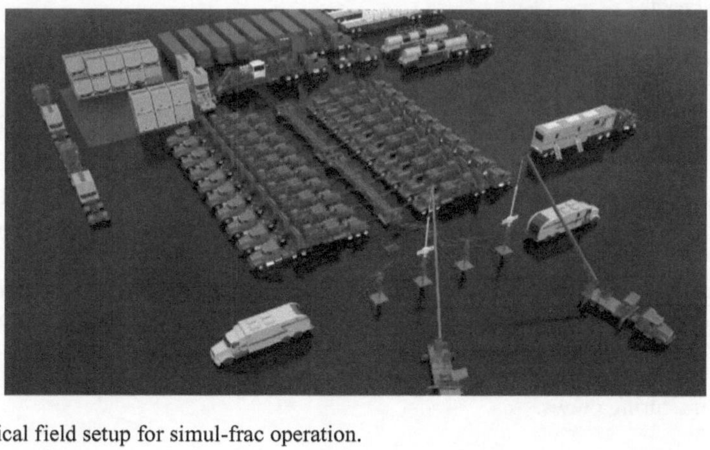

Figure 1. Typical field setup for simul-frac operation.

In terms of fracture design, the same proppant design is required for a two-well simul-frac because the same blender is used in both wells. During operation, if for any reason it is necessary to change the proppant concentration on one side, the other side shall be changed simultaneously. If there are indications that one of the wells is leaking, both wells need to be flushed and restarted.

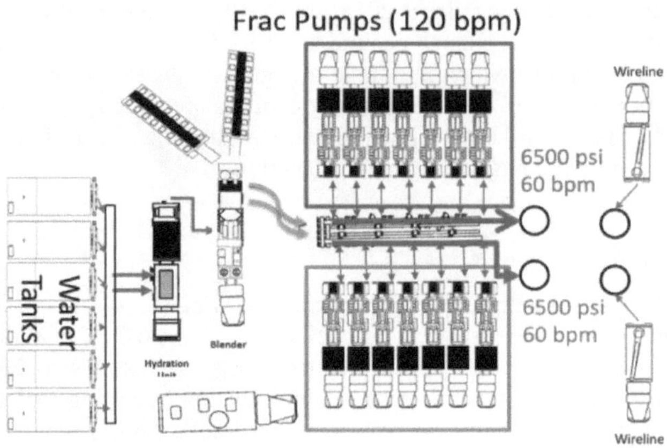

Figure 2. Simul-Frac schematic diagram.

2.2 *Simultaneous fracturing process features and challenges*

2.2.1 *Operation parameters*

For any particular field/formation, achieving the best completion treatment design often requires extensive modelling and several diagnostic methods for calibration and/or years of trial and optimization. The Simul-Frac technology is compatible with maintaining optimal treatment design, which can mostly follow the design philosophy and make adjustments to accommodate the slight rate decrease due to the split flow by taking into account objectives, economics, pad size, well spacing, and surface equipment constraints (Table 1).

Table 1. Operation parameters.

	Traditional (zipper)	Simultaneous fracturing	
Stage Length	30–60 m	60–75 m	
Number of clusters	6–10	8–12	
Pump Rate	10–16 m^3/min	22–25 m^3/min	11–12 m^3/min (Per well)
Proppant Intensity	2 m^3/m	1.7–2 m^3/m	
Fluid Intensity	25–30 m^3/m	17–26 m^3/m	

Reduced single-well pumping rates result in longer operation times for a single well compared to zipper fracturing, but simultaneous fracturing allows for two wells to be constructed at a time, reducing operation time by 25% overall.

The increased complexity of simultaneous fracturing increases the cost (equipment and extra wireline crew), but the overall pumping time and less fuel consumption reduce the overall cost.

In general, the simultaneous fracturing process provides significant increases in operational efficiency and thus a reduction in overall operating costs. However, the process also poses several challenges for fracturing crews.

2.2.2 *Challenges*

Several operational challenges need to be considered and plan ahead to successfully perform a simultaneous hydraulic fracture operation.

(1) Platforms with an odd number of wells and multiple designs create difficulties for simultaneous fracturing in each segment, requiring advanced planning and contingency strategies.
(2) There is a need for special manifolds to enable simultaneous fracturing.
(3) The instantaneous discharge requirement is increased from 100 bpm (16 cfm) to 140 bpm (22 cfm), and the pumps of the fleet need to be doubled to account for hydraulic horsepower (or replaced with the same number of high hydraulic horsepower pump, such as 5000HHP electronic frac skid.
(4) The high consumption of sand and liquid volumes on a single day (20,000 m^3 of liquid and 2,000 m^3 of sand on a single day) poses an extremely demanding logistical supply challenge.
(5) The increased workload on-site due to simultaneous wireline and fracturing activities required additional personnel for managing platform operations.

2.3 *Field application*

2.3.1 *Case study*

QEP has successfully applied simultaneous fracturing technology in several basins and continues to develop it today. Several years of successful operation have resulted in a 45% increase in efficiency over conventional completion methods. While early operation required two full fleets of personnel and equipment, and improvements to the split flow operation in the last two years have resulted in a significant increase in operation time and efficiency, as well as a 6.4% operational cost savings.

The first well completed at the simultaneous operation (in February 2017) was located in the Bakken field in North Dakota. The two wells, named FH 1 and FH 2, were both using sliding sleeve completions. In this case, simultaneous hydraulic fracturing was particularly beneficial because the inner diameter of the sliding sleeve system limited the surface fracturing rate. Since surface equipment can provide higher treatment rates than downhole, the potential efficiency gains of multi-well treatments are significant. This case study compares two skip-frequency wells with different completion techniques. Wells FH 1 and FH 2 were completed simultaneously, and wells FH 3 and FH 4 were completed separately. Since both wells use the same sliding sleeves completion systems, the fracture stage lengths and pumping rate limits are also the same.

To compare the different completion methods (single well, sliding sleeve, and zipper completions) and simultaneous fracturing, the overall efficiency was summarized by comparing the total daily pumping volume (Slurry volume per day). This measure of hydraulic efficiency accounts for many of the smaller operational metrics that might influence the operation and allows for the focus on the major differences in completion techniques.

According to the results of these four wells (as shown in Table 2), simultaneous completions of FH wells were much more efficient. The total daily production was almost twice as high as the wells completed individually. The two wells were completed in a total of 6 days, while the single well took 5 days per well (10 days total).

Table 2. Results of FH1, FH2, FH3, and FH4.

Well Name	Pump Rate (bpm)	Stage Length (ft)	Fluid Volume (bbl)	Prop Vol. (1000 lb)	Ops Days	Slurry per Day (bbl)
FH 1	35	200	126,838	9,945	3	42,279
FH 2	35	200	127,376	10,100	3	42,459
FH 3	35	200	120,071	9,584	5	24,014
FH 4	35	200	118,358	9,333	5	23,671

2.3.2 *Other implementations*

The Simul-Frac technology was also implemented by other major operators. Chevron presented its experimental results at the Unconventional Resources Technology Conference in 2021 (URTec 5340). Chevron piloted simultaneous fracturing as a cost and operation time reduction improvement program in early 2020, reducing total operation time by 29 percent and saving 3 percent in the first test.

According to Rystad Energy, another operator reported that the simultaneous fracturing program was completed 60 percent faster than the zipper fracturing program, noting that the improved efficiency reduced the average well cost by $400,000. In the four wells, the operator completed a total of 92 frac stages, averaging 14.2 stages per day, almost twice as many as the platform where zipper fracturing was performed (7.44 stages per day). Average lateral feet per day also increased by 46 percent, from 2,160 ft/day to 3,162 ft/day. Site operation time was reduced by 12 days, significantly controlling the cost of renting equipment on site.

3 WAY FORWARD

Simul-frac completion design provided a host of benefits. It helps to push for greater efficiency in fracturing operations and improve the financial and environmental impact of hydraulic fracturing with the various companies involved in a project.

As the oil and gas industry continues to work on energy transformation, to achieve the ESG, new ideas and innovation will be vital. By developing better equipment layout plans and helping operators implement simulated frac techniques, the industry is taking a step forward in ensuring that the transition is smooth and meeting the 2050 ESG goal.

4 CONCLUSION

This paper illustrates the simultaneous fracturing technology in detail and proposed the features and challenges in equipment, designs, and operations. The results from several operators further demonstrated the benefits of this novel technique, as it significantly improves the efficiencies of the fracture operations. Considering the Global Energy Transition, the simultaneous fracturing technique will possibly be a leading trend because of its controlled greenhouse gas emission.

ACKNOWLEDGMENTS

This work was supported by the Key Technologies R&D Program from China National Petroleum Corporation, "Daqing Gulong Shale Reservoir Stimulation Technologies Study" (2021ZZ10-04).

REFERENCES

Algadi, O., McLaughlin, N., Wilpitz, L., Farrell, T., & Gamadi, T. (2021). Simultaneous fracturing operations: Successful implementation and lessons learned. In: SPE Annual Technical Conference and Exhibition, 2021, Houton.

Jace Bullinger and Timothy Reist, (2021) 'Simul-frac' Technique Helps Shale Operator to Set Record. https://www.hartenergy.com/exclusives/simul-frac-technique-helps-shale-operator-set-record-194783

Kim, Amos S., Han, Seung, Ruhl, Andrew, Belcourt, Kristen, and Ross Cazenave. "Simul-frac Journey in the Permian Basin." In: the SPE/AAPG/SEG Unconventional Resources Technology Conference, Houston, Texas, USA, July 2021.

Mutalik, P. N., Gibson B., 2008. Case History of Sequential and Simultaneous Fracturing of the Barnett Shale in Parker County. In: the SPE Annual Technical Conference and Exhibition. Denver, Colorado. 21–24 September 2008.

Ogoke, V. C., Schauerte, L. J., Bouchard, G., & Inglehart, S. C. (2014). Simultaneous operations in a multi-well pad: A Cost-effective way of drilling multi wells pad and delivering 8 Francis a day. In: SPE Annual Technical Conference and Exhibition, Houton.

Russell, D., Stark, P., Owens, S., Navaiz, A., & Lockman, R. (2021). Simultaneous hydraulic fracturing improves completion efficiency and lowers costs per foot. In: SPE Hydraulic Fracturing Technology Conference and Exhibition 2021, Houton.

Waters, G., Dean, B., Downie, R., Kerrihard, K., Austbo, L., & McPherson, B. (2009). Simultaneous hydraulic fracturing of adjacent horizontal wells in the Woodford shale. In: SPE Hydraulic Fracturing Technology Conference 2009, Houton. 694–715.

Advances in Mineral Resources, Geotechnology and Geological Exploration – Rashid & Zhang (eds)
© 2023 Copyright the Author(s), ISBN 978-1-032-33377-9

Practical research on the application of inflow-control technology in desert oilfield

Dexiang Duan, Feng Qian & Ning Jing
China National Petroleum International Exploration and Development Company Limited, Beijing, China

Xiuyu Wang*
China University of Petroleum, Beijing, China

Kun Ning & Xi Wu
CNPC Engineering and Technology R&D Company Limited, Beijing, China
National Engineering Laboratory of Oil and Gas Drilling Technology, Beijing, China

ABSTRACT: Niger desert oilfield reservoir is characterized by strong non-homogeneity and large differences in permeability. It has caused the problems of uneven fluid production, rapid rise of water cuts in oil wells, and reduced recovery of oil reservoirs. Eight wells were successfully applied in the Niger oilfield, and good water control and oil production were achieved. This paper forms a set of inflow-control and oil recovery techniques applicable to desert oilfields by studying the operation process and application effects, which provides a reference for other oilfields to manage high water cut oil wells.

1 INTRODUCTION

The Niger oil field is located in the hinterland of the Sahara Desert in West Africa and has a tropical desert climate, making it one of the hottest countries in the world. Since 2008, CNPC has been conducting exploration and development operations in the Niger oil field. The Niger oil field has a variety of reservoir types, including gas-top reservoirs, bottom water reservoirs, and fault-block reservoirs. The gas-top reservoir has a high gas-to-oil ratio, the bottom-water reservoir has a rapid rise in water cut, the fault block reservoir has a rapid drop in formation pressure, and the reservoir properties, pressure systems, and fluid properties of multiple oil-bearing reservoirs vary greatly (Al-Enezi et al. 2012). The oilfield mainly adopts submerged electric pump oil recovery technology. As of 2021, the average water cut in the three main blocks of the oilfield is 25.2%, 37.5%, and 66.4% respectively. With the development of the oilfield, the oil wells have experienced uneven fluid production, rapid increase in water cut, and reduced oil recovery rate. In this situation, the problem of fast-rising water cuts in high water-bearing wells was successfully solved by successfully applying and promoting flow control technology. Not only has it increased the production of a single well, but it also provides a powerful technical tool for the economic development of the oilfield.

2 INFLOW-CONTROL FOR REDUCING WATER PRODUCTION TECHNOLOGY

2.1 *Process principle*

The flow control technology is designed to flow different fluids in differently shaped flow channels based on the different physical properties of density and viscosity of crude oil and water. Crude oil and water have different energy losses during the conversion of pressure energy to kinetic energy

*Corresponding Author: wangxiuyu@cup.edu.cn

 DOI 10.1201/9781003319412-4

in the cyclonic process. The flow pressure drop of water is higher, while the flow pressure drop of crude oil is lower, so as to control and limit the low viscosity of water and improve the passage of high viscosity crude oil. Compared to traditional flow control devices, adaptive flow control devices have an "active" flow control function. It is able to automatically adjust the additional resistance generated according to the changes in fluid production in the producing layer, to equalize the fluid production profile and control the bottom water cone (Zhao 2016; 2019).

A packer is used to separate the annulus between the flow control screen tube and the well wall into several relatively independent flow units so that the fluid does not flow through. The flow control screen pipe equalizes the flow rate of each flow unit. It prolongs the period of water-free oil production and water-free oil production and acts as a flow regulator (Ouyang & Liang 2009).

2.2 *Completion string design and downhole tools*

The flow and water control completion tubing column is designed from bottom to top with the following tubing column structure: pilot shoe + blind tubing + adaptive flow control screen tubing + isolation packer assembly + adaptive flow control screen tubing + blind tubing + suspension packer + drilling tool combination. The design of the tubular structure is based on the structure of the Niger well body and the characteristics of the submersible electric pumping process. The isolation packer is set in the middle of the completion tubular column, which enables the simultaneous extraction of the upper and lower main oil layers and avoids unnecessary subsequent layer transfer and workover operations. The mechanical water control valve enables the differentiation and control of fluids with different densities and viscosities of crude oil and water. The purpose of automatically adjusting the flow rate of different fluids is thus achieved (Song et al. 2018).

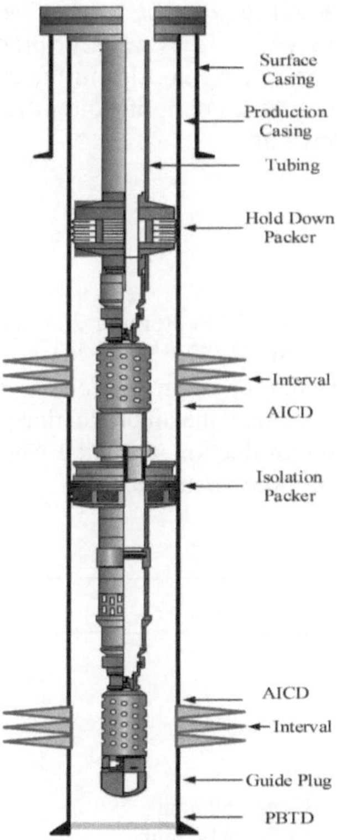

Figure 1. Schematic diagram of the water control completion string.

(1) Suspension packers

The suspension packer seating method is an in-tube ball drop hydraulic seating. When the ball falls into the ball seat position, pressure is applied from within the tubing, the anti-seating blocking mechanism is released, and the piston transmits force to shear off the seating pins, subsequently pushing the kava teeth out to be firmly stuck to the inner wall of the casing. Continued pressurization and pressure stabilization expand the rubber cylinder and the annular space of the casing is sealed, at which point the packer seating seal is completed. The casing can then be pressure tested and sealed. After the seal inspection, the hand-dropping form is selected according to the different sealing tools. There are three common types of hand drops, including tubing pressure drops, casing pressure drops, and mechanical positive rotation drops. The Niger field uses the mechanical positive rotation hand drop method. To unseal, the recovery tool is lowered to retrieve the salvage joint, the pipe column is lifted to the unsealing force, the unsealing pin is cut, and the packer is released.

(2) Isolation packers

The isolation packer is a hydraulic seat seal packer. It is mainly used for inter-layer isolation in multi-layer sections and can be used for flow control completions and general inter-layer isolation or stuck layers. Sealing is achieved by applying pressure to the rated working pressure in stages from the operating tubing. When unsealing, the top sealer is lowered into the top sealer recovery tool to lift the tubular column on the butt salvage joint, at which point the unsealing pin is cut and the kava teeth are automatically retracted by the kava spring, releasing the top sealer.

(3) Water control screen tubes

The body is set up with nozzles of different diameters and sizes in order to achieve uniform fluid production in each section of the well. As the fluid passes through the nozzles, the fluid will create different resistance to flow. Moreover, the nozzles can control the flow of high flow rate fluids, creating a uniform and effective production pressure differential profile and fluid production profile. The flow control screen tube is sensitive to the velocity of the fluid in the screen tube. If a section of the reservoir sees water or produces a rapid rise in water cut in the oil-water mixture, the flow regulating nozzles in the tube will create resistance to such high-velocity fluids. This reduces the amount of fluid produced in that section and achieves the purpose of regulating the flow rate.

3 OPERATION PROCEDURE

The S-11 well was put into production on 30 September 2011 using a submersible electric pump, producing at the following levels: 1768m–1772m and 1784m–1789.5m. The electric pump went down to a depth of 1611m. At the beginning of production, it was water-free and produced 1,700 bbl/d. In May 2012, it started to see water and the oil production gradually decreased. In June 2018, the water cut reached 68% and the oil production reached 295bbl/d. The well was shut on 25 June 2018. The basic drilling data of S-11 is shown in Table 1.

Table 1. Basic drilling data for S-11.

Well Name		Sokor-11
Well Classification		Production
Well Type		Vertical well
Surface Elevation		363.214m
Kelly Bushing		370.714m
Actual Vertical Depth		1900.0m
Casing Diameters	Casing 508mm	From surface to30m
	Casing 339.7mm	From surface to 297.14m
	Casing 244.5 mm	From surface to 1896.88m

The flow control tool string is connected on-site according to the design of the flow control completion project. The completion pipe column from bottom to top is: 4″ lead shoe + 1 blind 4″ pipe + 1 adaptive flow control screen pipe + 1 short section + 1 Φ211mm isolation packer assembly + 1 short 4″ section + 1 adaptive flow control screen pipe + 1 short section. This paper takes well S-11 as an example and describes in detail the main construction process on site.

(1) Pass-through well; lower 3 − 1/2″ outer thickened tubing bottom with φ213mm pass-through tool to explore the bottom of the artificial well.
(2) Scrape the pipe; lower 3 − 1/2″′ EU tubing bottom with a 9 − 5/8″ scraper to explore the bottom of the artificial well and repeatedly scrape the pipe 3 times at the sealer seating position.
(3) Lowering the sub-sectional flow control pipe column; (1) Preparation work before entering the well, surface inspection, connection of packer and water control screen pipe. (2) Lower the flow and water control completion tubing column. (3) Adjust the depth of the pipe column so that the packer is aligned with the seating position. (4) Throw φ38mm steel balls and pressurize the seating packer step by step. (5) Lift and lower the pipe column, if the pipe column does not move, the packer is successfully seated. If the pressure does not drop, the seal is qualified. (6) If the pipe column is lifted, turn the pipe column until the suspension weight of the pipe column drops, and the hand is successfully dropped. (7) positive pressure to ball seat action pressure, shear seat seal ball seat. (8) Seal the isolation packer with positive pressure in the oil pipe step by step. (9) Test the isolation seal. (10) Raise the lost pipe column in the well.
(4) Lower into the electric pump production column in accordance with the electric pump commissioning design.

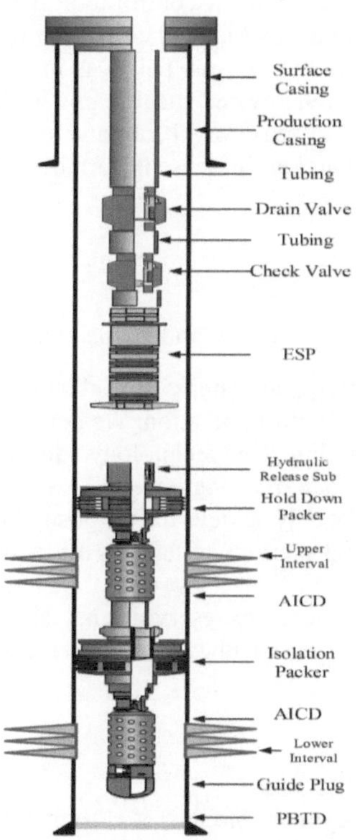

Figure 2. Schematic diagram of the water-control & completion string.

4 APPLICATION AND RESULTS

Up to now, the flow and water control technology has been applied on site for a total of 8wells. The flow and water control technology not only controls water and stabilizes oil, but also allows for 2–3 sections of layered oil recovery in one operation. The S-11 well has been in production for nearly one month using the flow and water control technology, and the water cut has decreased by 44% compared to the previous operation. The S-18 and S-19 wells were completed in one trip to achieve the completion of the two main oil sections. A future water plugging or layer change operation is saved. A comparison of daily oil production and water cut before and after the flow control as well as well operation is shown in Table 2.

Table 2. Oil production and water cut of inflow-control wells.

Well number	Daily fluid production bbl	Daily oil production bbl	Moisture content %	Remarks
S-11	922	295	68	Before Operation
	546	404	26	After Operation
S-18	1032	299	71	Before Operation
	620	618	0.4	After Operation
S-19	757	219	71	Before Operation
	585	582	0.6	After Operation

As a result of the blocking effect of the flow control tool, the fluid production capacity of well S-11 under the same working regime was reduced compared to the pre-operational regime. The dynamic fluid level was also reduced to 765 m. This was in line with design expectations. Five months after commissioning, the well's water cut remains at 28%. The daily oil production is 420bbl/d, with an average daily oil increase of 110bbl and a daily water production reduction of 513bbl. The effect of water control and oil increase is very obvious.

5 CONCLUSION

Based on the results and discussions presented above, the conclusions are obtained as below:

(1) By comparing the oil production and water cut production data of three high water-bearing wells before and after the flow control operation. We believe that the flow control process has a good "water blocking" capability. This technology can quickly reduce the water cut of the well and increase the recovery rate of a single well.
(2) The flow control process can not only achieve the purpose of water control and oil stabilization. It is also possible to achieve 2-3 sections of layered oil recovery in one operation, saving the cost of at least one operation.
(3) The core of the process lies in accurate reservoir analysis and finding the right water layer. Optimizing the parameters of the flow and water control tools is the key to the process working.

ACKNOWLEDGMENTS

This Work Was Supported by The Key Technologies r&d Program from CNPC Niger Petroleum S.A., "The Integrated Study On Improving The Quality And Efficiency Of Completion And Stimulation In Agadem Block".

REFERENCES

Al-Enezi, Khalaf, Das, Om Prakash, Aslam, Muhammad, Ziyab, Khaled, and Taher El-Gezeeri. (2012) Successful Case Histories of Smart Multilateral Well with Inflow Control Device and Inflow Control Valve for Life-cycle Proactive Reservoir Management in High Mobility Reservoir, Minagish Field West Kuwait. In: Abu Dhabi International Petroleum Conference and Exhibition, Abu Dhabi, UAE.

Ouyang Liang-Biao. (2009) Practical Consideration of an Inflow-Control Device Application for Reducing Water Production. In: SPE Annual Technical Conference and Exhibition, New Orleans, Louisiana.

Song Zhengcong, Li Qing, He Qiang. (2018) Research and application of adaptive flow control water and sand control process in Tahe oilfield. In: Drilling and Production Process. pp. 115–116.

Zhao Chongzhen. (2016) Development and application of adaptive flow control device for horizontal wells. In: Petroleum Drilling Technology, pp. 95–98.

Zhao X. (2019) Research and experiment on adaptive flow control technology. In: Petroleum Machinery China. pp. 93–94.

Advances in Mineral Resources, Geotechnology and Geological Exploration – Rashid & Zhang (eds)
© 2023 Copyright the Author(s), ISBN 978-1-032-33377-9

Research on the optimization of the re-open sequence of oil producers in the process of oilfield production resuming

Hengyu Lyu*
Research Institute of Petroleum Exploration and Development, CNPC, Beijing, China

Mengfei Liu
CNPC International Nile Limited, Khartoum, Sudan

Ran Wang
Petroleum Exploration and Production Research Institute, SINOPEC, Beijing, China

ABSTRACT: The Organization of the Petroleum Exporting Countries (OPEC+) began to restrict production as a response to the global pandemic. This resulted in the forced slowing down of numerous oilfield developments, as well as the closure of many oil operations. According to the reservoir's operators, a strenuous shut-in of the well will result in an imbalance in the subsurface fluid system of the reservoir, as well as a redistribution of the oil-water system and pressure potential field. Production curtailment, on the other hand, will be estimated to have reduced at long last. When wells are shut down and production is restarted in the future, it is necessary to efficiently recover production. It is also necessary to regulate the water cut of the reservoir and guarantee that the reservoir's development progresses in a balanced manner. This study provides a strategy for managing the sequence in which oil producers resume production, based on the use of essential indicators of single well development as evaluation criteria and the comprehensive evaluation score approach. The method is intended to instruct oil producers that are resuming production in terms of optimizing their production sequence. According to the findings of the prediction and comparison, it has been discovered that optimizing the sequence has a significant impact on oil output, with the water cut effectively decreasing by 5 percent under the same oil production scenarios.

1 INTRODUCTION

Many oil fields have been impacted by the pandemic and have adopted a policy of limiting production while maintaining pricing in order to maintain profits. Oil fields have no other option than to passively shut down wells in order to deal with this situation. When a product failure occurs or when remedial measures are required, an oil well will normally be shut down (Chiavico 2020; Pedullà 2019; Veil 2011). The abnormal shut-in of oil wells will have a significant impact on the steady production of oil wells, as well as the production and water cut of oil wells in the future. The redistribution of underground oil-water relationships in a reservoir caused by a large number of closed wells will have an uneven development effect, as well as cause the water cut of a single well to increase rapidly. Adverse consequences such as a significant and sudden decrease in productivity are possible (Belyakov 2021; Eltahan 2020; Jamie 2017; Han 2012; Sinha 2017). As a result, when an oilfield is experiencing the effects of production curtailment, it is aimed at optimizing the opening and closing of oil wells. This is also a challenge that must be confronted in order for the oilfield to maintain its long-term stability.

*Corresponding Author: herolhy@petrochina.com.cn

 DOI 10.1201/9781003319412-5

2 METHODOLOGY

Due to fluctuations in the international situation and the price of crude oil, an oilfield's production curtailment target must be a flexible and timely resumption strategy, capable of accommodating optimization of well opening and shut-in under a variety of production targets as well as operability for adjustment, is necessary.

2.1 *General principle*

Throughout the process of production limitation and production restart, the production control of oil wells must be carried out in line with the fundamental laws of reservoir development to ensure that the wells remain productive. In order to recover energy in an acceptable length of time while simultaneously ensuring steady crude oil output and keeping water cut within a tolerable range, a balanced injection and production strategy must be used.

- Ensure that the reservoir is operating at a stable production rate
- Optimize production wells with the ability to open wells
- Take treatment measures for synchronous high water cut oil producers
- Reasonably adjust the uneven recovery of reservoir pressure
- Control injection and production as per targeted water breakthrough type
- Conduct quantitative optimization of the artificial lift well system
- Keep reasonable sequence of re-open oil producers by stages

2.2 *Evaluation indexes*

Following the formulation of a policy for the resuming of reservoir production, it is required to plan the sequence in which single wells will be opened. The thorough evaluation approach is used extensively in this work. First and foremost, it is vital to identify the key indications that best reflect the parameters required for the reopening of a single well's production. According to development experience and current conditions, evaluation indicators such as water cut, cumulative oil production, well type (natural flow, ESP, GL), comprehensive well efficiency, pressure recovery condition, and daily production capacity are used to assess the possibility of oil wells to resume production, details are seen in Table 1.

Water cut is the most important indicator of a well's production performance. Looking at it from the standpoint of seepage while the water cut has been increased by more than 50%, the situation is quite dire. The flow of the water phase will significantly impede the flow of the oil phase. Therefore, in order to control the comprehensive water cut at the reservoir scale, efforts should be made to start oil wells with a low water cut while also taking advantage of the opportunity presented by production curtailment, countermeasures such as water plugging in wells with a high water cut that can be controlled, and then the production can be restarted after the countermeasures have been implemented. As a result, the water cut of the reservoir is necessary to be controlled within a reasonable range.

The accumulation of oil production is also an important factor to consider. Because of the severe heterogeneity of carbonate reservoirs, the production rates of oil wells vary widely, and the production scales are not evenly distributed over the reservoir. To allow for the cost-efficient extraction from the reservoir in a balanced planar and vertical direction, it is required to begin by opening the wells with the least amount of accumulated production initially.

The most intuitive metric of oil production capacity is the capacity to produce oil daily. Due to the limited production period of the reservoir, in order to achieve good economic benefits, the number of wells should be controlled as least as possible, and the cost of wells should be controlled under the conditions of adhering to certain production restrictions. The number of wells should be controlled as least as possible to achieve good economic benefits. As a result, it is vital to select oil wells with the right production capacity for the given situation. It is possible that there are many wells that produce at or over the required level of productivity. The location of the surrounding water injection wells, as well as the injection-production relationship, must be carefully considered at this point to select an oil well that will efficiently establish water flooding.

When considering the resumption of oil well production, it is also important to evaluate the production time rate of a single well, which can be a good indicator of the stability of oil well production. Priority should be given in the process of resuming production to selecting oil wells with stable production status in order to ensure steady production and to fulfill the production curtailment target as soon as possible.

Resumption of production is necessitated by a stable pressure situation, which can be generated by either the effect of water injection or the recovery of reservoir pressure caused by the shut-in of the well. Oil reservoir production is maintained by the reservoir pressure. Wells with pressure recovery are more capable of resuming production than wells without pressure recovery. Well regions with low pressure should be halted until the pressure has returned to normal.

In addition, the type of well is a key component to consider when determining whether or not to restart production. Under production conditions, the operating costs of different well types differ, and the operating costs of ESP wells will be higher than the operating costs of naturally flowing wells, for example.

Table 1. Evaluation scoring standards of indexes for producers re-opening potential.

WCT %	Score	Np Mbbl	Score	Type /	Score	WEFAC %	Score	PRE /	Score	Oil STB/d	Score
(0,5]	10	(0,10]	10	NF	10	(95,100]	10	Recovery by WI	7	>3000	10
(5,10]	9	(10,20]	9			(90,95]	9			(2000,3000]	8
(10,30]	8	(20,50]	8			(80,90]	8			(1500,2000]	7
(30,40]	7	(50,100]	7	ESP	3	(70,80]	7	Recovery due to shut-in	4	(1000,1500]	6
(40,50]	5	(100,200]	6			(60,70]	6			(800,1000]	5
(50,60]	4	(200,500]	5			(50,60]	5			(600,800]	4
(60,80]	3	(500,800]	4	GL	2	(40,50]	4	Positive response to water-flooding	10	(300,600]	3
(80,90]	2	(800,1000]	3			(30,40]	3			(200,300]	2
(90,100]	1	>1000	2			(0,30]	2			(0,200]	1

3 RESULTS AND DISCUSSIONS

The order in which production will resume and wells will be opened in the reservoir is determined by a complete scoring and evaluation of each well in the reservoir. When the value is higher, the priority of well opening is higher, which indicates that there is greater potential for the restart of production, which is more in accordance with the stated resumption strategy and evaluation criteria for resumption of production potential. As illustrated in Figure 1, oil wells with high priority are denoted by green color. Clearly, all wells with low water content, obvious pressure recovery, good productivity, and consistent production status are mostly natural flowing production wells. And oil wells with medium priority are denoted using the color orange. Most producing oil wells are featured as medium water cut, medium production, and relatively stable production status as can be seen in the diagram below. Oil wells having a lesser priority are highlighted in blue. Most oil wells have a larger water cut than normal, and the overall production is poor, as can be seen in the radar map, and well type is mainly artificial lift wells, ESP or GL, usually presenting pressure shortfalls and unpredictable production circumstances.

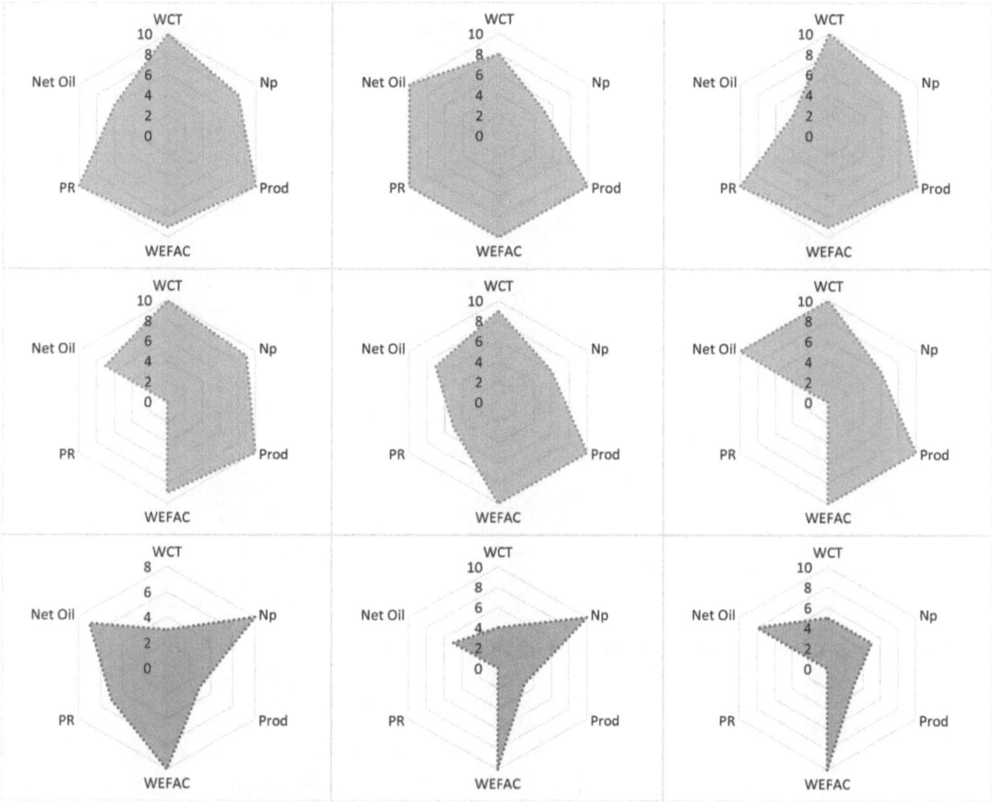

Figure 1. Radar map of the priority level of resuming production and reopening producers.

Due to the fact that the government's production curtailment directives are frequently issued on a temporary basis, the target is always shifting, increasing, and decreasing in a large range. All oil wells are divided into batches in order to achieve the various production restriction as per the target. The overall production of the first batch of oil wells is around 40,000 barrels per day, and the total production of each of the subsequent batches is approximately 20,000 barrels per day, respectively. Figure 2 shows the optimal well locations for each batch, which are clearly delineated.

A substantial pay zone thickness and good physical qualities are found in the crestal area of the reservoir where oil wells with the highest priority for restarting production are mostly located. And the oil and gas production regions are the ones in which the development effect of water injection is well and the reservoir pressure is restored. Because of their poor physical qualities, little pay zone thickness, and lack of effective water injection to supplement energy, wells at the edge and flank areas of the reservoir are often given lesser priority than wells in the crestal areas of the reservoir, for example.

Well opening should be done in stages to allow for changes in production curtailment targets to be accommodated. It is also recommended to make various preparations for the complete restoration of production and subsequent production of the oilfield. Treatment solutions for high water cut wells should be implemented simultaneously during the restart gap period to effectively minimize the comprehensive water cut and profit from the overall development benefits. Implementing high-pressure water injection zonal flooding and assessment, quantifying the injection-production ratios of the crestal and flank areas within waterflooding well groups, and allocating production and injection in a reasonable mode are all recommended.

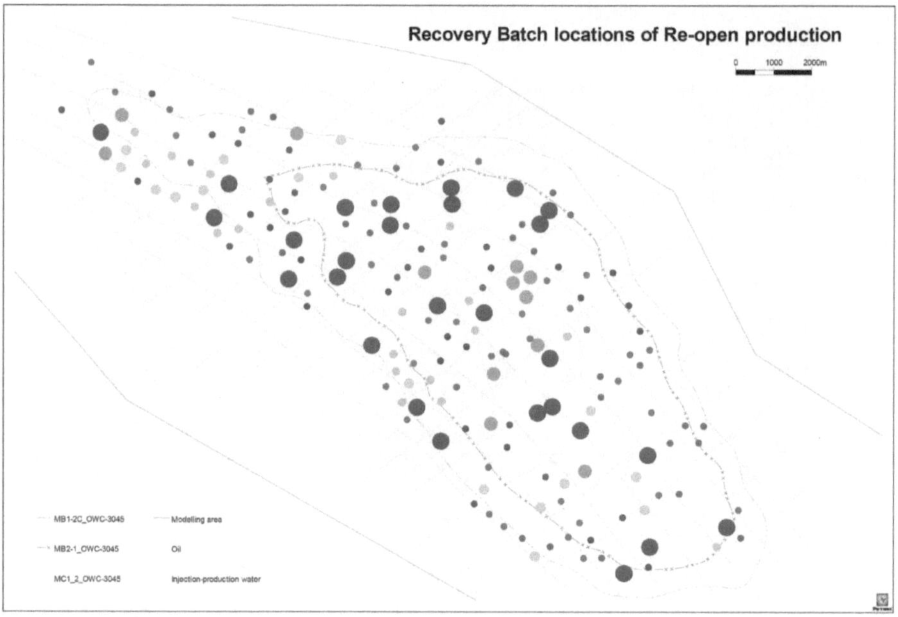

Figure 2. Optimized well location map for different batches.

4 CONCLUSION

Additional research into how to manage production curtail and change overall production by closing or opening wells is required in oilfield operations, but how to do it economically and effectively is still in the early stages. The research and analysis presented above indicate that it is prudent to develop a reasonable strategy for restarting production in response to changing curtailment instructions and that it is practical to propose an effective plan for optimizing the opening sequence to effectively respond to production curtailment. Using data on water cut, oil production, pressure condition, and production status as a strong foundation for a comprehensive evaluation of re-production potential, and in accordance with the batch strategy, oil wells with a high re-production potential are given priority over open wells, while also taking advantage of the production curtailment, resulting in a reduction in the amount of oil produced. Abnormal wells, wells with high water cut rates, and problematic wells will be addressed with the goal of suitably lowering the water cut and resuming consistent production throughout this period. A potential optimization process is used to determine the optimal sequence of re-production wells in a reservoir under the regulation of the overall re-production strategy until the objective of stabilizing reservoir production after the full revitalizing operation and managing the water cut within a suitable range is achieved.

ACKNOWLEDGMENTS

The authors express their gratitude to the Middle East Technical Team of the Research Institute of Petroleum Exploration and Development for their support.

REFERENCES

Belyakov, A.A., et al. The Efficiency of the Novel Technology Reservoir Pressure Reconstruction Without Well Shut-in and without Production Losses. Paper presented at the SPE Russian Petroleum Technology Conference, Virtual, October 2021.

Chiavico, M., et al. Comparative Analysis of Alternative Solutions for Produced Water Management in Offshore Environment Based on Life Cycle Assessment and Biodiversity Impact Assessment. Paper presented at the SPE International Conference and Exhibition on Health, Safety, Environment, and Sustainability, Virtual, July 2020.

Eltahan, E., et al. Impact of Well Shut-in After Hydraulic-Fracture Treatments on Productivity and Recovery in Shale Oil Reservoirs. Paper presented at the SPE Improved Oil Recovery Conference, Virtual, August 2020.

Han, G.Q., et al. Simulation of Multiphase Fluid-Hammer Effects during Well Shut-In and Opening. Paper presented at the SPE Asia Pacific Oil and Gas Conference and Exhibition, Perth, Australia, October 2012.

Jamie, S.A., et al. Optimizing Bean-Up Procedure after Well Shut-in. Simple Rock Mechanical Aspects and Operational Guidelines. Paper presented at the SPE Bergen One Day Seminar, Bergen, Norway, April 2017.

Pedullà, M., et al. Analysis and Comparison of Different Solutions for Produced Water Management - LCA and Biodiversity Impact Assessment. Paper presented at the Offshore Mediterranean Conference and Exhibition, Ravenna, Italy, March 2019.

Sinha, S., et al. Effect of Frequent Well Shut-In's on Well Productivity: Marcellus Shale Case Study. Paper presented at the SPE Eastern Regional Meeting, Lexington, Kentucky, USA, October 2017.

Veil, J. A., et al. Management of Water Extracted From Carbon Sequestration Projects: Parallels to Produced Water Management. Paper presented at the SPE Americas E&P Health, Safety, Security, and Environmental Conference, Houston, Texas, USA, March 2011.

Advances in Mineral Resources, Geotechnology and Geological Exploration – Rashid & Zhang (eds)
© 2023 Copyright the Author(s), ISBN 978-1-032-33377-9

Application and practice of compound dynamic disaster prevention technology in first mining face of deep mine

Jie Hu & Siqian Li*
CCTEG Chongqing Research Institute, Chongqing, China
State Key Laboratory of the Gas Disaster Detecting, Preventing and Emergency Controlling, Chongqing, China

ABSTRACT: The coupling disaster of coal and gas outburst and rockburst has become a difficult problem and bottleneck in the treatment of gas-dynamic disaster in deep mines. Taking the kilometer deep well as an example, the characteristics of coal and gas burst complex dynamic disaster in the first mining of coal seam 11-2 are studied and determined; based on the rockburst and the disaster causing factors of coal and gas outburst, the comprehensive index method and gas geological method are used to predict that there are four dynamic disaster risk areas in the working face. According to the principle of "regional prevention first, local treatment and supplement", the composite prevention measures such as large-diameter borehole pressure relief, coal seam water injection, and bedding borehole pre drainage are tested. The results show that the maximum amount of drilling cuttings in the mining area is 4.4kg/m, the gas concentration in the return airflow is maintained below 0.45%, and there are no accidents of exceeding the prediction index and gas exceeding the limit, which realizes the accurate prevention and control of zoning and grading of mine dynamic disasters and safe mining of the working face, and effectively prevents the occurrence of compound dynamic disasters.

1 INTRODUCTION

Coal and gas outbursts and rockburst are typical mine disasters in coal mining. In the process of shallow mining, coal and gas outburst and rockburst are typical mine disasters. In the process of shallow mining, these two dynamic disasters often occur independently and do not have the relationship of mutual coupling. With the increase in mine mining depth, complex coal and rock mass mechanical environments such as high ground stress, high heterogeneity, and low permeability have emerged (Cao 2014; Lan et al. 2016; Lei et al. 2017). Many mines show some characteristics of both coal and gas outburst and rockburst, showing a certain coupling (Yuan 2013, 2016). Their occurrence conditions may be lower and the disaster intensity may be greater than a single dynamic disaster. It has brought greater difficulties to mine safety production and management (Jiang et al. 2014; Li et al. 2019). The combined action of coal and gas outburst and rockburst has become the main factor restricting the safe and efficient mining of coal mines. Reliable and effective prevention and control technology of complex dynamic disasters in deep mines is urgently needed. In view of this, according to the classification of deep well composite dynamic disasters, taking the first mining face of the coal seam 11-2 as the engineering background, the author has carried out dynamic disaster characteristic analysis and regional risk prediction based on the test and analysis of coal seam physical mechanics and gas parameters, and formulated targeted prevention and control measures to realize safe and efficient mining of the working face. The technology of dynamic prevention and control of coal and gas dynamic disasters has an important reference for the prevention and control of dynamic disasters in deep mines.

*Corresponding Author: siqian_licqmk@163.com

 DOI 10.1201/9781003319412-6

2 ENGINEERING BACKGROUND

The mine is located in the north of Huainan coalfield and the south wing of the east section of Zhuji Tangji anticline. It is located in the footwall, which is opposite the South and North overthrust faults. The north side of the North overthrust group is accompanied by large tensile faults. The geological structure is complex due to the compression of North and South compressive stresses. The mine is mined on two levels, the first level elevation is –962m, and the second level elevation is tentatively-1150m. The first level 11-2 Coal seam is mined. The average thickness of the coal seam in the first mining face is 1.60m, the average dip angle is 5°, the buried depth is 980 ~ 1080m, the occurrence is stable, the whole area is minable, and the roof and bottom slate is composed of the commonly used mudstone powder sandstone. The inclined longwall one-time mining full height retreating mechanized mining technology is adopted, and the roof is managed by all caving methods. During the construction of the main roadway in the development and mining area, the roadway floor heave is up to 1m, the two sides move closer to 1.2m. The roadway deformation is serious, the support is difficult, and the repair cycle is short. The maximum lateral pressure coefficient is 1.55, and the stress concentration is relatively high, the coal roadway strip is adopted. After the outburst prevention measures in the pre-gas drainage area of the floor out of the layer drilling, obvious dynamic phenomena such as top drilling and coal blasting appear in the excavation process.

3 ANALYSIS OF DYNAMIC BISASTER CHARACTERISTICS OF THE FIRST MINING FACE

Coal and gas outburst is the result of the joint action of in-situ stress and gas pressure. Gas internal energy is the main energy of outburst (Zhang 2009, Liu 2018). Rock burst is caused by the instantaneous release of elastic properties of hard coal and rock mass. After the mine enters deep mining, the critical conditions for the occurrence of combined disasters of coal and gas outburst and rockburst are lower than those of a single disaster, and with the boost of one disaster, another disaster can occur under the low index of conventional risk prediction parameters. It is difficult to define some characteristics of both outburst and impact at some accident sites. Based on the instability failure mechanism of composite materials composed of coal rock solid and gas and the unified energy equation of coal rock dynamic disasters (Pan 2016; Zhang & Yang 2019), the deep mine dynamic disasters are divided into four categories, as shown in Table 1.

Table 1. Classification of compound dynamic hazards in deep mines.

Disaster type	Energy source	Coal seam and gas characteristics
Coal and gas outburst	In situ stress and gas are combined, and gas is the main energy source	No impact tendency and prominent risk, W>8m^3/t
rock burst	Elastic energy released by deformation and failure of coal and rock stratum	Have impact tendency and no prominent risk, W<6m^3/t
Coal and gas impact	The energy released by coal and rock mass deformation boosting gas	Have impact tendency and prominent risk, W>8m^3/t
Impact protrusion composite	The energy released by coal and rock mass promoted by gas	There is no impact tendency and prominent risk, and the local stress concentration is large, 6m^3/t<W<8m^3/t

In the development and mining area, the gas pressure of coal seam 11-2 is 1.2MPa, the maximum gas content is 8.63m^3/t, the failure type of soft stratification is IV, and the initial gas release velocity is 13, and the minimum firmness coefficient is 0.36. It is determined that coal seam 11-2 is an outburst coal seam. The dynamic failure time of the coal sample of the coal seam 11-2 is 360ms, the elastic energy index is 2.4434, the impact energy index is 1.52, and the uniaxial compressive strength is 12.264 MPa. It is determined that coal seam 11-2 has an impact tendency. According to the classification in Table 1, the disaster type of coal seam 11-2 is coal and gas impact.

4 PREVENTION AND CONTROL TECHNOLOGY OF COMPOUND DYNAMIC DISASTER IN THE FIRST MININGFACE

According to the characteristics of coal and gas impact composite dynamic disaster, on the basis of regional risk prediction, according to the principle of regional measures first and local measures supplementary, and according to the existing technical equipment of the mine, comprehensive prevention and control technologies such as bedding drilling pre pumping, pressure relief drilling and coal seam water injection are adopted.

4.1 *Prediction of dangerous areas of complex dynamic disasters*

According to the gas occurrence characteristics and measured gas parameters of the first mining face. The working face is divided into outburst risk area and non-outburst risk area by the gas geological method. At the same time, combined with geological conditions and mining technical conditions, impact risk areas of different degrees are divided by the multi-factor coupling method. The area where the outburst risk area overlaps with the serious impact risk area is called the key prevention area. It is predicted that there are two key prevention areas, one prominent danger area and one medium impact danger area in the working face, and other areas are regarded as normal areas, as shown in Figure 1.

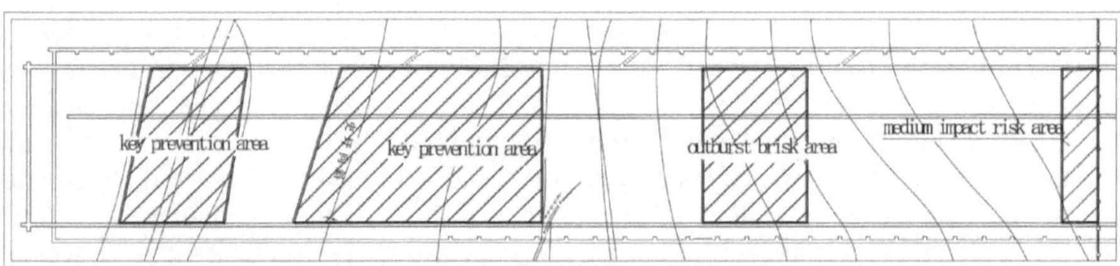

Figure 1. Prediction results of dynamic hazard risk area in the first mining face.

4.2 *Prevention and control measures for compound dynamic disasters in the first mining face*

4.2.1 *Pressure relief and pre pumping measures for large diameter drilling*

When large-diameter boreholes are constructed in the coal body, cracks are generated and damaged around the boreholes, forming a fracture area much larger than the diameter of the boreholes. After the fracture areas around the boreholes are connected with each other, a wider range of fracture areas is formed, which reduces the peak stress and transfers to the deep part of the coal body, which plays the role of pressure relief. The cracks in the coal body are developed and increase the air permeability of the coal body. It provides favorable conditions for gas drainage. In the key prevention area, 108mm diameter bedding pressure relief boreholes are constructed in the track roadway and transport roadway, with a spacing of 5m and a hole depth of 50m and 120m respectively. The boreholes with a length of 50m are sealed with yellow mud, and the boreholes with a length of 120m are sealed with two plugging and one injection, with a hole sealing depth of 20m and pre drainage of coal seam gas. In the coal and gas outburst risk area, only bedding pre-pumping boreholes are constructed, and in the medium impact risk area, only pressure relief boreholes are constructed. The schematic diagram of the borehole layout is shown in Figure 2.

4.2.2 *Coal seam water injection*

After coal seam water injection, the adhesion between coal particles is reduced, the surface friction is reduced, its physical and mechanical properties are greatly changed, the plasticity is improved, the accumulated elastic energy is reduced, the internal energy of water displaces gas and other gases is reduced, and the probability of compound dynamic disaster is reduced. During the recovery of the working face, the bedding borehole is used for water injection 100m ahead of the working face, the water injection pressure is 2.0MPa. The hole sealing depth is 10m, the water injection volume of a single hole is 48.86 ~ 65.68m^3, and the water content increment is not less than 2% as the inspection index. In the working face, the predicted borehole is used to inject water into the area beyond the control of the bedding hole, and the injection time is until the water is discharged from the coal wall.

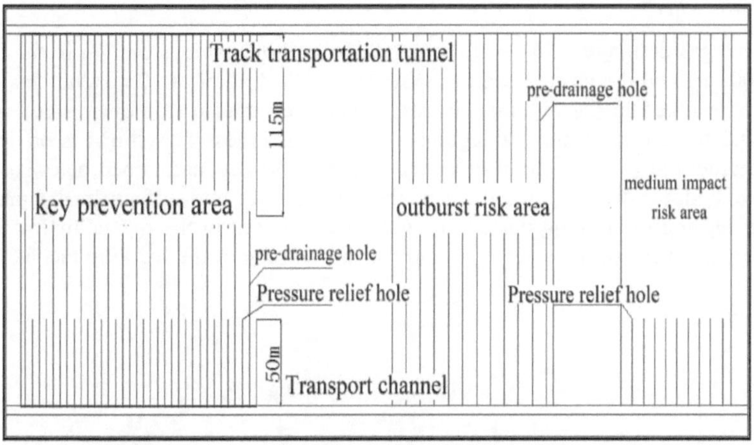

Figure 2. Layout of boreholes in the first mining face.

5 PREVENTION AND CONTROL EFFECT OF COAL ROCK GAS DYNAMIC DISASTER

5.1 *Prevention and control effect of rockburst*

During the mining of the working face, the amount of drilling cuttings is used for inspection. The changing trend of drilling cuttings within 500m of the mining of the working face is shown in Figure 3.

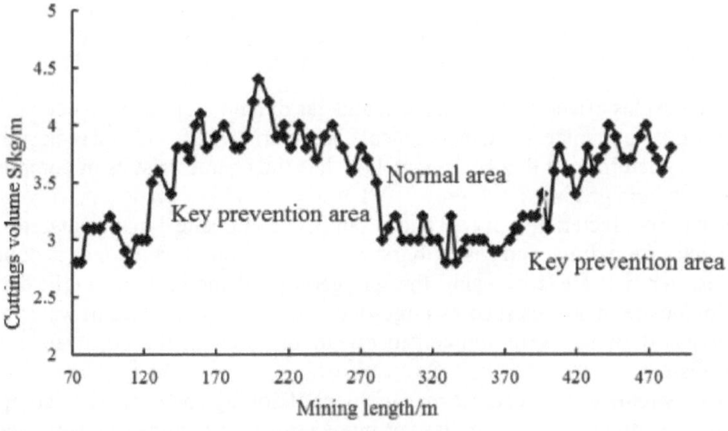

Figure 3. Trend of drilling cut quantity index in first mining face.

The measured S_{max} in the key prevention area is 3.8 ~ 4.4kg/m and S_{max} in the normal area is 2.8 ~ 3.4kg/m. There is no over-the-standard phenomenon and no abnormal dynamic phenomenon during drilling construction. The amount of drilling cuttings in the key prevention area is generally greater than that in the normal area, which verifies the accuracy of the prediction results in the dynamic disaster area. Several groups of coal samples were taken within 100m in front of the work to test. The average water content increment was 3.05%, and the water content increment was not less than 2%, which played a role in water injection and erosion prevention.

5.2 *Gas control effect*

The cumulative gas drainage volume of the bedding boreholes before mining in the first mining face is 896800 m^3, and the gas pre drainage rate is 35.9%. A total of 30 residual gas content test boreholes are arranged in the outburst risk area. The measured maximum residual gas content is 5.16m^3/t and the effect of pre drainage measures in areas less than 8m^3/t is obvious. During the mining period, the gas concentration in the return airflow is maintained below 0.45% by the combination of bedding drilling, buried pipe in the upper corner, and surface drilling pressure relief gas drainage. During the mining period, the gas concentration in the return airflow is shown in Figure 4. There is no gas overrun accident and the safe mining of the working face is realized.

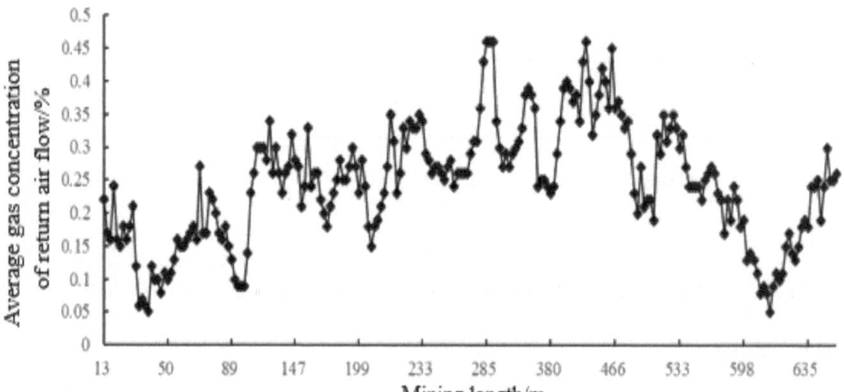

Figure 4. Trend of drilling cut quantity index in first mining face.

6 CONCLUSION

(1) According to the classification of coal, rock, and gas dynamic disasters in deep mines, through the analysis of gas parameters and mechanical characteristics of coal and rock strata in the coal seam 11-2, it is determined that coal seam 11-2 has the characteristics of composite dynamic disasters of coal and gas impact.
(2) According to the characteristics of coal and gas impact composite dynamic disaster, considering the two aspects of rockburst and gas outburst respectively, the working face is divided into four dynamic disaster risk areas by using the gas geological method and multi-factor coupling method. Composite measures such as large-diameter drilling, coal seam water injection, and drilling pre-gas drainage were adopted to effectively prevent the occurrence of composite dynamic disasters.
(3) This technology realizes the accurate management of zoning and grading of complex dynamic disasters, reduces the cost of mine disaster management, effectively prevents the occurrence of dynamic disasters, and realizes the safe mining of the working face.

ACKNOWLEDGMENTS

This work is financially supported by self-funded projects of CCTEG Chongqing Research Institute (Grant No. 2019ZDXM06)

REFERENCES

Cao Jian-jun. Research and application on prevention and treatment technology of deep mine disaster based on its dynamic property. [J]. Coal Science and Technology, 2014, 42(11): 50–54.

Jiang Yao-dong, Pan Yi-shan, Jiang Fu-xing, et al. The mechanism and prevention of rockburst in coal mining in China. [J]. Journal of China Coal Society, 2014, 39(2): 205–213.

Lan Hang, Chen Dong-ke, Mao De-bing. The current state of deep mining and disaster prevention in China. [J]. Coal Science and Technology, 2016, 44(01): 39–46.

Lei Shao-hua, Zhang Lin, Ma Zhen-kai. Analysis of the basic characteristics and present condition of dynamic disaster in mine deep mining. [J]. Shaanxi Coal, 2017, 36(06): 64–66+37.

Li Hong-yan, Mo Yun-long, Sun Zhong-Xue, et al. Research status and the prospect of coal mine rockburst disaster prevention and control technology. [J]. Coal Science and Technology, 2019, 47(01): 62–68.

Liu Xi-jun. Study on prevention and control of dynamic disaster of coal and rock gas in a deep mine. [J]. Coal Science and Technology, 2018, 46(11): 69–75.

Pan Yi-shan. An integrated study on the compound dynamic disaster of coal-gas outburst and rockburst.[J]. Journal of China Coal Society, 2016, 41(01): 105–112.

Yuan Rui-fu. Features of dynamic disasters combined rockburst and gas outburst in a deep coal mine and its preventive measures.[J]. Coal Science and Technology, 2013, 41(08): 6–10.

Yuan Liang. The strategic thinking of simultaneous exploitation of coal and gas in deep mining.[J]. Journal of China Coal Society, 2016, 41(01): 1–6.

Zhang Fu-wang, Li Tie. Cognizance of the compound dynamic disaster of coal and gas in deep mining[J].Zhongzhou Coal, 2009(4): 73–76.

Zhang Qing-he, Yang Ke. Mechanism analysis on continuous stress-relief mining for preventing coal and rock dynamic disasters in deep coal mines. [J]. Journal of Mining & Safety Engineering, 2019, 36(01): 80–86+102.

Advances in Mineral Resources, Geotechnology and Geological Exploration – Rashid & Zhang (eds)
 ISBN 978-1-032-33377-9

Reservoir protection strategy in drilling stage of Tarim oilfield

Hongtao Liu
CNPC Tarim Oilfield Company, Korla, Xinjiang, China

Yang Zhang
CNPC Engineering Technology R&D Company Limited

Shenglan Tang
CNPC Tarim Oilfield Company, Korla, Xinjiang, China

Yan Qiao
CNPC Engineering Technology R&D Company Limited

Lihu Cao
CNPC Tarim Oilfield Company, Korla, Xinjiang, China

Shiling Zhang
CNPC Engineering Technology R&D Company Limited

Hongjun Wu & Sun Tao
CNPC Tarim Oilfield Company, Korla, Xinjiang, China

ABSTRACT: Tarim Oilfield has developed five major oil and gas fields in Tazhong, Lunnan Lungu, Donghe Yaha Yingmai, Kuqa, and Southwest Tarim. The reservoir types can be divided into medium porosity and medium (high) permeability sandstone reservoir, low porosity, and low permeability sandstone reservoir, fracture pore sandstone reservoir, fracture pore cave carbonate reservoir, and cave carbonate reservoir. The current reservoir protection design and construction are mainly based on acid soluble weighting materials, low water loss, and shielding temporary plugging technology. Through the analysis, it is found that the reservoir protection in the current drilling stage is not targeted, the whole process of reservoir protection has not been effectively implemented, and the evaluation method of reservoir protection is unreasonable. Different types of reservoir protection schemes are established through research. The reservoir with medium porosity and medium permeability is shielding temporary plugging technology, low water loss, appropriate salinity, and acid-soluble weighting material; the protection of low porosity and low permeability reservoir is temporarily blocked by shielding of small particles; the damage mechanism of fractured porous sandstone reservoir is fracture leakage, basic fast phase trap, salt\water\alkali sensitivity. The protection measures are low damage drilling fluid system, leak prevention and plugging technology while drilling, acid-soluble plugging materials, membrane temporary plugging, etc; The fractured vuggy carbonate reservoir mainly needs leakproof, and the cavernous carbonate reservoir is less damaged. It is recommended not to add ultra-fine calcium carbonate for shielding and temporary plugging. The research also established the prediction software to predict the fluid sensitivity and the ideal filling software to design the fine reservoir protection formula. The results provide a preliminary idea for establishing an effective reservoir protection system in the drilling stage of Tarim Oilfield, which is conducive to providing the pertinence and effect of reservoir protection in the drilling stage of Tarim Oilfield.

 DOI 10.1201/9781003319412-7

1 INTRODUCTION

The research on reservoir damage evaluation and protection technology is not only an important content in the process of oil and gas field exploration and development, but also an important link to improving the quality of oil and gas exploration and development (Lucas 1918, Washburn 1921). Reservoir protection in exploration is conducive to the discovery and correct evaluation of oil and gas reservoirs; Reservoir protection during production is conducive to improving oil and gas production and economic benefits of oil and gas field development (Bennion et al. 1994; 1996).

The reservoirs in Tarim Oilfield cover five types: medium porosity and medium (high) permeability sandstone reservoirs, low porosity and low permeability sandstone reservoirs, fracture pore sandstone reservoirs, fracture pore carbonate reservoirs, and cave carbonate reservoirs (Bennion et al. 2000). Different types of reservoirs have different damage mechanisms and different protection measures. However, at present, the design and construction of reservoir protection measures in the drilling stage of the oilfield are basically the same, there are many problems such as poor pertinence, and the reservoir protection effect is poor (You et al. 2006; Li et al. 2007). Therefore, based on the summary of the damage mechanism, cause, and degree of different types of reservoirs at home and abroad, as well as the technical achievements of reservoir protection in the drilling stage, combined with the actual situation of reservoir protection in the drilling stage of Tarim Oilfield, this paper analyzes the similarities and differences and causes of different types of reservoir damage in the drilling stage and puts forward the characteristic protection scheme, so as to provide guiding opinions for the design of reservoir protection scheme in the future (Li et al. 2007).

2 BASIC STATUS AND PROBLEM ANALYSIS OF RESERVOIR PROTECTION IN TARIM OILFIELD

Tarim Basin has developed five major oil and gas field groups in Tazhong, Lunnan-Lungu, Donghe-Yaha-Yingmai, Kuqa, and the Southwest block of Tarim Basin. The development horizons are Paleogene, Cretaceous, Jurassic, Triassic, Carboniferous, Silurian, Ordovician, and Cambrian. According to the reservoir damage mechanism, cocoa divides the reservoirs into five types: medium porosity and medium (high) permeability sandstone reservoir, low porosity and low permeability sandstone reservoir, tight sandstone reservoir with fractures, fracture pore carbonate reservoir, and cave carbonate reservoir. See Table 1 for typical blocks and representative strata.

Table 1. Classification of reservoir characteristics in Tarim oilfield.

Reservoir type	Typical reservoir sequence	Typical block
Medium porosity and medium(high) permeability sandstone reservoir	Carboniferous	Tazhong 4, Donghe, hade
	Jurassic and Triassic	Lunnan
	Paleogene and Cretaceous	Kela
Low porosity and low permeability sandstone reservoir	Silurian	Tazhong 10
Fractured tight sandstone gas reservoir	Cretaceous	Dina, Keshen, Dabei
Fractured vuggy carbonate reservoir	Ordovician Cambrian	Tazhong, Lungu, Yingmai, Yaha 7
Cave carbonate reservoir	Ordovician	Harahatang

2.1 *Design and construction of reservoir protection in the drilling stage*

During drilling, the drilling fluid system used in the reservoir section is mainly water-based and oil-based drilling fluid. The key points of the designed reservoir protection formula are as follows. The field implementation is basically based on the design, adding a certain amount of reservoir protective agent to ensure that the permeability recovery value of drilling fluid is qualified. The test results of drilling fluid permeability recovery value of carbonate and sandstone reservoirs in the recent five years are shown in Figures 1 and 2.

Reservoir protection measures designed for water-based drilling fluid:

a. Near balanced pressure drilling, the pressure difference should be kept as small as possible;
b. The rheological property of drilling fluid is stable, the filtration loss of drilling fluid is low, the quality of mud cake is better, and the solid phase control is strengthened;
c. Acid soluble or oil soluble temporary plugging materials shall be used, and limestone powder shall be used for weighting in the whole well section (less than 1.35).
d. Shielding temporary plugging: 2 ~ 3% oil soluble temporary plugging agent, 1 ~ 2% shielding temporary plugging agent type 1, 1 ~ 5% shielding temporary plugging agent type 2, 0.02 ~ 0.05% surfactant;
e. Permeability recovery value meets relevant regulations.

Oil-based drilling fluid reservoir protection measures:

a. For near-equilibrium pressure drilling, the density of drilling fluid should be strictly controlled;
b. The rheological property of drilling fluid is stable, the filtration loss is low, the mud cake quality is good, and the solid phase control is strengthened;
c. Acid-soluble or oil-soluble temporary plugging materials are used to prohibit the use of permanent plugging materials;
d. Permeability recovery value meets relevant regulations.

Reservoir protection measures during site construction:

a. Near balanced pressure drilling is adopted.
b. Limestone powder shall be used for weighting when drilling;
c. Good performance of drilling fluid is maintained to ensure rapid drilling;
d. Solid control equipment well is used to remove inferior solid phase in time and keep drilling fluid low solid phase;
e. 1% TYZ-8, 1.5% YX-1, and 1.5% YX-2 are timely added to protect the oil and gas reservoir.

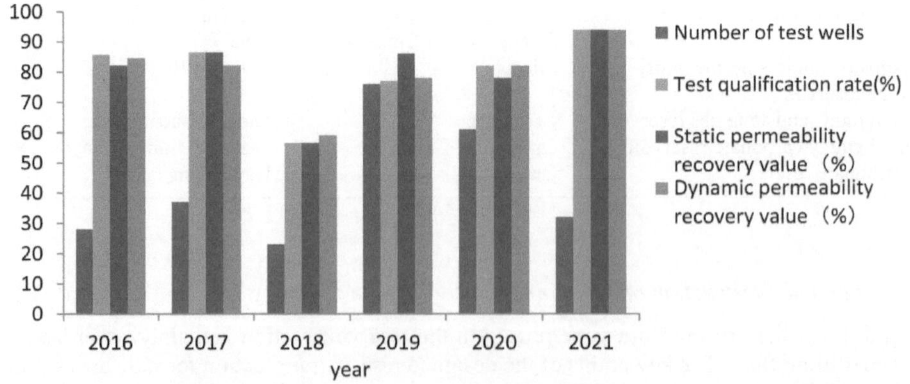

Figure 1. Detection of permeability recovery value of clastic rocks in Tarim Oilfield in recent five years.

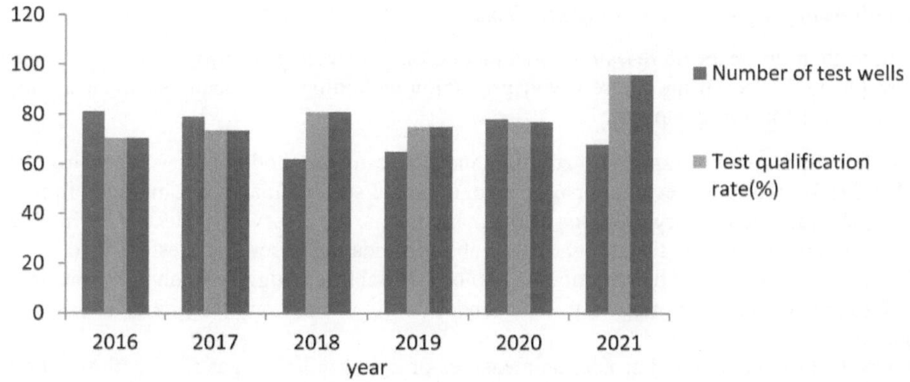

Figure 2. Detection of carbonate permeability recovery value in Tarim Oilfield in recent five years.

2.2 *Construction effect of reservoir protection in the drilling stage*

Figures 3 and 4 show the reservoir protection effect of development wells drilled in the recent three years. The data show that 100 clastic rock development wells have been drilled, and 71 wells have been directly put into production, accounting for 71%; 152 carbonate development wells have been drilled and 39 wells have been directly put into operation, accounting for 25.66%. The protection effect is poor.

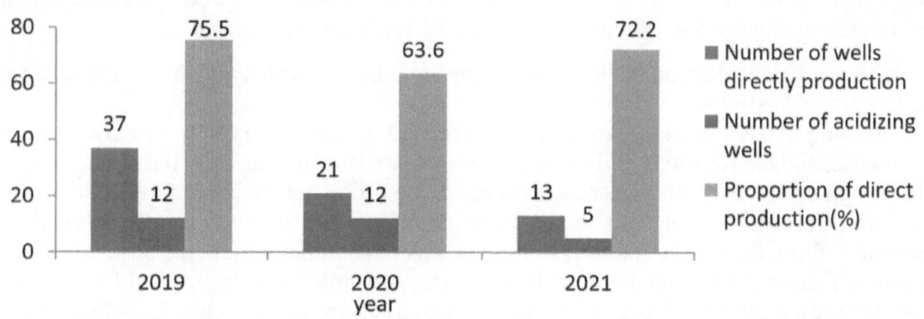

Figure 3. Statistics of the proportion of clastic rock development wells directly put into production in Tarim Oilfield in recent three years.

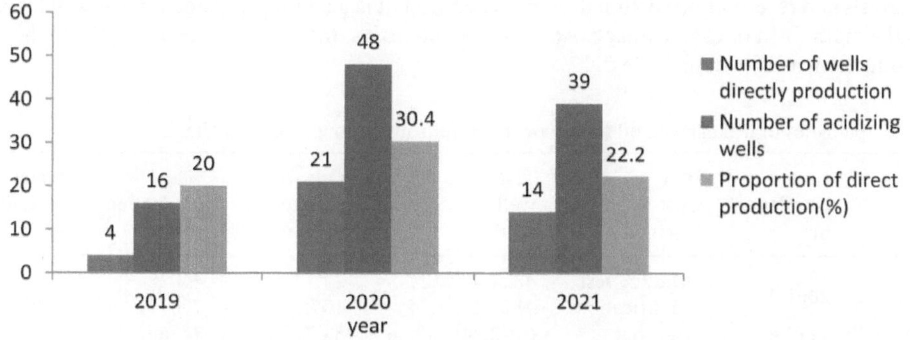

Figure 4. Statistics of the proportion of carbonate development wells directly put into production in Tarim Oilfield in recent three years.

2.3 *Analysis of progress and existing problems*

After more than 20 years of research and Practice on reservoir protection in the drilling stage of Tarim Oilfield, a set of unique reservoir protection technology has been established, which is embodied in the following aspects:

1) Based on the industrial standards "experimental evaluation method of reservoir sensitivity flow (SY / t5358-2010)" and "evaluation method of dynamic/static damage of completion fluid", and taking shielding temporary plugging as the main means, the reservoir protection technology of medium porosity and medium (high) permeability sandstone reservoir is established.
2) The evaluation method and protection technology of fractured vuggy carbonate reservoir based on fracture modeling formation fracture and shielding temporary plugging technology are established.
3) The evaluation method and protection measures of tight sandstone gas field with fractures are explored and established based on the evaluation methods such as pressure attenuation method or fracture making experiment, and the concept of total acid dissolution of oil-based drilling fluid, nitrogen drilling fluid, and water-based drilling fluid.

Through research and implementation, the oilfield has issued many documents, such as operating instructions for sandstone reservoir protection technology in Tarim Oilfield and management measures for clastic rock reservoir protection and development of Tarim Oilfield Company, to standardize the reservoir protection technology in the drilling stage. However, with the deepening of exploration and development and the implementation of the concept of reservoir protection in the whole process, A series of problems have been exposed in reservoir protection during oilfield drilling, which restricts the in-depth development of reservoir protection and the realization of the goal of improving quality and efficiency (Li et al. 2011; Zhang et al. 2007).

1) There is a lack of evaluation methods for unconventional reservoirs, and the results of different studies are inconsistent.
2) The existing shielding temporary plugging technology and reservoir protection have poor pertinence, and the formulas of different types of reservoirs are basically the same.
3) The design of the reservoir protection scheme in the drilling stage is not fully combined with subsequent processes. For tight sandstone reservoir, several research results reveal that the depth of drilling fluid polluting the reservoir is tens of centimeters, and the perforating charge is easy to penetrate this damage zone. It is necessary to protect the drilling fluid reservoir. The carbonate reservoir framework and interstitial materials are acid-soluble, and acidification is generally required before production. It is necessary to use acid-soluble limestone for weighting and shielding in the drilling stage.
4) The damage to the cave-type reservoir is controversial and its application is confusing. The size and scale of underground caves are not clear, which is not conducive to reservoir damage analysis and reservoir protection design. In addition, it is generally considered that well leakage is the main form of cave damage. According to statistics, the larger the leakage, the higher the production, as shown in Table 2.

Table 2. Statistics of leakage and oil testing of some wells in Tazhong, Tarim Oilfield.

Well code	Lost circulation m^3	Oil test method	Test well section	Oil production m^3/d	Gas production m^3/d	Water production m^3/d
ZG511	2061.4	Midway test	4824-4836.27	24.72	44828	–
		acidification	4824-5022.74	107.77	70677	5.46
ZG103	1914.4	Blowout	6148.446233.46	53.77	74366	
ZG11-H7	1027	Blowout	6263-6302.7	31.68	35758	2.64

(continued)

Table 2. Continued.

Well code	Lost circulation m^3	Oil test method	Test well section	Oil production m^3/d	Gas production m^3/d	Water production m^3/d
ZG111	155.3	Blowout	6008-6250	62.86	105395	1.94
ZG21-H5	99.6	Blowout	5816-6210	18.18	14974	82.44
ZG22	121	Blowout	5605-5736.66	40.78	70260	1.4
	3191.55	Midway test	6059.58-6279	19.44	166164	45.56
ZG14-2H	162.2	Blowout	6240.2-6550.5	/	29556	/
		acidification	6240.2-6550.5	15.69	20520	/
ZG431-H1	950.8	Blowout	5350.01-5567	39.86	82179	/
ZG433C	49.1	Blowout	6125-6221.71	77.84	12688	/
ZG45	388.1	Blowout	5637.2-5650.17	52.96	34044	11.84
ZG44-H2C	545.8	acidification	5350-6755	37.62	59670	1.32

3 DAMAGE MECHANISM AND PROTECTION STRATEGY OF DIFFERENT TYPES OF RESERVOIRS

The general idea of the protection strategy is to divide the reservoir design scheme, design a strategy according to the reservoir type, refine the strategy according to the sub-blocks, and establish a protection strategy for each block. The establishment of a protection strategy is designed according to the ideas of reservoir identification, damage mechanism, evaluation method, and protection measures (Zhang et al. 2009, et al. 2015).

3.1 *Medium porosity and medium(high) permeability sandstone reservoir*

1) Basic characteristics of the reservoir. This kind of reservoir has good physical properties and a large pore size. Generally, the formation contains a certain amount of expansive clay minerals, resulting in strong water/salt sensitivity insensitivity test evaluation. Due to the large pore size, weak velocity sensitivity, and stress sensitivity, see Tables 3 and 4. Reservoir protection should mainly prevent liquid and solid phases from invading the reservoir.

Table 3. The whole-rock analysis and clay mineral content of medium porosity and medium(high) permeability sandstone reservoir in the oilfield.

		The mineral content of whole-rock analysis %							
Strata	Typical block	Clay mineral	Quartz	Potassium feldspar	Plagioclase	Calcite	Dolomite	Anhydrite
Carboniferous	Donghe	8.0	80					
	Hade	6.56	68.3	5.23	11.33	7.52	2.56	0.11
Triassic	Lunnan	5.84	44	16	26			

		Clay mineral content %					
Strata	Typical block	Montmorillonite	Illite	Kaolinite	Chlorite	Illite/ montmorillonite	Interlayer ratio%
Carboniferous	Donghe		8.0	60	22	10	15
	Hade	–	31	–	38	31	20
Triassic	Lunnan		17.0	55.7	17.9	7.4	

Table 4. The sensitivity of medium porosity and medium(high) permeability sandstone reservoir in oilfield.

Strata	Typical block	Porosity %	Permeability mD	Pore thickness size μm	Velocity sensitivity	Water sensitivity	Salinity sensitivity	Alkali sensitivity	Differential pressure sensitivity
Carboniferous	Tazhong4	19.2	932	46	Nothing	Medium-strong	Medium-strong	Medium-weak	Medium-weak
	Donghe	11	222	22	Nothing	Medium-strong	Medium-strong	Weak	Weak
	Hade	10.6	221	19	Nothing-weak	Medium-strong	Medium-strong	Weak	Weak
Triassic	Lunnan	18.7	155	15	Nothing-weak	Medium	Medium-strong	Medium-weak	Nothing

2) Reservoir damage mechanism: the damage mechanism and causes of drilling fluid to the reservoir are analyzed as follows:
   a. The size of reservoir pore roar is larger than the size range of solid particles in drilling fluid, which provides a channel for solid invasion of drilling fluid. Due to the non-uniformity of pore roar, it is easy to block in the roar and reduce permeability. Attention should be paid to protection.
   b. The reservoir contains a certain amount of expansive clay minerals with strong water/salt sensitivity. Drilling fluid enters the reservoir through filtration, causing sensitivity damage. Attention should be paid to protection.
   c. The content of kaolinite in the reservoir is small, the migration of formation particles is not serious, and the structural characteristics of pore roar determine that it can be eliminated in time after migration, so it is not easy to block the pore roar.
   d. The pore structure size of the reservoir is large, the capillary force is small, the aqueous phase is not easy to stay in the reservoir, and the water lock has little damage to the reservoir.
   e. There is weak stress sensitivity in the reservoir. The difference between the drilling fluid column pressure and the formation pressure is small in the drilling stage, and the permeability decreases under the action of the pressure difference, which is conducive to preventing the filtration and leakage of drilling fluid from entering the reservoir, resulting in damage, that is, the drilling pressure difference will not cause stress-sensitive damage to the reservoir.
   f. The pressure difference between the drilling fluid column and the formation pressure will not cause stress sensitivity damage in the drilling stage, but the excessive pressure difference acting on the reservoir will increase the risk of drilling fluid leakage and cause drilling fluid leakage damage. Therefore, attention should be paid to reservoir protection.
3) Reservoir protection strategy: the adopted reservoir protection strategy is as follows:
   a. Shielding temporary plugging technology is adopted to prevent solid particles and liquid drilling from invading the reservoir and reduce solid and liquid damage. The grading of the shielding temporary plugging agent designed according to the pore size of this kind of reservoir is 800 mesh: 600 mesh: 200 mesh = 20:25:55.
   b. Try to control the low drilling fluid filtration, reduce the amount of liquid entering the reservoir due to filtration and reduce the liquid phase damage.
   c. Try to control the salinity of drilling fluid within the critical salinity range to reduce salt sensitivity damage;
   d. Adopt leakage prevention and plugging technology while drilling, control tripping speed, adjust drilling fluid performance, reduce circulating pressure consumption and reduce drilling fluid leakage risk;
   e. Acid soluble weighting materials and leak-proof and plugging materials may be used, which is conducive to the injury and can be relieved in the subsequent process.

3.2 *Low porosity and low permeability sandstone reservoir*

1) Basic characteristics of reservoir: this kind of reservoir has poor physical properties, small pore size, and large capillary force, and the reservoir contains a certain amount of clay minerals, including a high content of kaolinite, serious velocity sensitivity in sensitivity evaluation, and weak other sensitivities, as shown in Tables 5 and 6. Reservoir protection should focus on preventing the liquid phase from invading the reservoir.

Table 5. The whole-rock analysis and clay mineral content of low porosity and low permeability sandstone reservoir in oilfield.

		The mineral content of whole-rock analysis %							Clay mineral content %				
Strata	Typical block	Clay mineral	Quartz	Potassium feldspar	Plagioclase	Calcite	Dolomite	Anhydrite	Montmorillonite	Illite	Kaolinite	Chlorite	Illite/ montmorillonite
Silurian	Tazhong10	7.6	66	8	6	4	1	7.4	–	28.9	38.4	5.6	27.1

Table 6. The sensitivity of medium porosity and medium(high) permeability sandstone reservoir in oilfield.

Strata	Typical block	Typical block	Porosity %	Permeability mD	Velocity sensitivity	Water sensitivity	Salinity sensitivity	Alkali sensitivity	Differential pressure sensitivity	Water lock damage
Silurian	tazhong10	11.34	17.7	1.24-2.77	strong	Weak	medium-weak	medium-weak	medium	medium-weak

2) Reservoir damage mechanism: the damage mechanism and causes of drilling fluid to the reservoir are analyzed as follows:
   a. The pore size of the reservoir is small, and the solid particle size of the drilling fluid is large. It is not easy to enter the reservoir and will not have a great impact on the reservoir.
   b. The reservoir contains a certain amount of clay minerals. The invasion of drilling fluid will cause water and salt sensitivity damage to a certain extent, which should be paid attention to in the protection.
   c. The content of kaolinite and illite in the reservoir is high, the formation particles are easy to fall off and migrate, and the formation pore roar is small. Once the reservoir particles migrate, they will block the pore roar and it is not easy to eliminate, so the velocity sensitivity is strong. However, in the drilling stage, there will be no large-area liquid flowing in the reservoir, so the velocity sensitivity influence in the drilling stage is small.
   d. The pressure difference between the drilling fluid column and the formation pressure will not cause stress sensitivity damage in the drilling stage, but the excessive pressure difference acting on the reservoir will increase the risk of drilling fluid leakage and cause drilling fluid leakage damage. Therefore, attention should be paid to reservoir protection.
3) Reservoir protection strategy: the adopted reservoir protection strategy is as follows:
   a. The shielding temporary plugging technology is adopted to form a better mud cake and reduce the invasion of drilling fluid into the reservoir. The grading of the shielding temporary plugging agent designed according to the pore size of the reservoir is 2000 mesh: 1000 mesh: yx-2 = 30:45:25.
   b. Try to control the low drilling fluid filtration, reduce the amount of liquid entering the reservoir due to filtration and reduce the liquid phase damage.
   c. Adopt leakage prevention and plugging technology while drilling, control tripping speed, adjust drilling fluid performance, reduce circulating pressure consumption, and reduce drilling fluid leakage risk;
   d. Acid soluble weighting materials and leak-proof and plugging materials may be used, which is conducive to the injury and can be relieved in the subsequent process.

3.3 *Fractured tight sandstone gas reservoir*

1) Basic physical properties of reservoir: the matrix of this kind of reservoir is dense, the pore size is very small, the capillary force is very large, the solid particles of drilling fluid are not easy to enter the reservoir, and the liquid phase is easy to enter. Once entering, it is not easy to eliminate, and all kinds of water lock and fluid sensitivity are very strong, as shown in Tables 7 and 8. Reservoir protection is mainly to prevent liquid phase invasion and fracture leakage.

Table 7. The whole-rock analysis and clay mineral content of fractured porous sandstone reservoir.

Strata	Typical block	The mineral content of whole-rock analysis %							
		Clay mineral	Quartz	Potassium feldspar	Plagioclase	Calcite		Dolomite	Anhydrite
Cretaceous	Keshen	4.2	53	7.6	29	0		5.82	0
	Dabei	5.2	26.5	3.1	7.5	3.1		0	0
	Bozhi	5.0	58	8	14	8		2	0
Strata	Typical block	Clay mineral content %							
		Montmorillonite	Illite	Kaolinite	Chlorite	Illite/ montmorillonite	Interlayer ratio%		
Cretaceous	Keshen	–	61	11	13	15	15		
	Dabei	–	67	0	23	10	15		
	Bozhi	–	72	12	7	9	20		

Table 8. The sensitivity of fractured porous sandstone reservoir.

Strata	Typical block	Porosity %	permeability mD	Pore thickness size μm	Velocity sensitivity	Water sensitivity	Salinity sensitivity	Alkali sensitivity	Differential pressure sensitivity	Water lock damage
Cretaceous	Dina	8.97	0.99	0.5–1.5	Weak	Medium-strong	Medium-strong	Medium-weak	Medium-strong	Medium
	Keshen	5.77	0.1	0.03–0.66	Nothing	Medium-strong	Strong	Strong	Medium-strong	Medium-strong
	Dabei	4.29	0.02–0.2	0.02–0.13	Weak	Strong	Medium-strong	Strong	Medium-strong	Medium-strong
	Bozhi	3.0–6.0	0.01–0.8	0.15–0.5	Weak	Medium-strong	strong	Medium-weak	Strong	Strong

2) Reservoir damage mechanism: the damage mechanism and causes of drilling fluid to the reservoir are analyzed as follows:
   a. The fast pore roar of the reservoir base is very small, and the solid particles of drilling fluid and formation particles are large, so they will not migrate and block the roar channel, and the solid damage is very small relative to the fast pore roar of reservoir base.
   b. The reservoir is in the gas reservoir environment, with a fast foundation, special wettability, small pore roar, large capillary force, easy liquid phase to enter the reservoir, and difficult retention and discharge, resulting in great water (liquid) lock damage. Attention should be paid to reservoir protection in the drilling stage.
   c. The reservoir foundation contains 4-6% clay minerals, but due to the small seepage channel and the great influence of clay mineral expansion on the seepage channel, the sensitivity of various fluids is medium to strong, which should be paid attention to in the drilling stage.
   d. There are a large number of fractures and microfractures in this kind of reservoir, which facilitates the leakage of drilling fluid into the reservoir. After the leakage, the solid phase

of drilling fluid will block the fractures, and the liquid phase will cause a fast foundation, strong fluid sensitivity, and water lock.

e. The pressure difference between the drilling fluid column and the formation pressure will not cause stress sensitivity damage in the drilling stage, but the excessive pressure difference acting on the reservoir will increase the risk of drilling fluid leakage and cause drilling fluid leakage damage. Therefore, attention should be paid to reservoir protection.

3) Reservoir protection strategy: the adopted reservoir protection strategy is as follows:
   a. This kind of tight reservoir is seriously damaged, so it is necessary to build two shielding and plugging layers. The first one is to block the fractures, and the leakage prevention and plugging technology while drilling is adopted to reduce the reservoir differential pressure and leakage; The second way is to block the basic fast hole roar and prevent the entry of the liquid phase. Technologies such as isolation membrane and nonpermeable membrane can be adopted;
   b. Try to control the low drilling fluid filtration, reduce the amount of liquid entering the reservoir due to filtration and reduce the liquid phase damage.
   c. Adopt leakage prevention and plugging technology while drilling, control tripping speed, adjust drilling fluid performance, reduce circulating pressure consumption, and reduce drilling fluid leakage risk;
   d. Try to control the salinity of drilling fluid within the critical salinity range to reduce salt sensitivity damage;
   e. Acid soluble weighting materials, leakproof and plugging materials, and unblocking technology shall be used as much as possible;
   f. Select low damage process technology (nitrogen) and drilling fluid (oil-based, etc.)

3.4 *Fractured vuggy carbonate reservoir*

1) Reservoir damage mechanism: the damage mechanism and causes of drilling fluid to the reservoir are analyzed as follows:
   a. Carbonate reservoir is calcareous cementation, basically free of clay minerals, and the porosity of the base block is very small (1-2%). Generally, water/salt sensitivity damage will not be caused after the entry of the drilling fluid liquid phase;
   b. There are fractures in the reservoir. Solid particles such as weighting materials will enter the reservoir fractures and block the fractures, resulting in certain damage;
   c. The pressure difference between the drilling fluid column and the formation pressure will not cause stress sensitivity damage in the drilling stage, but the excessive pressure difference acting on the reservoir will increase the risk of drilling fluid leakage and cause drilling fluid leakage damage. Therefore, attention should be paid to reservoir protection.
   d. The reservoir is temporarily blocked by fine carbonate particles in the dolomite, which is easy to block the water in the reservoir.
2) Reservoir protection strategy: the adopted reservoir protection strategy is as follows:
   a. Adopt leakage prevention and plugging technology while drilling, control tripping speed, adjust drilling fluid performance, reduce circulating pressure consumption, and reduce drilling fluid leakage risk;
   b. Acid soluble weighting materials and leakproof and plugging materials may be used, which is conducive to the injury and can be relieved in the subsequent process.
   c. For the dolomite reservoir, the shielding temporary plugging technology is adopted to form a better mud cake and reduce the invasion of drilling fluid into the reservoir. The grading of the shielding temporary plugging agent is designed according to the size of the reservoir pore roar.

3.5 *Cave carbonate reservoir*

1) Reservoir damage mechanism: the damage mechanism and causes of drilling fluid to the reservoir are analyzed as follows:

a. The matrix is dense, the solid particles are difficult to enter the formation, and the formation particles cannot migrate, so the particles have little impact on the reservoir damage;
b. Leakage easily occurs in the drilling stage, and a large number of drilling fluid phases will enter the karst cave with rich formation fluid. However, whether the drilling fluid leakage causes damage to the cave in the reservoir and the degree of damage cannot be experimentally evaluated.
c. The pressure difference between the drilling fluid column and the formation pressure will not cause stress sensitivity damage in the drilling stage, but the excessive pressure difference acting on the reservoir will increase the risk of drilling fluid leakage and cause drilling fluid leakage damage. Therefore, attention should be paid to reservoir protection.

2) Reservoir protection strategy: the adopted reservoir protection strategy is as follows:
a. Because the reservoir matrix is dense and acid-soluble, shielding temporary plugging technology cannot be used;
b. A large amount of drilling fluid leakage will increase the economic cost. Therefore, leakage prevention and plugging technology should be adopted in the drilling stage to reduce leakage.

Through the above analysis and research, the reservoir protection schemes of different reservoir types are established. In the actual drilling, the damage mechanism of drilling fluid in specific blocks shall be determined according to the drilling experiments in the block, and the reservoir protection strategies and formulas in different blocks shall be designed by means of software and experiments to guide the on-site construction, and according to the on-site implementation effect, it is needed to constantly adjust the reservoir protection strategy of each block. In order to improve the depth of reservoir protection in the drilling stage, improve the effect of reservoir protection and ensure the smooth implementation of the reservoir protection scheme, the following work needs to be carried out at the same time.

1) Based on the experiment of drilling fluid permeability recovery value, according to different damage types, a variety of means are used to establish the guidelines for drilling fluid damage evaluation and standardize the evaluation of drilling fluid damage
2) Make full use of seismic and adjacent well data to analyze reservoir characteristics and determine reservoir types.
3) Make full use of the fluid sensitivity prediction software to help increase the experimental efficiency.
4) The shielding temporary blocking formula is finely designed to solve the problem that the formula is not targeted. The shielding temporary plugging particles shall be designed according to the formation hole roar through the ideal filling software; The formula of shielding temporary plugging is verified by indoor experiment, and the formula used on site is finally determined;
5) Enrich and improve oil conservation materials. Add 0.5-10 on the basis of existing ultrafine calcium carbonate μM ultrafine calcium carbonate; Develop a variety of variable particles with softening points according to the formation temperature; R & D, the introduction of isolation membrane and nonpermeable membrane, and experimental application.
6) Carry out the research on the damage mechanism and protection methods of dolomite reservoir and cave carbonate reservoir, establish and study the models of drilling fluid intrusion depth and damage depth in different reservoirs, study the field evaluation method of drilling fluid reservoir protection effect, improve the theoretical depth of reservoir protection and improve the basis of supporting the site.

4 CONCLUSION

1) Tarim Oilfield has many reservoir types and great changes in reservoir characteristics. The existing reservoir protection measures in the drilling stage have some problems, such as poor pertinence, inconsistent reservoir damage evaluation standards, unclear understanding of the damage mechanism of some types of reservoirs, and so on;

2) It is required to study and design different reservoir protection schemes according to different reservoir types. The medium porosity and medium (high) permeability sandstone reservoir protection scheme takes shielding temporary plugging (800, 600, and 200 mesh ultra-fine calcium carbonate particles) as the core, and the low porosity and low permeability sandstone reservoir protection takes shielding temporary plugging (2000, 1000, and 800 mesh ultra-fine calcium carbonate particles) as the core; the core of fracture pore tight sandstone reservoir is to optimize low damage process technology (nitrogen) and system (oil-based), leak-proof and plugging technology while drilling to block fractures and reduce leakage, isolation membrane, nonpermeable membrane and other technologies to block fast pore roar and prevent liquid phase from entering the reservoir; the protection of fractured vuggy carbonate reservoir adopts leakage prevention and plugging technology, with reducing drilling fluid leakage as the core.
3) In order to improve the depth of reservoir protection in the drilling stage and improve the effect of reservoir protection, the main work idea at this stage is to prepare and establish the guide for drilling fluid damage evaluation, standardize the drilling fluid damage evaluation, make full use of seismic and adjacent well data, analyze reservoir characteristics, determine reservoir types, and make full use of fluid sensitivity prediction software to assist in increasing experimental efficiency. Ideal filling software is used to design fine shielding and temporary plugging formula, enrich and improve oil conservation materials, carry out research on reservoir damage mechanism and protection methods such as dolomite reservoir and cave carbonate rock, establish reservoir damage depth model and field evaluation method of drilling fluid reservoir protection effect, so as to improve the theoretical depth of reservoir protection and enhance the basis of supporting the site.

REFERENCES

Aifeng, Li, Tianyou Fan, Lin Zhao, Experimental study on spontaneous imbibition of low permeability cores in fractured reservoirs[J]. Journal Petroleum Geology and Recovery Efficiency, 2011, 18 (5): 67–70.

Bennion D B, Bietz R F, Thomas F B, et al. Reductions in the Productivity of Oil and Low Permeability Gas Reservoirs due to Aqueous Phase Trapping[J]. Journal of Canadian Petroleum Technology, 1994, 33 (9): 45–54.

Bennion D B, Thomas F B, Bietz R F, et al. Water and Hydrocarbon Phase Trapping in Porous Media-Diagnosis, Prevention and Treatment[J]. Journal of Canadian Petroleum Technology, 1996, 35 (10): 29–36.

Bennion D B, Thomas F B, Ma T. Recent Advances in Laboratory Test Protocols to Evaluate Optimum Drilling, Completion and Stimulation Practices for Low Permeability Gas Reservoirs[C]. SPE 60324, 2000.

Fengdong Zhang, YIli Kang, Lijun You, et al. Damage model of water phase trap in fracture pore tight sandstone gas reservoir[J]. Natural Gas Geoscience, 2009, 20 (5): 812–816.

Fengdong Zhang, YIli Kang, Yu Yang, et al. Research Progress on damage mechanism and prevention of water phase trap in tight gas reservoir development[J]. Natural Gas Geoscience, 2007, 18 (3): 457–461.

Gao Li, Yingfeng Meng, Hongming Tang, et al. Experimental study on Countercurrent self-absorption effect of water-based underbalanced drilling in sandstone gas reservoir[J]. Natural Gas Industry, 2007, 27 (1): 75–77.

Jie Zhang, Xueqing Teng, Hui Yang, et al. Study on damage mechanism of oil-based drilling fluid to gas reservoir[J]. Drilling & Production Technology, 2015, 38(3): 91–94.

Lijun You, Yili Kang, Yijian Cheng, et al. Experimental study and application of water phase trap damage in tight sandstone gas reservoir[J]. DrillingFluid & Completion Fluid, 2006, 23 (2): 4–7.

Lucas R. Rate of Capillary Ascension of Liquids. Kolloid-Zeitschrift, 1918, 23: 15–22.

Shikui Li, Weidong Liu, Haiqing Zhang, et al. Experimental study on spontaneous imbibition displacement in low permeability reservoir[J]. Acta Petrolei Sinica, 2007, 28 (2): 109–112.

Washburn E W. Dynamics of Capillary Flow. Phys, 1921, 17 (3): 273–283.

Advances in Mineral Resources, Geotechnology and Geological Exploration – Rashid & Zhang (eds)
© 2023 Copyright the Author(s), ISBN 978-1-032-33377-9

Study on construction technology of super long and super deep retaining wall structure with prestressed anchor cable

Yinan Wang & Guangxiu Fang*
Department of Civil Engineering, College of Engineering, Yanbian University, Yanji, Jilin, China

Dingguo Wang
Jilin Shuangli Construction Project Management Co., Ltd., Yanji, Jilin, China

Wei Jang
Jilin Province Transportation Construction Group Co., Ltd., Yanji, Jilin, China

Huiquan Wang
Yanji Construction Project Quality Supervision Station, Yanji, Jilin, China

ABSTRACT: Taking the construction of an ultra-long and ultra-deep retaining wall with prestressed anchor cable in the project of Guantang Hot Spring Resort Hotel in Yanji City, Jilin Province as an example, the geological and hydrological conditions are analyzed, the construction process, construction method, and key control points are put forward, and the construction technical measures are deepened. The type selection, supporting scheme, and technical measures for strengthening the construction of the slope retaining wall support project are feasible, meet the requirements of the current Technical Specification for Retaining and Protection of Building Foundation Excavations (JGJ 120–2012), and shorten the construction period and reduce the project cost, which provides a reference for the construction organization and application of similar retaining wall structures.

1 INTRODUCTION

Although the supporting structure of the deep foundation pit is a temporary retaining structure during construction, its selection, calculation, and construction are correct and reasonable, which has a significant impact on the safety, duration, and economic benefits of the project (Wang & Yin 2015). Row piles (Mao 2015) are usually used to support the foundation pit when the excavation depth is between 6 and 14 m. The row piles are composed of a series of adjacent piles, and their function is to resist the earth's pressure. Prevent slope instability. At present, for the super-long and super-deep retaining wall with a specific structure, when the retaining wall with prestressed anchor cable is designed, the calculation method of its supporting mode is not uniform, and the research on deepening construction technology is scarce (Ying et al. 2019), so it is of great significance to carry out research on it. Based on the engineering case, The feasible construction technology and method of prestressed anchor cable retaining wall (Xu et al. 2019) are analyzed and studied, and the key points of quality control and strengthening construction technical measures are put forward, which can be used for reference for similar projects.

*Corresponding Author: fangjian888888@163.com

DOI 10.1201/9781003319412-8

2 PROJECT OVERVIEW AND HYDROGEOLOGICAL CONDITIONS

2.1 *Project overview*

Kwun Tong Hot Spring Resort Hotel Project is located in the south of Yanji City, Yanbian Korean Autonomous Prefecture, Jilin Province, east of the intersection of Yannan Road and Yanlong Road, near Yanji Dinosaur Museum and Maoershan National Forest Park. The supporting object is the soil mass on the west side of the Guantang Hot Spring Resort Hotel Project, and the project plan is shown in Figure 1.

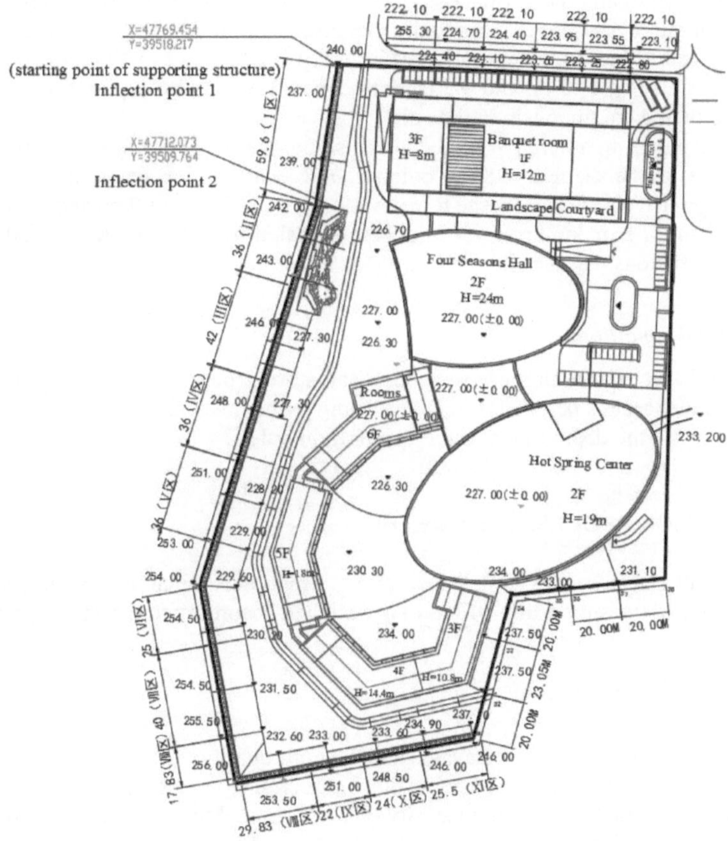

Figure 1. Project plan.

The length of the retaining wall is 415.10m, the elevation of the ground line at the top of the slope of the retaining wall is 237.00m~256.00m, and the elevation of the ground line at the bottom of the slope of the retaining wall is 225.00m~237.11m, with a large fluctuation. The minimum distance from the east and north sides of the retaining wall to Kwun Tong Resort Hotel is 19.50m, and there are no other buildings on the west and south sides of the retaining wall.

2.2 *Formation conditions*

Layer ①: plain fill, the layer thickness revealed by exploration is 0.70~1.80m, and it is continuously distributed in the site.

Layer ②: silty clay with sand, plastic and hard, medium compressibility, medium-dry strength, and toughness. The top buried depth is 0.70~1.80 m, the top elevation is 236.22~261.89 m, and the bottom buried depth is 3.40~12.80 m. The layer thickness revealed by exploration is 2.40~11.10 m. Continuously distributed on the site. The thickness of the sand layer is 0.10 ~ 0.40 m, and it is in lump shape locally.

Layer ③: Completely weathered argillaceous siltstone, the rock structure is basically destroyed, with residual structural strength, drillable by dry drilling, and the rock core is cohesive soil mixed with coarse sand or coarse sand mixed with cohesive soil. The top buried depth is 3.40 ~ 12.80 m, the top elevation is 228.02~255.79 m, and the bottom buried depth is 13.50~17.20 m. The layer thickness revealed by exploration is 4.40~11.60 m, which is distributed continuously in the site.

Layer ④: intensely weathered argillaceous siltstone, most of the rock structure is damaged, and the strength of the rock stratum is gradually increased from top to bottom. The buried depth of layer top is 13.50~17.20 m, the elevation of layer top is 222.72~247.99 m, the buried depth of layer bottom is 16.50~21.0 m, and the layer thickness revealed by exploration is 1.0~5.10 m, which is distributed continuously in the site.

Layer ⑤: moderately weathered argillaceous siltstone, the rock structure is partially damaged, the rock core is in the shape of a short column, and locally in the shape of the fragment. The top buried depth is 16.50~21.0 m, the top elevation is 219.42~246.09 m, and the drilling depth is 22.0~30.0 m. The maximum layer thickness revealed by exploration is 12.60 m, which is distributed continuously at the site.

2.3 *Hydrological conditions*

The cohesive soil layer in the proposed site area can store groundwater in wet years or seasons, and the groundwater type is perched water. According to the investigation report, no underground water is found within the depth range of the exploration hole.

3 CONSTRUCTION PROCESS AND METHOD

The retaining wall structure of cast-in-place pile + prestressed anchor cable is selected as the supporting form, and the supporting structure is designed according to the safety grade.

3.1 *Pile foundation construction*

3.1.1 *Drilling*

① Mud preparation: the mud pit is excavated and the high-quality clay is taken to prepare the mud. According to the geological conditions of this bid section, the mud performance indicators are the relative density of 1.1–1.2, the viscosity of 19–22 S, and the sand ratio of less than 4%.

② Drilling speed: the sand layer shall be controlled at about 1 meter per hour, the clay layer shall determine the drilling speed according to the actual situation, and decompression drilling shall be adopted.

3.1.2 *Fabrication of reinforcement cage*

① The reinforcement cage is fabricated by the support forming method: the support is composed of a fixed body and a movable body. Select 3~4 cm thick wood board, and make it into semicircular fixed support according to the design size of the skeleton. At its peripheral edge, the groove for supporting the main reinforcement is determined according to the position of the main reinforcement. The fixing bracket is fixed to the ground with two 4*10 cm struts. A semicircular movable bracket is arranged above it, and several wooden boards are selected with a thickness of 3–4 cm and nail them on the two wooden bars on the outward inclined surface at the lower end. The two ends of each batten of the movable support are also provided with grooves for supporting the main reinforcement according to the position of the main reinforcement. The lower end of the oblique batten of the movable support is fixed on the fixed support by a bolt. The two semicircular supports are connected together to form concentric circular support. According to the length of the skeleton, Set one every 2 meters or so. Each support shall be parallel to the other, and the center of the circle shall be on the same horizontal line. The main reinforcement is put into the groove one by one, and then the stirrup is put on the periphery of the skeleton according to the design position, and then it is bent into a hoop and bound with the main reinforcement. After binding the stirrups, it is required to remove the connecting bolts of the movable support and the fixed support, and draw out the movable support from both ends of the framework. The whole skeleton can be taken down from the fixed bracket.

② The reinforcement cage shall be processed into a reinforcement framework in advance according to the design requirements, which shall be put in place as a whole or sections, and lifted into the drilling hole. Before hoisting the reinforcement cage, it is needed to check whether the hole depth meets the design requirements and whether the hole wall has the conditions that hinder the cage hoisting in places, such as diameter reduction and hole deviation. The reinforcement cage shall be hoisted by a crane, and the cage shall not collide with the hole wall during hoisting. In order to ensure the thickness of the concrete protective layer outside the framework, cement mortar protective layer cushion blocks shall be set every 2 meters in the longitudinal direction and every 1/4 of the circumference in the transverse direction, and the framework position shall be corrected at any time. After the framework reaches the design elevation, the framework shall be firmly fixed at the orifice, and concrete shall be poured immediately.

(a) Fabrication of reinforcement cage (b) Hoisting of reinforcement cage by crane

(a) Fabrication of reinforcement cage (b) Hoisting of reinforcement cage by crane

Figure 2. Fabrication and hoisting of reinforcement cage.

3.1.3 *Concrete pouring*

① Lowering the conduit: it is required to number the conduit and mark the scale, put the conduit into the hole in sections with a crane, leave a distance of 20~40cm from the bottom of the hole at the lower end, connect a funnel to the upper opening of the conduit, and set a water plug at the interface to isolate the contact between the concrete and the water in the conduit.

② Storage bin: In order to facilitate a certain impact force during concrete pouring, a storage bin shall be set above the funnel. The storage bin shall be 2~2.5 meters higher than the funnel mouth, and its capacity shall meet the requirement that the volume of the buried pipe shall not be less than 1 meter.

③ Mixing concrete: to make the cast-in-place pile concrete have good fluidity, the slump shall be controlled at 18~22cm. The concrete shall be mixed in the mixing plant, transported continuously by concrete transport vehicles, and pumped by delivery pumps.

④ Concrete pouring: firstly, it is needed to pour and store enough concrete in the storage bin and funnel, then cut off the lifting rope of the water stop, open the valve of the storage bin, so that the stored concrete together with the water stop falls to the bottom of the hole, and at the same time, the water level in the hole rises suddenly and overflows, indicating that the concrete has been poured into the hole. The first buried depth of the conduit shall be greater than 1.0m, and the buried depth of the conduit shall be calculated at any time during the concrete pouring process. The buried depth of the conduit shall be controlled within 2~6 meters. When lifting the conduit each time, the minimum buried depth of the conduit in the concrete shall not be less than 2 meters to prevent the pile from being broken.

In order to make the quantity of the first batch of poured concrete meet the requirements of the first embedment depth ($\geq$ 1.0m) of the conduit and filling the bottom of the conduit, the formula for calculating the quantity of concrete required is as follows:

$$V \geq \pi \mathrm{D}^2/4(H_1 + H_2) + \pi \mathrm{d}^2/4h_1 \tag{1}$$

Where:
V— note the quantity required for the first batch of concrete (m^3);
D — pile hole diameter (m);
H_1— spacing from pile hole bottom to tremie bottom, generally 0.4m;
H_2— initial burial depth of conduit (m);
D— inner diameter of conduit (m);
H_1— the height (m) required for the concrete column in the tremie to balance the external pressure when the concrete in the pile hole reaches the embedment depth H_2, i.e., $H_1 = H_w\gamma_w\gamma_c$;
H_w —refers to the depth of water or mud in the borehole (m);
γ_w— the gravity of water or mud in the borehole (kN/m^3);
γ_c— the unit weight of concrete mixture ($24kN/m^3$).

3.2 *Earth excavation*

Open excavation is adopted. As the soil is mainly argillaceous siltstone and the excavation depth is large, in order to maintain the stability and safety of the slope, it is proposed to adopt the excavation method from high to low, mechanical excavation, manual cooperation, section by section, and layer by layer. The side slope is protected by shotcrete.

3.3 *Slope shotcrete protection*

① Reinforcement of foundation pit slope shall be φ6, with a spacing of 200×200. Vertical additional reinforcement shall be 2φ18@3000, and horizontal additional reinforcement shall be 2 φ18@3000. The reinforcing mesh shall be laid after a layer of concrete is sprayed, and shall be firmly anchored and connected with soil nails.

② Key points of shotcrete construction: the thickness of net shotcrete is 80mm; The strength grade of waterproof concrete is C20; drain holes shall be made of galvanized steel pipes with a diameter of 100mm, the exposed part shall not be less than 100mm, and the spacing shall be 3000mm×2000mm.

③ Shotcreting shall be carried out in sections and sections in turn, and the shotcreting sequence in the same section shall be from bottom to top. When spraying, the lower part of the reinforcement and soil nail shall be sprayed first, and then the upper part.
④ Concrete slope protection curing: 0.5mm thick plastic film shall be used for curing.

3.4 *Anchor cable construction*

3.4.1 *Drilling*

① In the process of drilling, the formation change, drilling status, groundwater and other conditions of each hole shall be recorded on site.
② When the drilling depth exceeds 6m, it is required to remove the rock powder in the hole once every one meter of drilling. The diameter of the anchor hole is 275mm. In order to ensure the depth of the anchor hole, the actual drilling depth is 0.5m greater than the design depth (which should meet the standard value of the design uniaxial compressive strength).

3.4.2 *Fabrication of anchor cable*

The prestressed steel strand is $\varphi^s 15.2$, the length of the anchor cable is made according to the design hole depth length plus the reserved tensioning length of 1.2m, the anchoring section is a composite type, which is set as two sections of six bundles of steel strands L1 and L2, each section is connected with the bearing body by extrusion head (P-type anchorage), and the steel bearing plate is made of Q235 steel. One stringing ring and tightening ring shall be set every 3m. In the middle of the steel strand, a Φ25PVC grouting pipe passes through the isolation ring and the steel bearing body and is preset in the middle of the anchor cable to the bottom of the hole.

3.4.3 *Anchor cable installation*

The anchor body shall be installed manually, and a 1.5m long steel strand shall be reserved outside the anchor hole.

3.4.4 *Grouting*

① M35 cement mortar shall be used, with a cement-sand ratio of 1:1, the water-cement ratio of 0.45 ~0.5, and grouting pressure of not less than 1.0MPa. During mortar mixing, mortar strength confirmation test blocks shall be prepared on-site and submitted to the supervising engineer for signature after the curing period expires.
② The anchor rod shall be grouted to fill the anchoring section for the first time, and shall be grouted from the reserved orifice of the anchor head for the second time after being tensioned and locked. Grouting for the second time shall be repeated several times until it is full. M30 cement mortar is used for the primary grouting, and cement paste is used for the secondary grouting.

3.4.5 *Installation of anchorages*

After the concrete reaches 50% of the design strength after curing, tighten the nut on the end rod screw and press the bearing plate on the anchor head. The anchor cable locks the anchor backing plate and the lockset.

3.4.6 *Tensioning*

① The frame and anchorage section shall be tensioned after the concrete reaches 80% of the design strength. Before formal tensioning, the anchorage body shall be pre-tensioned twice, and the pre-tensioning load shall be 0.1 times the design load.
② The tensile stress of the anchor rod (cable) is applied in six levels, which are 0.1, 0.25, 0.5, 0.75, 1.0, and 1.1 of the design tensile force respectively, and are increased to the over-tensile load step by step. The stabilization time of the first five levels of the load is 5 minutes, and that of the last level is 15 minutes, and the elongation of each level of load is recorded respectively.

In order to reduce the loss of prestressing, the total tension elongation value shall include the overtension value, and the overtension value shall not exceed 1.15 times.

The anchor cable construction process and quality control key points are shown in Figure 3.

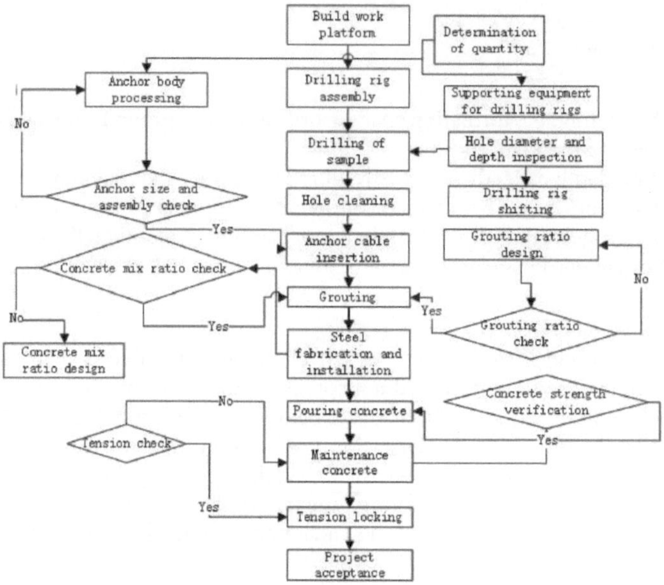

Figure 3. Anchor cable construction process.

③ For the stress dispersion type prestressed anchor cable, the unit anchor cables are required to be tensioned respectively first, and then the unit anchor cables are tensioned at the same time after the elastic elongation difference caused by the unequal length of the free section of each unit anchor cable under the same load condition is compensated. The prestress value of the anchor cable is 1.1 times of the design value, Tensioning shall be carried out twice in the sequence of the middle, upper, and lower (the two anchor cables shall go up first and then down). The first tensioning load shall be half of the design tension value, and the tensioning shall be carried out in three stages, which are 0.1, 0.25, and 0.5 times the design tension value respectively. The second tensioning shall be performed in the sequence of middle, upper, and lower to 1.1 times the design load. 0.75, 1.0, and 1.1 times of the design pulling force respectively.

④ After tensioning in place, lock and mechanically cut the redundant steel strand, and the remaining steel strand of the anchor head shall be ≥10 cm to prevent slipping. The anchor bolt shall be grouted first and then sealed with C30 fine aggregate concrete. The thickness of the concrete shall not be less than 100mm, and the curing shall be strengthened.

3.5 *Construction of waist beam and top beam*

The waist beam is connected with the anchor cable by I 20 B I-steel. The joints are welded with 300*150*8mm thick steel plates. The size of the top beam is 1300×800, and the concrete strength grade is C40.

4 CONCLUSION

Taking the construction of prestressed anchor cable type extra-long and extra-deep retaining wall structure of Guantang Hot Spring Resort Hotel project in Yanji City, Jilin Province as an example, the construction techniques and work methods of the pile foundation construction, such as drilling, steel cage fabrication and concrete filling, were analyzed according to the engineering characteristics of the construction project and the geological and hydrological conditions of the location.; earth excavation and slope shotcrete protection; anchor cable construction, including drilling, anchor cable making, and installation, grouting, anchor installation, tensioning, etc. were studied, and five key points for quality control of support construction were proposed, while measures to strengthen construction technology and quality were deepened. After construction verification, it meets the requirements of the current construction specifications and regulations, satisfies the quality inspection and evaluation standards, and can be used as a reference for the construction of similar projects.

ACKNOWLEDGMENT

This paper was supported by the Key Tackling Project of the Science and Technology Development Plan of the Jilin Provincial Science and Technology Department (Project No. 20170204032SF).

REFERENCES

Mao H.Q. Civil engineering construction (Fifth edition) [M]. (2021). Wuhan: Wuhan Science and Technology Press

Wang B.T, Yin D.Sh. Theoretical research and engineering application of earth pressure of retaining wall and foundation pit retaining structure [M]. (2015). Beijing: Science and Technology Press.

Xu J.L, Fang G.X, Zhao Sh. F, Han W.Q. (2019). The design of Excavating Piles Bored with Long Auger for Deep Foundation and Anchor Cable and Reinforced Methods. [J]. Building Technology, 50(03): 335–337.

Ying R.P, Fang G.X, Zhao Sh. F, Han W.Q. (2019). Design and deeper construct to artificial digging hole pile as supporting structure of deep foundation pit [J]. Building Technology, 50(01): 88–91.

Advances in Mineral Resources, Geotechnology and Geological Exploration – Rashid & Zhang (eds)
© 2023 Copyright the Author(s), ISBN 978-1-032-33377-9

Geochemical characteristics and genetic types of deep natural gas in dehui fault depression, Songliao basin

Weiqiang Gao*
Liaohe Oilfield Exploration and Development Research Institute, Panjin, Liaoning, China

ABSTRACT: The deep layers of the Dehui fault depression in the Songliao Basin have great potential for natural gas resources. With the continuous development of oil and gas accumulation theory, the understanding of the geochemical characteristics and genesis of deep natural gas in the Dehui fault depression is relatively backward, which seriously restricts the exploration and development and requires further detailed evaluation. Based on the collection of 122 natural gas samples, this paper analyzes the hydrocarbon and non-hydrocarbon gas components and carbon isotope characteristics of the samples by analyzing and testing experiments and identifying the genetic type in combination with the relationship between hydrocarbon gas components and carbon isotopes. The results show that the hydrocarbon components of the natural gas in the Dehui fault depression are mainly alkane gas. Among them, the average content of methane is 86%, the average content of ethane is 4.2%, and the average content of propane is 1.7%. The distribution of the drying coefficient of natural gas is 0.75~0.99, which reflects that the natural gas is mainly wet. Non-hydrocarbon components mainly include N_2 and CO_2, and the content is mainly concentrated in 2%~6% and 0.02%~1%. The carbon isotope characteristics of alkanes show that the distribution of $\delta^{13}C_1$ ranges from -50‰ to -26‰, the distribution range of $\delta^{13}C_2$ ranges from -35‰ to -20‰, and the distribution of $\delta^{13}C_3$ ranges from -33‰ to -24‰. Natural gas is mainly coal-derived, mixed with a small amount of oil-cracking gas and mixed gas. The results of the research can provide theoretical help for the further exploration and development of deep natural gas in the Dehui fault depression.

1 INTRODUCTION

Deep natural gas resources in the Dehui fault depression are abundant. In recent years, with the continuous improvement of exploration and development technology, three oil and gas fields, namely Nong'an, Buhai, and Xiaohelong, have been discovered successively, which shows wide exploration prospects. At present, there are some controversies about the characteristics of natural gas accumulation in this study area (Di 2007; Li et al. 2011). However, the factors restricting the accumulation of natural gas are the identification of the geochemical characteristics and genetic types of natural gas. Based on this, this paper collects natural gas samples in the oil field and tests them. The hydrocarbon, non-hydrocarbon gas components, and carbon isotope characteristics of natural gas samples are analyzed. Combining the experimental results, the genetic type of natural gas is identified. It is expected that this study can provide support for the exploration and development of deep natural gas in the Dehui fault depression.

2 GEOLOGICAL BACKGROUND

The Dehui fault depression is located in the middle of the eastern fault depression in the southeastern uplift area of the southern Songliao Basin, with an area of about 4000 km^2 (Figure 1). The Dehui

*Corresponding Author: gaowq01@petrochina.com.cn

 DOI 10.1201/9781003319412-9

fault depression belongs to the secondary structural unit and can be divided into Xiaohelong-Buhai structural belt, Xiaohelong subsag, Central anticline structural belt, Nong'andong subsag, and Nong'an-Wanjinta structural belt. The same as the entire southeastern uplift area and even the entire Songliao Basin, the Dehui fault depression has an obvious composite double-layer structure of fault-depression (Di 2007). The study area mainly develops Jurassic, Cretaceous, Neogene, and Quaternary strata. On this basis, three sets of source rocks are developed in the study area, whose lithology includes dark mudstone, carbonaceous mudstone, and coal. Reservoirs include sedimentary rocks and volcanic rocks. The sedimentary rock reservoirs are silt-fine sandstone and glutenite, and the volcanic rock reservoirs are volcanic rocks and pyroclastic rocks. At present, the discovered oil and gas reservoirs in the Dehui fault depression are dominated by structural traps, including fault block traps, fault nose traps, and anticline traps.

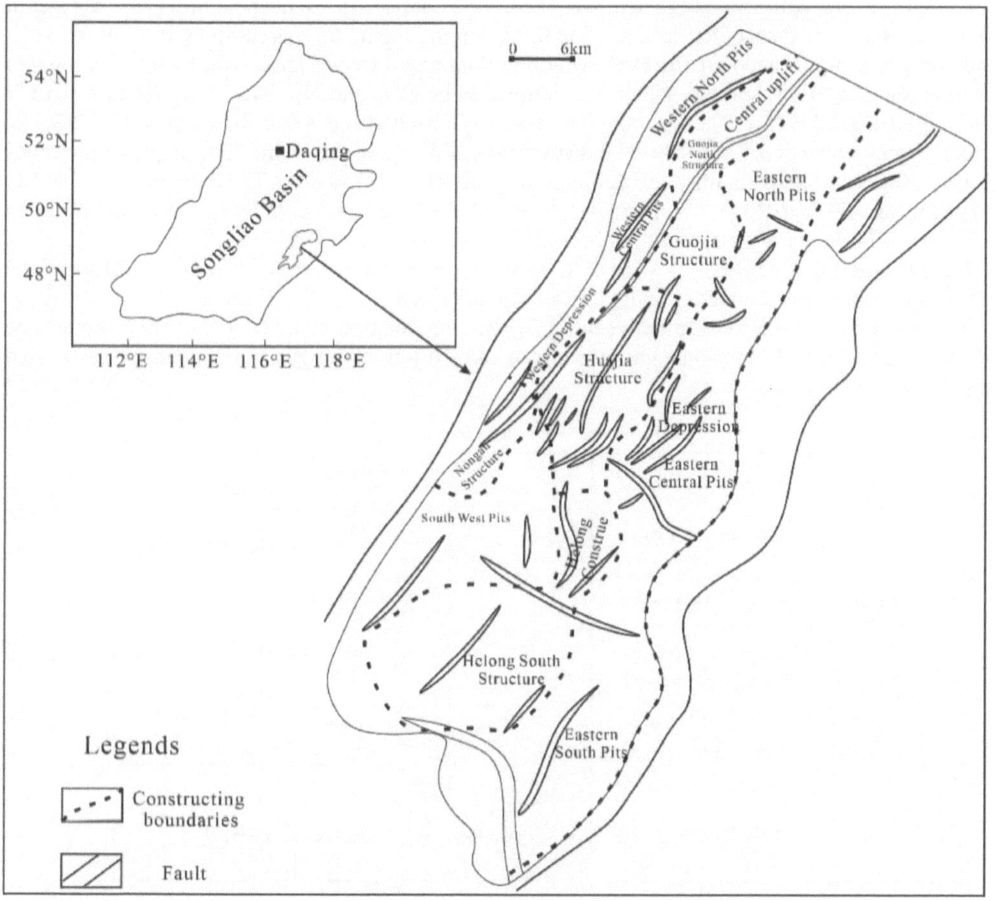

Figure 1. Geographical location and structural development characteristics of Dehui fault depression.

3 METHODS

In this study, a total of 122 natural gas samples from 45 wells were collected for component and carbon isotopic analysis. Component analysis was performed on an Agilent Model 6890N Gas Chromatograph instrument. The experiment was initially set at 25°C and heated at a constant temperature for 10 minutes. After 10 minutes, the temperature rate was set to 10°C/min, the experimental temperature was increased from 30°C to 180°C, and finally, the temperature was held at 180°C for 20 to 30 minutes. The carbon isotope analysis of the gas was done on GC/IRMS. The experimental temperature was first increased from 25°C to 80°C at a rate of 5°C/min, then increased to 260°C at a rate of 5°C/min, and finally held constant for 10 minutes.

4 RESULTS AND DISCUSSIONS

The natural gas in the Dehui fault depression is mainly alkane gas, of which CH_4 is absolutely dominant, and the content is mainly between 78% and 99%, with an average value of 86%. The C_2H_6 content is between 0.7% and 14%, with an average of 4.2%. The C_3H_8 content is between 0.07% and 13.2%, with an average of 1.7%. The drying coefficients of deep natural gas in the Dehui fault depression are distributed from 0.75 to 0.99. Among them, dry gas samples account for 31%, and wet gas samples account for 69%, which is dominated by wet gas. Non-hydrocarbon gases include N_2, CO_2, H_2S, and He, which are dominated by CO_2 and N_2. The CO_2 content is mainly distributed in 0.02%~1%. The N_2 content is relatively enriched and is distributed in 0.27%~23.5%, which is mainly between 2%~6%. The distribution of $\delta^{13}C_1$ in the Dehui fault depression ranges from −50‰ to −26‰, mainly in the range of −40.0‰ to −30.0‰. The distribution of $\delta^{13}C_2$ ranges from −35‰ to 20‰, mainly in −29‰ ~ −23‰. The $\delta^{13}C_3$ content is distributed between −33‰ ~ −24‰, mainly concentrated in −25‰ ~ −22‰.

Dai proposed that natural gas with $\delta^{13}C_1$<-30‰ is of organic origin (Dai 1992). Most of the natural gas samples in the Dehui fault depression have $\delta^{13}C_1 < -30$‰, which are of organic origin. Combined with the relationship between $\delta^{13}C_1$ and composition characteristics, the genetic type of natural gas is mainly coal-type gas, accompanied by oil-cracking gas and condensate associated gas (Figure 2).

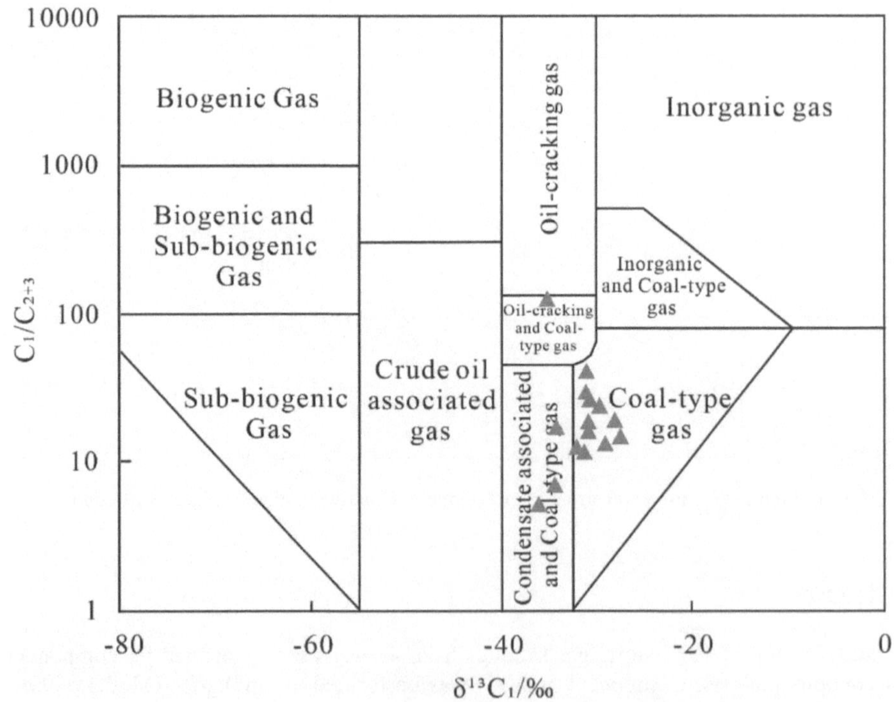

Figure 2. Identification of genetic types of deep natural gas in Dehui fault depression.

5 CONCLUSIONS

The natural gas in the Dehui fault depression is dominated by alkane gas, with relatively high methane content, and is dominated by wet gas. The non-hydrocarbon gases are relatively high in N_2 and CO_2 content. Combined with the characteristics of natural gas components and carbon isotopes, the genetic type of deep natural gas in the Dehui fault depression is dominated by coal-type gas, accompanied by oil-cracking gas and associated gas.

ACKNOWLEDGMENT

This work was financially supported by the National Natural Science Foundation of China.

REFERENCES

Dai, J.X. (1992) Identification of various alkane gases. Scientia Sinica. 2: 185–193.

Di, J.X. (2007) Research on the charge times stages of deep natural gas in the Dehui depression, the south Songliao Basin. Jilin University Publishing.

Li, H.G., Xue, H.T., Huang, W.B., et al. (2011) Hydrocarbon source rock exploration potential of the deep layer in Dehui fault depression. Science Technology and Engineering. 27: 6578–6582.

Advances in Mineral Resources, Geotechnology and Geological Exploration – Rashid & Zhang (eds)
© 2023 Copyright the Author(s), ISBN 978-1-032-33377-9

Research on flotation process of copper-nickel sulfide ore

Qi Nie, Ping Lu*, Lei Ma*, Yong Cheng, Cong Liu & Yiming Wen
Kunming Metallurgy College, Kunming, Yunnan, China

ABSTRACT: With rapid economic development, low-grade nickel-containing polymetallic ore has become an important source for the production of nickel concentrate, copper concentrate, and other metals. According to process mineralogy, Jinping ore is a typical refractory nickel ore, containing 0.634% nickel and 0.329% copper. The ore contains a large number of gangue minerals, such as chlorite, mica, and others, which have a complicated relationship with valuable minerals. The complexity of mineral composition, distribution, and intercalation characteristics results in low nickel and copper recoveries. In this paper, a new process is researched and developed. The ore is ground to 75% -0.074 mm, and then mixed flotation process is carried out to obtain mixed concentrate, which undergoes two roughings, three beneficiations, and two sweeps. The obtained mixed concentrate contains 3.03% nickel, 73.03% recovery, 2.01% copper, 92.58% recovery, 0.24% cobalt, and 77.62% recovery.

1 INTRODUCTION

Nickel sulfide ore has complex mineralogical characteristics, which is one of the key factors that lead to substandard flotation treatment index of such ore (Hu et al. 2021). These ores are often associated with gangue minerals, such as chlorite, serpentine, and feldspar. The presence of serpentine minerals in the ore can interfere with the flotation performance of nichrome ore (Bo et al. 2012). It forms an "anisotropic agglomerate" or slime coating on the surface of nickel sulfide ore, which is not conducive to the recovery of nickel.

The flotation is the most widely used method for handling such ores. This is because of its advantages, for instance low product cost, easy-handful for tailings, and more (Bruckard et al. 2011; Edwards et al. 1980; Merve Gen et al. 2012). In traditional flotation research, stage grinding is the main grinding method used; however, most of the research focuses on concentrate regrinding or sweeping regrinding, and few researchers are engaged in conducting research on medium ore regrinding, particularly the flotation of nickel-bearing ores (He Hao et al. 2020; Witney & Yan 1997). In this study, a new process and agent were studied in view of the complex composition and fine particle size of refractory nickel ore in Jinping.

2 MATERIALS AND METHODS

Representative samples were taken from the Jingping nickel mine. Ore was prepared according to laboratory test preparation procedures, and samples were ground in a wet ball mill to obtain specified fineness. The chemical composition and the analysis results of the nickel phase and copper phase are shown in Tables 1, 2, and 3.

*Corresponding Authors: 370461393@qq.com, 22035052@qq.com

 DOI 10.1201/9781003319412-10

Table 1. Results of chemical composition analysis (by XRF).

Element	Ni	Cu	S	Fe	Sn	Co	Al_2O_3
Content (%)	0.634	0.329	7.6	21.6	0.047	0.047	4.54
Element	CaO	Ag*	SiO_2	Zn	Pb	MgO	
Content (%)	2.23	<10	20.18	0.022	0.02	29.72	

(*: ppm)

Table 2. Results of nickel phase analysis (by XRD).

Phase	Nickel sulfide	Nickel sulfate	Nickel silicate	Total
Content%	0.544	0.060	0.031	0.635
Distribution%	85.67	9.45	4.88	100.00

Table 3. Results of copper phase analysis.

Phase	Copper sulfide	Free copper oxide	Combined copper oxide	Total
Content%	0.316	0.11	0.004	0.331
Distribution%	95.47	3.32	1.21	100

3 RESULTS AND DISCUSSIONS

3.1 *Mineralogy of the ore*

The main metallic minerals found in the ore are pentlandite, pyrrhotite, magnetite, and chalcopyrite, and the main gangue minerals are chlorite, mica, and quartz, accounting for 48.498%, 13.474%, and 5.494%, respectively. Chlorite is mainly associated with pentlandite, pyrrhotite, magnetite, chalcopyrite, mica, and quartz; hence, it is important to remove or prevent the effect of chlorite on valuable minerals flotation. Observing the mineral dissemination characteristics by mineral liberation analysis (MLA), most minerals in the ore are unequal and complexly distributed, i.e., pentlandite disseminated with chalopyrite and pyrrhotite; finely grained, micro-grained, acerous grained chalcopyrite are coated by chlorite, which is associated with other valuable minerals. The selected scanning electron microscopic (SEM) images are shown in Figure 1.

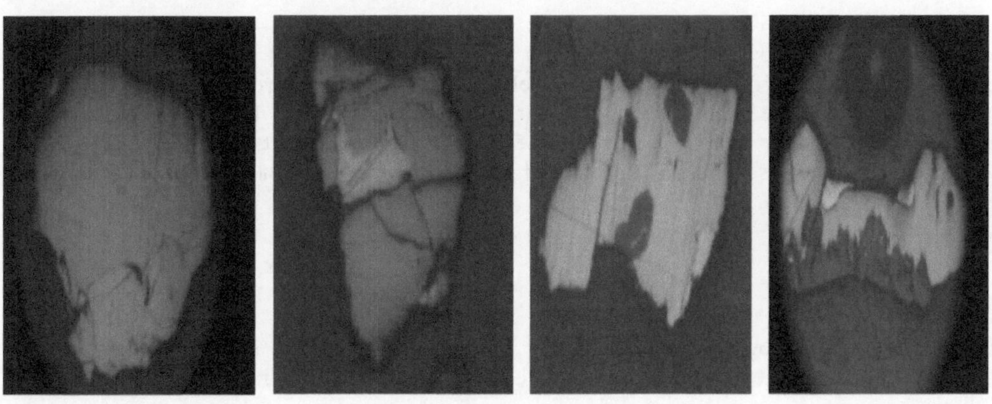

Figure 1. SEM image: typical occurrence of main minerals.

3.2 *Effect of grinding fineness*

Tests for grinding fineness on the roughing flotation were conducted. The flowsheet is depicted in Figure 2 and the results are shown in Table 4.

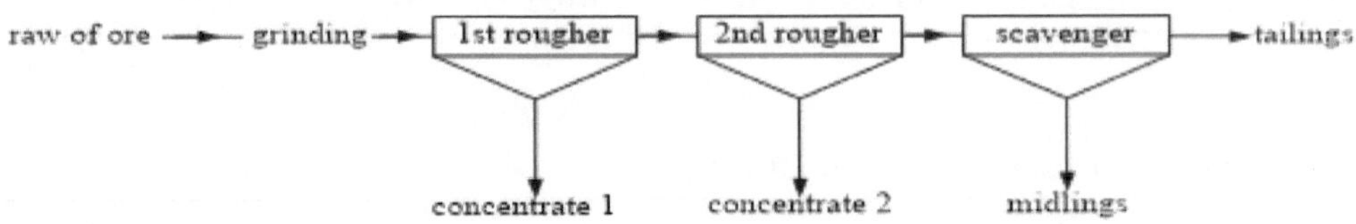

Figure 2. The flowsheet of grinding fineness tests.

Table 4. The results of grinding fineness tests.

Grinding fineness	Products	Yield (%)	Ni grade (%)	Ni recovery (%)
70%–0.074 mm	Concentrate 1	11.21	2.39	41.93
	Concentrate 2	7.00	1.22	13.44
	Middlings	3.50	0.74	4.08
	tailings	78.29	0.63	40.55
75%–0.074 mm	Concentrate 1	12.32	2.98	57.30
	Concentrate 2	8.60	1.08	14.54
	Middlings	4.00	0.65	4.07
	Tailings	73.08	0.51	24.09
80%–0.074 mm	Concentrate 1	12.75	3.01	59.99
	Concentrate 2	8.99	1.00	13.95
	Middlings	4.25	0.60	3.96
	Tailings	74.01	0.53	22.12
85%–0.074 mm	Concentrate 1	12.20	3.09	58.77
	Concentrate 2	8.80	1.03	14.39
	Middlings	4.00	0.65	4.12
	Tailings	75.60	0.51	22.72

The crystal size of pyrite is fine. To obtain a better flotation effect, proper grinding fineness is essential. However, if the liberation of nickel chromite is realized through primary grinding, a large amount of gangue minerals and coarse nickel chromite particles will be slimed together, and a large amount of slime will lead to an increase in the dosage of reagents and the recovery rate of valuable metals decreases. Finally, the optimal grinding fineness is 75% -0.074 mm, as shown in Table 4.

3.3 *Flotation tests*

To obtain optimum flowsheet and satisfying flotation indicators, the middling regrinding and flotation test was carried out. The flowsheet is shown in Figure 3 and the results are shown in Table 5.

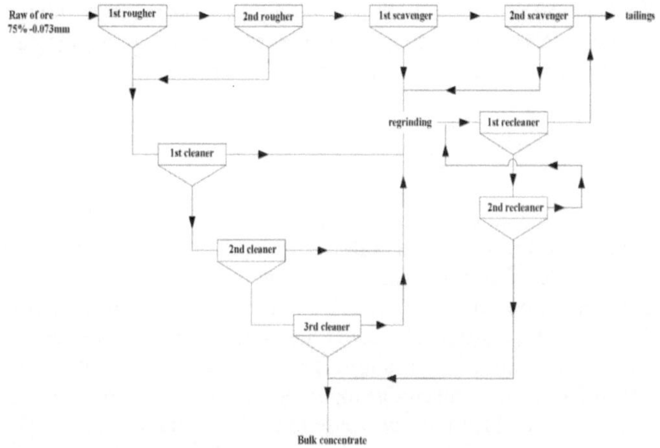

Figure 3. The middling regrinding and flotation flowsheet.

Table 5. The results of the middling regrinding and flotation test.

		Grade (%)			Recovery (%)		
Product	Yield%	Ni	Cu	Co	Ni	Cu	Co
Concentrate	14.89	2.80	2.01	0.18	66.21	92.62	66.31
Tailings	85.11	0.25	0.028	0.016	33.79	7.38	33.69
Raw of ore	100.00	0.634	0.331	0.047	100.00	100.00	100.00

As shown in Table 5, compared with the regrinding and flotation test of the scavenging concentrate, the obtained bulk concentrate shows little change in the grades of nickel, copper, and cobalt, and the recovery rate is improved. The rate increased by 5.31%.

3.4 *Effect of middling regrinding fineness*

To obtain better results by middling regrinding flowsheet, the tests of middling regrinding fineness were carried out. The results are shown in Table 6.

Table 6. The results of middling regrinding and flotation tests.

			Grade (%)			Recovery (%)		
Fineness	Product	Yield%	Ni	Cu	Co	Ni	Cu	Co
85% –0.074mm	Concentrate	14.80	2.89	2.02	0.19	67.40	92.28	71.50
	Tailings	86.20	0.24	0.029	0.013	32.60	7.72	28.50
90% –0.074mm	Concentrate	15.20	3.03	2.01	0.240	73.03	92.58	77.62
	Tailings	84.80	0.20	0.029	0.012	26.97	7.42	22.38
95% –0.074mm	Concentrate	15.76	3.01	1.97	0.22	72.84	92.47	76.00
	Tailings	82.24	0.21	0.03	0.013	27.16	7.53	24.00

As presented in Table 6, when the middling regrinding fineness is 90% –0.074 mm, both Ni and Cu recovery increased due to the regrinding of intermediate products, significantly increasing in particle floatability, independently of the liberation of the particles.

4 CONCLUSIONS

The research results show that a large number of gangue minerals are inhibited under the condition that sodium carbonate is the modifier and water glass and carboxymethyl cellulose (CMC) are combined inhibitors. The role of pulp is conducive to the improvement of concentrate quality.

The particle size of nickel chromate in the ore sample is wide, and the pyrite and pyrrhotite, chlorite, chalcopyrite, hematite, quartz, calcite, actinite, and other minerals are complex symbiotic. If monomer dissociation is achieved by a single grinding before flotation, the easily recovered coarse-grained nickel-iron ore will form sludge together with magnesium minerals, and the increase in sludge content in the slurry will lead to a decrease in flotation rate and chemical consumption. At the same time, due to lightweight, high surface energy and charge, and large specific surface area of the fine-grained nickel-iron ore particles, the fine-grained ore particles are easy to agglomerate with the fine-grained gangue ore particles, which is the main reason for the agglomeration of the fine-grained gangue ore particles. The flotation effect is poor. Therefore, stage grinding and stage flotation are beneficial to the monomer dissociation of fine nickel-cobalt ore particles, and finally, the grade and recovery of nickel concentrate are improved.

ACKNOWLEDGMENTS

1. This work was financially supported by Yunnan Provincial Department of Education Scientific Research Fund Project No. 2021J0945.
2. This work was financially supported by Kunming Metallurgical College Research Fund Project No. 2020XJZK06

REFERENCES

A. Merve Gen, I.kilickplan, J.S.Laskowski. Canadian Mertallurgical Quarterly, vol.51, n.4, pp: 368–375(2012).

Bo Feng et al. The effect of conditioning methods and chain length of xanthate on the flotation of nickel ore. Minerals Engineering 39, pp. 48–50(2012).

C.R.Edwards, et al. The effect of slime coatings of the serpentine minerals, chrysotile, and lizardite, on pentlandite. International journal of mineral processing, vol.7, pp. 33–42(1980).

He Hao, Shao Yanhai, Wu Haixiang, et al. Research status of flotation technology and reagents of copper-nickel sulfide ore in my country [J]. Mineral Protection and Utilization, 40(2):5(2020).

Hu Cong, Chen Wei, Xu Pengyun. Inhibition of serpentine slime in the acidic system and its effect on flotation of copper-nickel sulfide ore [J]. Chinese Journal of Nonferrous Metals, 31(1): 11(2021).

J.Y.Witney and D.S. Yan Reduction of magnesia in nickel concentrates by modification of the froth zone in column flotation. Minerals Engineering, vol.12, n.2, pp. 139–154. (1997).

W.J.Bruckard et al. A review of the effects of the grinding environment on the flotation of copper sulphides. International journal of mineral processing 100, pp.1–13(2011).

Advances in Mineral Resources, Geotechnology and Geological Exploration – Rashid & Zhang (eds)

Study on deformation characteristics of soft soil subgrade treated by plastic drainage board stacking preloading technology

Kexin Zhao
South China University of Technology, Guangzhou, China

ABSTRACT: When building roads on soft soil foundations, problems such as embankment instability or excessive settlement are commonly seen. Therefore, soft foundation treatment should be carried out. This paper mainly introduces the treatment of the soft soil subgrade of Wanhuan East Road in the Pearl River Delta of China. Using the plastic drainage board stacking and preloading treatment technology, through the engineering site monitoring of two sections, the surface settlement value and level of the soft soil foundation after treatment were analyzed. The variation law of displacement value and pore water pressure value was analyzed and studied. The results showed that the subsidence after the soft soil foundation treatment was obviously different. The subsidence of the road was larger than that of the shoulder of the road, and the lateral displacement mainly occurred at the elevation of −4 to −7m. When the depth was greater than 7m, it began to decrease with the increase of the depth. Due to the effect of the plastic drainage board, the soil drainage speed was accelerated, the pore water pressure dissipation speed was accelerated, and the strength of the soft soil foundation was also rapidly increased, which was conducive to the construction safety of the foundation treatment. The plastic drainage board was used to treat the soft soil foundation. Most of the settlement of the soft soil foundation was basically completed during the filling period. The post-construction settlement calculated by the layered sum method was small, and the stability coefficient met the requirements and could meet the requirements of various indicators.

1 INTRODUCTION

China has a long coastline, and most of the river estuaries are alluvial deltas. Such deltas are mainly composed of water-saturated soft cohesive soils and are called soft soil areas (Arulrajah 2011; Roy 2009; Huang 2020; Londo 1999; Shanks 1981). Soft soil includes water-saturated soft cohesive soil and silt (Amundson 2008; Park 2015). When building roads on soft soil foundations, problems such as embankment instability or excessive settlement are likely to occur (Muthukkumaran 2016; Yan 2011). The stability and bearing capacity of the roadbed, as well as the treatment of soft soil foundation, should be based on the physical and mechanical properties of soft soil and floating mud, depth of buried layer, embankment height, material site conditions, road grade, and other factors, plastic drainage board, powder spray pile, gravel pile, overload pre-processing and other measures (Karatzetzou 2021; Jiang 2015).

The surcharge preloading method to control settlement technology is also one of the foundation treatment technologies widely used at home and abroad (James 2002; Liu 2014; Wang 2018). In the surcharge preloading method, the sand well load preloading method first appeared, which was proposed by Moran in 1925, and then covered a layer of horizontal sand cushion on the surface of the foundation, which greatly improved the efficiency of drainage and consolidation (Chai 2015; Zheng 2016). The pile load preloading method is to reduce the moisture content of the soil body and improve the shear bearing capacity of the foundation by stacking the load on the ground and using the pile load greater than or equal to the design load to promote the drainage and consolidation of the foundation soil, the load is removed, and the main consolidation settlement of the foundation soil is

DOI 10.1201/9781003319412-11

completed in advance, which plays a role in strengthening the foundation (Ivanova 2017; Pal 2010; Yang, 2007). At first, the stacking preloading method only used heavy objects to pressurize the foundation but found that the treatment efficiency was low. The engineers added vertical drainage bodies to the foundation and found that the foundation reinforcement effect was obvious.

The piling load preloading method usually consists of two parts: a drainage system and a pressurized system. The drainage system is generally composed of a horizontal drainage cushion and a vertical drainage channel (Rujikiatkamjorn 2011). The horizontal drainage cushion is generally a sand cushion, and it is also formed by a combination of a sand cushion and a geosynthetic cushion. Vertical drainage channels are often formed by ordinary sand wells, bagged sand wells, and plastic drainage boards. The pressurization system usually adopts the following methods: stacking method, vacuum preloading method, vacuum preloading combined stacking method, and groundwater level reduction method. If the permeability coefficient of the foundation soil is large and the construction period is long, the vertical drainage body can be installed only through the pile load pressure, thereby reducing the construction cost. Due to the simple construction, low technical construction requirements, and flexible approach, this method can be combined with other methods to treat the foundation, such as the vacuum combined stacking preloading method, the stacking preloading combined dynamic compaction method, etc., without the need for special construction machinery and foundation treatment. low cost. However, using this method to treat the foundation has a long construction period, and now the construction period is tight. It is often used in conjunction with other methods to treat the foundation to improve reinforcement efficiency. Saturated soft clay is widely distributed in coastal areas and has the characteristics of high-water content, large void ratio, low permeability, and high compressibility. Under the action of load, the soft soil foundation will produce an excessive settlement and settlement difference, which has always been a major problem to be solved in the construction process. The law of settlement and deformation and the situation of the settlement are predicted. In this paper, the treatment process of soft soil subgrade of a highway in Nansha, Guangzhou, located in the Pearl River Delta region. During the construction of the embankment, predict the possible unfavorable conditions, control the loading rate of soft soil foundation, and ensure the safety of construction. By collecting on-site monitoring data, after sorting out, the soil properties indicators such as the degree of consolidation in the project area can be calculated, and finally, the effect evaluation can be made. The soft foundation treated by the combined method of stacking and preloading of plastic drainage board is studied in order to obtain the settlement law and propose an effective prediction method, which can be used as a reference for similar projects.

2 DEFORMATION SITUATION OF SOFT SOIL FOUNDATION TREATED WITH PLASTIC DRAINAGE BOARD STACKING PRELOADING

2.1 *Project overview*

The design starting point of Wanhuan East Road is WHK0+000, the coordinates of the starting point are (X=182448.533, Y=70193.552), the design endpoint number is WHK4+000, the coordinates of the endpoint is (X=179679.077, Y=72980.361). The length of the roadbed is 3.559km.

Among them, the length of the soft foundation treatment is 3083m, the plastic drainage board is 2753695 linear meters, the amount of lead holes is 92904m, the two-way geogrid is 184285m^2, the bottom sand cushion is 99022m^3, the backfill sand is 29837m^3, the pre-pressed soil is 300434m^3, and the settlement and wide filling are 151077m^3. The pond subgrade is 600m long, the sandbag is 921m long, and the cofferdam volume is 5865m^3.

The whole line of the subgrade soft foundation project adopted a plastic drainage board and equal load preloading construction for consolidation treatment. The plastic drainage was arranged in a regular triangle with a spacing of 1.2m and the average height of the fill was 3.5m. Considering that the elevation of the centerline of the flood control and drainage road was greater than 6.45m, the slope was edging with a viscosity of 1m, and the grass was sprayed for protection. The foot of the slope was used as a temporary drainage ditch and a water-retaining cofferdam was added.

2.2 *Soft foundation treatment construction monitoring scheme*

During the construction of foundation reinforcement, various instruments embedded in the foundation soil could reflect the change of the preload load of the foundation, the consolidation, settlement, and displacement of the foundation with time and space. Through the processing, analysis, and calculation of these observation data, the corresponding evaluation and dynamic design could be carried out, the stability of the embankment during filling was guaranteed, and a reliable basis for reasonably determining the preloading time and unloading time was provided. The layout of the monitoring points of the project was shown in Figure 1, where I was the monitoring device of the surface settlement points and ⊙ was the side piles. The settlement marks were buried before the cushion laying construction, and the elevation data were read after the cushion laying was completed.

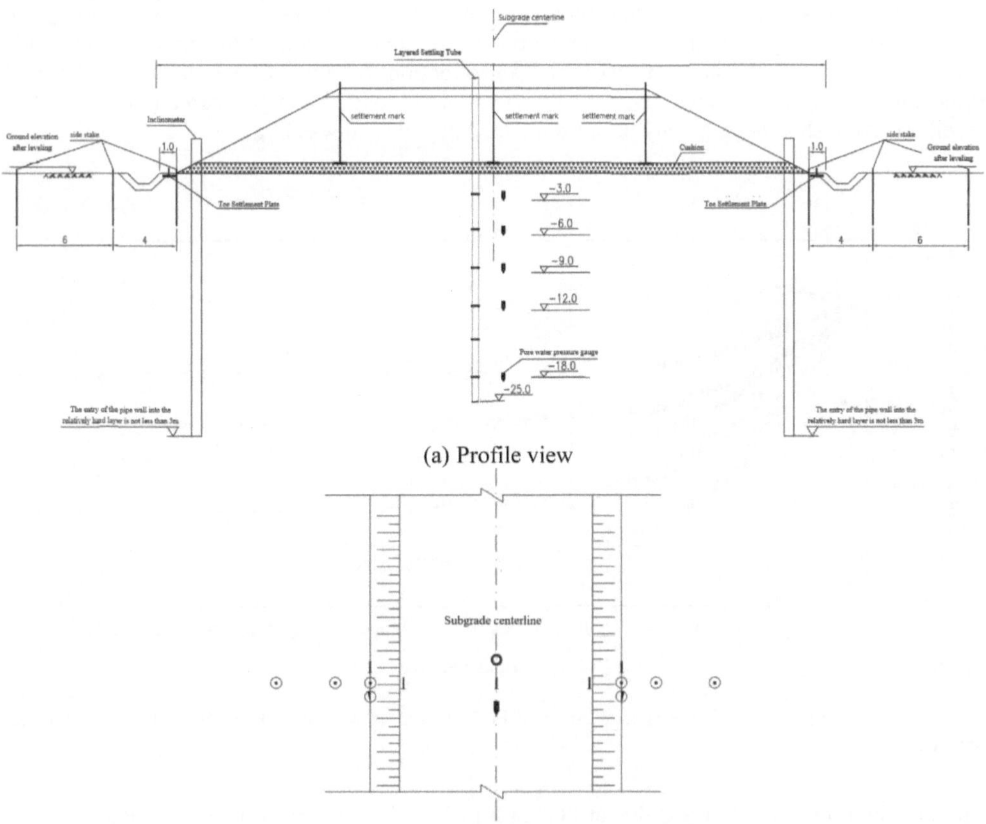

(a) Profile view

(b) Top view

Figure 1. Layout of monitoring points in the process of stacking and preloading plastic drainage panels.

In this paper, two monitoring sections, WHDK1+660 and WHDK1+860, were studied, and measuring points were arranged at the toe, shoulder, and middle of the slope in the two sections, as shown in Table 1.

Table 1. Monitoring section and monitoring point distribution.

Monitoring section	Monitoring points				
WHDK1+660	9CJ1	9CJ2	9CJ3	9CJ4	9CJ5
	Toe of the slope	Shoulder	Middle	Shoulder	Toe of the slope
WHDK1+860	10CJ1	10CJ2	10CJ3	10CJ4	10CJ5
	Toe of the slope	Shoulder	Middle	Shoulder	Toe of the slope

2.3 *Analysis of the law of surface subsidence*

2.3.1 *Typical soil settlement law*

Figure 2 and Figure 3 were the relationship diagrams of settlement-load-time in the process of layered filling of embankments at K1+560 and K1+860 sections, respectively. It could be seen from the figure that the settlement curve was slow at the initial stage of loading, indicating that the settlement rate was small and the consolidation rate is slow. In the later stage of preloading, the curve becomes flat again, and the sedimentation rate became smaller. At this time, the subgrade of this project entered the stage of secondary consolidation. It could be seen that the differential settlement exists and is obvious, and the settlement on the road was larger than the settlement on the left and right shoulders, mainly because the load on the road was larger.

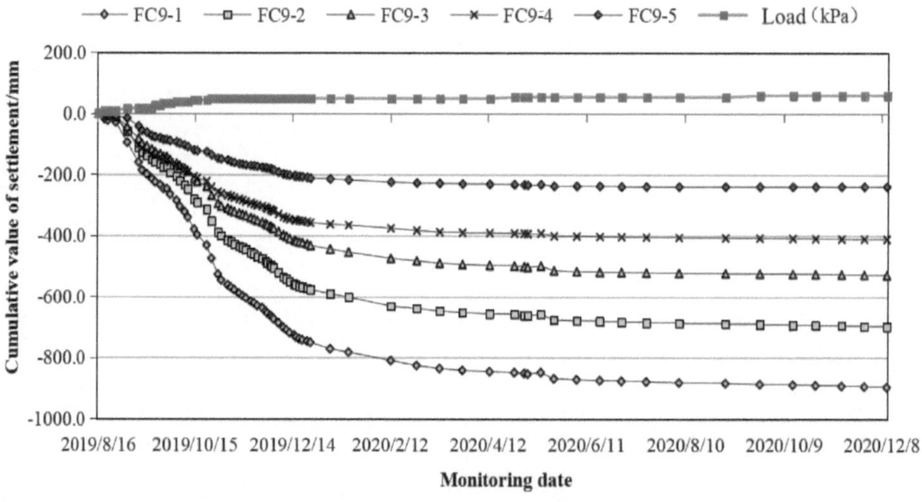

Figure 2. Cumulative change-time change curve of FC5 layered settlement at K1+560 section of Wanhuan East Road.

It could be seen from Figure 2 that at FC5 at K1+560, the settlement of each measuring point increased rapidly in the first 4 months, the subsidence in the road was as high as 500mm, and then the subsidence gradually became smaller and the area was stable, and the final settlement value reached 896mm. One characteristic of using this method to treat soft soil foundation was that the settlement could be basically completed during the construction period, and the settlement after construction was small, which was beneficial to the construction of the highway. It could be seen from Figure 3 that the settlement of FC6 at the position of K1+860, the settlement of each measuring point increased rapidly in the first 4 months, the settlement in the road was as high as

437mm, and then the settlement gradually became smaller and the area was stable, and the final settlement value reached 750 mm.

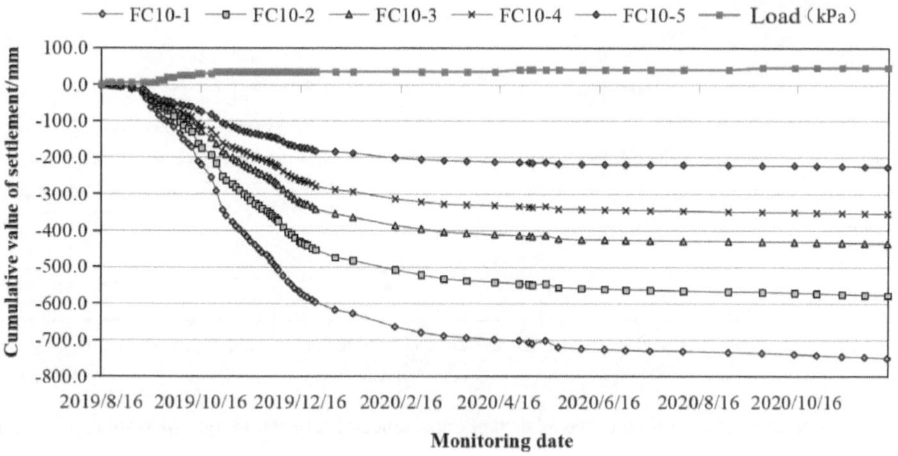

Figure 3. Cumulative change-time change curve of FC6 layered settlement at K1+860 section of Wanhuan East Road.

It could be seen from Table 2 that the subsidence in the middle of the road was the largest, and the subsidence of the road shoulder was small. The subgrade settlement laws of different stakes were roughly the same.

Table 2. Statistical values of layered settlement of each section.

Cumulative value of layered settlement (mm)	WHDK1+560				
	FC9-1	FC9-2	FC9-3	FC9-4	FC9-5
2020/12/11	−896.0	−698.0	−529.0	−412.0	−240.0
Cumulative value of layered settlement (mm)	WHDK1+860				
	FC10-1	FC10-2	FC10-3	FC10-4	FC10-5
2020/12/11	−750.0	−578.0	−437.0	−355.0	−227.0

2.3.2 *The law of settlement of the surface soil layer*

Consistent with the layered settlement curve, it could be seen from Figures 4 and 5 that the settlement curve was slow at the initial stage of loading, indicating that the settlement rate was small and the consolidation rate was slow. With the increase of the load, the drainage rate of the plastic drainage board was accelerated, the settlement rate was significantly increased, and the soil consolidation was accelerated. In the later stage of preloading, the curve became gentle again, and the settlement rate became smaller. At this time, the subgrade of the project entered the stage of secondary consolidation, and the settlement in the middle of the road was larger than that of the left and right shoulders.

It could be seen from Figure 4 that the settlement of each measuring point in K1+560 increased rapidly in the first 4 months, the subsidence in the road was as high as 800mm, and then the subsidence gradually became smaller and the area was stable, and the final settlement value reached 1000mm. The settlement of each measuring point in K1+860 increased rapidly in the first 4 months,

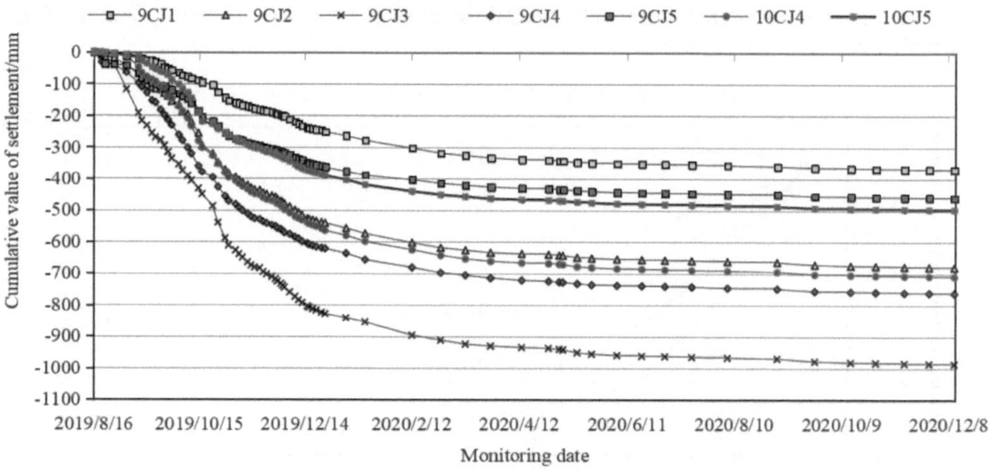

Figure 4. Cumulative change-time curve of surface subsidence of K1+560 section of Wanhuan East Road.

the subsidence in the road was as high as 850mm, and then the subsidence gradually became smaller and the area was stable, and the final settlement value reached 1000mm.

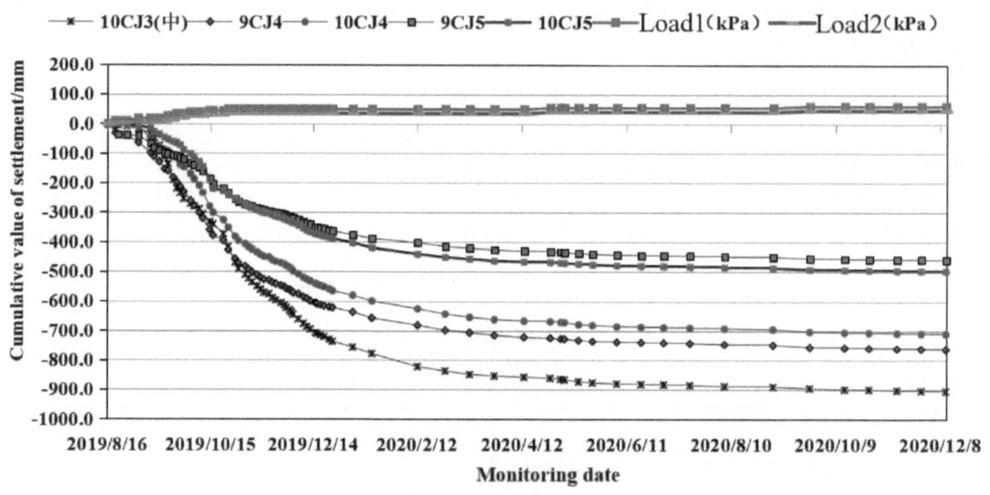

Figure 5. Cumulative change-time curve of surface subsidence of K1+860 section of Wanhuan East Road.

It could be seen from Figure 5 that the settlement of different sections was slightly different, but the rules were roughly the same.

The drainage board was completed around August 10, 2019, and the geogrid was laid on August 15, 2019. The earthwork piling construction began on August 20, 2019; the cumulative pile load preloading construction was about 480 days, and the cumulative dead load during the preloading construction is about 380 days; the current elevation of the subgrade top center before unloading is currently measured: the K1+560 section is 8.226m, the K1+860 section is 7.554m, and the accumulated filling thickness in the road is about 2.6~3.5m. The design elevation of the drawing is: K1+560 is 7.832m, K1+860 is 7.518m, and the road median elevation measured before dead load unloading is basically in line with the design drawings.

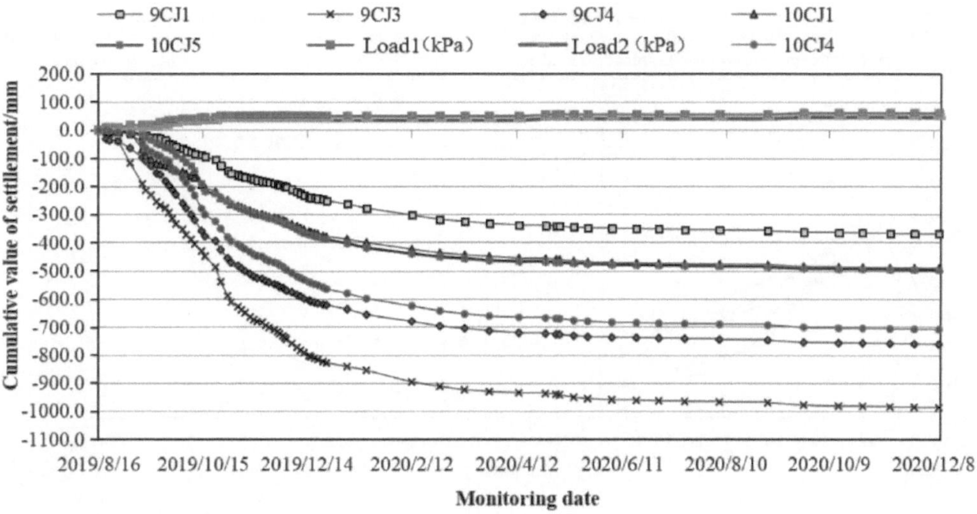

Figure 6. Cumulative change-time curve of surface subsidence of K1+560 and K1+860 sections of Wanhuan East Road.

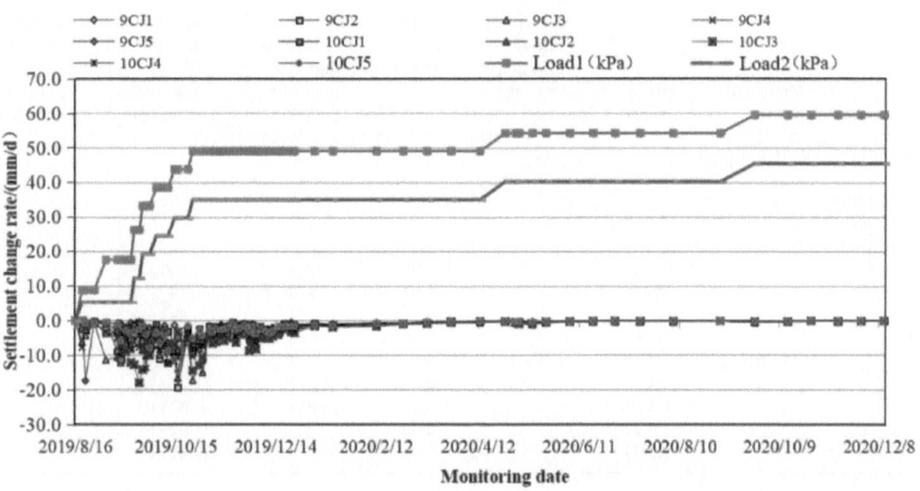

Figure 7. Curve of surface settlement change rate of K1+500~K1+900 section of Wanhuan East Road.

The drainage board was finished around August 10, 2019, the geogrid was laid on August 15, 2019, and the earthwork stacking construction began on August 20, 2019. The cumulative stacking preloading construction was about 480 days, and the cumulative dead load of the preloading construction was about 380 days. The design elevation of K1+560 was 7.832m, and the design elevation of K1+860 was 7.518m. Before unloading, the current measured elevation of the subgrade top of the K1+560 section was 8.226m, and the subgrade top of the K1+860 section was 8.226m. The currently measured elevation of the center was 7.554m, and the accumulated thickness of the road in the K1+860 section was about 2.6~3.5m. At present, the middle road elevation was measured before the dead load unloading, which basically conformed to the design drawing.

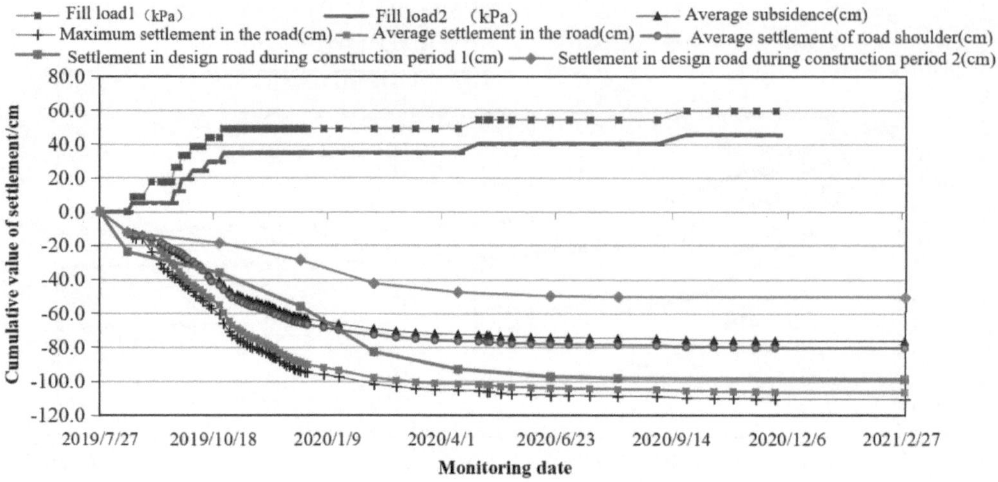

Figure 8. The cumulative curve of the settlement and design settlement of the K1+500~K1+900 section of Wanhuan East Road.

Table 3. The latest monitoring settlement statistics of general surcharge preloading sections.

Monitored road	Monitored section	Monitoring points	Average Settlement During Insertion of Drainage Board (mm)	Current latest settling rate (mm/d)	Monitoring accumulated settlement during the preloading period (mm)	Total settlement value during the construction period (mm)	Design settlement during the preloading period (mm)
Wanhuan East Road K1+500~K1+900 Road (General stacking preload)	WHDK1+660	9CJ1	−120	−0.04	−369.8	−489.8	−989
		9CJ2		−0.08	−680.3	−800.3	
		9CJ3		−0.07	−987.0	−1107.0	
		9CJ4		−0.08	−762.4	−882.4	
		9CJ5		−0.06	−460.3	−580.3	
	WHDK1+860	10CJ1		−0.04	−491.3	−611.3	−506
		10CJ2		−0.07	−590.9	−710.9	
		10CJ3		−0.08	−940.4	−1024.4	
		10CJ4		−0.06	−709.5	−829.5	
		10CJ5		−0.07	−497.9	−617.9	

2.4 *Analysis of horizontal displacement law*

After the foundation soil was under the action of filling pressure, when vertical consolidation and settlement occurred, horizontal displacement and deformation would also occur. Although the plastic drainage board accelerated the soil drainage when the loading speed was too fast, the shear stress had a great influence on the soil, and lateral deformation would occur. However, the lateral restraint of the foundation soil would be reduced during the loading process to ensure that foundation slippage was avoided.

It can be seen from Figures 9 and 10 that due to the increase of the load, the lateral displacement would increase accordingly. When the elevation depth was less than 5.0m, the displacement increased significantly with the change of depth; when the depth was greater than 5.0m, as the elevation decreased, the lateral displacement also decreased, and the displacement could be ignored at

20m. It could be seen from the observation that the lateral displacement mainly occurred in the silty soft soil area, and the maximum value occurred at 3.0m to 7.0m below the surface. The maximum displacement of these two sections was 210.2mm and 32.0mm, respectively.

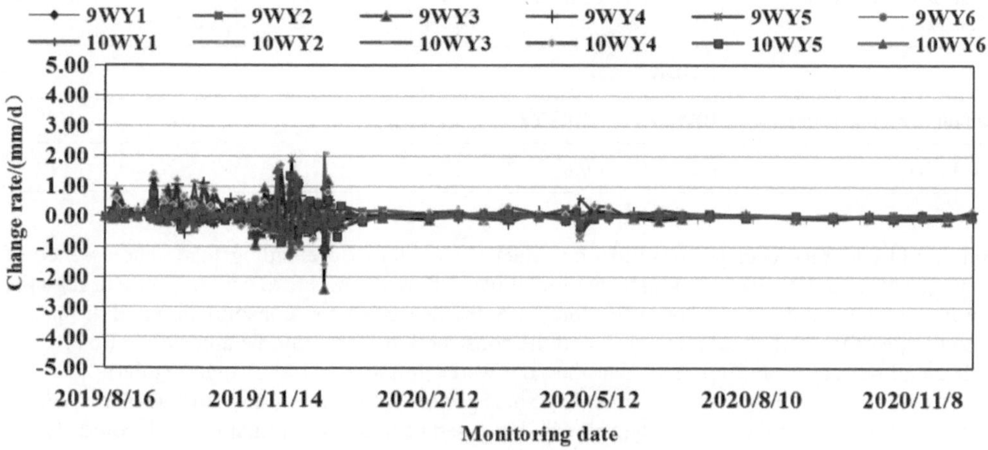

Figure 9. The cumulative curve of layered settlement of K1+500~K1+900 section of Wanhuan East Road.

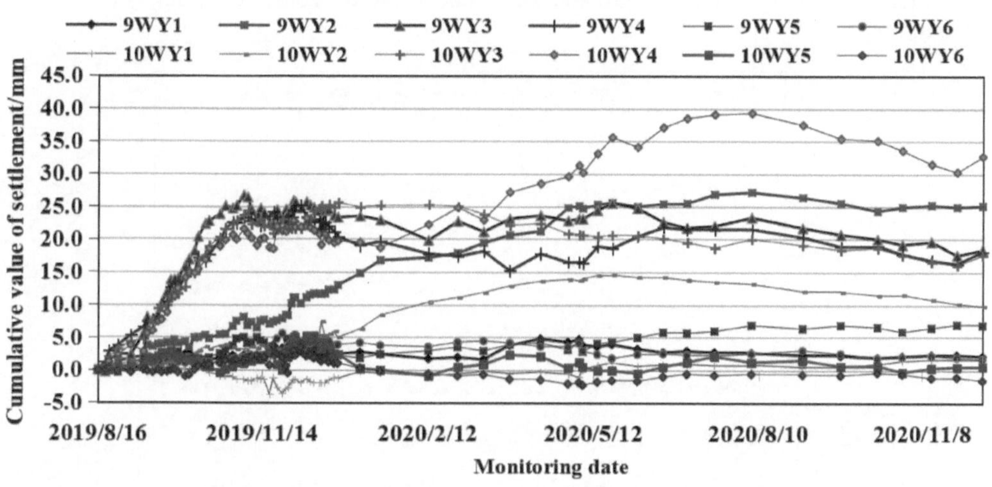

Figure 10. The cumulative curve of displacement side piles in the K1+500~K1+900 section of Wanhuan East Road.

In the early stage of filling, the lateral displacement developed slowly but with time. In the middle stage of filling, the load increased steadily, and the lateral displacement began to develop rapidly. At the end of filling, soil consolidation was basically completed, and the effective stress of soft soil foundation had grown, the soil strength has increased, and the resulting lateral displacement was changing slowly, which was similar to the three stages of vertical displacement.

2.5 *Analysis of pore water pressure law*

In order to more comprehensively understand the law of consolidation and settlement of soft soil, the change of pore water pressure was also analyzed during the construction of this project. Saturated

Table 4. K1+560 and KA+860 displacement values.

Horizontal displacement cumulative value (mm)	WHDK1+560					
	9WY1	9WY2	9WY3	9WY4	9WY5	9WY6
2020/12/11	2.1	25.1	18.4	18.2	6.9	1.7
Horizontal displacement cumulative value (mm)	WHDK1+860					
	10WY1	10WY2	10WY3	10WY4	10WY5	10WY6
2020/12/11	1.7	9.8	17.6	32.7	0.6	−1.6

soil would undergo seepage consolidation under the action of preloading load. The essence of seepage consolidation was the transformation of two different stress forms in the soil. Through the observation and study of the pore water pressure, the degree of soil consolidation and the strong growth rate of the soil could be judged, and the construction plan could be adjusted in time.

It could be seen from Figure 11 that the pore water pressure would change regularly with the increase of load and the passage of time. In a short period of time after the load was increased, the value increased significantly, and then slowly decreased back to the original value. It could also be seen from the figure that the pore water pressure tended to increase with the increase of the buried depth of the measuring point. Due to the effect of the plastic drainage board, the soil drainage speed was accelerated, the water pressure dissipation speed was accelerated, and the strength of the soft soil foundation also increased. It is fast and was conducive to the construction safety of soft soil treatment.

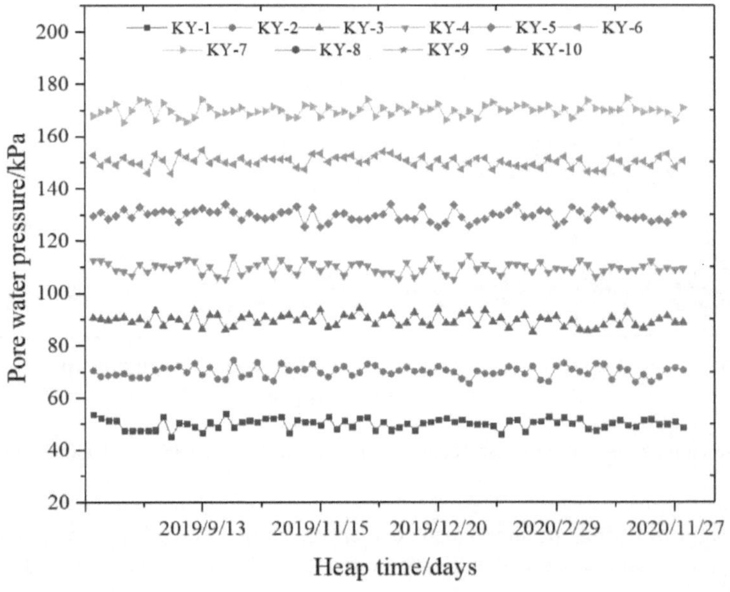

Figure 11. Pore water pressure-time course diagram of K5+650 section.

In the early stage of filling, the soil skeleton bore the external force, resulting in low excess pore water pressure and small drainage, so the settlement rate was slow. In the middle stage of filling, the foundation soil skeleton structure had been destroyed, and the load Most of it was borne by water in the soil, the excess pore water pressure increased significantly, and the settlement rate of soft soil increased rapidly. At the end of filling, due to the discharge of pore water, the soil skeleton structure was rearranged, and the effective stress of the soil increased. Obviously, under the main external load, the excess pore water pressure still increased, but the growth rate was small, and the settlement rate of soft soil also decreased, and showed a convergence trend. As the water pressure dissipated and decreased, the settlement also converged to a stable value.

3 CONCLUSION

(1) During the construction of the soft soil foundation, the surface settlement was roughly represented in three stages. In the first stage, the settlement of the embankment was small in the early stage of filling, and the soil body undergoes elastic deformation, which was the stage of settlement. In the second stage, the settlement rate was large in the middle stage of filling. At this time, with the continuous increase of the load, the soft soil foundation will undergo plastic deformation, and the soil skeleton would be rearranged after the frame was damaged. In the third stage, at the end of embankment filling, settlement continued to occur, but the settlement rate gradually slowed down. With the continuous removal of pore water, the effective stress of the foundation increased until the settlement rate approached zero.

(2) In this project, the lateral displacement mainly occurred in the elevation range of −4 to −7m. When the depth was greater than 7m, the value began to decrease with the increase of the depth. In the early stage of embankment loading, the lateral displacement rate was slow, indicating that in the early stage of construction loading, the soil mainly underwent elastic deformation, and the shear displacement generated by the soil was very small. In the middle stage of construction, the lateral displacement grew rapidly. At the end of the filling stage, the effective stress of the soft foundation was close to stable, the soil strength was gradually strengthened, and the lateral displacement was gradually reduced.

(3) In this project, the pore water pressure increased significantly within a short period of time after the load was increased, and then the pore water pressure slowly decreased back to the original value. The pore water pressure tended to increase with the increase of the buried depth of the measuring point. Due to the action of the plastic drainage board, the soil drainage speed was accelerated, the pore water pressure dissipation speed was accelerated, and the strength of the soft soil foundation also increased rapidly, which was beneficial to the treatment of the foundation construction safety.

(4) The soft soil foundation was treated with a plastic drainage board. This treatment method made most of the settlement of the soft soil foundation basically completed during the filling period. The post-construction settlement calculated by the layered sum method was less, and the stability coefficient met the requirements. Advantageously, the soft soil treatment method adopted was feasible and could meet the requirements of various indicators.

REFERENCES

Arulrajah A, Bo M W, Piratheepan J, et al. In Situ Testing of Soft Soil at a Case Study Site with the Self-Boring Pressure meter[J]. Geotechnical Testing Journal, 2011, 34(4):355–363.

Amundson R, Ewing S, Dietrich W, et al. On the in situ aqueous alterations of soils on Mars[J]. Geochimica Et Cosmochimica Acta, 2008, 72(15):3845–3864.

Chai J, Rondonuwu S G. Surcharge loading rate for minimizing lateral displacement of PVD improved deposit with vacuum pressure[J]. Geotextiles & Geomembranes, 2015, 43(6):558–566.

D Roy, Robinson K E. Surface settlements at a soft soil site due to bedrock dewatering[J]. Engineering Geology, 2009, 107(3–4):109–117.

Huang W, Wen K, X Deng, et al. Constitutive Model of Lateral Unloading Creep of Soft Soil under Excess Pore Water Pressure[J]. Mathematical Problems in Engineering, 2020, 2020(2):1–13.

Ivanova T V, Albert I U, Kaufman B D, et al. The load-bearing capacity of hanging piles by the strength criterion of a pile or soil material[J]. Magazine of Civil Engineering, 2017, 67(7):3–12.

James S C, Jr W W K. Evaluation of demonstrated and emerging technologies for the treatment and clean-up of contaminated land and groundwater[J]. Land Contamination & Reclamation, 2002, 10(4):239–245.
Jiang N J, Du Y J, Liu S, et al. Multi-scale laboratory evaluation of the physical, mechanical, and microstructural properties of soft highway subgrade soil stabilized with calcium carbide residue[J]. Canadian Geotechnical Journal, 2015:373–383.
Karatzetzou A, Pitilakis D, Karafagka S. System Identification of Mosques Resting on Soft Soil. The Case of the Suleiman Mosque in the Medieval City of Rhodes, Greece[J]. Geosciences (Switzerland), 2021, 11(7):275.
Liu Z J, Guo Y, Yuan Q W, et al. On the Construction Technology and Quality Control of Vacuum Combined Surcharge to Foundation Reinforcement[J]. Advanced Materials Research, 2014, 1030–1032:827–829.
Londo A J, Messina M G, Schoenholtz S H. Forest Harvesting Effects on Soil Temperature, Moisture, and Respiration in a Bottomland Hardwood Forest[J]. Soil Science Society of America Journal, 1999, 63(3):637–644.
Muthukkumaran K, Raaj S K, Kumar M V. Assessment of pile failures due to excessive settlement during pile load test[J]. Japanese Geotechnical Society Special Publication, 2016, 2(73):2520–2524.
Park D, Lee J. Interaction effects on load-carrying behavior of piled rafts embedded in clay from centrifuge tests[J]. Canadian Geotechnical Journal, 2015:1–12.
Pal M, Deswal S. Modelling pile capacity using Gaussian process regression – ScienceDirect[J]. Computers & Geotechnics, 2010, 37(7):942–947.
Rujikiatkamjorn C, Indraratna B. Analytical solutions and design curves for vacuum-assisted consolidation with both vertical and horizontal drainage[J]. Canadian Geotechnical Journal, 2011, 44(2):188–200.
Shanks D H, Barrett R V. PERFORMANCE OF AIRCRAFT PNEUMATIC TYRES IN SOFT SOIL[J]. Aeronautical Journal -New Series-, 1981, 85(840):20–28.
Wang J, Fang Z, Cai Y, et al. Preloading using fill surcharge and prefabricated vertical drains for an airport[J]. Geotextiles and Geomembranes, 2018, 46(5):575–585.
Yan S W, Zhang L L, Sun L Q. Research on settlement and wavy failure of the road on very soft soil foundation[J]. Advanced Materials Research, 2011, 261–263:1370–1373.
Yang M, Yao L K, Wang G J . Study on the effect of width and space of anti-slide piles on soil arching between piles[J]. Yantu Gongcheng Xuebao/Chinese Journal of Geotechnical Engineering, 2007, 29(10):1477–1482.
Zheng G, Liu J, Lei H, et al. Improvement of the very soft ground by a high-efficiency vacuum preloading method: A case study[J]. Marine Geotechnology, 2016:631–642.

Advances in Mineral Resources, Geotechnology and Geological Exploration – Rashid & Zhang (eds)
© 2023 Copyright the Author(s), ISBN 978-1-032-33377-9

Research and application of deep well drilling speed-up technology in the Bayan area

Zhen Qin, Yongsheng Ye, Xinxin Wang & Bo Zhang
Engineering Technology Research Institute of Huabei Oil Company, Renqiu City, Hebei Province, China

Aimin Qiu
Bayan Exploration and Development Branch of Huabei Oil Company, Bayan Nur city, Inner Mongolia Nei Mongol, China

Junjie Jia, Yan Guo & Jingjing Zhang
Engineering Technology Research Institute of Huabei Oil Company, Renqiu City, Hebei Province, China

Xiaowei Zhang
Exploration and Development Research Institute of Huabei Oil Company, Renqiu City, Hebei Province, China

Yan Wang
Engineering Technology Research Institute of Huabei Oil Company, Renqiu City, Hebei Province, China

ABSTRACT: Under the current drilling technology conditions, in deeply buried reservoirs with poor mechanical drilling speed, complex formations such as hard, highly abrasive, and special well sections such as high inclination are drilled, the drilling speed is slow, and the drill and drilling tool assembly need to be tripped frequently to replace, resulting in low drilling efficiency, long drilling cycle, high drilling cost, and high safety risk. In this paper, the torsional impactor speed-up tool for hard formations is optimized by the impact vibration in the circumferential direction of the drill. The reciprocating motion of the piston drives the valve to generate a shut-off pressure pulse to optimize the high inclination. The hydraulic oscillator speed-up tool is used for horizontal wells and a static offset rotary steering system is used for directional acceleration of push-back deep wells. Combined with the characteristics of slow drilling rate in deep wells in the bayan area, field tests were carried out on wells Jihua x ping x, Xinghua x, and Linhua x ping x, and the drilling efficiency was increased by 38 %, 65 %, and 62 % compared with the adjacent wells, and the drilling cycle significantly shortened, which is of great significance for efficient drilling of deep wells in the Bayan area and improving the effectiveness of exploration and development.

1 INTRODUCTION

At present, the exploration and development of the Bayan area have entered a new stage. With the in-depth promotion of the exploration field, the Bayan oilfield is facing the following difficulties in drilling speed up. First, under the influence of geological tectonic movement, the formation dip angle in the block changes greatly, the drilling process of stabilizing the inclination is difficult, and it is easy to drop the inclination. In this paper, through the reciprocating motion of the piston, the valve is driven to generate a shut-off pressure pulse, and the hydraulic oscillator speed-up tool for high inclination and horizontal wells is selected, The torsional impactor speed-up tool for hard formation is optimized through the impact vibration of the bit circumferential direction, which solves the problems of difficult drilling and low mechanical penetration rate caused by large formation angle; Second, the horizontal well reservoir in Linhua block is deeply buried and the reservoir is thin. And the well trajectory is difficult to control. In this paper, the implementation difficulty is reduced during drilling through the static offset push-pull rotary steering system for directional speed-up of deep wells, and well trajectory is controlled.

2 OPTIMIZATION OF NEW TOOLS FOR SPEEDING UP

2.1 *Speed-up test of torsion impactor*

The torsional impactor can be used in combination with the PDC drill to greatly improve the drilling speed in difficult-to-drill formations in deep wells. The tool generates shock vibration in the circumferential direction of the drill through a special structure, which converts the fluid energy of the drilling fluid in the drill string into torsional, high-frequency, uniform, and stable mechanical shock energy, and directly transmits it to the PDC drill, which allows the bit to cut through the formation without waiting for enough torsional energy to accumulate. The vibration frequency of the tool is adjusted by the flow area of the internal variable flow nozzle and the displacement of the drilling pump. The amount of torque produced is determined by the pressure loss of the drilling fluid flowing through the tool (Li et al. 2016).

(a) Working principle: When the drilling fluid sent by the drill collar flows through the filter of the deflector, due to the special structure of the filter as shown in Figure 1, the large particles in the drilling fluid will move down along the central pipe string. It continues to flow under the action of inertia and enters into the hydraulic hammer. Drilling fluid that does not contain large particles will flow out from the diverter hole of the deflector to open the hydraulic hammer in the guide cavity (control room); the rest will enter from the diverter in the center and pass through the drilling of the double-layer spiral filter. The hydraulic hammer is driven by the fluid to drive the hydraulic hammer; the flow of drilling fluid and the pressure drilling speed generated by the nozzle set the torsional impact force of the hydraulic hammer; the torsional impact energy generated by the hydraulic hammer is transmitted to the drill through the transmission joint, thereby overcoming the stick-slip effect of the PDC drill.
(b) Tool failure criteria and post-failure work effects: The main phenomenon after tool failure is a decrease in drilling speed and a small amount of footage. After the tool fails, it will not have any impact on the drilling. Because of its strong tensile, torsion, and compression resistance, its function is similar to the drill pipe short joint, and the drilling can continue.

Figure 1. Torsion impactor.

(c) The tool performance parameters are shown in Table 1:

Table 1. Torsion impactor performance parameters.

Project	Parameter	Project	Parameter
Wellbore size	9-1/2"	Maximum tensile load	>100 t
Maximum compressive load	>300 t	Rotating speed	50 rpm–70 rpm
Material yield limit	930 MPa	Flow L/s	26–32
Pressure drop	0.6–0.8 MPa	Drilling pressure range	8 kN–16 kN
Top make-up torque / kN-m	20	Lower end make-up torque / kN-m	18

2.2 *Speed-up test of hydraulic oscillator*

(a) Working principle

The hydraulic oscillator is divided into pulse subsections and oscillationsubsectionsn. The drilling fluid passes through the jet element of the pulse subsection to produce a wall effect, and the flow of the drilling fluid in the element is continuously reversed, which pushes the piston to reciprocate, and drives the valve to generate shut-off pressure. pulse. The oscillating short joint is mainly a mandrel that is sealed to the outside, and spring is installed axially on the periphery of the sealing mandrel. When the internal pressure acts on the lower end face of the mandrel, under the action of the pressure, the mandrel moves downward and compresses the spring. When the pressure is released, the mandrel returns to the original position under the action of the spring. The pressure pulse can make the oscillating sub joint repeat the above action continuously so that the pipe string reciprocates in the direction of its own axis so that the static friction resistance becomes the dynamic friction resistance.

(b) The main performance parameters are shown in Table 2

Table 2. The main performance parameter table of hydraulic oscillator.

Project	Parameter	Project	Parameter
Recommended working displacement	21–30 L/s	Rated pressure drop	2.3–3.5 MPa
Maximum operating temperature	150°C	Joint	41/2 "
Working frequency	6–9 Hz	Maximum pulling force	200 t
Working torque	<35 kN-m	Tightening torque	40 kN-m

3 OPTIMIZATION OF NEW TECHNOLOGY FOR SPEEDING UP

3.1 *Torsion impactor process*

Torsion impactors are suitable for both conventional drill tool assemblies and screw drill fittings. It can change the working state of the bit and prolong the life of the bit as shown in Figure 2. For the formation with poor drillability, the drill is selected according to the characteristics of the formation to give full play to the auxiliary effect of tool speed and efficiency improvement, using special the drill, the speed-up effect is better. The working life of the turbo-driven torsion impactor can be increased by up to 180 h. It is not recommended to use this tool for seriously lost circulation sections; If there is serious block loss or reaming, this tool cannot be used; a vibrating screen of not less than 30 meshes is used; drill pipe filters must be added to the drilling tools.

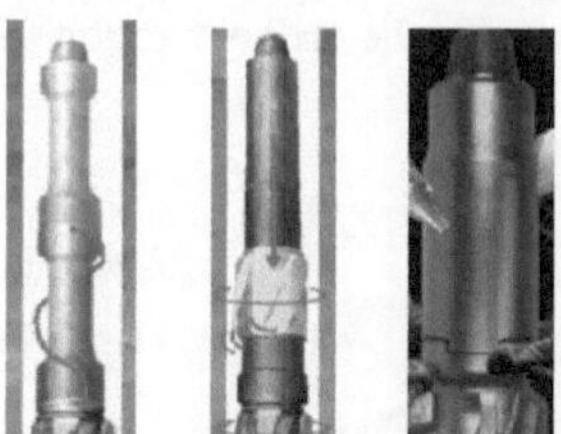

Figure 2. Torsion impactor combination.

3.2 *Hydraulic oscillator technology*

As shown in Figure 3, the hydraulic oscillator can effectively eliminate the accumulation effect of the weight of the drilling tool in a certain section of the well wall during the drilling process with a small drilling pressure; If the drilling weight is too large, the spring of the oscillation sub joint will be compressed, which reduces the usefulness of the tool. The effect of the hydraulic oscillator is mainly determined by the displacement flowing through, but each tool is set according to the specific application conditions, so the displacement during the construction process is as consistent as possible with the displacement of the setting tool so that the tool can be used optimally (Zha et al. 2017).

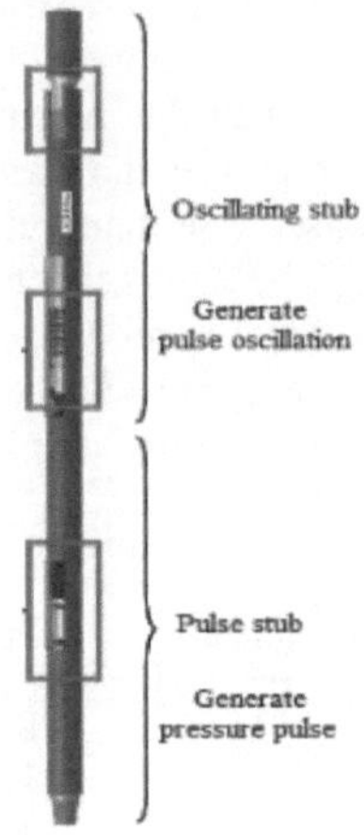

Figure 3. Hydraulic oscillator.

3.3 *Rotary steering drilling technology*

Rotary steering technology is an advanced wellbore trajectory control technology in the field of oil drilling at home and abroad. This technology enables the drilling tool to drill according to the preset wellbore trajectory during the process of rotary drilling (Li et al. 2000).

(a) Working principle

The rotary steerable drilling system currently used in the Bayan area is shown in Figure 4 as a "static offset push-back type". The downhole offset mechanism of the system consists of two parts, a non-rotating outer casing and a rotating shaft, which are connected by upper and lower bearings For the relatively rotatable structure, the upper end of the rotating shaft is connected with the drill string, and the lower end is connected with the drill, which can not only transmit rotational pressure and torque, but also transport drilling fluid. The support rib is placed at the position close to the drill without rotating the casing. During the steering process, the support rib pushes against the well wall to give the drill a reaction force so that the drill drills in a given direction.

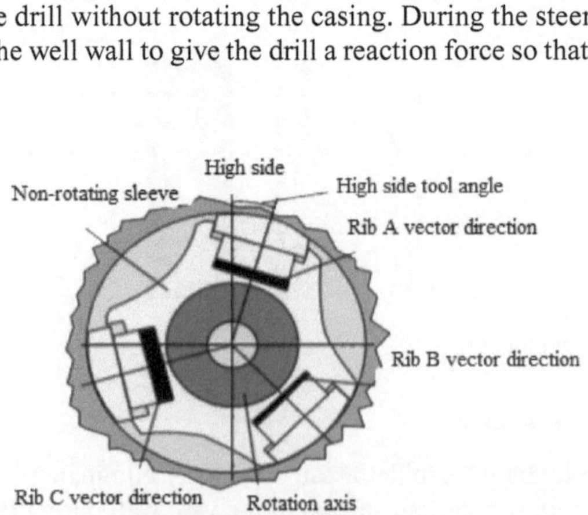

Figure 4. Internal structure of the rotary guide tool.

(b) The main performance parameters are shown in Table 3:

Table 3. Rotary guide tool performance parameters.

Project	Parameter	Project	Parameter
Applicable wellbore size	8 1/2"–9 7/8 "	Maximum temperature	150°C
Build rate (Max)	9.3°/30 m	Highest compressive strength	140 Mpa
Near drill zero-length	1.5 m	Trouble-free working time	≥ 200h
Inclination zero length	10.7 m	Mud displacement range	2041 L/s
Maximum weight on bit	250 kN	Sand content	≤ 0.3 %
Maximum speed	240 rpm	Tilt angle	±0.1°(0–180°)
Maximum torque	21 kN·m	Azimuth	±1.0°(0–360°)

4 APPLICATION AND ANALYSIS OF DRILLING SPEED-UP OPTIMIZATION TOOLS

4.1 *Hydraulic oscillator*

(a) Horizontal well Jihua x ping x

For the horizontal well Jihua x ping x, the service life of the drilling tool reaches 119 hours, and the service life of the horizontal gneiss section can meet the requirements of the PDC drill (Li et al. 2015). The hydraulic oscillator tool cooperates with the inverted drilling tool to drill at a weight of 120 kN, which effectively reduces the friction between the drilling tool and the well wall, relieves the drag pressure downhole, and the effective transmission of the weight on the bit can protect the drill, improve the drilling speed and the drill. The footage, as shown in Table 4, only 2 drills are used in the test section of Jihua x ping x well, and 3 PDC are saved compared to offset wells. The drilling speed reaches 6.07m/h, the drilling speed is increased by 38 %, the drilling depth reaches 2126m, extending the length of the horizontal section, and the footage of gneiss is 1366 m. A new record is set for the longest footage of gneiss in domestic horizontal wells, and the task of construction of the horizontal section of the well has been successfully completed (Zhao 2003).

Table 4. Comparison of field drilling data of hydraulic oscillator.

Hashtag		Footage m	Pure drill time/h	Drill size mm	Mechanical drilling speed/m/h	Speed up drilling ratio
Test well	Jihua x ping x	722	119	215.9	6.07	
Contrast well	Jihua x ping x	500	113.2	215.9	4.42	38 %

4.2 *Rotary steering technology*

(a) Xinghua x

Well, Xinghua x is a vertical well. The field data is shown in Table 5. In the 3497–4473 m well section, frequent orientation is required due to the formation dip. In rotary steerable drilling, the drilling speed is 20.76 m/h, which is 65% faster than the offset well, which ensures the drilling speed.

(b) Linhua x ping x

Linhua x ping x is a horizontal well, and the rotary steerable drilling technology is used during the drilling of the horizontal section of the reservoir. The geo-steering team uses the angle difference to find the oil layer and conducts fine stratigraphic comparison with the surrounding wells in real-time to ensure the accurate landing of the horizontal well. After entering the horizontal section, the well seismic calibration, the logging response in the oil layer,

Table 5. Comparison of field drilling data.

Hashtag	Application well section	Footage/m	Speed/m/h	Compared
Xinghua x	3497-4473	976	20.76	65% faster
Xinghua x	3420-4422	1002	12.53	

and the gamma-ray of the rotary steering direction are fully utilized to accurately determine the formation occurrence and the top and bottom positions of the oil layer, ensuring that the oil layer drilling rate is as high as 90.79%; after the geological task of the section, the inclination was continuously increased upward to find the oil layer of the overlying horizontal well, which is equivalent to the pilot hole of the overlying horizontal well. As shown in Table 6, the drilling speed is 10.4 m/h, which is 62% higher than that in the upper section of the same well, ensuring the drilling speed (Li et al. 2018).

Table 6. Comparison of field drilling data.

Hashtag	Well deep	Application well section	Footage m	Speed m/h	Drill tool assembly	Compared	Reservoir encounter rate
Linhua x ping x	4314	2802–3347 build-up section	545	6.42	Compound drilling tools	62%	90.79%
		3347–4314 deflection and stabilization section	967	10.4	Rotary drill		

5 CONCLUSION

The characteristics of low drilling speed in deep wells in the Bayan area, torsional impactor speed-up tools for hard formations, hydraulic oscillator speed-up tools for high-angle and horizontal wells, and rotary steering systems for directional speed-up in deep wells are selected. Field tests were carried out on Jihua x ping x, Xinghua x, and Linhua x ping x wells, and the drilling efficiency was increased by 38%, 65%, and 62% compared with adjacent wells. It significantly increased the drilling speed of deep wells in the Bayan area, laying a solid foundation for increasing production in the Bayan area.

REFERENCES

Li Ruiying, Wang Feng, Chen Shaoyun, et al. ROP improvement in deep formations in the Daqing Oilfield[J]. Petroleum Drilling Techniques, 2015,43(1):39–43.

Li Shunping, Zhao Tong, Zhu Renfa, et al. Research on ROP improvement technology for Tazhong block at Tarim Basin and its application[J]. Drilling and Production Technology, 2018, 41(6):34–36.

Li Songlin, Su Yinao, Dong Haiping. Structural principle and characteristics of American automatic rotary steering drilling tool[J]. Petroleum Machinery, 2000, (28): 22–27.

Li Wei, Li Zhuolun, Liu Weiqing. Field application of torsion impact ROP-improvement tool in block Wenan[J]. Special Oil and Gas Reservoirs, 2016, 23(4): 144–147.

Zha Chunqing, Liu Gonghui, Li Jun. Development and field application of a compound percussive jet[J]. Petroleum Drilling Technique, 2017, 45(1): 57–61.

Zhao Jingshan. Horizontal well drilling technology in the thin oil layer of Shengli Oilfield[J]. Oil Drilling Technology, 2003(10).

Advances in Mineral Resources, Geotechnology and Geological Exploration – Rashid & Zhang (eds)
© 2023 Copyright the Author(s), ISBN 978-1-032-33377-9

Study on characteristics of the conglomerate reservoir in Shahezi Formation of XJWZ

Wang Xu* & Kai-Yan Wang
School of Earth Sciences, Northeast Petroleum University, Heilongjiang Province Daqing, China

ABSTRACT: With the deepening of XJWZ fault tight conglomerate gas exploration, the Shahezi Formation shows great exploration potential. To strengthen the deep natural gas exploration in the northern Songliao Basin and undertake the prediction of the conglomerate reservoir in the Shahezi Formation, this paper studies the prediction of a well-to-seismic integration reservoir in the study area and predicts the favorable reservoir development area. We carry out sequence division and comparative interpretation of the Shahezi Formation on the basis of the whole XJWZ fracture tertiary sequence division scheme, and the characteristics of the wave group, to synthetic seismic records as a bridge, combined with logging, seismic profile characteristics, and accurately identify the three, four, and five sequence interfaces. Unidirectional well analysis, seismic phase analysis, and sedimentary phase analysis were conducted by combining core observation, logging data, and logging curve feature identification. The reservoir characteristics and physical distribution were also analyzed.

1 INTRODUCTION

The XJWZ depression is a large depression in the northern part of Songliao Basin, 90 km from north to south, 55 km in the widest part, and an area of 4300 km^2. It is divided into XX depression, XD depression, XD depression, XD slope belt, and AD-SP uplift belt. According to previous studies, XJWZ fracture is a west-east supertype skip fracture, which is an anti-oblique structure controlled by the reverse thrust fault of Lingxi. After the deposition of Shahezi Formation, the fracture is strongly active, forming a broken fold, where the top of the fold is strongly denudated. In the Shahezi period, XJWZ subsidence mainly developed deep lake-half deep lake, shore shallow lake, braided river delta plain, braided river delta front, fan delta plain, fan delta front, and other types. In general, the central and western regions are dominated by the lake phase, the west steep slope belt develops the fan delta, and the braided river belt develops the gentle slope delta.

The study area is located in the southwest of the XJWZ depression and the southeast of the ancient central uplift belt. In the northeast of the study area, it extends into the main area of XX depression, and the Shahezi Formation enters the lake deposition, mainly with the deep lake-semi-deep lake phase, and the central and western part is located in the XX fault slope zone, and the Shahezi Formation enters the fan delta deposition, mainly with the fan delta plain and the leading edge of the fan delta. Since Shahezi period is the main period of fracture development, the fracture angle is large, the rock debris alluring from the west uplift quickly rushes into the lake, and the fan delta developed along the fault steep slope is deposited. The single sand body is thick and the thickest place can reach hundreds of meters.

In recent years, with continuous development of seismic exploration technology, deep natural gas exploration has achieved high-yield industrial airflow in B1 Wells and S9 Wells in XJWZ. However,

*Corresponding Author: 752228696@qq.com

DOI 10.1201/9781003319412-13

due to complex geological conditions of deep earthquake and deep burial, it has been difficult to image seismic data, challenge to ensure the signal-to-noise ratio and resolution, and difficult to carry out the fine characterization of targets. To strengthen the deep natural gas exploration in the northern Songliao Basin and tackle the prediction of the conglomerate reservoir in the Shahezi Formation, this paper studies the prediction of a well-to-seismic integration reservoir in the study area and predicts the favorable reservoir development area.

2 SHAHEZI FORMATION ORDER DIVISION AND COMPARATIVE INTERPRETATION

2.1 *Three-level sequence division and comparative interpretation*

In the study area, 10 exploration wells were drilled in Shahezi Formation; however, due to the huge thickness of the Shahezi Formation, few exploration wells were drilled through multiple sets of sequence strata. Therefore, the research area of three sequence division is mainly the whole XJWZ fracture. On the basis of the three sequence division schemes, based on the earthquake integration, super cutting sequence interface features, and the different wave characteristics, to determine the position of the seismic profile, to synthetic seismic records as a bridge, combined with logging, accurately identify three sequence interface, to the line by surface, to surface, in the sand group strata three sequence comparison tracking. Three three-level sequence interfaces were identified in the study area, namely, $T_{4\text{-}1c}$, $T_{4\text{-}1b}$, and $T_{4\text{-}1a}$, which subdivided the Shahezi Formation into four three-level layers, from bottom to SQ_1, SQ_2, SQ_3, SQ_4, as shown in Table 1.

Table 1. Shahezi formation three-level sequence division scheme.

	Three-level sequence division scheme	Earthquake reflection interface
Denglouku Formation		$T_{4\text{-}1}$
	SQ_4	$T_{4\text{-}1a}$
Shahezi Formation	SQ_3	$T_{4\text{-}1b}$
	SQ_2	$T_{4\text{-}1c}$
	SQ_1	$T_{4\text{-}2}$
Huoshanling Formation		

$T_{4\text{-}1a}$ reflection layer is equivalent to the Shahezi Formation tertiary sequence SQ_3 top reflection. In the seismic profile, the overall reflection characteristics are clear, more reflected in the strong amplitude than continuous reflection, its upper layer gradually enhanced, the upper layer gradually weakened from bottom to top (Figure 1), which is a clear sequence interface, easy to track and contrast, its reflection time between 2400 ms and 3120 ms.

$T_{4\text{-}1b}$ reflection layer is equivalent to the Shahezi Formation tertiary sequence SQ_2 top surface reflection. In the seismic profile, the overall reflection wave characteristic is clear, mostly as strong amplitude continuous reflection, its upper stratum is mostly more strong amplitude than continuous reflection, the lower layer is more weak amplitude than continuous reflection (Figure 1), is a clear sequence interface, easy to track and contrast, its reflection time is between 2400 ms and 3300 ms.

$T_{4\text{-}1c}$ reflection layer is equivalent to the reflection of SQ_1, its features in the fracture depth are clear, mostly more strong amplitude than continuous reflection; however, but near the slope, area characteristics are worse, mostly weak amplitude more continuous reflection or blank reflection, can only track the envelope surface (Figure 1), its reflection time between 2400 ms and 3400 ms.

2.2 *Division and comparative interpretation of four and five levels*

Because the research area structure is located in the central uplift zone and XX depression junction, Shahezi Formation is mainly developed in the XX fracture slope zone. The development range of

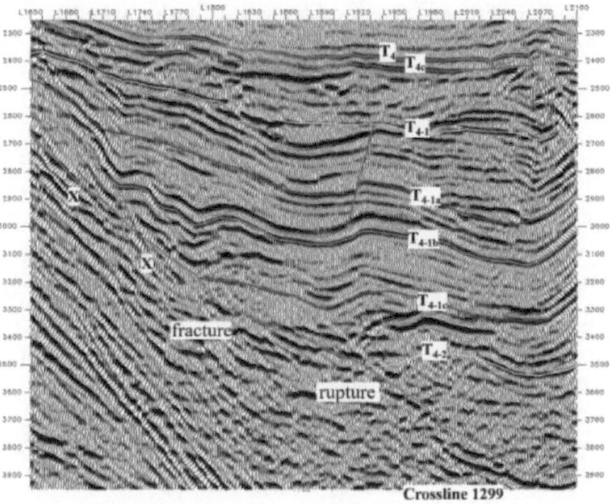

Figure 1. $T_{4\text{-}1a}$, $T_{4\text{-}1b}$, and $T_{4\text{-}1c}$ wave group characteristics.

the sequence with depth (SQ_1 due to the old formation area increases), and the porosity is gradually smaller due to physical deterioration, see Table 2. Via comprehensive consideration, SQ_4 and SQ_3 are mainly selected as two three-level sequences to carry out four-level and five-level sequence subdivisions.

Table 2. Brief comparison of each three-level sequence of Shahezi Formation.

Class Sequence	Area (km^2)	Deep altitude(m)	Maximum thickness (m)	General thickness (m)	Porosity (%)
SQ_4	100.43	−3740∼−4580	1150	500∼1000	>4
SQ_3	57.76	−3800∼−5500	750	250∼600	2∼6
SQ_2	39.11	−4700∼−5900	650	200∼525	<4
SQ_1	215.90	−2975∼−6250	850	200∼650	<4

a. Four levels of sequence division

The tertiary order is a secondary sedimentary order with an internal structure similar to the tertiary order. However, limited by the shallow group drilling depth of the research area Shahezi, the revealed four-level sequence well cannot be used for logging curve and core data identification, and can only function on the basis of the seismic profile features (seismic integration, super, cutting, etc.) for four-level sequence interface identification and division.

According to the above four-level sequence division method, SQ_4 is subdivided into three four-level sequence categories: $SQ_{4\text{-}1}$, $SQ_{4\text{-}2}$, and $SQ_{4\text{-}3}$. The four-level sequence division scheme and the corresponding seismic reflection interface are shown in Table 3 below.

b. Five-level sequence division

The fifth sequence is the secondary sedimentary sequence within the fourth hierarchy, which inherits the internal structural characteristics from the fourth hierarchy. The division and identification of the study area are mainly carried out by the uniaxial identification and comparison of the four-level sequence based on the characteristics of the seismic profile, the sedimentation law, and the thickness relationship.

According to the above five-level sequence division method, $SQ_{4\text{-}1}$ is required to be divided into $SQ_{4\text{-}1\text{-}1}$, $SQ_{4\text{-}1\text{-}2}$, and $SQ_{4\text{-}1\text{-}3}$, then $SQ_{4\text{-}2}$ into $SQ_{4\text{-}2\text{-}1}$ and $SQ_{4\text{-}2\text{-}2}$, while $SQ_{4\text{-}3}$ is too thin to

Table 3. Shahezi Formation four-level and five-level sequence division scheme.

Stratigraphic unit	Three-level sequence	Four-level sequence	Five-level sequence	Earthquake reflection interface
Denglouku Formation				$T_{4\text{-}1}$
		$SQ_{4\text{-}3}$		$SB_{4\text{-}2}$
		$SQ_{4\text{-}2}$	$SQ_{4\text{-}2\text{-}2}$	$SB_{4\text{-}2\text{-}1}$
			$SQ_{4\text{-}2\text{-}1}$	$SB_{4\text{-}1}$
	SQ_4		$SQ_{4\text{-}1\text{-}3}$	$SB_{4\text{-}1\text{-}2}$
		$SQ_{4\text{-}1}$	$SQ_{4\text{-}1\text{-}2}$	$SB_{4\text{-}1\text{-}1}$
			$SQ_{4\text{-}1\text{-}1}$	$T_{4\text{-}1a}$
Shahezi Formation			$SQ_{3\text{-}1\text{-}3}$	$SB_{3\text{-}1\text{-}2}$
	SQ_3	$SQ_{3\text{-}1}$	$SQ_{3\text{-}1\text{-}2}$	$SB_{3\text{-}1\text{-}1}$
			$SQ_{3\text{-}1\text{-}1}$	$T_{4\text{-}1b}$
		SQ_2		$T_{4\text{-}1c}$
		SQ_1		$T_{4\text{-}2}$
Huoshanling Formation				

divide; SQ_3 is divided into $SQ_{3\text{-}1\text{-}1}$, $SQ_{3\text{-}1\text{-}2}$, and $SQ_{3\text{-}1\text{-}3}$. The five-level sequence division scheme and the corresponding seismic reflection interface are shown in Table 3.

2.3 *The explains the results*

a. SQ_1 sequence results

In the SQ_1 sedimentary stage of the early stage of the Shahezi Formation, the lake basin is shallow and in the overall shallow lake environment, and the center of the basin receives debris from the western boundary, forming the fan delta and lake deposition. The development area of SQ_1 is 215.90 km^2, the maximum deposition thickness is 850 m, the general thickness is between 200 and 650 m (Figure 2), the shallowest buried depth of the top surface is -2975 m, and the maximum elevation is -6250 m (Figure 3). The lithology is mainly coarse-grained conglomerate and fine-grained dark mudstone. The SQ_1 formation at the edge of the XX fault slope zone is partially denuded, and the front edge of the sand body is mostly developed separately.

b. SQ_2 sequence results

During the SQ_2 sedimentary period, the lake basin was still in the shallow lake environment, but the water body of the lake basin gradually deepened, inheriting the characteristics from the SQ_1 sedimentary period, and developing the fan delta phase. The development area of SQ_2 is 39.11 km^2, the maximum deposition thickness is 650 m, the general thickness is between 200 m and 525 m, the top surface is −4700 m, the maximum altitude is −5900 m, and the general depth is between −5300 m and −5700m.

c. SQ_3 sequence results

During the SQ_3 deposition period, the scope of the lake basin gradually expanded outward, and the water was further deepened. The deep lake-semi-deep lake environment appeared in the northeast of the study area, and the scope of the lake basin further reduced the sand body. The western XX fault slope zone develops the fan delta plain and the leading edge of the fan delta. The development area of SQ_3 is 57.76 km^2, the maximum deposition thickness is 750 m, the general thickness is between 250 m and 600 m, the shallowest buried depth of the top surface is −3800 m, the maximum altitude is −5500 m, and the general depth is between −4500 m and −5300 m. SQ_3 layers are $SQ_{3\text{-}1\text{-}1}$, $SQ_{3\text{-}1\text{-}2}$, and $SQ_{3\text{-}1\text{-}3}$, with an area greater than 45 km^2. The top buried depth is generally greater than-4500 m, and the maximum buried depth is −5775 m, as shown in Table 4.

d. SQ_4 sequence results

In the early stage of SQ_4 deposition, the water body continued to deepen, the scope of the lake basin continued to expand, and the sedimentary environment was a deep lake-semi-deep lake. In

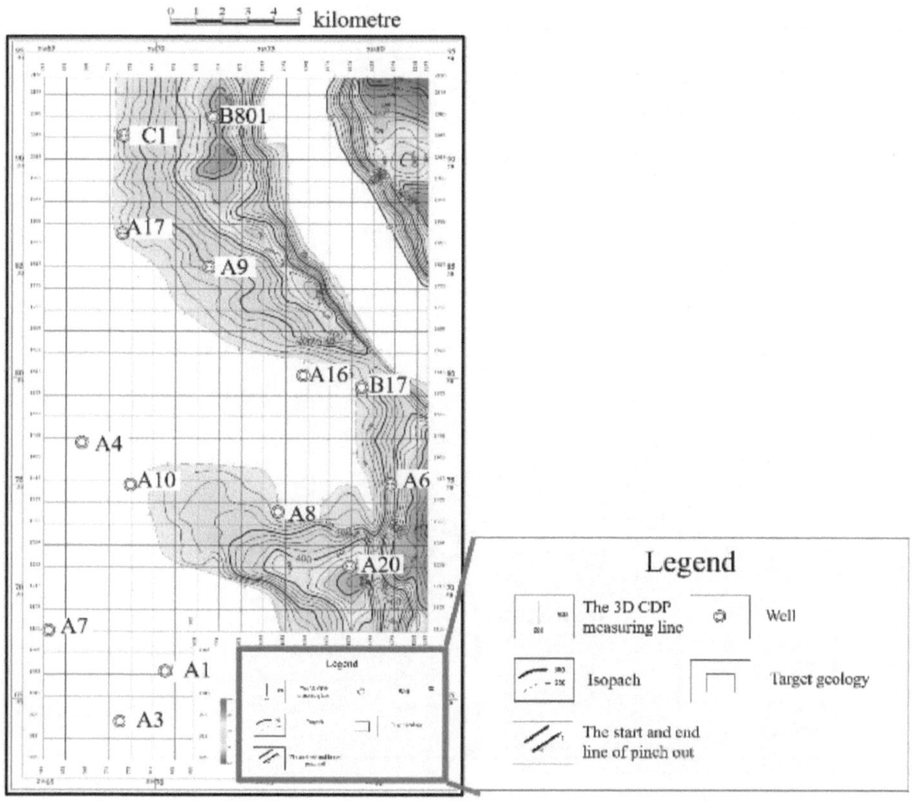

Figure 2. SQ_1 thickness diagram of the tertiary sequence of Shahezi Formation.

Table 4. Brief table of internal sequence development in SQ_3.

Class Sequence		Area (km^2)	Maximum thickness (m)	General thickness (m)	Maximum deep altitude (m)	General depth elevation (m)
$SQ_{3\text{-}1}$	$SQ_{3\text{-}1\text{-}3}$	57.76	375	100~300	−5500	−4500~−5300
	$SQ_{3\text{-}1\text{-}2}$	51.34	220	100~200	−5650	−4750~−5500
	$SQ_{3\text{-}1\text{-}1}$	45.55	220	75~150	−5775	−4750~−5600

the later stage, the water of the lake basin retreated, the edge of the western XX fault slope zone was serious, and the leading subphase of the fan delta was developed. The development area of SQ_4 is 100.43 km^2, the maximum sedimentary thickness is 1150 m, the general thickness is between 500 m and 1000 m, the top surface is −3740 m, the maximum altitude is −4580 m, and the general depth is between −4000 m and −5250 m. The SQ_4 sequence develops $SQ_{4\text{-}1}$, $SQ_{4\text{-}2}$, and $SQ_{4\text{-}3}$. Among them, $SQ_{4\text{-}1}$ developed three-fifth layers: $SQ_{4\text{-}1\text{-}1}$, $SQ_{4\text{-}1\text{-}2}$, and $SQ_{4\text{-}1\text{-}3}$, and the area is greater than 60 km^2. The buried depth of the top surface is generally greater than the -4500 m altitude. The maximum buried depth is -5,225 m in altitude; $SQ_{4\text{-}2}$ developed $SQ_{4\text{-}2\text{-}1}$ and $SQ_{4\text{-}2\text{-}2}$. Area is greater than 80 km^2. The buried depth of the top surface is generally greater than -4100 m altitude. The maximum buried depth is −5,025 m in altitude. The $SQ_{4\text{-}3}$ top suffered from denudation. The small developmental area is 54.45 km^2. The buried depth is generally greater than the altitude of −4000 m. The maximum buried depth is at an altitude of −4,380 m, see Table 5.

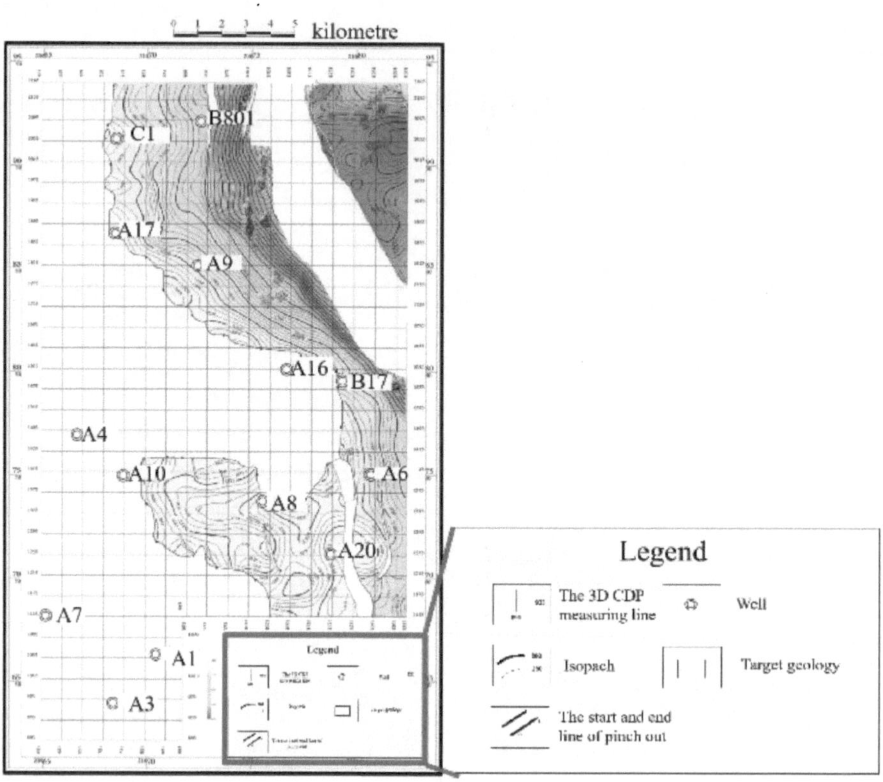

Figure 3. Top surface depth of SQ_1.

Table 5. Brief table of internal sequence development in SQ_4.

class sequence		Area (km^2)	Maximum thickness (m)	General thickness (m)	Maximum deep altitude (m)	General depth elevation (m)
$SQ_{4\text{-}3}$	$SQ_{4\text{-}3}$	54.45	375	100~200	−4380	−4000~−4300
$SQ_{4\text{-}2}$	$SQ_{4\text{-}2\text{-}2}$	85.95	475	100~375	−4575	−4100~−4500
	$SQ_{4\text{-}2\text{-}1}$	82.76	475	100~350	−5025	−4300~−4800
$SQ_{4\text{-}1}$	$SQ_{4\text{-}1\text{-}3}$	70.02	390	50~125	−6250	−4400~−4700
	$SQ_{4\text{-}1\text{-}2}$	63.64	140	50~125	−5150	−4500~−4900
	$SQ_{4\text{-}1\text{-}1}$	66.42	475	150~375	−5225	−4500~−5100

3 GRAPHIC PICTURE

According to the typical structural characteristics reflected by the seismic profile information, the seismic phase analysis is conducted and combined with lithology, and other data revealed by drilling the sedimentary phase mode are finally established.

3.1 *Single well phase analysis*

Combined with core observation, logging data, logging curve characteristics identification method, first of the representative well deposition, and microphase division, on the basis of the total sediment phase and micro acquaintance logging curve standard, to complete the drilling in sand group

10 drilling of sediment phase division, and study the situation of a single well and microphase development.

(1) Lithological mark

According to core observation, the gray conglomerate (such as B13 well 4323 m–4330 m, etc.), gray conglomerate (such as B401 well 4191 m–4532 m, B141 well 4124 m–4213 m, etc.), dark gray siltstone (such as B401 well 4429 m–4442 m, etc.), and black mudstone (such as B141 well 4226 m–4228 m, etc.), gray mudstone (such as B401 well 4479 m–4486 m, etc.), indicating the deep, mainly reducing environment.

Drilling in the research area revealed that the Shahezi Formation mainly developed coarse-grained conglomerate, fine-grained siltstone, and mudstone, and a small number of pyroclastic rocks were developed locally, indicating that the hydrodynamic conditions of the sedimentary environment were weak with extensive development of fan delta and lake phases.

(2) Logphase sign

The well logging data and natural gamma logging curve data in the study area are complete and well corresponding. Using the above single well phase analysis method, the natural gamma curve and resistivity curve characteristics are fully utilized in the study area.

The upper part of well B8 developed a large set of conglomerates, natural gamma curve box low value, resistivity curve high bell value, fan delta plain subphase, gray mudstone, gray clay siltstone, natural gamma curve serrated median, and tooth low value, for deep lake-semi-deep lake subphase (Figure 4, left).

The upper part of Well B141 developed a large set of gray conglomerate, natural gamma curve is box median, resistivity curve is bell high, fan delta plain subphase, lower black mudstone, and gray conglomerate interformation, natural gamma curve is tooth high value, low resistivity curve, for the leading edge of the fan delta (Figure 4, right).

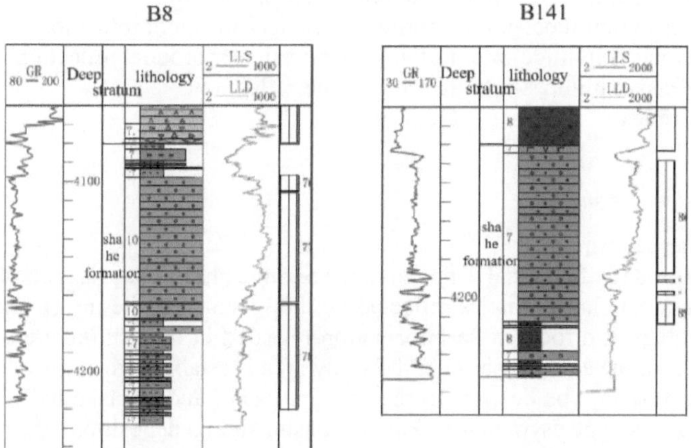

Figure 4. B8, B141 Single well phase analysis.

The upper part of well B13 has interbedded gray conglomerate and black mudstone, the natural gamma curve is serrated low, and the resistivity curve is serrated high, which is the leading edge of the fan delta (Figure 5, left).

The upper part of well B801 developed gray conglomerate and dark gray mudstone, natural gamma curve showed box low value and tooth high value, which is the fan delta leading subphase, the lower developed black mudstone and natural gamma curve showed tooth high value and low resistivity value, which is the deep lake-semi-deep lake subphase (Figure 5, right).

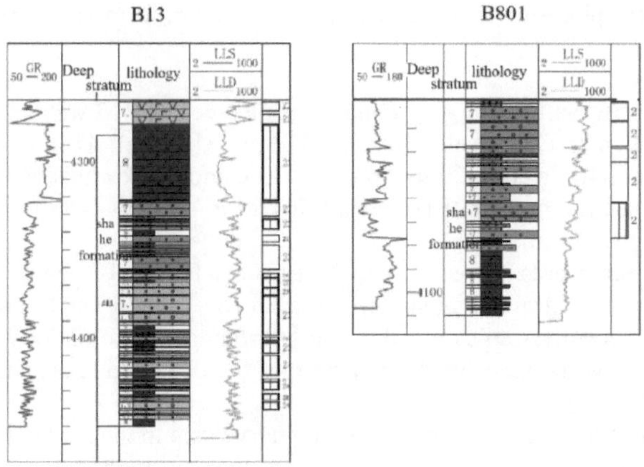

Figure 5. B13, B801 Single well phase analysis.

3.2 *Seismic facies analysis*

The amplitude, continuity, and internal reflection structure were selected as the main parameters for seismic phase analysis, and the geometric appearance was used as auxiliary parameters. According to the principle of seismic stratigraphy and the previous research results on the deep seismic reflection characteristics in the northern Songliao Basin, there are mainly six seismic phases of Shahezi Formation in this area:

Strong amplitude-continuous-mat-like parallel reflection
Medium and strong amplitude-continuous-mat parallel reflection
Medium and weak amplitudes-more continuous-wedge-divergent reflection
Medium and strong amplitude-intermittent-mound anterior product reflection
Medium and weak amplitudes-intermittent-hillock reflection
Moundy blank reflex

3.3 *Analysis of deposition phase*

(1) Sediment phase pattern

Under the background of regional deposition, the seismic phase analysis is conducted according to the typical structural characteristics reflected by the seismic profile information, in combination with the lithology and rock phase information revealed in the drilling area, and finally, the sedimentary model of the seismic phase in the study area is established.

The XJWZ deep basin type belongs to the fracture basin, and the lake basin type belongs to the continental fracture type asymmetric skip lake basin, and its deposition effect is controlled by the tectonic boundary and activity. The steep slope zone of fracture in the basin is the location of controlled fault development, which is characterized by a steep slope, near the material source, with great fluctuation and strong tectonic activity. Sedimentary systems such as alluvial fan and fan delta are developed, which advance closer to the lake. Moreover, the coarse debris such as alluvial fan and fan delta is directly connected to the deeper lake, which shows the characteristics of fast phase band change and discontinuity in the plane. The gentle slope zone has a slow slope, relatively weak tectonic activity, relatively fine sedimentary particles, and advances toward the center of the lake. It mainly develops braided river delta, river delta, and fan delta. Settlement-sedimentation center is often linear along the control fault zone, and the deposition pattern shows large deposition thickness and deep-water center of the lake basin. The sedimentation thickness is thin and shallow, vulnerable to later denudation, resulting in loss of phase zone. The sedimentary pattern of XJWZ fracture is shown in Figure 6 (Lu Pengji et al. 2014).

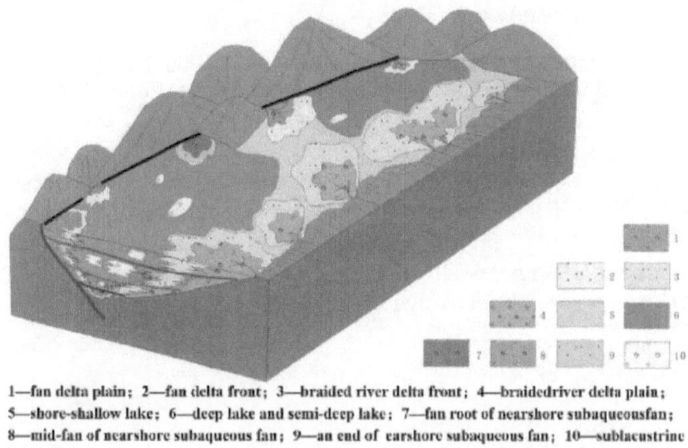

Figure 6. Sedimentary pattern of XJWZ.

The study area is mainly located in the southwest steep slope zone of XJWZ depression, due to the influence of XX controlled subsidence fracture, the main development of fan delta phase, and lake phase deposition. The fan delta phase develops in the western steep slope zone of the fracture, the thickness is large, and the lake phase mainly develops in the central part of the depression near the western depression.

Due to the large angle of the steep slope zone and the rapid material source accumulation in the study area, the front fan delta has entered the deep lake-semi-deep lake area, which is not easy to distinguish from the lake phase deposition. Therefore, the subphase is only divided into the fan delta plain and the leading edge of the fan delta in this study.

In the study area, the lake phase was widely developed, mainly by the deep lake-semi-deep lake subphase. The deep lake-semi-deep lake subphase is a relatively quiet part of the water body below the wave base surface, located in a weak reduction-reduction environment, and has a relatively single sedimentary structure. Mainly the horizontal layer, block layer, and deformed bedding are formed by differential compaction. Its lithology is mainly dark ash, black-gray mudstone, shale, and oil shale, occasionally mixed with thin layers of ash, marl, and argillaceous siltstone. The seismic profile shows more mat-like parallel and sub-flat line reflection characteristics.

(2) Interpretation of seismic phase sedimentology

Through single well phase and seismic phase analysis, combined with the deposition background in the study area, the corresponding relationship between seismic phase and deposition in the study area is summarized in Table 6.

Table 6. Deep seismic phase and deposition phase conversion in XJWZ fracture B-BN area.

Type of sedimentary phase		Type of seismic phase
Phase	Parfacies	Main seismic phase type
Lakes	Deep Lake-Half Deep Lake	Strong amplitude-continuous-mat-like parallel reflection
		Medium and strong amplitude-continuous-mat-like parallel reflection
Delta	Fan delta plain	Medium and weak amplitude-intermittent-mound-shaped disorderly granite reflection
		Glossary blank reflection
	Fan delta leading edge	Medium and weak amplitudes-more continuous-wedge-shaped divergent reflection
		Medium and strong amplitude-intermittent-mound anterior product reflection

3.4 *Results of sediment phase analysis*

This interpretation focuses on the seismic phase and sedimentary phase of two sequence units, SQ_{4-3} and SQ_{4-2-2}, aiming to find the favorable sedimentary phase zone of the Shahezi Formation and lay a good foundation for providing drilling targets.

For SQ_{4-3}, B401, B13, B8, and B141, for example, single-well phase and seismic phase analysis of B401 Well, see Figure 7. SQ_{4-3} sequence has western and eastern sources, western sources from near XX fault belt, and eastern sources from SJ-CYG uplift belt. The phase zone mainly develops near Wells B401, B8, and B141, and the leading edge of the fan delta extends along the deep depression of the fan delta plain near well B13 and develops a deep lake-semi-deep lake between Wells B401 and B13 and B141, respectively (Figure 8).

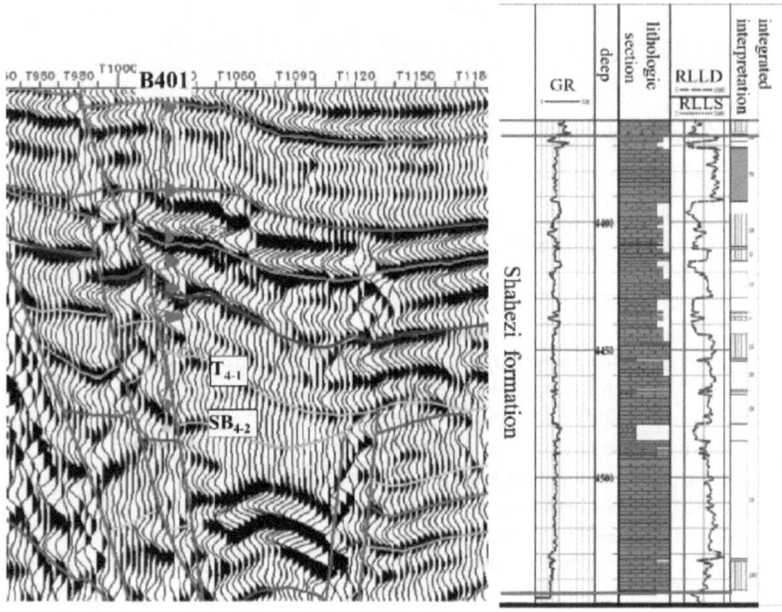

Figure 7. B401, B13, B8, B141 Single-well phase and seismic phase analysis.

Two exploration wells are B401 and B141 in the SQ_{4-2-2} sequence. The SQ_{4-2-2} sequence source is mainly from the XX fault zone in the west of the study area. The fan delta plain phase zone mainly develops in the northeast of well A9 and near well B141. The leading edge of the fan delta extends along the deep depression of the fan delta plain and develops from well B401 to well B141, and the deep lake-semi-deep lake phase develops in the east of the well B401 in the northeast of the study area (Figure 9).

4 RESERVOIR CHARACTERISTICS AND PHYSICAL DISTRIBUTION

Shahezi Formation develops during the strong fracture period, which is affected by the XX fault of the main control fault, and the material source mainly comes from the east-west short axis direction. The parent rock component is from volcanic rocks, including tuff, rhyolite, rough rock, andesite, and early sedimentary conglomerates, breccia, and metamorphic rocks. Due to the large change in structural activity and the small width of the short axis, the rapid unloading of debris material results in poor sorting and grinding circle, different gravel sizes, particle support or matrix

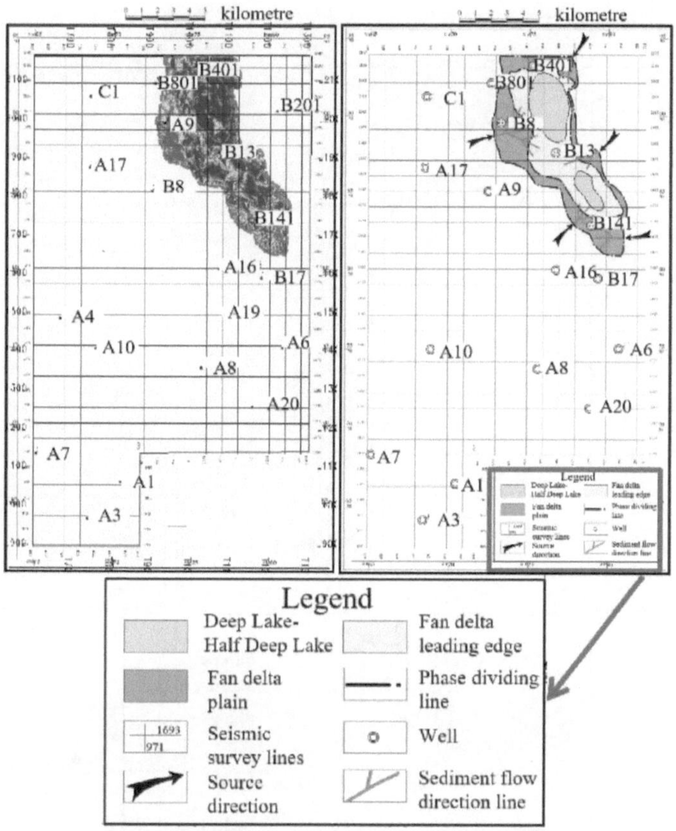

Figure 8. SQ_{4-3} property plan and sediment phase diagram of Shahezi Formation.

support, and angular gravel. The sandstone type is mainly feldspar rock chip sandstone, followed by rock chip sandstone, and rock chip feldspar sandstone, which reflects the characteristics of fast sedimentation speed, strong handling capacity, rapid hydrodynamic change, and close lithology in the study area.

4.1 *Storage space type*

The research results reveal that the important reservoir space types of the sand reservoir in the study area are primary pores, secondary pores, and a small number of microcracks, mainly secondary pores (Leon 2015).

The Shahezi Formation in the study area is a dense conglomerate with high primary intergranular pore content in the reservoir, mainly the remaining native intergranular pore modified by diagenesis with poor connectivity (Figures 10a, b). The most common particle dissolution in the reservoir is the dissolution hole, and feldspar often solves along the disconnection joints, making the particles show a grid shape, partially dissolving from the inside, and leaving dissolved residues at the side. The dissolution hole of rock debris is mostly dissolved from the particles (Figure 10c). The reservoir crack pores are mainly pressure cracks, which rupture some particles of rock chips and feldspar along the joints, forming microcracks (Figure 10e). In addition, structural cracks are also common. These cracks generally do not cut through particles, and the crack surface is curved (Figure 10f).

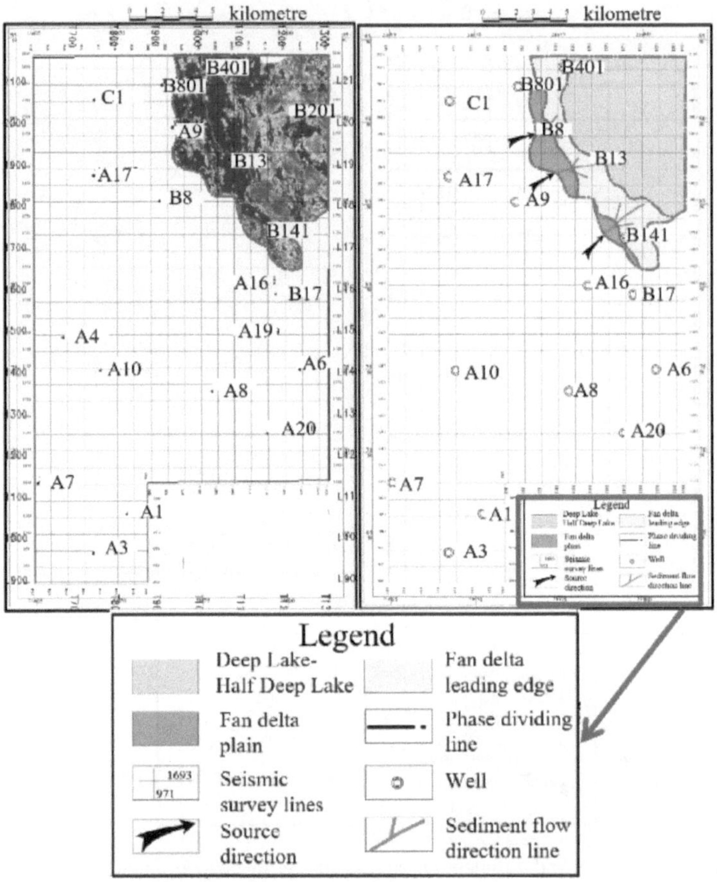

Figure 9. $SQ_{4\text{-}2\text{-}2}$ property plan and sediment phase diagram.

4.2 *Reservoir physical characteristics*

Drilling revealed the conglomerate and bedrock weathering shell of the XJWZ deep reservoir. The main reservoir is the Shahezi Formation conglomerate.

Table 7. Evaluation criteria of reservoir physical properties (CNPC standard).

Evaluation project	Very good	Good	Medium	Inferior	No avail
Porosity (%)	>20	15~20	10~15	5~10	<5
Penetrance ($\times$10-3 μm^2)	>1000	1000~100	100~1	1~0.01	<0.01
Carbonate (%)		3~5	10	>10	

The rock types of Shahezi Formation are conglomerate, sandstone, and siltstone, mainly conglomerate and generally coarse. In lithology, 55% with 4.0% above 4.0%, 25% above 6.0%, 40%, 40% with 0.1m D (88%), and 6% over 6.0% (Liu Chao 2015). It can be seen that the physical property of conglomerate reservoir is better than that of sandstone, and siltstone is the worst. From the perspective of the phase zone, the porosity of sand (gravel) rock is greater than 4.0%, the porosity of sand (gravel) is between 2% and 4%, the shallow lake sand (gravel) rock is between

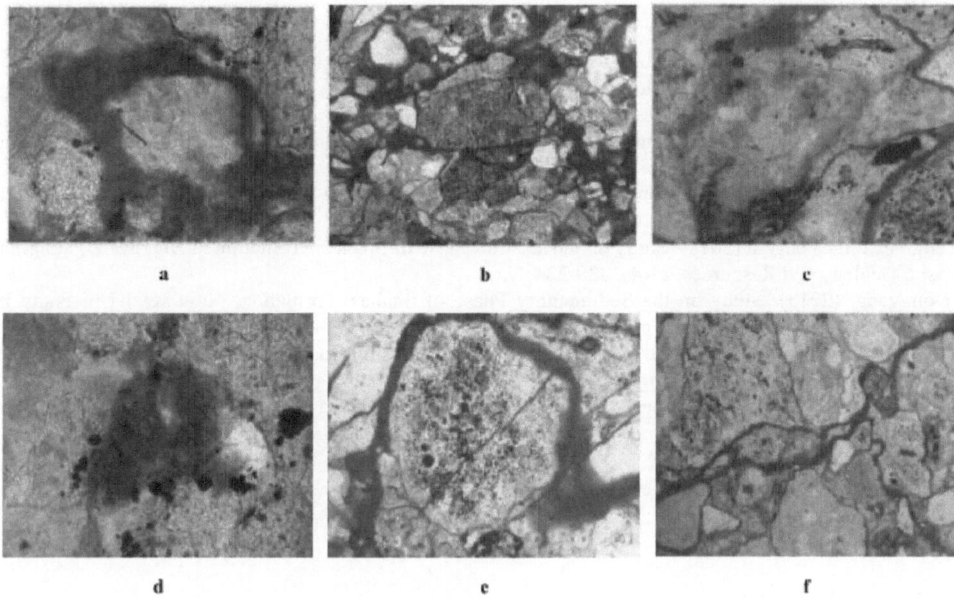

Figure 10. Type of Shahezi subsystem storage space (take well A6 as an example).

2.5% and 3%, the reservoir is good, which is the best SQ_4 reservoir, the porosity is more than 4%, SQ_3 porosity is 2%–6%, and SQ_2 and SQ1 is less than 4%. According to the physical property evaluation standard of CNPC reservoir, the porosity and permeability indexes of dense sand conglomerate are generally low, especially the low permeability index and the reservoir is poor, which belongs to the very low hole and very low permeability reservoir. It is analyzed that the gas reservoir of the Shahezi Formation is mainly a lithic gas reservoir.

5 CONCLUSION

Through the interpretation of the Shahezi Formation, the following conclusions and understandings are drawn:

(1) The Shahezi Formation mainly develops coarse-grained conglomerate, conglomerate, fine-grained siltstone, and mudstone, and a small number of pyroclastic rocks are developed locally, indicating that the hydrodynamic conditions of the sedimentary environment are weak, and the fan delta and lake facies are developed widely.
(2) The overall performance of XJWZ fracture is the western superskip fault, and the study area is in the western fault area. XX fault is the whole fault, and the fault is most developed near the central uplift zone. The deep fault is mainly north, north, west, and west.
(3) The gas reservoir type of Shahezi Formation is mainly conglomererial lithological gas reservoir, and the combined form of reservoir layer is mainly self-generated and self-storage type with lower and upper storage.
(4) Shahezi Formation mainly develops in the northeast of the working area, mainly deposited by the fan delta phase and deep lake-semi-deep lake phase. The sand body at the leading edge of the fan delta extending into the lake area is the most favorable reservoir in Shahezi Formation.

REFERENCES

Cui Fenglin & Cao Guoyin (2014). Study on Prediction Technology of Dense Conglomerate Reservoir in Shahezi Formation of Song Station. Daqing Drilling Engineering Company Geophysical Exploration Company I, 11

Li Ang. (2015). Study on the deposition and reservoir Characteristics of Shahezi Formation in Sunan Zhaozhou Block. Northeastern Petroleum University, 5

Liu Chao. (2015). Study on the conditions and characteristics of the reservoir. Northeastern Petroleum University, 6

Lv Pengji & Liu Chengzhi (2014). Study on the seismic phase of Shahezi Formation in Xujiaweizi, Songliao Basin. Geology and Resources, 23(4): 330-334

Pei Dongyang. (2013). Study on the Sedimentary Phase of Shahezi Formation, Xujiaweizi University in Songliao Basin. China University of Geosciences (Beijing), 5

Xiali. (2009). Sedimentary system and prediction of hydrocarbon source rock. China University of Geosciences (Beijing), 5

Zhou Chuanchen. (2013). Study on the sequence characteristics of Shahezi Formation. Northeastern Petroleum University, 5

Advances in Mineral Resources, Geotechnology and Geological Exploration – Rashid & Zhang (eds)

Application of mine bore-hole transient electromagnetic method in rapid excavation roadway

Yuteng Li* & Jiajia Zhao
Xi'an Research Institute of China Coal Technology & Engineering Group, Xi'an, China

ABSTRACT: Rapid excavation of coal mine roadways has become a hot topic, but there has been a lack of ideal pre-detection means for geological disasters in rapid excavation roadways. In order to achieve the goal of accurately and remotely detecting the low resistivity geological body in front of the excavation roadway in advance, a dynamic source and dynamic receiving transient electromagnetic detection device constructed in the directional borehole is adopted. The 0-704 m section of the fracturing borehole in a coal mine is detected by the push of the drilling rig and analyzed by three-dimensional imaging. The results show that the borehole transient electromagnetic detection technology is an organic combination and mutual supplement of mine geophysical exploration, drilling, and roadway rapid excavation technology. It can effectively identify the resistivity geological body within 30 m of the drilling hole diameter, and can be effectively applied to the geological prediction of rapid excavation roadway.

1 INTRODUCTION

Coal is important fossil energy in China. With the continuous advancement of the intelligent coal mine, the rapid excavation technology of roadways tends to be mature, and the maximum tunneling footage of single-day roadway has been greatly improved. In the mining area of northwest China, with better geological conditions and surrounding rock conditions, the roadway excavation speed can reach 1500–2000 m per month (Cheng et al. 2021; Kang et al. 2021). The rapid excavation system in Daliuta Mine has created a record of 158 m daily maximum footage and 3088 m monthly maximum footage (Ma et al. 2015). Although the geological conditions in this region are simple, with few geological conditions such as faults, collapse columns, old goaves, burning areas, and magmatic intrusions, there are situations such as small coal mines crossing the boundary and goaves in local mining areas in this region, which requires accurate long-distance geophysical exploration technology to detect the strata in the region in advance. Some scholars pointed out that the current geophysical prospecting technology has the bottleneck problems of insufficient detection accuracy and low intelligence of detection technology and equipment, which seriously restricts the full play of the performance of high-end mining equipment. It can be seen that the degree of advanced identification of geological conditions of Excavation Roadway is an important reason affecting the rapid excavation of coal mine roadway. One of the tasks of advanced geological detection for rapid excavation of underground roadways in coal mines is to identify the potential hidden disaster-causing geological factors of excavation face and ensure the geological safety of excavation. In terms of the advance of the water hazard in excavation roadway, the traditional roadway DC method and transient electromagnetic method can only realize the detection of low resistivity anomaly area in the range of 0–100m ahead of tunneling. In the advance drilling of roadway excavation, the method of advance drilling is mainly used for water exploration and drainage. Due to the effective

*Corresponding Author: liyuteng@cctegxian.com

DOI 10.1201/9781003319412-14

area exposed by drilling can only control the geological anomalies within the limited radius around the borehole, and the economic cost and time cost of drilling are high, more and more coal mining enterprises are more willing to use geophysical methods to detect abnormal geological bodies in advance (Han et al. 2010).

As an important geological safety guarantee technology in roadway excavation, the traditional geophysical prospecting method has been unable to meet the requirements of rapid roadway excavation, which puts forward new requirements for coal mine geophysical prospecting technology. Some scholars have proposed the detection method of comprehensive application of mine drilling and mine geophysical prospecting technology. Drilling and drilling are used to provide a construction environment for geophysical prospecting to carry out geophysical prospecting technology in drilling. This method makes full use of the advantages of drilling and geophysical prospecting technology. In the coal mine borehole geophysical prospecting, because the coal mine industry focuses on whether there is a long-distance geological anomaly body and whether it will pose a threat to coal mine safety production, the conventional logging method cannot achieve ideal application effect in coal mines. So, it is necessary to carry out new borehole geophysical prospecting technology and equipment suitable for the coal mine environment (Fan et al. 2021; Zhao et al. 2020). The large depth geophysical exploration technology of the surface is applied to the underground drilling of the coal mine by changing the measuring device, which provides ideas for solving the problems.

Based on this, this paper puts forward the transient electromagnetic detection method in the borehole, designs the three-component receiving coil in the borehole, and uses the drilling rig to push the geological anomaly body in the rapid excavation roadway of a coal mine to carry out the detection experiment and three-dimensional imaging of the detection results.

2 PRINCIPLE OF BOREHOLE TRANSIENT ELECTROMAGNETIC DETECTION

The borehole transient electromagnetic method (BH-TDEM) is a time-domain electromagnetic detection method. The transmitting coil excites the step waveform electromagnetic pulse. After the excitation source is shut off, the secondary field signal generated by the formation eddy current begins to be collected. The purpose of effectively identifying the geological body is achieved by analyzing the variation difference of the secondary field signal with time. In transient electromagnetic detection, Maxwell's differential expression is:

$$\nabla \times E = -\frac{\partial B}{\partial \mathrm{t}} \tag{1}$$

$$\nabla \times H = j \tag{2}$$

$$\nabla \bullet B = 0 \tag{3}$$

$$\nabla \bullet E = 0 \tag{4}$$

$$B = \mu H \tag{5}$$

$$j = \sigma E \tag{6}$$

In the formula, B, H, and E are magnetic induction intensity, magnetic field intensity, and electric field intensity, σ and μ are magnetic permeability and conductivity of the medium, and j is the density of conduction current.

BH-TEM principle of coal mine borehole is shown in Figure 1, where T is the transmitter coil and R is the three orthogonal receiving coils. It is composed of Z component receiving coil Rz coaxial with the transmitter coil, X component receiving coil Rx orthogonal with the transmitter coil, and Y component receiving coil Ry orthogonal with the transmitter coil. The transmitting coil adds step excitation to send pulse electromagnetic signals to the surrounding rock, and collects them by the three-component receiving coil to realize the range detection and azimuth recognition of all-space geological anomalies.

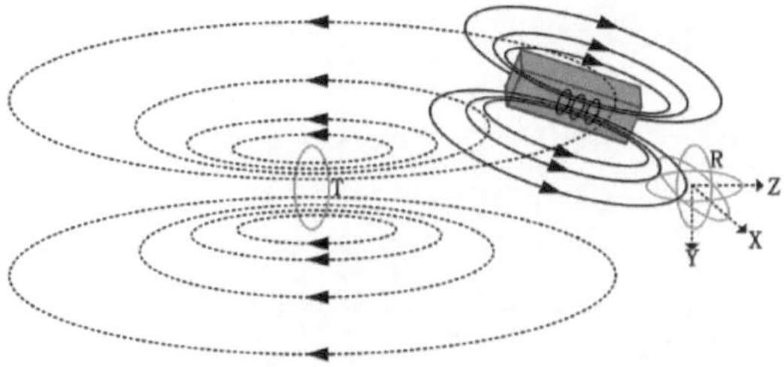

Figure 1. Three-component BH-TEM principle.

3 STRUCTURE OF DETECTION DEVICE

The designed hole detection device is shown in Figure 2, which is composed of the probe tube in the hole and the synchronizer outside the hole. The borehole probe consists of three modules: battery host, transmitting circuit, and receiving probe, which contains the core circuit of the instrument. Out-hole synchronizer is mainly used to issue measurement parameters and clock synchronization in the measurement process. During the detection, the drilling rig pushes the probe to measure the secondary field signal according to a certain distance along the drilling hole extension direction.

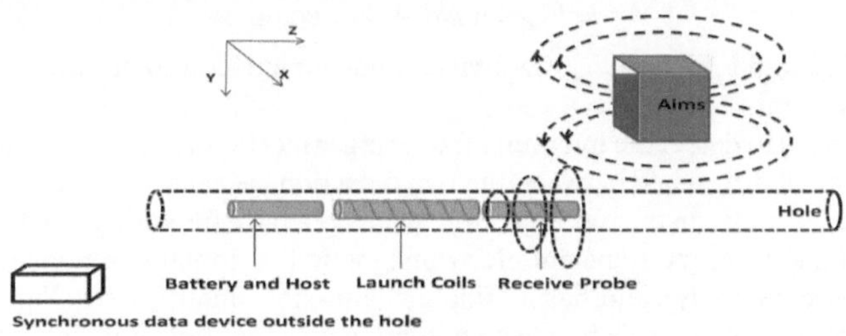

Figure 2. Three-component drilling transient electromagnetic detection device.

4 DATA PROCESSING METHOD

4.1 *Vertical component data processing*

The curve shape of the vertical component data of the borehole transient electromagnetic is basically the same as that of the mine transient electromagnetic detection data (Figure 3). Therefore, the data processing method can refer to the mine transient electromagnetic, and the pre-treatment technology is used to correct the influence of inductance. The corrected data can be processed and interpreted by various inversion imaging methods.

4.2 *Horizontal component data processing*

Based on the three-component BH-TEM device pushed by the drilling rig, the device will inevitably rotate in the borehole. Therefore, the actual formation of horizontal component data needs to be

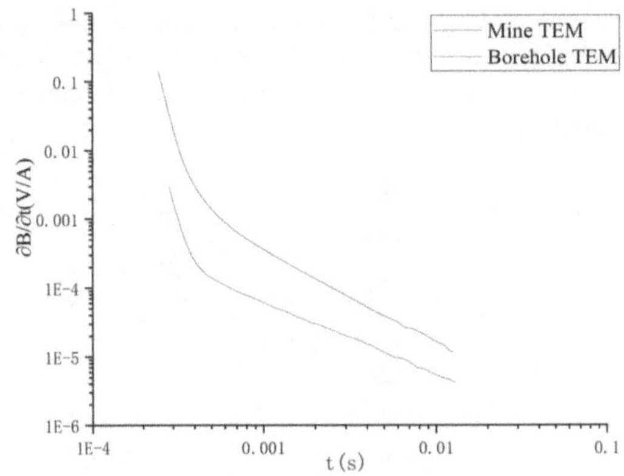

Figure 3. Comparison of measured curves of mine equipment and drilling equipment.

obtained through the corresponding transformation of the tool surface angle. The rotating angle of the instrument in the borehole is α_0, the tool face angle at the i measuring point in the hole is α_i, and the rotation angle of the instrument in the borehole is

$$\Delta\alpha = \alpha_i - \alpha_0 \tag{7}$$

The corrected induced electromotive force of X and Y components at the i measuring point is

$$V_{i,x} = V_{i,x'} \cos\Delta\alpha + V_{i,y'} \sin\Delta\alpha \tag{8}$$

$$V_{i,y} = V_{i,x'} \sin\Delta\alpha + V_{i,y'} \cos\Delta\alpha \tag{9}$$

In the formula, $V_{i,x'}$ and $V_{i,y'}$ are the horizontal component induced electromotive force measured at the M measuring point.

The BH-TEM is used to detect the hydraulic fracturing effect before and after fracturing. Because the main environmental difference between the two detections is caused by fracturing, it is considered that the difference between the two detection results is a pure fracturing field effect. According to the numerical simulation, the transient electromagnetic horizontal component detection curve generated by the pure anomaly field has a 'sine' or 'anti-sinusoidal' shape (Figure 4). The curve characteristics of pure anomalous field when simulating different quadrant anomalies can be summarized as the combination of 'sine' and 'anti-sinusoidal' shapes, as shown in Figure 5, so the quadrant where the anomalous body is located can be judged according to Figure 5.

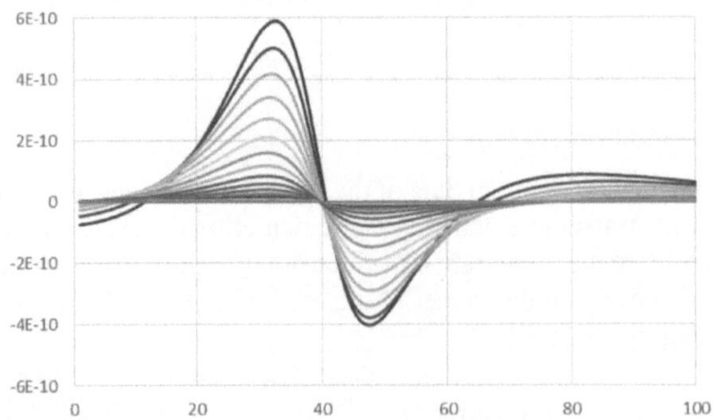

Figure 4. Pure abnormal field pattern.

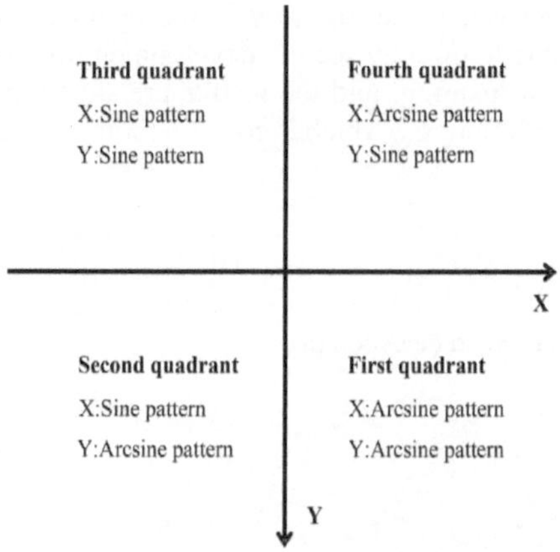

Figure 5. Horizontal component abnormal field pattern combination.

The trigonometric function relationship between the amplitude of the horizontal component anomaly field (Figure 6) is used to calculate the deflection angle θ of the anomaly body within this quadrant according to the following formula.

$$\theta = \arcsin\left(\frac{\left|V_n^{yA}\right|}{\sqrt{(V_n^{xA})^2 + (V_n^{yA})^2}}\right) \tag{10}$$

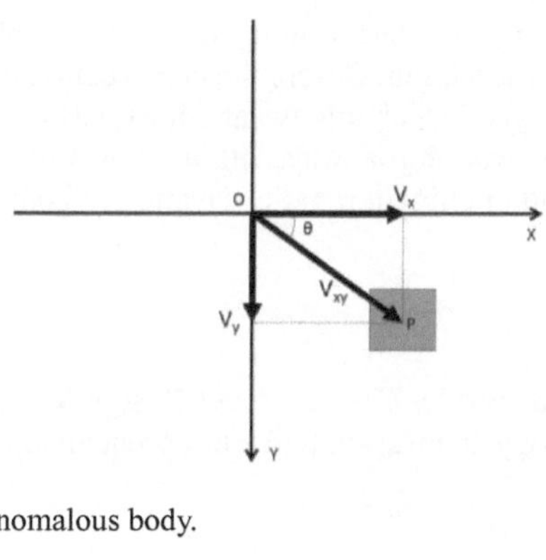

Figure 6. Deflection angle of anomalous body.

5 ENGINEERING APPLICATIONS

In a coal mine in Shaanxi, borehole transient electromagnetic was used for advanced long-distance detection in fast excavation roadway. The test project is set to be implemented in the 4700 m chamber of the main haulage gateway of the 122110 working face, and the drilling trajectory plan is shown in Figure 8. The horizontal length of SY1 and SY2 boreholes designed is 792 m, extending along both sides of the roadway. In the actual detection, due to the collapse of the borehole, the actual detection distance is 15-705 m for SY1 and 18-507 m for SY2.

In order to display the detection results more intuitively, the results are shown in Figure 7. The horizontal coordinate is the drilling depth, and the ordinate is the radial distance of the drilling. It can be seen that the overall visual resistance value is low in the range of 0 – 20 m in the radial distance of the borehole. However, since Figure 7 is the result of the overall superposition of abnormal fields, it is impossible to directly see the development trend of the low-resistance body. The detection results are stereo-imaging, and the results are shown in Figure 8. Figure 8 clearly shows the spatial distribution of fractures, which is convenient for evaluating the effect of hydraulic fracturing.

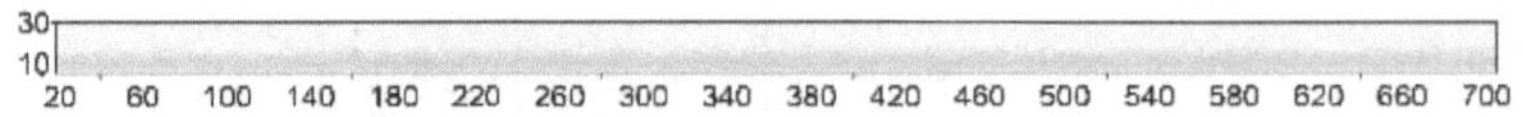

Figure 7. Results of pure anomaly field detection plane.

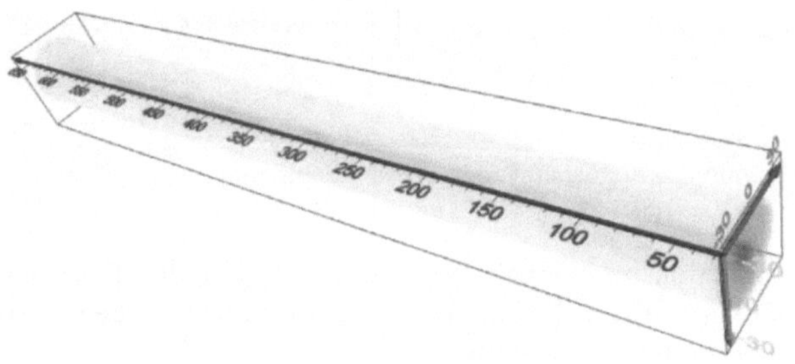

Figure 8. 3D results display.

6 CONCLUSIONS

The borehole transient electromagnetic detection in coal mines combines the advantages of mine drilling and mine geophysical prospecting. Before the construction of a rapid excavation roadway, the borehole transient electromagnetic technology can effectively detect the geological anomaly of the central axis radius of the excavation roadway within 30 m, which provides a geological safety guarantee for safe, efficient, and intelligent rapid excavation of coal mine.

ACKNOWLEDGMENTS

This work was financially supported by The Top-level Design Key Project of Tiandi Technology Co., Ltd. Science and Technology Innovation and Entrepreneurship Fund Special Project (2020-TD-ZD003).

REFERENCES

Cheng J.Y., Lu Z.Q., Jiang B.C. (2021) A novel technology set of "long excavation/ long detection" for rapid excavation in the coal mine roadway. J. Journal of China Coal Society. 1–10.
Fan T., Zhang Y.Z., Zhao R. (2021) Advance detection method of rapid excavation based on borehole TEM intelligent stereo imaging. J. Journal of China Coal Society,46(2):578–590.
Han D.P., Li D., Cheng J.L. (2010) DC method of advanced detecting disastrous water-conducting or water-bearing geological structures along the same layer. J. Journal of China Coal Society,35(4):635–640.
Kang H.P., Jiang P.F., Gao F.Q. (2021) Analysis on stability of rock surrounding heading faces and technical approaches for rapid heading. J. Journal of China Coal Society, 2023–2045.
Ma C., Dai G.S., Cao G.M. (2015) Application of efficient Fast driving system in Daliuta coal mine. J. Coal Engineering, 47(12): 34–37.
Zhao R., Fan T., Li Y.T. (2020) Application of borehole transient electromagnetic detection in the test of hydraulic fracturing effect. J. Coal Geology & Exploration. 41–45.

Advances in Mineral Resources, Geotechnology and Geological Exploration – Rashid & Zhang (eds)

Research and application of production technology by layers in Agadem oilfield, Niger

Tiancheng Yang*, Hongwei Zhang, Wentao Wang, Chang Shu, Jintong Hu, Panfeng Liu & Shiju Ren
Petroleum Engineering Institute of Dagang Oilfield, Tian Jin, China

ABSTRACT: The Agadem oilfield in Niger has various types of reservoirs and multiple sets of oil-bearing layers. There are great differences in reservoir physical properties, pressure system, and fluid properties in different layers, and the contradiction of interlayer interference is prominent, which requires layered production. Therefore, we have studied and tested the technology of electric switch tools controlled separate layer oil production suitable for the characteristics of desert oilfield, which uses the pressure wave to control the electric switch tools of each layer, and realizes the safe and rapid layered production without moving the string. This technology has been successfully applied in four wells in the Agadem oilfield, realizing one-time completion, continuous control of unmoved string, multi-layer exchange, and separate production. In addition, it can avoid interlayer interference, improve oil production, reduce water cuts, reduce operation workload and save operation cost, and realize the innovation of oil production technology in overseas oilfields.

1 INTRODUCTION

The Agadem oilfield is located on the southern edge of the Sahara Desert. There are various types of reservoirs, and there are multiple sets of oil-bearing layers. The physical properties, pressure systems, and fluid properties of different layers are quite different. They generally do not have the ability to self-spray and need ESP artificial lift. Some blocks do not have the conditions for layered development, and need to be replaced between layers during the development technology and returned layer by layer. At present, electric pump wells in Niger Oilfield generally use Y-tube electric pump oil production technology to achieve stratified oil production (Zhen 2002). This technology uses steel wires or cables to switch multi-stage sliding sleeves to control the production and shutdown of each layer. It is inconvenient to change layers and requires high equipment configuration. The structure of the pipe string is complex, and it is only suitable for oil wells with well inclination <45° and 9-5/8" production casing, which is obviously not suitable for Agadem oilfield application. In addition, some oilfields carry out layered oil production by means of a mechanical switch controlled by a pressure spring (Fu 2018; Li 2014; Meng 2013; Zhang 2006), which has poor reliability and can only separate two layers. Therefore, the research on layered oil production technology of electric pump well in Agadem oilfield is carried out to realize one-time multi-layer completion, pump down and layered production, reduce operation volume and operation cost, and achieve the purpose of cost reduction and efficiency increase.

2 RESEARCH THE TECHNOLOGY OF ELECTRIC SWITCH TOOLS CONTROLLED SEPARATE LAYER OIL PRODUCTION

In view of the prominent production contradiction between single well-combined production layers and the low production efficiency, the separate layer oil production technology of electric pump

*Corresponding Author: yangtcheng@petrochina.com.cn

DOI 10.1201/9781003319412-15

well without moving string is studied and formed. This technology uses packers to seal the oil layers and electric switch tools to control the production and shutdown of each oil layer, so as to realize one-time completion, find and block water layers, and control the separate layer oil production of multiple layers by layer replacement, which can reduce operation cost and oil recovery, and improve oil recovery.

2.1 *Research and experiment on tools of control separate layer oil production*

An electric switch tool to control separate layer oil production is developed, which is mainly composed of two parts, the integration of mechanical and electronic components. The mechanical part is composed of a working cylinder, a central tube, a transmission mechanism, switchgear, etc.; the electronic components are integrated by a pressure sensor, a data storage, a detection circuit, a motor, and lithium batteries (Figure 1).

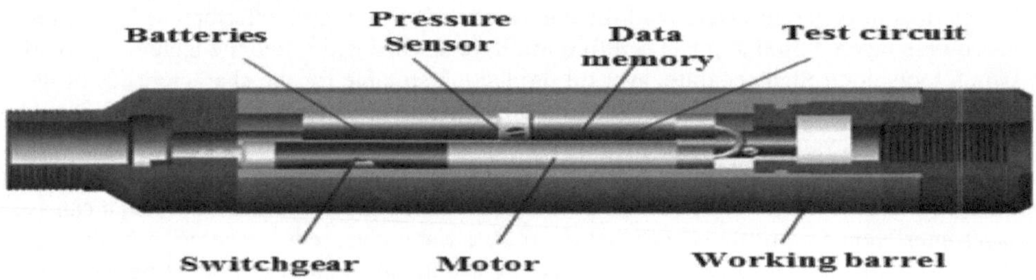

Figure 1. Schematic diagram of the structure of the electric switch tool.

2.1.1 *Working principle of electric switch tools*

Before going down the well, electric switch tools are initialized with the setting parameters with the software on the ground. After going down the well, the electric switch tools will open or close at the set time to automatically adjust the production state of the oil layers. When the production plan is adjusted, and a certain layer needs to be closed or opened manually, it is necessary to use a pump truck to transmit a set of specific pressure waves from the oil casing annulus to the downhole. After the electric switch tools of each layer receive the pressure waves, they are sent to the detection circuit for analysis. When the target layer electric switch tool recognizes the received pressure wave, it will command to adjust the switch to the required production state to achieve the purpose of layered production.

2.1.2 *Features of electric switch tool*

(1) The electric switch tool is powered by high-capacity, high-temperature-resistant lithium batteries, with a service life of three years.
(2) Software initialization to set parameters such as the layer number, pressure wave type, timing switch status, etc.
(3) Each layer of the electric switch tool has a unique pressure wave corresponding to the opening and closing states.
(4) The specific pressure wave is transmitted downhole with the pump truck, and the relevant electric switch tool is instructed to open or close to control the stratified production.
(5) The formation or string pressure fluctuation does not affect the opening and closing state of the electric switch tools.
(6) A one-flow valve is added at the liquid inlet of the electric switch tool, which solves the situation that the pump truck cannot quickly boost the pressure to convey the pressure wave when the oil layer pressure is low.

(7) The electric switch tool adopts high-temperature resistant seals and electronic components, and the surface is treated with anticorrosion to improve the temperature resistance and anticorrosion performance of the tool.

2.1.3 *Technical parameters of electric switch tool*
The technical parameters of the electric switch tool are shown in Table 1.

Table 1. Technical parameters of electric switch tool.

Max. outer size (mm)	Length (mm)	Min. inner size (mm)	Work pressure (MPa)	Work temperature (°C)
Φ114	850	Φ45	60	125

2.2 *The string of electric switch tools controlled separate layer oil production*

The string of electric switch tools controlled separate layer oil production is composed of layered control releasing string and electric pump production string, which are mutually independent. The separate layer oil releasing string comprises a hydraulic releaser, removable packers, electric switch tools, safety subs, a guiding plug, etc. The electric pump string comprises ESP, oil drain, etc. (Figure 2).

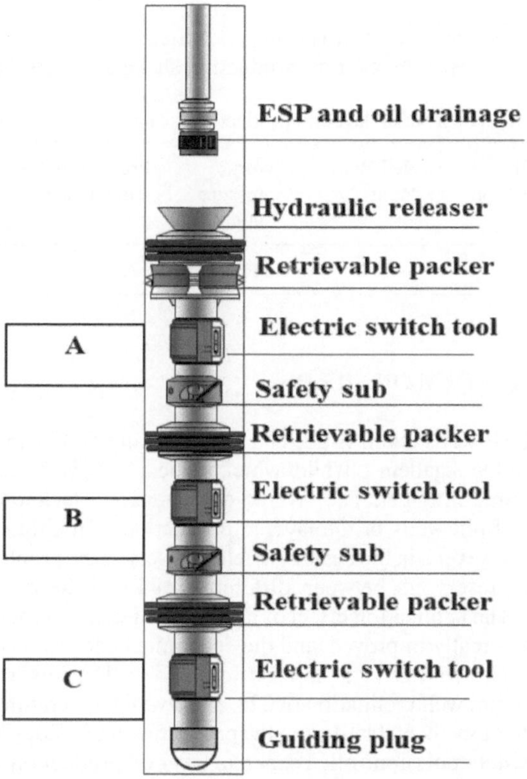

Figure 2. String of electric switch tools controlled separate layer oil production.

2.2.1 *Technology principle of separate layer oil production string*
Firstly, the RIH electric switch tools are initialized in advance to the designed depths with the layered control releasing the string. Then packers are set by pressurization, and the releaser is released by dropping the ball. After that, the upper string is pushed. Finally, the RIH electric pump string performs layered production according to the production plan. When the production plan is adjusted, the downhole is pressurized from the tubing casing annulus with the pump truck to change the electric switch tool of the target layer to the state suitable for production, thus achieving the aim of changing the production layer.

2.2.2 *Technology characteristics of separate layer oil production string*
(1) The electric pump production string and the layered control string are mutually independent, RIH and POOH, respectively, and the pump inspection does not affect the layered control releasing the string.
(2) The removable packers with high temperature and high-pressure resistance are used to separate each oil layer, and the electric switch tools are used to control the production and shutdown of each oil layer.
(3) Adding pressure wave from the annulus can produce any layer in layers without moving the string, which avoids the operation of steel wire or cable and reduces the safety risk and construction operation cost.
(4) The production of any one or more layers of multiple layers (more than two layers) can be adjusted with no adjustment times limit.
(5) The separate layer of oil production string can be salvaged, disconnected, and flushed from the safety sub after the sand is buried, reducing the risk and cost of construction operations.

2.2.3 *Technical parameters of separate layer oil production string*
The technical parameters of separate layer oil production string are shown in Table 2.

Table 2. Technical parameters of separate layer oil production string.

Applicable casing (mm)	Well deviation (°)	Well Work depth (m)	Work pressure (MPa)	Production temperature (°C)	Validity layer section (layer)	period (year)
5-1/2″, 7″, 9-5/8″	≤ 90	≤ 4000	65	125	2~6	3

3 APPLICATION IN AGADEM OILFIELD

In October 2019, four ESP wells (Well S-20, G-11, A S-5, and G-17) applied separate layer oil production technology in the Agadem oilfield, which made a historic breakthrough in layered oil production for the first time in Niger. The layered oil production technology realized the layers replacement production of old wells or the layered production of multiple sets of oil formations perforated at one time in new wells in the desert oilfield and solved the problems of large differences in physical properties of reservoirs between different strata and the rapid rise of water cut in commingled production. The application effect of layered production in four wells was remarkable, the production index was greatly improved, and the economic benefit was prominent.

Application example: The Well G-11 is an electric pump well in Agadem Oilfield, with a well depth of 2995 m, a maximum well inclination of 32°, and a well temperature of 121.3°C. The water cut gradually increased to more than 70% during the production technology. In November 2019, the technology of electric switch tools controlled separate layer oil production was applied to produce two layers. The upper layer separately was produced first, the oil production was increased from 125 bbl /d to 397 bbl / d, and the water cut was decreased from 71% to 0%. After a period of production, due to a shortage of the formation fluid supply, the pressure of pump intake gradually decreased until the well was closed. The upper and lower layers were put into commingled production using special pressure waves to control the electric switch tools. The average daily oil production was 281.27 bbl/d, and water cut was decreased by 40.7% on January 15, 2020. Two months later, after closing the upper layer and replacing the lower layer for production, the average daily oil production dropped to 53.07bbl/d, and the water cut to 89.6%. The upper and lower layers were put into commingled production again on June 8, 2020, the average daily oil production rose to 246.7bbl/d, and the water cut was dropped to 24.2%. (Figure 3).

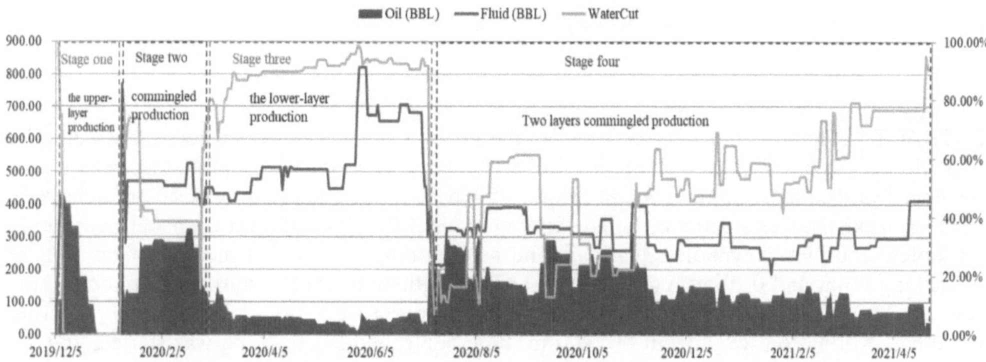

Figure 3. The production curve of well G-11.

4 CONCLUSION

(1) The research and application of the technology of electric switch tools controlled separate layer oil production is easy to change the layer production operation without the need for wire fishing or cable lifting operations.
(2) The technology of electric switch tools controlled separate layer oil production realizes one-time completion and the exchange and separate production of multiple layers controlled by the non-moving string, which can reduce the operation volume, shorten the operation period, continuously improve the oilfield development effect, and finally achieve the purpose of reducing costs and increasing efficiency.
(3) The separate layer oil production technology solves the problem of prominent contradiction between multi-layer system combined production layers, fully excavates the potential of each oil layer, and improves the development effect.
(4) The technology of electric switch tools controlled separate layer oil production needs to be further studied to solve the problem of downhole and surface information feedback.

REFERENCES

Fu Yarong, Li Yangmin, Ma Yongzhong, et al. (2018) Technology of combined production of stratified oil production and sand control in inclined wells. Oil Field Equipment, 47(04): 80–83.

Li Zhiguang, Yan Yongwei, Li Fengtao, et al. (2014) Experimental study on the stratified oil recovery technology of protected oil layers. Oil Field Equipment, 43(11): 57~60.

Meng Yong. (2013) Research and application of new layered oil recovery technology. Guangzhou Chemical Industry, 41 (23): 140 ~ 141.

Zhang Lei, Ren Houyi, Zheng Jinzhong, et al. (2006) Research and application of integrated string of mechanical sand control and layered oil production. China Petroleum Machinery, 34 (05): 51~53.

Zhen Baosheng, Hu Zhongtai. (2002) Testing technology of Y-shaped tubing string for submersible electric pump wells. Well Testing, 11(03): 37~38.

Advances in Mineral Resources, Geotechnology and Geological Exploration – Rashid & Zhang (eds)
© 2023 Copyright the Author(s), ISBN 978-1-032-33377-9

Discussion on long open-hole sidetracking technology based on an ultra-deep well in the Northern Tarim Basin

Xin Qiao*, Renmin Zhang*, Cong Xiong, Jintao Wu, Dabin Fan & Fangyi Xu
Supervision Center of PetroChina Tarim Oilfield Company, Korla, China

ABSTRACT: The well in the northern part of the Tarim Basin is a four-open horizontal well. The completion depth of the third opening is 7538.5 m. The well was drafted before the electric test, but the well wall collapsed in the section below 7450 m, so it was decided to backfill and sidetrack. After two backfills, five sidetracks failed. Aiming at the difficulties of ultra-deep well, ultra-long open holes, complex downhole situations, and serious drag pressure, a method of optimizing sidetracking point and drilling tool assembly and strengthening control during sidetracking was proposed. The drilling tool combination was a 215.9 mm PDC bit and a 172 mm rotary steering tool. When drilling to a well depth of 7200 m, time-controlled sidetracking was started, and the sidetracking was successful. This study can provide a reference for long open-hole sidetracking in ultra-deep wells in the same block.

1 INTRODUCTION

A well in the northern part of the Tarim Basin is a four-open horizontal well. After the second opening is completed, the 244.5 mm casing is lowered to the well depth of 5443 m. The third opening is a 215.9 mm wellbore, with a planned completion depth of 7609 m and a build-up point of 7190 m. When drilling to the 7500 m interval, oil and gas shows are active, and the risk of well control is high. Therefore, the intermediate completion depth is 7538.5 m ahead of schedule. During the drilling operation before the electrical test, the well wall collapsed in the section below 7450 m, and the drilling tool assembly was changed. The drilling tool assembly for drilling the well was repeated five times, but the drilling could not be completed smoothly, and the reaming torque fluctuated greatly. Taking into account the simultaneous existence of lost circulation, collapse, formation of brine, and active oil and gas at the bottom of the well, the decision was finally made to backfill and sidetrack.

2 SIDETRACKING PROCESS

2.1 *The first backfill and sidetracking*

The first backfill and cement plug to the well section from 7100 m to 7400 m, the sidetrack point is selected at 7231 m, corresponding to the well deviation of 6.94°, the drill plug is 7231 m, and the static pressure is 10 t without displacement.

The drilling tool combination for the first drilling trip was a 215.9 mm roller cone bit, a 172 mm straight screw, and a 2.5° bend joint. Time-controlled sidetracking was adopted, and the

*Corresponding Authors: 695433263@qq.com, 342026140@qq.com

 DOI 10.1201/9781003319412-16

sidetracking was up to 7240 m. The tool face was unstable, and the drilling tool combination was replaced after tripping. The drilling tool combination of the second drilling is a 215.9 mm multiple drill bit and a 1.75° single-bend screw. Time-controlled drilling is adopted to sidetrack to 7261 m at a speed of 3h/m. During the sidetracking process, the supporting pressure is serious, and the tool face is always high. The sidetracking track follows the direction of the old eye and stops the sidetracking.

Figure 1. The debris carried by the second sidetracking.

2.2 *The second backfill and sidetracking*

Through the sidetracking analysis, the cement plug is not well cemented, and the strength is not enough, so it is decided to re-plug the cement plug, and plug the well section 6961 m to 7261 m. The sidetracking point is selected in the vertical well section, and the well depth is 7150 m. Considering the cement plug strength, the sidetrack point is changed at 7121 m, and the static pressure is 20 t without displacement.

The third round of drilling increases the rigidity of the sidetrack drill and adds a 159 mm drill collar. The drill tool combination is based on a 215.9 mm roller cone bit, a 172 mm straight screw, and a 2.5° elbow joint, and then three 159 mm drill collars are added. Time-controlled drilling is adopted, sidetracking at 3 h/m, the tool face is placed at 90–120°, the support pressure is still serious, the maximum support pressure is 14 t, and the maximum lifting force is 35 t. When the sidetracking reached 7135 m, it was decided to trip out due to the time of use of the roller cone bit. In order to solve the problem of supporting pressure in the fourth drilling trip, an attempt was made to connect a hydraulic oscillator to the drilling tool assembly. The drilling tool assembly is a 165 mm hydraulic oscillator based on the 215.9 mm hybrid drill bit, 172 mm straight screw, and 2.5° elbow joint. During the time-controlled sidetracking process, the support pressure is still serious. Glass microbeads (0.3%) are added to the well slurry. The support pressure has been significantly improved, but the cost of glass beads is high, and it is not continuously replenished. After a period of time, the support pressure situation is serious, and drilling reaches 7144 m. The sand samples are lumps and cement, the inclination data are consistent with the old wellbore, and it is judged that the drill bit is in the old wellbore. Then, we trip out to re-run the well and handle drilling fluid performance. On the basis of the fourth trip, the drilling tool assembly of the fifth trip removed the hydraulic oscillator and added a 206 mm stabilizer to further enhance the lateral force. In order to solve the problem of supporting pressure in this drilling, the well team installed torsion pendulum software in the top drive and cooperated with the torsion pendulum when sidetracking. However, this measure still failed to solve the supporting pressure problem, and the sidetracking was stopped when drilling to 7153 m.

3 DIFFICULTIES IN SIDETRACKING

3.1 *Well depth is too deep*

The sidetracking point of this well is more than 7100 m deep, and the sidetracking point is in a state of high temperature and high pressure, which seriously affects the working state and service life of the directional tool and cannot guarantee the accuracy of the tool data during the long-term sidetracking process (Wang 2006). At the same time, due to the deep sidetracking point well, it is difficult to center the drilling tool, and the drilling tool is easy to stick to the edge. In addition, the existence of virtual and thick mud cakes causes the WOB to be added to the drilling tool instead of the bit, resulting in the supporting pressure of the drilling tool. This increases the frequency of activities is increased, resulting in the continuous creation and destruction of steps during the sidetracking process. During the five times of sidetracking in this well, there was pressure support in each drilling. During the sidetracking process, the drilling pressure rose to 120–140 kN for every 0.3–0.4 m drilled on average, requiring movable drilling tools.

It is difficult to drive cement plugs in ultra-deep wells because the construction displacement is limited, and the replacement efficiency is poor. High temperature and high pressure have high requirements on the performance of cement slurry, which may cause the cement slurry to retard setting, poor cement bonding, and low strength. It is difficult for sidetracking tools to form effective support, and the success rate of sidetracking is low (Wang 2017). Ultra-deep wells make it more difficult for directional engineers to position the tool face, and it takes a lot of time to adjust the appropriate tool face. At the same time, the tool face is also prone to instability. Two of the sidetracking runs were terminated due to the instability of the tool face.

3.2 *Open-hole is too long*

Due to the influence of lithology and formation dip during the drilling process, the wellbore diameter changes greatly, and the step surface and well diameter expansion will inevitably exist. The sidetracking open-hole section of this well is nearly 2000 m long, and there are many points of resistance downhole. During the sidetracking process, due to the time-controlled drilling (3–4 h/m), the drilling tool has a small moving distance for a long period of time, there are risks such as key-seating sticking and adhesion sticking, and the downhole safety risk is extremely high.

3.3 *Downhole situation is complicated*

During the drilling process of this well, the formation of brine is produced, and each time the well is drilled down to the bottom, the brine mixed slurry is more than 20 m^3. The direct consequence of the brine production is to destroy the performance of the drilling fluid, form a virtual thick mud cake, and increase the frictional resistance. Saltwater immersion destroys the stability of the wellbore, the formation of the unstable well section expands the diameter of the well, which makes tripping difficult, and the time-controlled sidetracking affects the effective transmission of the drill pressure. At the same time, it is also accompanied by lost circulation, resulting in a small density window, and it is difficult to suppress the brine by increasing the density. Once the pressure fluctuates greatly, there is a great risk of lost circulation and overflow.

In addition, the lithology of the lower strata in the block where the well is located is shown in Table 1.

The main lithology of the lower strata in that area is mudstone and sandstone. Due to the good physical properties of sandstone, the mud loses water remarkably, and the mud cake is thick, which will lead to sticking. The open-hole sections of the well are all mudstone and sandstone strata, and the long-term active section of the drilling tools caused by time-controlled drilling is short, which is prone to sticking and sticking.

Table 1. Main lithology of lower strata in the block where the well is located.

Initial well depth/m	End well depth/m	Lithology
7098	7113	Grey mudstone
7113	7119	Grey calcareous mudstone
7119	7126	Brown gray argillaceous sandstone
7126	7155	Grey mudstone
7155	7192	Grey mudstone
7192	7218	Grey calcareous mudstone
7218	7245	Grey mudstone

4 THINKING AND PRACTICE OF LONG OPEN-HOLE SIDETRACKING IN ULTRA-DEEP WELLS

4.1 *Selection of sidetracking points*

Sidetracking points should be selected in stable formations with uniform lithology, small dip angle, good drillability, and regular well diameter, avoiding faults and well sections where the formation is soft and hard (Liu 2006). Generally, mudstone formations with better formation stability should be selected in the Yueman block.

4.2 *Requirements for sidetracking cement plugs*

Generally, a high-density cement slurry should be used for the cementing plug before sidetracking to ensure that there is a cement plug with high strength of 50-100 m in the sidetracking section (Cui 2015). In the process of drilling the cement plug, we should always pay attention to the drilling pressure, torque, and the situation of returning cement blocks. After drilling to the vicinity of the sidetrack point, if there is any phenomenon such as venting, it is necessary to re-fill the cement to seal and ensure the sidetracking requirements (Cao 2012). Due to the presence of oil and gas shows and brine overflow in the well, the poor quality of the cement plug was one of the reasons for the failure of the three sidetracking. The quality of the cement plug can be guaranteed by controlling the pressure.

4.3 *Selection of sidetrack drill assembly*

The roller cone bit has the function of self-cleaning, is not easy to produce a sudden change of reverse torque, is less affected by supporting pressure, and has a strong lateral cutting ability (Jia 2018). It is the first choice for open-hole side drilling.

At present, there are three types of open-hole sidetracking tools commonly used: rotary steering tool, curved screw, and straight screw with curved joint (Chen 2020). Rotary guide tools are safe and efficient, but are often limited by cost control (Wu 2021). The bending point of the bent screw is close to the drill bit, the lateral displacement is small, and the lateral force is weak. The bending point of the drilling tool combined with the elbow joint and the straight screw is far away from the drill bit, the lateral displacement is large, the lateral force is strong, and the sidetracking success rate is high. It is usually used for sidetracking in hard formations with poor drillability. If it is too low, it cannot be rotated. After sidetracking, it is necessary to replace the steerable drilling tool assembly to continue construction (Tan 2017). The commonly used sidetracking tool in the Yueman block is an elbow and a straight screw to ensure the success rate of sidetracking in hard formations.

The rigidity of the drilling tool can ensure the centering degree of the drilling tool, and secondly, it can stably provide a large lateral force without changing the orientation. Therefore, the sidetrack drill assembly must ensure sufficient rigidity, and the rigidity can be improved by adding drill collars and centralizers (Liang 2005).

4.4 *Downhole preparation before sidetracking*

After drilling the cement plug, it is necessary to fully circulate for more than two weeks to adjust the mud performance, and it is strictly forbidden to re-enter the well with contaminated mud to avoid the instability of the wellbore caused by the decrease of mud performance.

For long open-hole sidetracking, it is necessary to pass the well downhole before sidetracking, and the downhole step surface and the virtual thick mud cake are destroyed by drawing up and down and reaming, so as to reduce the supporting pressure during the sidetracking process and ensure the safety of the sidetracking process.

4.5 *Control during sidetracking*

After adjusting the tool face to the sidetracking direction before sidetracking, by repeatedly and quickly drawing the well wall and pulling out the groove surface within a range of 5 m, the drilling tool stabilized to a certain extent, the fluctuation of the tool face and the lifting of the drilling tool were reduced. The frequency of the tool face was re-adjusted to avoid the tendency of frequent back-and-forth pulling and scratching, which was conducive to the formation of the initial step face.

The downward drilling of the drill bit is toward the soft side of the formation, which means that the drill bit tends to drill in the direction of the cement plug during the sidetracking process. It is necessary to control the drilling time to ensure that the drilling direction of the drill bit is consistent with the sidetracking direction. Generally, the time-controlled drilling time is 3 to 5 times or even more than that of the original drill.

During the sidetracking process, the formation of a new wellbore should be judged according to the increase in the percentage of cuttings in the returned sand sample. When the content of cuttings in the formation reaches 80%, the drilling time can be gradually shortened until the new wellbore is drilled.

4.6 *Judgment basis for drilling to new wellbore*

The only basis for judging the new wellbore is that the percentage of cuttings in the sand sample gradually increases to 100%, and the particle size of the returned cement block becomes smaller and smaller until there is no cement at all. The success of sidetracking cannot be judged by changes in torque and drill pressure.

4.7 *Sixth sidetracking*

The main reason for the failure of sidetracking with conventional directional drilling tools in the first five trips of the well was severe underpinning, which resulted in the inability of continuous time-controlled sidetracking. In order to solve the problem of supporting pressure, it was decided to use a rotary guide tool for side drilling. The construction idea is to use the rotary guide tool to drill vertically through the deflecting point of the old wellbore, and then orientate after entering the new formation.

The drilling tool combination of the sixth pass is a 215.9 mm PDC bit plus a 172 mm rotary steering tool. Time-controlled sidetracking started when drilling to a well depth of 7200 m. The forced high-side tool face is 180°, the speed of time control is 3 h/m, and the sidetracking is 4 m to observe that the proportion of cuttings in the sand sample is 60-70%. The drilling parameters are changed, the time is controlled to 2 h/m, the stratum cuttings account for 100% of the sidetracking to a well depth of 7210 m, the near-bit well deviation is 1.2°, and the well deviation is 4.07° in the old wellbore at the same depth. So, it is judged that the sidetracking is successful.

5 CONCLUSION

Based on the results and discussions presented above, the conclusions are obtained as below:

(1) Before sidetracking, a good sidetracking point is selected, and the well section is avoided with large hole diameter change and serious sticking.
(2) The quality of the cement plug is ensured, the static pressure of 5–10 t is verified before sidetracking, and the cement is re-filled to seal or replace the sidetrack point if there is a void phenomenon.
(3) The lithology of the Yueman block is relatively hard, and it is best to use a rotary steering tool and then choose a combination of a high lateral force elbow and a straight screw.
(4) Before sidetracking, the mud properties should be treated well, and the wellbore should be drilled to destroy the steps on the well wall and the virtual thick mud cake to ensure that the sidetracking can be inactive for a long time.
(5) The only basis for judging the new wellbore is that the percentage of cuttings in the sand sample gradually increases to 100%, and the particle size of the returned cement block becomes smaller and smaller until there is no cement at all.

In terms of future work, these experiences can provide a reference for long open-hole sidetracking in ultra-deep wells in the same block.

REFERENCES

Cao, Y.L. (2012) Discussion on the technology of drilling and drilling cement plugs in horizontal wells. J. Science and Technology of West China., 11(08):38–39+37.

Chen, J. (2020) Analysis of drilling technology for open-hole sidetracking in ultra-deep wells. J. China Petroleum and Chemical Standard and Quality. 40(14):253–254.

Cui, L., Luo, W.L., Zeng, Y.S., et al. (2015) Construction Practice of Sidetracking in Well Huadong 12x in Huachang Structure in Fushan Sag. J. Unconventional Oil & Gas. 2(05):66–71.

Jia, X.J. (2018) Research on Tooth Profile of Single Cone Bit. D. Northeast Petroleum University.

Liu, Y.H., Fu, J.H., Liu, M.G., et al. (2006) Trajectory prediction method for sidetracking horizontal Wells considering the interaction between drill bit and formation. J. Drilling & Production Technology. 01: 9–11+121.

Liang, G.H., Guan, H.J., Mao, Y. (2005) Application of screw-type centralizer and drill collar combination full-hole anti-deviation drilling tool in deep water well construction in bedrock strata. J. Hydrogeology and Engineering Geology. 01:98–101.

Tan, F.L. (2017) Theoretical analysis and application selection of screw drilling tools in oil and gas drilling. D. Southwest Petroleum University.

Wang, Z.S., Zou, Q. (2006) Flexible application of several sidetracking methods in directional wells in hard formations. J. Drilling & Production Technology. 06: 19–20+142.

Wang, Z.G. (2017) Practice of cement plug placement for deep oil base drilling fluid. J. Inner Mongolia Petrochemical Industry. 000(001):102–104.

Wu, X.X. (2021) Development status of foreign commercial rotary steerable drilling tools. J. Inner Mongolia Petrochemical Industry. 47(11):39–41+66.

Advances in Mineral Resources, Geotechnology and Geological Exploration – Rashid & Zhang (eds)
© 2023 Copyright the Author(s), ISBN 978-1-032-33377-9

Development and application of hole integrated blowout prevention system

Zunyu Xu* & Zhanjin Lu
China Coal Technology Engineering Group Chongqing Research Institute, Gas Prevention and Control Branch, Chongqing, China
State Key Laboratory of the Gas Disaster Detecting, Preventing and Emergency Controlling, Chongqing, China

ABSTRACT: During the construction of geological drilling in high gas mines, coal and gas outburst mines, due to the use of drilling water to discharge slag and the pressure of the gas or rock powder mixture in the drilling, it is easy to cause instantaneous ejection to the outside. The construction personnel were unable to respond immediately, resulting in an instant gas gushing out of the tunnel or other mixtures in the hole spewing out and hurting people. The blowout prevention system is mainly made into a targeted treatment of the mixture sprayed from the orifice through careful exploration and research of the orifice, reducing gas and personal accidents caused by drilling construction.

1 INTRODUCTION

As one of the main disasters in coal mines, gas disasters seriously threaten the health of underground workers and the safety of coal mine production (Cheng 2010). At present, gas control adopts drilling to extract coal seam gas, and then mining coal seams after the gas content reaches the standard (Xie 2013). However, due to the complex geology of the coal mine, the gas condition in the coal seam is unknown, and a large amount of gas is easily ejected suddenly during drilling, flooding into the working space of the workers, causing the gas in the roadway to exceed the limit, seriously affecting the safety of operators and the safety of coal mine production (Yuan 2013, 2019). At present, it is mainly used to install the blowout preventer at the borehole, but the existing blowout preventer is troublesome to install, not easy to fix, and is limited by the size of the borehole. Thus, it is easy to cause the blockage of the nozzle preventer, which affects the effect of gas extraction and sediment discharge. Therefore, there is an urgent need to develop a drilling integrated blowout prevention system.

The gas exceeding the limit in the drainage roadway is mainly caused by the gas gushing out instantaneously and sharply from the borehole. Although measures have been taken to strengthen the blowout prevention and strong extraction of the orifice, there are many easy gas leaks in the part where the gas is released due to the conventional blowout prevention devices and systems. So, the anti-overrun effect is not ideal (Zhang 2007). During the drilling and hydraulic punching period, the reasons for the over-limit on-site and the easy gas gushing points are as follows: 1) The casing and the hole wall are not tightly sealed after the casing is installed in the blowout preventer, and a large amount of gas gushing out. 2) The seal between the drill pipe and the blowout preventer is not strict, which is another channel for the gas to flow out. Conventional blowout preventer drill pipe seals often use leather pads, polyurethane, and other materials that are easily deformed under the condition of drilling friction of the drill pipe, resulting in the continuous expansion of the gap. 3)

*Corresponding Author: 837623270@qq.com

 DOI 10.1201/9781003319412-17

The outlet of the drainage ballast of the coal-water separation device is the channel for gas gushing. If the control is not timely, there will be a large amount of gas gushing. 4) When dismantling the drill pipe, it is necessary to remove the water tail. At this time, the drill pipe channel is connected with the borehole and becomes the channel for gas gushing.

2 OCCURRENCE CHARACTERISTICS OF COAL (ROCK) LAYERS

The mine is about 6 km long from east to west and 5 km wide from north to south, covering an area of about 22 km^2, with an approved production capacity of 2.9 million tons/year. The mine development mode is a vertical shaft, main crosscut, multi-level, and centralized transportation development. There are three vertical shafts (main shaft, auxiliary shaft, and air shaft), all of which are arranged within the scope of the industrial square. It is divided into three levels: the first level elevation is – 550 m, the second level elevation is – 750 m, and the third level elevation is tentative – 1000 m.

The coal-bearing strata of the mine are Carboniferous and Permian. The Carboniferous Taiyuan formation contains 5 $\sim$ 7 coal layers, which are poorly developed and have no industrial value. The Permian coal-bearing strata are Shanxi Formation and the lower and upper Shihezi Formation, which are divided into seven coal-bearing sections. The coal seam in the mine belongs to high-quality 13 coking coal and gas coal, with low phosphorus, low sulfur, high calorific value, and high volatile content. All coal seams mined in the mine have the tendency of natural ignition. Coal dust in each coal seam has explosion risk. The minefield belongs to a high geothermal anomaly area, with an average geothermal gradient of 3.4°C/HM. Among them, the average ground temperature at – 550 m is above 31°C, and the highest is 37°C, belonging to the first-class heat damage area.

The main coal seams of the mine are one coal, six coal, eight coal, 11 coal, and 13 coal, of which one coal, six coal, eight coal, and 13 coal are outburst coal seams, and 11-2 coal is non-outburst coal seams within the current mining and development scope. The natural gas composition in the coal seam of the minefield is mainly methane, up to 95.94%, while the gas content in the coal seam is mostly below 6 ml G.R. and up to 12.46 ml G.R. In addition, the occurrence of gas in coal seams is relatively regular. With the increase of buried depth, the gas content in the same coal seam also increases gradually.

A coal floor roadway is constructed in the sandy mudstone layer between 1 coal group and one ash. Local sections may expose ash due to the influence of geological structure and changes in the inclination of the coal (rock) layer. The roadway distance is one coal method distance of 10.8$\sim$ 15.0 m.

The features of the mudstone are as follows: gray, sandy mudstone structure, massive structure, parallel bedding, well-developed joints, flat ports, slightly rough, easy to break, locally containing several layers of siderite nodules, 0.1-0.5 m thick siderite nodules. The thickness of this layer is 0$\sim$17.9 m, with an average of 7.3 m.

The features of the ash are light gray, cryptocrystalline structure, thick layered, mainly composed of calcium carbonate, with high argillaceous content at the top, dense and hard, with fossil fragments, developed fissures, filled with calcite, and violent exposure to dilute hydrochloric acid bubble. The thickness of this layer is 1.7–4.6 m, with an average of 2.5 m.

3 CURRENT STATUS OF DRILLING CONSTRUCTION

During the mining of the first coal seam group, the floor roadway constructed during the mining of the 1st coal seam group was affected by the floor limestone water. There are often gas emissions and coal (rock) mixture ejection events during the construction of a geological exploration drilling hole, and the maximum mixed pressure ejected from the drilling hole is more than 2 MPa. This not only endangers the safety of drilling workers, but also brings unfavorable factors to mine safety.

In order to effectively control the abnormal gas emission during drilling, a pumping device was installed at the hole of the drilling hole. During the drilling construction process, the device was used for drainage to reduce the impact of the drill pipe holder and extractor with the drill pipe and prevent the gas from leaking from the gap.

3.1 *The feasibility of an integrated blowout prevention system*

The system integrates "blocking," "drainage," and "buffering" into one blowout prevention device, which greatly solves the problem of gas gushing in boreholes or high-speed ejection of coal (rock) particles, water, and other mixtures.

Among them, the "plugging" system mainly plays the role of sealing and preventing backflow, blocking the chance of gas in the borehole from contacting the outside. The diversified design of the "drainage" system mainly focuses on the persistence of drainage to deal with drilling problems. The principle is to implement comprehensive drainage of boreholes, increase the drainage volume of boreholes, and improve the drainage capacity; when the above two measures can meet the needs of drilling borehole management, the "buffer" system will be used as an auxiliary system or as an auxiliary system. The channel for the mixture of coal (rock), water, etc., sprayed from the borehole reduces the possibility of spraying into the roadway.

4 THE COMPONENT COMPOSITION AND ANALYSIS OF THE INTEGRATED BLOWOUT PREVENTION SYSTEM

The drilling integrated blowout prevention system is mainly composed of gas extraction pipes, connecting hoses, reverse skirts, air-water separators, buffer airbags (skeleton air ducts), rubber gaskets, control gate valves, and other devices. The device is mainly composed of "blocking," "drainage," and "buffer" systems, and is centrally controlled by a gas-water separator connected to the main extraction pipeline. The components of the blowout prevention system are shown in Figure 1.

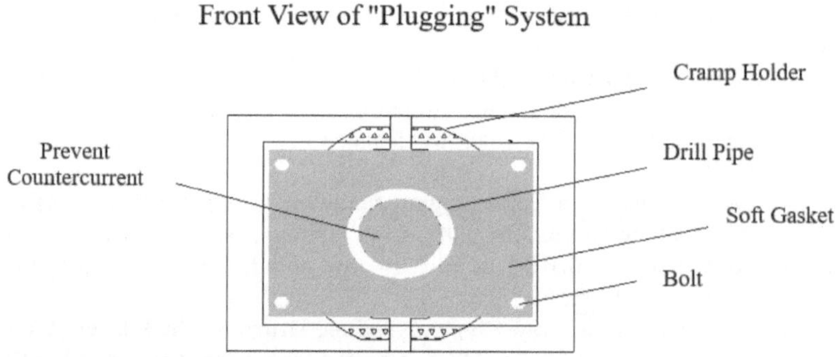

Figure 1. Components of the BOP system.

4.1 *Blocking system*

The "blocking" system is mainly installed inside and outside the drilling rig slips. The rubber gasket installed on the outside is mainly used for sealing, and the skirt device installed on the inside plays the role of preventing backflow. When the drilling is under normal construction, the skirts and rubber gaskets installed on the inside and outside of the slip are in a state of screwing into the hole between the drill rod and the barrier. When gas is ejected from the borehole, the skirt device installed on the inner side forms a reverse flow state, and the outer barrier forms a completely sealed state, which prevents the gas in the borehole from gushing out and cuts off the channel between the gas in the hole and the outside world.

4.2 *Drainage system*

The "drainage" system mainly comprises the pumping device above the extractor and the switching device at the end of the drill pipe to centrally extract the boreholes. The pumping and extracting device is sealed by the extractor during normal drilling. The negative pressure of the extraction can be appropriately controlled due to the influence of the state; the tail of the drill pipe is installed as a switching device connected to the wind (water) pipeline during drilling. When gas gushing occurs in the borehole, increase the negative pressure of extraction while drilling to above 10 kPa. At the same time, the switching device connected to the tail of the drill pipe needs to close the control air (water) pipeline gate valve and open the gas and water separation. The control gate valve connected to the gas generator enhances the gas drainage in the borehole and reduces the possibility of gas gushing out of the borehole.

4.3 *Buffer system*

The "buffer" system is used as the auxiliary system of the blowout prevention system. The "buffer" system is processed by a skeleton air cylinder and connected to the air-water separator. When the borehole is under normal construction, the "buffer" system is affected by the negative pressure of the extraction and is in a closed state of adsorption. "When the system cannot meet the needs, it will enter the buffer airbag along the multi-pass device. At this time, the gas and coal (rock) gushing from the hole will cause the airbag to bulge, and the sprayed coal (rock) powder can directly enter the buffer." The system and the extraction hose connected to the gas-water separator will work together with the orifice to draw and extract simultaneously, which enhances the ability to deal with the gas gushing from the borehole and the coal (rock) dust in the hole. The specific design is shown in the figure below).

In addition, during the drilling construction process, the drainage pipe is connected with a ¢76 mm hose. The bottom of the drill site or roadway must be slightly upwards to prevent gas leakage from the channel.

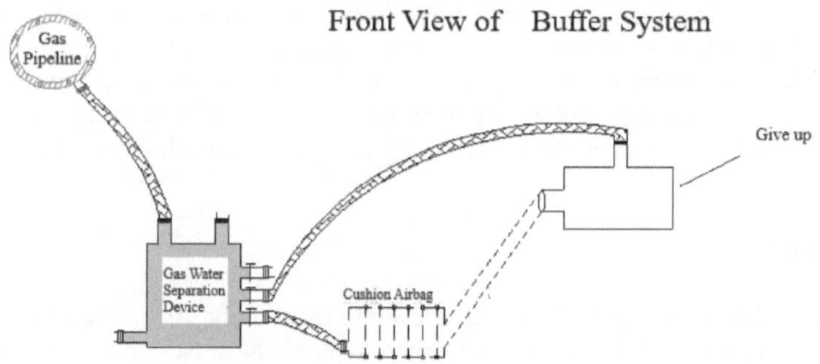

Figure 2. Schematic diagram of the buffer system.

5 TECHNICAL COMPARISON

5.1 *Production cost comparison*

The proposed blowout prevention device has the following advantages: convenient production (all kinds of connecting devices and processed parts are common parts), low material cost (skeleton air duct and control gate valve are commonly used downhole parts), simple installation, flexible use, convenient operation, with rubber gasket, good performance in reverse skirt barrier, high safety, with drilling construction and drainage being carried out at the same time, and strong controllability.

5.2 *Technical comparison*

The original blowout prevention device has a poor sealing effect. The drill pipe holders, orifices, etc., can form channels for gas or other mixtures to spray out. In addition, it can only be applied to small gas pressures (below 0.6 MPa). It can be used when the mixture in the hole sprays out slowly. If the gas pressure is high (over 0.6 MPa) or the mixture in the hole sprays faster, it cannot meet the needs.

The drilling integrated blowout prevention system comprehensively considers the various parts of the drilling rig and seals all exposed parts to ensure that the drilling is separated from the outside. In addition, the distance between the air-water separator and the drilling channel is increased (1.5–2 m). It reduces the occurrence of gas pressure or other mixtures that are not released when the gas pressure is sprayed at high speed.

6 ANALYSIS OF ECONOMIC AND SOCIAL BENEFITS

6.1 *Low material cost*

Both the gas-water separator and the integrated box can be welded by ⌀3 mm steel plate. The skeleton air duct is an air duct commonly used for underground roadway turning installation. The control gate valve is a ⌀50 mm borehole gate valve. The rubber gaskets are all underground. Using commonly used parts, the production cost is low. The welding process of the connecting device of each part is ordinary oxygen welding, which is simple to manufacture.

The installation is simple and convenient. The connecting part of the device is fixed with 10–12# bolts, and the whole device can be fixed on the tunnel side and the top (bottom) slab of the drilling site by using 8# iron wire coiling, without affecting the safety space of drilling construction.

6.2 *Social benefit analysis*

After the integrated system 1 was put into use during the construction of the floor tunnel of the coal seam group, it solved the problem that the tunnel could not be constructed due to the large gas in the drill hole when the construction geology advanced detection hole. In the future, we will gradually use this system to replace the original blowout preventer in the drilling holes of the various construction geology and pre-draining pits of our mine. A large amount of coal (rock) ejected from the borehole hurts people, ensuring the safety of the borehole construction personnel.

7 CONCLUSION

(1) A new type of drilling integrated blowout prevention system has been successfully developed, which mainly consists of gas extraction pipes, connecting hoses, reverse skirts, gas-water separators, buffer airbags (skeleton air ducts), rubber gaskets, control gate valves, etc.
(2) The drilling integrated blowout prevention system can separate the drilling hole from the outside world, which increases the channel distance between the gas-water separator and the drilling hole, and has the characteristics of low material cost and easy installation.
(3) Drilling integrated blowout prevention system can effectively reduce the occurrence of gas exceeding the limit or the mixture hurting people when the gas pressure or other mixtures are ejected at high speed due to the unreleased pressure, which has good social benefits.

ACKNOWLEDGMENTS

This work was supported by the key projects of the joint fund for regional innovation and development (U21A20110).

REFERENCES

Cheng Yuanping. Coal Mine Gas Prevention Theory and Engineering Research, Xuzhou: China University of Mining and Technology Press, 2010.

Xie Ping.Study on the migration and accumulation of pressure relief gas in the cracked space caused by overburden mining[D]. Hunan University of Science and Technology, 2013.

Yuan Liang et al. Technical Manual of Chief Coal Mine Engineer[M]. Beijing: Coal Industry Press, 2013.

Yuan Liang. Theory of pressure relief mining and gas extraction and technology system of coal and gas co-excavation[J]. Journal of China Coal Society, 2019.01, 34(1).

Zhang Guoshu, etc. Ventilation Safety[M]. Xuzhou: China University of Mining and Technology Press, 2007.

Advances in Mineral Resources, Geotechnology and Geological Exploration – Rashid & Zhang (eds)
© 2023 Copyright the Author(s), ISBN 978-1-032-33377-9

Distribution characteristics of Neogene remaining oil in M oilfield

Lu Yang*
Liaohe Oilfield Exploration and Development Research Institute, Panjin, Liaoning, China

ABSTRACT: M oilfield has a history of exploration and development for more than 30 years. The oil and gas reservoir has entered a period of medium and high water cut, with few high-quality reserves and difficult to exploit the remaining oil. Therefore, it is very hard to stabilize production, and the task of stabilizing production and increasing production in the long term is increasingly difficult. At present, it is tough to determine the remaining oil-gas enrichment area of the main oil-gas producing strata in the study area, and there is not much room for adjustment. In order to deeply understand the spatial characteristics of reservoir sand body, remaining oil distribution law, and main control factors in the M oilfield, and help tap the remaining oil potential in the old area. Based on the establishment of 3D geological modeling and combined with oilfield performance data, this paper studies the distribution law of remaining oil by integrating various methods for the Neogene of M gas field. The conclusions of this study are as follows: (1) The remaining oil is mainly distributed in the form of the remaining oil in the layer affected by heterogeneity and the remaining oil in the local enrichment controlled by the barrier and interlayer, and the remaining oil is mainly distributed in the X52-2 sub-layer. (2) The irreducible water saturation of reservoir rock in the M gas field is high and has obvious hydrophilic characteristics. (3) The controlled and recoverable reserves of the M gas field are 6.668 million tons and 2.7 million tons, respectively, and the cumulative oil production and cumulative gas production show a linear relationship. (4) The remaining oil distribution has four characteristics: enriched in the high part of the structure, controlled by the development degree of barrier and interlayer, controlled by interlayer difference, and formed by the imperfect well pattern.

1 INTRODUCTION

M oilfield is located on the structural belt of a basin in Western China. It is a nearly symmetrical short-axis anticline with a near east-west strike, a slightly steep north wing, and a slightly gentle south wing. After decades of exploration and development, it has entered the production decline stage of medium and high water cut, with few high-quality reserves and difficult to exploit the remaining oil. The development of the following oil fields is becoming more and more difficult, and the oil fields have put the remaining oil tapping into the primary position. During the Sixth Five-year Plan period, domestic research on remaining oil was carried out, and the variation of remaining oil characteristics overtime on different geological scales was studied. In order to further improve oil recovery and excavate remaining oil, it is necessary to accurately describe and quantify remaining oil and study its distribution law.

At present, studies on remaining oil at home and abroad mainly focus on three aspects: remaining oil distribution characteristics, remaining oil saturation measurement and detection technology, and remaining oil exploration technology (Chen 2000; Han 1995; Liu 2004; Lu 2001; Xie 2003). Various international conferences have been held to describe the distribution of remaining oil after reservoir displacement in macro and micro aspects. The determination of the distribution characteristics and laws of remaining oil in the reservoir has always been an important subject to be discussed and explored in the field of petroleum research (Li 2001; Yan 2016).

*Corresponding Author: yangl09@petrochina.com.cn

 DOI 10.1201/9781003319412-18

In view of the existing problems in the study area, this paper will start with the collected geological, logging, and seismic data, and use Petrel 3D geological modeling software to carry out 3D deterministic modeling, focusing on the delineation of the barrier and interlayer. On this basis, the remaining oil distribution characteristics are analyzed by integrating various methods. Through the above research work, we hope to promote the technical development of the remaining oil research and help tap the potential of the remaining oil in the study area.

2 METHODS AND MATERIALS

In this paper, a 3D geological model of the study area was first established. Then, according to the reality of high water cut and scattered and complex types of remaining oil in the late development period, the research ideas of comprehensive application of multiple methods, multi-angle description, and semi-quantitative and quantitative characterization were determined.

Petrel 3D geological modeling software was used for geological modeling, and the target layer of modeling was Neogene. The modeling method is single sand body deterministic modeling, and the emphasis is on the characterization of barrier and interlayer.

2.1 *Data preparation*

Five types of data are prepared for the establishment of a 3D geological model, namely well location data, stratified data, seismic interpretation data, reservoir parameter data, and geological understanding data. Among them were a total of 112 pieces of well location data, including wellhead coordinates, bolting altitude, drilling depth, and deviation data. There were 41 stratified data sets. Seismic interpretation data include fault data and seismic interpretation X52 top surface structure data. Reservoir parameter data includes interpretation data of vertical lithology, sedimentary facies, porosity, permeability, and water saturation physical parameters of each well. Geological data include the plane distribution range of a single sand body and the distribution range of barrier and interlayer.

2.2 *Modelling procedure*

First of all, we should focus on the description of the distribution of a single sand body, barrier, and interlayer. Secondly, on the basis of phase control, the deterministic method is adopted to increase the continuity of data distribution and facilitate numerical simulation. Finally, the geological model is constantly revised and improved in combination with the trial calculation of numerical simulation and the results of reservoir performance analysis (Figure 1).

2.2.1 *Structural modeling*

Overall consideration, the step size of the plane grid of this fine geological model is 20 m×20 m. In the process of establishing the fault model, the dip, dip angle, and distance of the fault are considered, and the fault model is determined by seismic interpretation of the fault.

2.2.2 *Attribute modeling*

The lithofacies model is established before attribute modeling. The lithofacies model of each single sand layer is established by the deterministic modeling method based on the sand body plan of each single sand layer in geological research. The 3D attribute modeling adopts the Kriging algorithm under the phase-controlled condition to establish porosity, permeability, and water saturation models. The spatial distribution characteristic of data is that the parameters of the main range, secondary range, and vertical range are fitted by the variation function.

2.2.3 *Calculation of reserves*

The reserve calculation method of the model is the volume method, and the reserve calculation parameters are the parameters used in the reserve calculation.

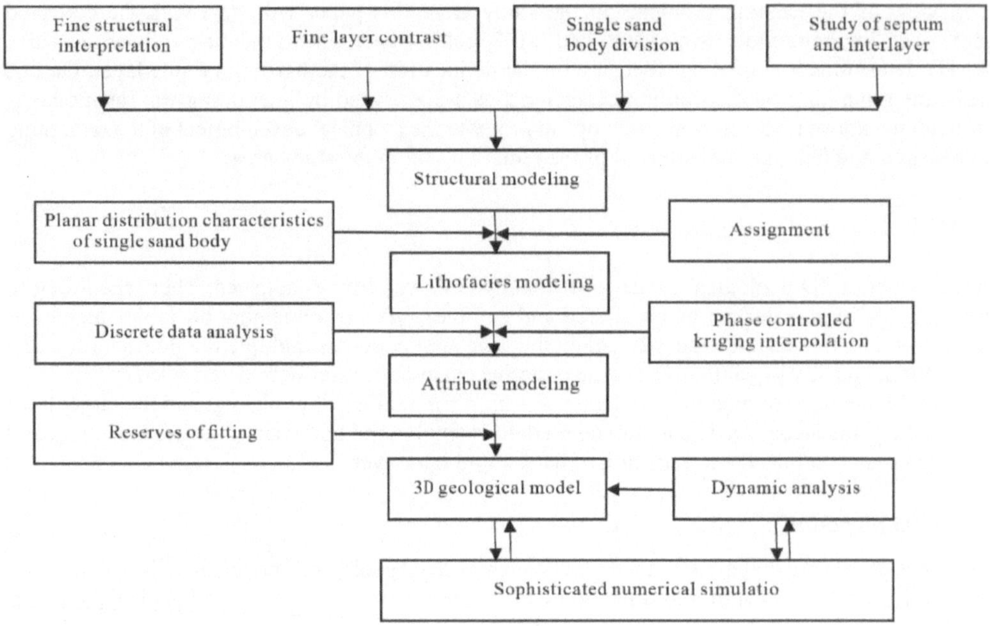

Figure 1. 3D Geological modeling process.

2.3 *Model coarsening*

The plane grid design of the coarsening model is 50 m ×50 m. The coarsening of the sand layer is based on the division of a single sand body, and the single layer with a large thickness is split in equal proportion to ensure the accuracy of numerical simulation.

2.4 *Remaining oil study*

Due to the lack of closed coring well data, the remaining oil analysis mainly adopts the residual oil detection method, reservoir engineering method, and numerical simulation method to study the remaining oil (Figure 2). There are two types of remaining oil distribution: one is the remaining oil that has not been displaced by edge water and bottom water, that is, the remaining oil that has not been flooded between layers. The other is the remaining oil in the unwashed part of the formation that has been displaced by edge water and bottom water, that is, the remaining oil in the formation that has not been flooded.

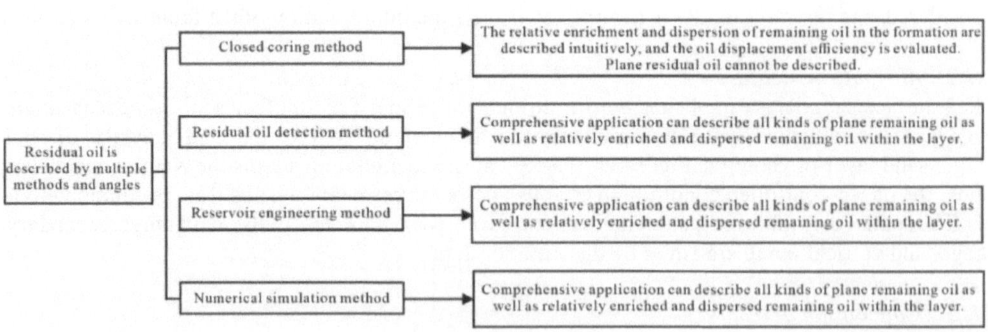

Figure 2. Remaining oil research ideas.

3 RESULTS AND DISCUSSIONS

3.1 *Remaining oil enrichment form*

The remaining oil is mainly distributed in the form of the remaining oil in the layer affected by heterogeneity and the locally enriched remaining oil controlled by the barrier and interlayer, and some remaining oil exists in the upper and lower parts of the barrier and interlayer. There are three evaluation units in total, including the upper and lower units (X52-2-1 and X52-2-2) and X52-4 within the X52-2 substratum, and the remaining oil and gas are mainly distributed in the X52-2 substratum.

3.2 *Remaining oil reservoir characteristics and reserves*

The irreducible water saturation of reservoir rocks in the M gas field is relatively high, generally in the range of 0.3-0.4, and increases with the deterioration of the reservoir. The remaining oil saturation is generally 0.2-0.3, and with the increase in reservoir variation, the water saturation at the intersection of the curve is 0.65, showing obvious hydrophilic characteristics.

According to the material balance method and decline analysis method, the controlled reserves and recoverable reserves of the M gas field are 6.668 million tons and 2.7 million tons, respectively. Using cumulative oil production and cumulative gas production as the relationship curve, the relationship between them is linear in the later stage of development.

3.3 *Distribution rule of remaining oil*

3.3.1 *Remaining oil is enriched in the high part of the structure*

The structure is not only the basic geological factor controlling the generation, migration, accumulation, and preservation of oil and gas, but also has a great influence on reservoir waterlogging and distribution of remaining oil. In the north and east of the block with more production wells in two small zones, the production wells show water penetration to high water cut from the structural low point to the structural high point, and the structural influence is obvious. At present, the flooding degree is weak only in the middle and south of the block, and the remaining oil is enriched. The numerical simulation results show that the remaining oil is enriched in the high part of the structure.

3.3.2 *Remaining oil is controlled by the development degree of barrier and interlayer*

The neutron lifetime logging results show that the remaining oil saturation is high in the area where the barrier and interlayer are relatively developed, and the water flooding degree is low in this area. In well K241, the X52-2 subzone was produced at a relatively low structural position. Due to the development of barriers and interlayers, the edge water advances slowly. Therefore, in the later period, the daily fluid volume is basically less than 10 tons, the overall water cut is less than 10%, and the anhydrous oil recovery period is 5.8 years.

3.3.3 *Remaining oil is controlled by the interlayer difference*

The difference in interlayer remaining oil saturation is mainly affected by the injection-production relationship and interlayer heterogeneity (permeability). The reservoir of natural gas water flooding in the work area is mainly affected by permeability, and the production layer with high permeability at the well point has a high degree of washing. The upper part of X52-2 is a low permeability layer with a low flooding degree, and the remaining oil is distributed widely and relatively enriched. The permeability of the lower 2-2 small layer is obviously better than that of the upper 2-1 small layer. The production degree is high, and the remaining oil mainly accumulates in the relatively high point area.

3.3.4 *Remaining oil enrichment area formed by the imperfect well pattern*

Remaining oil accumulation can result from the undrilled or unperforated original well pattern, sparse well pattern and uncontrolled existing well pattern. The well pattern of the KS101C-K331 well area is sparse. From the remaining oil reserves abundance of small formation 2, it can be seen that the area has a low production degree, and the nearby wells producing small formation 2 have low water cut. Well K8004 only shoots small formation X52-2-1 in X52 formation. X52-2-1 is co-produced with X51 and X8-7 formation, with a daily output of 1.4 tons of oil and no water cut. In addition, the X52-2 subzone electrical survey explained the 9.5 m oil and gas zone in the upper part of the well, and the X52-2 subzone was not fully utilized at the well site.

4 CONCLUSION

The research on remaining oil is very important for tapping the potential of old oil fields. It is the top priority to clarify the distribution characteristics of the remaining oil in the actual exploration and development process. After completing the relevant research work, this paper draws the following conclusions:

(1) The remaining oil is mainly distributed in the form of the remaining oil in the layer affected by heterogeneity. The remaining oil in the local enrichment is controlled by the barrier and interlayer. The remaining oil is mainly distributed in the X52-2 sub-layer.
(2) The irreducible water saturation of reservoir rock in the M gas field is high and has obvious hydrophilic characteristics.
(3) The controlled and recoverable reserves of the M gas field are 6.668 million tons and 2.7 million tons, respectively, and the cumulative oil production and cumulative gas production show a linear relationship.
(4) The remaining oil distribution has four characteristics: enriched in the high part of the structure, controlled by the development degree of barrier and interlayer, controlled by the interlayer difference, and formed by the imperfect well pattern.

ACKNOWLEDGMENTS

This work was financially supported by the National Natural Science Foundation of China.

REFERENCES

Chen, C., Jia, A.L., Sun, Y.M. (2000) Internal facies structure model and remaining oil distribution characteristics of thick reservoir. Acta Petrolei Sinica., (05): 99–102.
Han, D.K. (1995) Discussion on enhanced oil recovery in deep and high water cut oilfield. Petroleum Exploration and Development., (05):47–55.
Liu, B, J., Xie, J., Zhang, J.L. (2004) Research status and progress of residual oil technology in China. Northwestern Geology., (04):1–6.
Li, S.X., Chen, Y.M., Feng, Q.H., et al. (2001) Method for determining remaining oil saturation by using cross well tracer. Petroleum Exploration and Development., (02): 73–75.
Lu, J.L., Li, G, Q., Fan, Z.H., et al. (2001) Study on remaining oil distribution in oil field in high water cut period. Acta Petrolei Sinica., (05): 48–52.
Xie, J., Zhang, J.L. (2003) Description and prediction of remaining oil. Petroleum Industry Press Beijing., pp.1–13.
Yan, W.C., Sun, J.M. (2016) Analysis of microscopic remaining oil research status. Progress in Geophysics.,31(5):2198–2211.

Advances in Mineral Resources, Geotechnology and Geological Exploration – Rashid & Zhang (eds)
© 2023 Copyright the Author(s), ISBN 978-1-032-33377-9

The application of sensor-based logging technology in logging

Haixu Li*
No. 5 Oil Production Plant of Daqing Oilfield Company, Daqing, China

ABSTRACT: The global logging industry has developed continuously and steadily in recent years. The technical progress is mainly reflected in the improvement of the performance of traditional technologies and the exploration and research of cutting-edge technology. Significant progress has been made in imaging logging, nuclear magnetic resonance logging, formation testing, and reservoir monitoring. The accuracy of logging measurement has been greatly improved; the oil logging technology is mainly used to find oil and gas reservoirs. This paper introduces the present situation of oil logging technology in China and its shortcomings, expounds on the common logging technology and its characteristics, and finally discusses the application of sensor-based logging technology.

1 INTRODUCTION

The main task of logging technology is to find oil and gas reservoirs and monitor the dynamic of oil and gas reservoirs. In a word, logging technology is an indispensable part of oil exploration and exploitation. Logging technology is a kind of dynamic logging technology used in the process of modern oilfield exploitation. State monitoring technology and logging technology can be used to monitor the oil field physically and solve and deal with problems in quality and reservoir engineering. The following is to place the downhole equipment and equipment into the wellbore and then to the bottom of the well and the well, measure the data information and shaft structure of the surrounding bottom layer. After the data is transmitted back to the ground, analyze and sort out the data.

2 THE PRESENT SITUATION AND EXISTING PROBLEMS OF LOGGING TECHNOLOGY IN CHINA

2.1 *Site production is more difficult*

With the continuous exploitation of oil resources for many years, the oil resources gradually dried up, and the exploration and exploitation degree of oil and gas reservoirs is becoming more and more difficult. Therefore, it is the only way to improve the oil logging technology and survey the foreland basin and carbonate layer. For example, the internal structure of the carbonate layer is complex, and the oil and gas reservoir has the characteristics of low permeability and low resistivity (Fang Chaoliang). This requires high logging technology and requires higher accuracy and reliability of logging equipment. In addition, various special geological conditions and geological guidance will also put forward new requirements for oil logging technology, and the exploration and exploitation of oil and gas reservoirs will be more difficult.

2.2 *Logging environment is more complex*

Logging is a passive physical measurement method, and its application is limited by environmental factors. Therefore, logging workers should actively learn solid geological knowledge to master the

*Corresponding Author: lhaixu@petrochina.com.cn

DOI 10.1201/9781003319412-19

causes of various geological phenomena and give full play to the subjective initiative of human beings, so that logging technology can meet the requirements of the environment.

2.3 *Higher requirements for logging technology and equipment*

With the development of modern drilling technology, the environment of logging has changed greatly, so it puts forward higher requirements for the relevant logging technology and logging equipment. At present, the logging technology in China is relatively low as a whole, and the quality of logging equipment is uneven. The imported foreign equipment in the early stage cannot adapt to the present production because of the improvement of production requirements, and the price of the excellent logging equipment abroad is constantly rising. This situation greatly hinders the development of China's petroleum exploration and exploitation industry. Therefore, we must strengthen the research and development of logging technology and equipment and improve the technical level of the whole industry.

3 COMMON OIL LOGGING TECHNOLOGY

3.1 *Nuclear logging technology*

Nuclear logging, also known as radio logging, is a technology that applies nuclear technology to logging. It mainly studies the geological profile of the well by using the physical properties of the rock layers or inter stratum fluids to survey oil and natural gas resources. The specific way is to add an isotope tracer into the fluid in the rock gap and then use the natural gamma tester to track and measure the fluid, so as to judge the direction, path, and flow of the fluid to evaluate the well condition.

3.2 *Acoustic logging*

Acoustic logging is a technique that can record acoustic properties of rock by acoustic measurement to achieve logging purposes. In the underground, the sound velocity of various rocks is different. For example, the sound velocity of sandstone is much faster than that of mudstone. Therefore, different rock strata can be distinguished by detecting the sound velocity of underground rock (Zhu Guiqing). In addition, when sound waves pass through the oil and natural gas, their propagation speed and frequency will change. Therefore, acoustic logging can also be used to distinguish the rich oil, water, or gas in the reservoir of rock formation.

3.3 *Well temperature logging*

Well temperature logging is a technique to judge the flow state of a production well or injection well by using the abnormal ground temperature step line in the well. Generally speaking, in the place where the ground temperature is normal, the ground temperature line is a smooth straight line. But in the abnormal place of the geo-temperature line, the geo-temperature line will change, which can be clearly distinguished.

3.4 *High precision near-bit gamma imaging logging technology*

Near-bit gamma imaging logging is to install the measuring sensor directly above the drill bit. During drilling, it can scan and detect the radioactive intensity of the stratum encountered during drilling and monitor the change of well deviation and orientation. According to the logging data, it can judge the lithology and permeability layer, detect the formation boundary, calculate the formation inclination and thickness, control the borehole trajectory, etc. to guide the drilling and evaluation of the reservoir and improve the drilling speed of high slope well and horizontal well

(Xia Kewen). Compared with conventional gamma logging, the formation information obtained by near-bit gamma imaging logging is more real, timely, and precise. The blind area is small, which can significantly improve the drilling encounter rate of high-quality reservoirs.

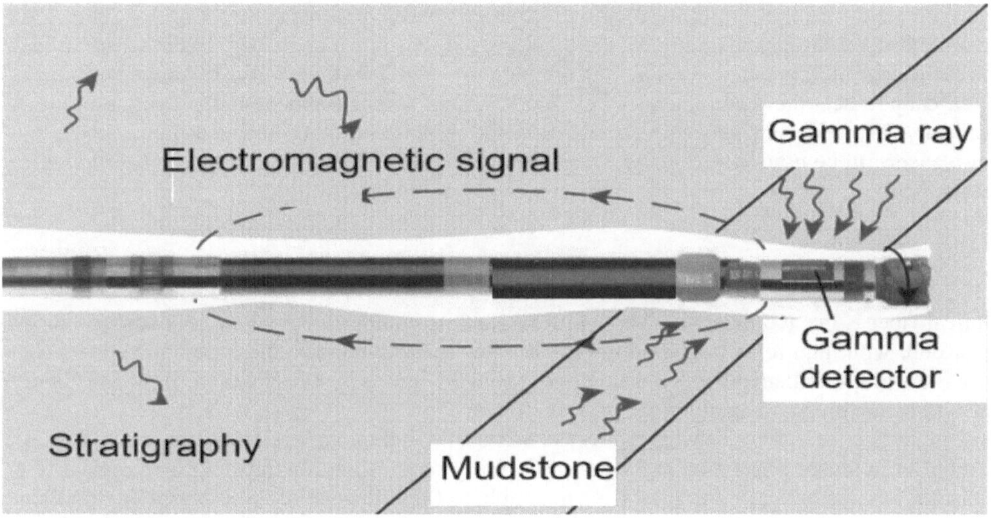

Figure 1. Gamma imaging logging of near-bit.

3.5 *Micro resistivity scanning imaging logging technology*

Micro resistivity scanning imaging logging technology uses a large number of button electrodes to deal with the change of formation resistivity, combined with image recognition and processing technology. The two-dimensional or three-dimensional borehole wall images taken by the operation are recorded. Micro resistivity scanning imaging logging technology can easily realize underground fractures and holes, the identification and evaluation of the hole, the structural interpretation and rock stratum properties near the bottom of the well, the judgment of the in-situ stress data analysis, etc. Micro resistivity scanning imaging logging technology can effectively solve the problem of bottom hole fracture identification, evaluation and well side structure, so this technology is relatively simple. It is applicable to different types of lithologic strata. But micro resistivity scanning imaging technology is measured by an electrode plate sticking to the well, which can be detected. The depth and range are limited, and the results obtained after measurement will be affected by error due to the influence of the borehole, so it should be guaranteed when using this technology. The expansion rate of well diameter is low, and the well wall is regular.

3.6 *Natural gamma measurement*

Natural gamma-ray logging is based on the natural radioactivity of the formation. With the development of the oil field, radioactive materials are continuously transported and accumulated, resulting in radioactive anomalies in the water flooded layer of the oil well, the water injection horizon of the water injection well, and the channeling outside the casing. In the exploration and development of oil and gas fields, the natural gamma logging curve is mainly used to divide lithology, determine the shale content of reservoir, stratigraphic correlation, track and locate perforation, evaluate reservoir water flooding, judge the channeling outside the casing of oil and water wells, and determine the water absorption of water injection wells. Laser optical fiber nuclear sensor is produced based on optical fiber transmission and optical fiber sensor. It uses the physical effects such as optical fiber photo loss and photoluminescence. It can develop a sensitive probe for the energy level range of nuclear detection and prolong the service life of the photomultiplier tube.

4 APPLICATION OF OIL LOGGING TECHNOLOGY BASED ON SENSOR

With the development of science and technology, logging technology has been developing. In addition to the above-mentioned logging technology, there are also some logging technologies based on sensor technology. In the logging technology based on sensors, two kinds of sensors are commonly used. One is the traditional electronic-based sensor, which belongs to the sensor used in the early stage and cannot be used in high temperature, high pressure, and corrosion environment. So, it cannot meet the requirements of the modern logging environment and is being eliminated gradually. The other is fiber optic sensor, network sensor, and other new sensors. Because this new sensor can be used in various harsh environments, it is widely used in sensor-based logging technology.

4.1 *Application of optical fiber sensor in logging technology*

Optical fiber is the abbreviation of optical fiber, and its component is silica. It is composed of a fiber core with high refraction, cladding with a low refractive index and a protective layer. The core is a fine glass filament with a diameter of about 0.1 mm, which encloses the light and Guided wave structure propagating along the axial direction.

A general optical fiber sensing system consists of a light source, sensor, optical cable, and photoelectric sensor conversion and signal processing composition. In the process of optical fiber transmission, the optical fiber sensor is vulnerable to the influence of the external environment, such as temperature, pressure, magnetic field, electric field, displacement, resulting in the change of light wave characteristic paramcters, such as intensity, phase, frequency and polarization state of the transmitted light. The corresponding physical quantity, chemical quantity, or biomass can be obtained by detecting the change in these quantities.

The fiber Bragg grating sensor is a new type of fiber sensor made by using the wavelength sensitivity of the fiber Bragg grating to temperature and stress parameters. Fiber Bragg grating is formed by using single-mode Germanium Doped Fiber irradiated by ultraviolet light. The fiber core, after grating, presents periodic distribution. The changes in temperature, strain, and pressure will cause changes in grid distance and refractive index of fiber grating. Thus, the reflection spectrum and transmission spectrum of fiber Bragg grating are changed. The corresponding temperature, strain, or pressure information can be obtained by detecting the changes in the reflection spectrum or transmission spectrum of fiber Bragg grating.

The special optical cable for logging comprises optical fiber core wire and its tensile wire, protective material, coating, and other armor, and the special connector for connecting optical fiber and instrument (i.e., bridle). The optical fiber data transmission rate is greater than 10 Mbit / s.

Fiber optic sensor technology is a new sensing technology developed with the development of optical fiber technology. Because of its high anti-interference to an electromagnetic wave, it can be used in high temperature and high-pressure environments and can measure various environmental parameters under the well with high precision. Recently, the new laser fiber nuclear logging technology is a technology that combines laser technology with optical fiber technology in logging. This technology is generally used in the well logging of non-transparent fluid. Because the laser fiber optic core sensor is based on optical loss and photoluminescence, it is more sensitive and accurate than the general nuclear detector.

4.2 *Application of network sensor in logging technology*

The essence of network logging is to integrate and control the sensors in the well. At present, the development of this technology has three main directions: first, the array of downhole sensors; second, the image of the ground acquisition system; third, information sharing and real-time implementation of various solutions. According to the development direction, the combined fast platform, including MRI, array induction, and acoustic and electrical imaging, will be established and added to the network sensor logging system after improvement. At present, logging technology is undergoing a major change, and the network has become the trend of logging technology development. The new generation of logging technology based on Internet technology is becoming the mainstream of logging technology, which is more rapid, reliable, and excellent for information sharing. This new logging technology can improve the accuracy and immediacy of downhole real-time observation information. After years of development, logging technology has made many breakthroughs. Now it is developing in the direction of high precision, high efficiency, high reliability, and networking to meet the requirements of the new geological environment. In the near future, with the development of downhole permanent sensor technology and multi-sensor integration technology, logging technology will be further developed to promote the development of oil exploration and production in China.

REFERENCES

Fang Chaoliang, Wu Mingde, Feng Qining. Prospect of key logging technologies [J] Petroleum Science and technology.

Xia Kewen, Song Jianping. Laser optical fiber nuclear logging technology.

Zhu Guiqing. The latest progress and development trend of logging while drilling technology abroad [J] Survey Well technology.

Advances in Mineral Resources, Geotechnology and Geological Exploration – Rashid & Zhang (eds)
© 2023 Copyright the Author(s), ISBN 978-1-032-33377-9

Research on the application of carrier phase difference technology in geological exploration

Yilin Dong*
College of Resource and Environment, Shandong Agricultural University, Shandong, China

ABSTRACT: As a new type of GPS measurement technology, GPS-RTK technology has the advantages of high measurement efficiency, high measurement accuracy, and less workload. It has a good application prospect in geological exploration engineering measurement activities. Applying this technology to practice can effectively break through the limitations of traditional measurement methods, reduce errors and workload, and enhance the effectiveness of geological exploration activities. This paper mainly gives an overview of GPS-RTK technology, discusses its specific application function and application strategy in geological exploration engineering measurement, and analyzes the main causes of errors and influences on the accuracy as well as the relevant practical points of the location setting of reference stations and mobile stations.

1 INTRODUCTION

GPS technology is an important measurement technology with the advantages of high precision, all-weather, automation, and high efficiency. With the continuous development of GPS dynamic positioning function, it also plays an increasingly important role in the field of measurement. GPS-RTK technology is a new measurement method based on the GPS satellite positioning system. This technology is applied in the field of geological exploration engineering measurement. It can obtain centimeter-level positioning accuracy in real-time without the need for post-settlement settlement. It is of great practical value to clarify the composition, positioning mode, and application points of GPS-RTK technology (Lv 2020).

2 GPS-RTK TECHNOLOGY OVERVIEW

2.1 *GPS-RTK system composition*

GPS-RTK system is mainly composed of three parts: base station, rover, and application software. Among them, the base station mainly includes a receiver, antenna, antenna transmitting station, and power supply; the rover mainly includes an electronic handbook, antenna, receiver, and receiver radio; application software is mainly composed of engineering measurement software and real-time dynamic differential software.

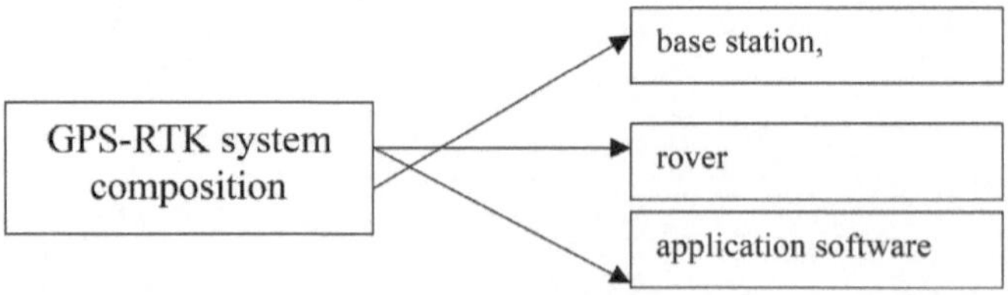

Figure 1. GPS-RTK system composition.

*Corresponding Author: 975424130@qq.com

DOI 10.1201/9781003319412-20

2.2 *GPS-RTK positioning mode*

GPS-RTK technology mainly includes three types: dynamic measurement, quasi-dynamic measurement, and fast static measurement. In fact, the normal GPS measurement method itself includes these three types, but the RTK using the carrier phase dynamic real-time differential technology can obtain the measurement accuracy of centimeter-level in real-time according to the carrier phase observation, thus greatly improving the development of the measurement work. level. (Li 2020).

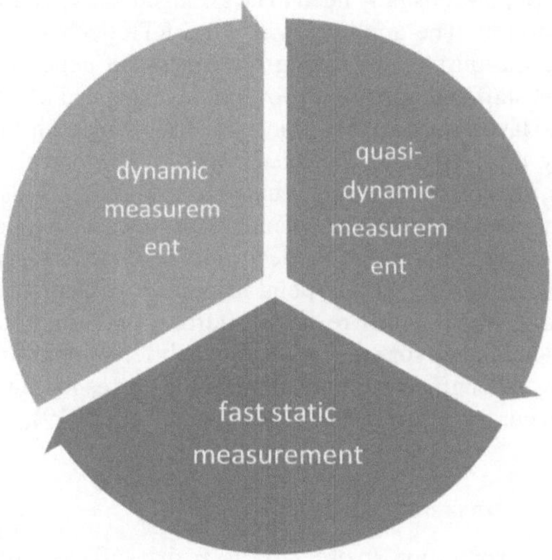

Figure 2. GPS-RTK positioning mode.

3 PRACTICAL APPLICATION OF GPS-RTK TECHNOLOGY IN GEOLOGICAL EXPLORATION ENGINEERING SURVEY

3.1 *Control measurements*

Conventional control measurement mainly includes two steps: overall control and partial encryption control. In the process of overall control, it is necessary to prepare for encryption work, and fully consider the overall situation to avoid contradictions and conflicts. Conventional methods are usually limited by the need for local encryption control.

The first-level wire must be measured, and on this basis, the graph root control is carried out, which is not only difficult to obtain a good control effect but also easily causes a lot of manpower and material costs. In this regard, GPS-RTK has obvious advantages: when performing overall control, there is no need to consider the line-of-sight point and encryption control. It can smoothly complete data collection and obtain accurate coordinates, so it provides great convenience for the first-level control point selection activities. After the practical application of GPS-RTK technology, the point selection activities in the first-level control only need to be safe for the base station. It is only necessary to consider the practicability and practicality of the point, which is conducive to improving the overall work speed and efficiency and reducing the workload, so as to reduce the burden on the staff (Liu 2019). In practice, the choice of positioning mode needs to be determined according to the side length of the control network: the side length is 5~10km, and the fast static positioning mode can be selected first; the side length is 10~15km if a sufficient number of satellites are supported and external observation conditions are obtained. If the standard is met, the fast static positioning mode can be applied. If it is in a relatively open plain area, the dynamic positioning mode can be selected; if the side length is less than 5km, it is necessary to reasonably select the fast static positioning mode or dynamic positioning mode according to the specific environmental conditions and exploration requirements.

3.2 *Layout of engineering points*

When applying the traditional surveying method represented by the total station, in order to reduce the shading and blocking of the exploration line of sight by various ground objects, a large number of high-level surveying control points usually need to be arranged, thus resulting in a lot of manpower and material cost consumption. The application of GPS-RTK technology can effectively solve this problem and achieve the purpose of saving resources and expenses. At the same time, it can reduce the workload of staff and improve the overall layout speed of engineering points. When applying this technology to layout engineering points, the following points need to be implemented: First, it needed to survey the first-level control network of geological exploration engineering projects, and on this basis, analyze the working routes and geographical distribution of engineering points, and make a reasonable determination; secondly, it is needed to enter the coordinates of the design engineering point in the mobile station electronic handbook, apply the lofting function of RTK technology to complete the engineering point layout activities on the spot, and calculate the coordinates of the location of the antenna in real-time; thirdly, for the engineering point, verification and calculation should be conducted for the error between different points (usually using formulas) to ensure that the accuracy of engineering points meets the corresponding standard requirements and further improve the accuracy of the overall measurement (Li 2019).

3.3 *Survey line profile measurement*

In the practice of geological exploration, in order to ensure that sufficient basic data can be obtained for exploration design, engineering layout, reserves calculation, comprehensive research, and other activities, when surveying the exploration line profile, it is necessary to strictly follow the relevant standards and specifications, and at the same time, it should be fully considered. Requirements are designed for geological exploration engineering. When using conventional surveying methods to carry out survey line profile survey work, it is usually necessary to use a large number of substations to complete the survey, while when GPS-RTK technology is used, with the help of receivers and electronic handbooks, real-time collection of profile three-dimensional coordinate data can be achieved. It is required to effectively reduce the overall workload and enhance the intuitiveness of exploration work, quickly observe the profile status and engineering field topography, and reduce unnecessary work. With the support of RTK technology, only one person can complete all the stakeout and measurement work, which can effectively reduce the accumulation of errors and further improve measurement efficiency and accuracy.

3.4 *Geoengineering point location measurement*

Traditional engineering point positioning methods such as handheld GPS method and semi-instrument method are not only time-consuming and labor-intensive, but also difficult to ensure that the accuracy meets the corresponding standard requirements. Because GPS-RTK technology has centimeter-level real-time measurement accuracy, it can fully meet the requirements of related technologies and ensure high-quality implementation of geological exploration projects. When carrying out field operations, the first-level control points (the distance between points should not exceed 15km) pre-laid in the engineering project should be used first, and the work should be carried out around them as reference points. If the distance between the first-level control point and the engineering point that needs to be positioned is relatively far, RTK can be used to conduct a pilot survey, select favorable terrain to set up a base station, and then use the mobile station to carry out survey work. To ensure sufficiently accurate data, the number of engineering points needs to be reasonably determined.

3.5 *Fragment measurement*

Traditional fragmentary surveying usually uses a total station or flat panel instrument to measure the existing map root control points in the target area. This measurement method requires multiple inputs of the feature codes of the map root control points and the application of mapping software on this basis. When drawing the map, higher requirements are put forward for visibility: the features and landforms around the north point must be visible. At the same time, an instrument usually requires at least 2 to 3 people to participate in the operation at the same time, so the work efficiency is low. RTK technology not only has no requirement for visibility, but also only one person can complete the relevant operations when carrying out the measurement work. During the official operation, the staff uses the instrument that has been initialized and the measuring rod to align the fragment point on the topography. After a few seconds, the coordinates of the point can be accurately obtained. After the accuracy meets the relevant standard requirements, the point can be input feature encoding and save and submit. After measuring the topographical features in a specific area, with the help of professional data transmission and processing software, the obtained measurement points can be output and processed centrally.

4 GPS-RTK TECHNOLOGY APPLICATION NOTES

4.1 *Measurement error control*

Although GPS-RTK technology has a series of advantages that traditional technologies and methods do not have, it is subject to a series of external factors and lacks necessary verification conditions. Therefore, errors will also occur in practice. If effective measures cannot be taken to control the errors, then the accuracy of the measurement results cannot be effectively guaranteed. To this end, we must strengthen the ideological understanding, and pay attention to do a good job in error control. The main measurement errors in RTK technology include two types of errors related to the same distance and those related to the same station: First, for the errors related to the same distance, as the distance between the mobile station and the reference station increases, such errors will also follow. Therefore, the distance between the mobile station and the base station must be reasonably controlled, and the operating radius should be scientifically limited; secondly, the error related to the same station can be mainly controlled by applying correction techniques and methods. In addition, it is also necessary to avoid the accumulation of errors caused by multiple operations and eliminate the network closure difference and network cable offset (Li 2019).

4.2 *Base station and mobile station settings*

The settings of the base station and the mobile station are directly related to the accuracy of the measurement results, so they must be set reasonably. The following points need to be paid attention to when setting up the base station: First, the base station should be kept away from strong electromagnetic interference sources such as substations, microwave stations, TV stations, wireless transmission towers, etc., to avoid the base station being interfered by them; second, the interference between the base station and the mobile station should be avoided. It is blocked by high-rise buildings or mountains, and the antennas of the base station and mobile station should be set as high as possible; thirdly, the base station should be avoided to be set around large buildings, large areas of water and other signal reflectors. The following points should be paid attention to when setting the mobile station: firstly, it is needed to control the coordinate parameters reasonably to ensure that the coordinate parameters are consistent with the base station; secondly, it is important to ensure that the data link between the mobile station and the base station is always connected; thirdly, it is of significance to avoid the distance from the base station to the mobile station is too far, otherwise, the accuracy will not meet the standard requirements; thirdly, when setting up the mobile station, it is necessary to check the relevant data to ensure the accuracy of its input; fourthly, due to the elevation of different locations in the mountainous area, it is necessary to ensure that the number of high-level control points fully meet the practical requirements, and to strengthen the corresponding verification work, so as to reduce the effect of the result accuracy by the elevation anomaly.

5 CONCLUSION

GPS-RTK technology has a wide range of applications in geological exploration engineering. When using this technology to carry out exploration and measurement, it is necessary to implement various practical points according to specific measurement objectives and requirements, so as to break through the limitations of traditional methods and effectively improve the results. accuracy. In addition, the error control work should be done well to avoid the negative impact of the excessive error on other work. When setting up the base station and mobile station, it needs to be carried out according to the actual situation of the project to ensure the scientificity and rationality of the location.

REFERENCES

Li Donghui. (2019) Application of GPS-RTK in Geological Profile Survey [J]. Western Prospecting Engineering, 31(11): 118–121.

Li Gang. (2019) Application of GPS-RTK measurement technology in geological exploration [J]. World Nonferrous Metals, (14): 188+190.

Li Qian. (2020) Research on the application of GPS fast static method in geological exploration engineering survey work [J]. World Nonferrous Metal (15): 148–149.

Liu Zhiyuan. (2019) Discussion on the application of RTK technology in geophysical surveys [J]. Engineering Construction and Design, (16): 270–271.

Lv Xiaofeng. (2020) Application of GPS measurement technology in geological exploration [J]. Heilongjiang Science 11(4): 112–113.

Research on geological structure characteristics and disaster prevention and control

Advances in Mineral Resources, Geotechnology and Geological Exploration – Rashid & Zhang (eds)

Analysis of blockage characteristics and exploration of unblocking processes in unconsolidated sandstone reservoirs

Taiwei Yang, Yu Yan*, Qingqin Cai, Shuzhao Guo & Yuqiao Du
Petroleum Engineering Research Institute of Dagang Oilfield, Dagang, Tianjin

Wei Zhang
Downhole Operation Company of Dagang Oilfield, Dagang, Tianjin

ABSTRACT: Consideration was given to the characteristics of high porosity and permeability, poor rock formation, loose cementation, high mud content, low dissolved gas-oil ratio, and low flow rate in Unconsolidated Sandstone reservoirs. Based on the analysis of blockage material characteristics and production characteristics with injection wells and production wells, this paper put forward two ways for unblocking processes: degradation and acidification unblock technology and integration technology for stable production, which have been applied and the results have shown to be effective.

1 RESERVOIR CHARACTERISTICS

The Unconsolidated Sandstone reservoirs are widely spread in many oil fields in Bohai Bay, mainly developed in Minghuazhen Group and Guantao Group. In unconsolidated sandstone reservoirs in the G oil field, lithology is mainly medium-fine Unconsolidated Sandstone, reservoir space is mainly primary intergranular pore space, oil formation cementation is loose, and permeability is very high. The measured formation temperature is about 45°C, the average porosity is 30.6%, the average permeability is 5.2D, and the median pore radius is 9.0. The surface crude oil is relatively dense, with high viscosity, 5.4% wax content, and 21.5% colloidal asphalt content, which is a normal temperature-pressure gradient system. The lithology and rock combination are mostly "sand-covered mud", with more sand and less mud, and coarse lithology, and the main rock types are feldspathic Unconsolidated Sandstone and feldspathic fine Unconsolidated Sandstone.

Table 1. The content of mineral rock.

	Clastic constituents (%)						Cement constituents (%)			
Formation	Quartz	Flint	Feldspar	Cuttings	Heavy mineral	Total	Argillaceous	Carbonate	Other	Total
G1	32.3	1.9	42.2	19.4	1.7	84.9	14.5	2	0.4	16.7
G2a	26.8	2.1	47.1	21	0.8	89.7	9.3	1.1	1.8	10.3
G2b	38.5	4.8	34.9	20.7	0.5	80.1	16.6	2.1	1.3	19.9
G3	32	6.5	31.3	28.9	0.2	80.3	17	1.6	1.1	19.7
Average	31.9	3.5	41.4	21.3	0.9	84.6	13.3	1.6	1.4	15.4

*Corresponding Author: dg_yanyu@petrochina.com.cn

DOI 10.1201/9781003319412-21

2 ANALYSIS OF BLOCKAGE MATERIAL CHARACTERISTICS

2.1 *Analysis of blockage composition in the injection wells*

The source of injection water in the Unconsolidated Sandstone reservoir in the G oilfield area is mainly stratigraphic recovered water, the main water type is $NaHCO_3$ type, while the mineralization is high, and there is a large amount of $CaCl_2$ type water in the reinjection sewage, which is more likely to produce scaling pollution; at the same time, the reinjection sewage contains a large amount of SRB, CO_2, and chloride in the joint erosion of oil casing, resulting in a large amount of iron oxide corrosion in the production well.

Table 2. Scaling analysis of injection pipeline.

Source of the scale	Fe_2O_3 %	P_2O_5 %	MgO %	CuO %	Insoluble of acid %	Other %	550°C Ignition loss %	550°C–950°C Ignition loss %
Well G1	60.86	0.008	0.31	0.085	11.36	0.007	25.94	1.43
Well G2	62.21	0.27	0.32	0.06	10.50	0.34	24.08	2.22
Well J3	59.14	0.05	0.27	0.03	11.70	0.33	25.40	3.11

About 32% of the injected wells in the G oilfield are injected with polymer. The microstructure of the blockage in the high seepage block after the rejection of polymer oil washing is observed by scanning electron microscopy (Figure 1), the blockage has an internal spatial mesh structure with a large number of pores on the surface, and some of the pores are filled by inorganic scale. The strength of the blockage in the high seepage block is low, and the main components are polymer, oil, and scale flocs. The plug contains a certain amount of Ca and Fe elements, indicating that the plug contains inorganic scale components such as calcium carbonate and iron hydroxide. The polymer is absorbed and retained in the near-production well zone, and the molecular chain is curled to form a colloid when it comes into contact with the formation fluid and clay minerals. The inorganic scale and oil scale are encapsulated to block the formation (An 2014; Li 2017).

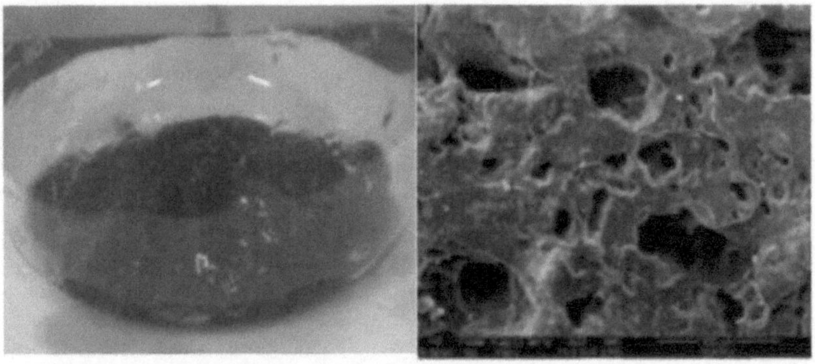

Figure 1. The reversed-out polymer microstructure of well G4.

2.2 *Analysis of the composition of the blockage in the production wells*

The blockage of production wells is mostly in the form of black and fine particles gummy (Figure 2), generally formed by the mixture of polymer gel, oil, inorganic scale, reservoir minerals, corrosion products cross-linked, and complex structure. During the movement of polymers, molecular chains will intertwine and wrap clay particles for transportation, which will narrow the seepage pore channel or even block the pore channel, resulting in the inability to effectively release the capacity of a single production well. At the same time, mostly the crude oil in the G oilfield belongs to heavy oil, and it is very easy to cause asphaltene and gum in the crude oil to precipitate and aggregate during the development process due to the change in reservoir temperature and pressure, resulting in the flow resistance.

Figure 2. Wellbore blockage of Well X1.

Table 3. Analysis of production well blockage composition.

		Inorganic matter			
Sump oil	Organic matter	Fe	Mg	Na	$Al_2O_3 \cdot SiO_2$
22.5	40.3	10.6	4.3	14.2	8.1

3 ANALYSIS OF PRODUCTION CHARACTERISTICS

3.1 *Production characteristics of injection wells*

G oil field carries out Triple oil mining at an earlier time. Most of the production wells have a rapid pressure rise of 2-4 MPa at the beginning of polymer drive, accompanied by the phenomenon of section reversal. The reason for this is the dynamic adjustment process from imbalance to balance and then to an imbalance of the resistance in the high and low permeability layers, in order to achieve the effect of enhancing the overall reservoir utilization.

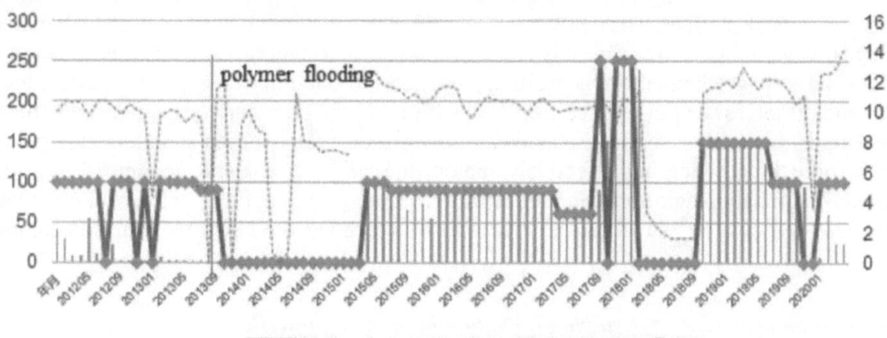

Figure 3. The blockage feature of Well B4.

After the long-term injection of polymer, the concentration of the aggregates will be formed at the throat of the reservoir in the near-production well zone, forming a filter-cake-like substance, and as the injection volume increases and the injection pressure increases, the blockage becomes denser and denser, making it very difficult to inject aggregates (Wu 2020). Besides, because of the discontinuity of the injection process, the free metal ions in the wellbore cause the polymer cross-linking in the wellbore, and when the injection is resumed after the stoppage, the injection volume is also very likely to drop.

3.2 *Production characteristics of production wells*

Because of the good porosity, permeability, and connectivity of the Unconsolidated Sandstone reservoir, it is common for the water to flee and gather in the production wells. Owing to the high waxy and heavy oil nature of the reservoir fluid, it is easy for the wax and colloidal asphalt to precipitate in the production fluid, resulting in a decrease in fluid supply to the wellbore. Furthermore, most of the Unconsolidated Sandstone production wells have been filled with gravel and apply sand control, and the blockage will further block the seepage channel due to the complementary effect of the screen tube, resulting in a decrease in fluid production. On the power diagram, the unloading is slowed down and jaggedness is obvious (An 2017; Zhou 2020).

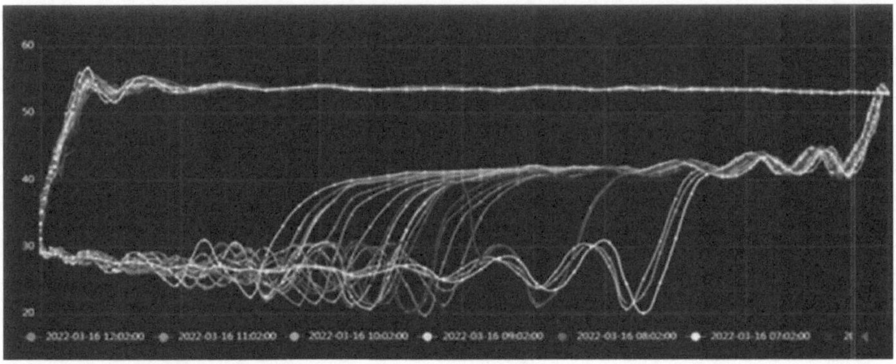

Figure 4. Indicator diagram changes of Well X2.

4 SUGGESTIONS FOR UNBLOCKING PROCESSES

4.1 *Unblock technology of degradation and acidification in the injection well*

It has been proved that the single performance unblocking technology cannot cope with the complex and changing blockage types. The degradation acidification technology of injection well uses non-oxidizing polymer unblocked and weakly corrosive solid acid to synergize and use degradation fluid to unblock the seepage channels in the wellbore and near-production well zone, which degrades the polymer and dissolves the polymer colloid while stripping the polymer encapsulation on the rock and mechanical impurities, so that the acid can fully dissolve the exposed sand and clay minerals, expand the pore throat space, and effectively unblock the carbonate and silicate minerals (He 2017). In 2021, the degradation acidification unblocking technology was applied twice, and the injection volume increased by 60m^3/day after the measure, which shows the injection plan was completed stably.

4.2 *Integration technology for stable production in production wells*

Due to the existence of pore liquefaction type sand out of seeing production wells, the output sand particle is fine, but due to the polymer and heavy components in heavy oil aggregation, the particle size is increasing, which makes the seepage channel be gradually blocked. Degradation unblocking measures can briefly unblock the near-production well area, but due to the distal sand particles and polymer constantly moving in the direction of the wellbore, the unblock and sand proofing effect are not ideal. This kind of measure production wells should learn from the concept of cold extraction of sand and choose the process model of "large aperture anti- sanding in sieve tube & sand discharge pump & periodic production well washing" to form a stable seepage channel composed of formation sand and transport sand, while using the degradation and decongestion system to periodically clean the deposited impurities in the near-production well zone and the pump mouth to ensure the liquid injection capacity of the wellbore. The annual fluid production of the measure production wells using this process is 70% higher than that of the neighboring production wells.

REFERENCES

An Kun (2014). New remove congestion and increase injection technology research in polymer injection Wells. Master's Thesis. Southwest Petroleum University,19.

He Peng; Feng Juan; Xue Hui (2017). Application of chlorine dioxide composite blockage removal technology in Zibei Oil Production Plant. Petrochemical Industry Technology.10, 89–90.

Li Feng (2017). Research of heavy oil field polymer injection well-plugging method research. Master's Thesis. Southwest Petroleum University, 22.

Wu Guang; Du Xun. et al. (2020). Study and application on a chelates removal technology for polymer flooding response wells: a case of Bohai S oilfield. Chemical Engineering of Oil & Gas. 187–92.

Zhou Baifeng (2020). Performance Analysis of Helical Axial Multiphase Pump Based on ANSYS. Coal Mine Machinery.7, 83–85.

Advances in Mineral Resources, Geotechnology and Geological Exploration – Rashid & Zhang (eds)
© 2023 Copyright the Author(s), ISBN 978-1-032-33377-9

Rheological optimization of drilling fluid in the upper salt formation of ultra-deep wells in Tarim Basin

Lijuan Pan
SINOPEC Northwest Company of China Petroleum and Chemical Corporation, Urumchi, Xinjiang, China
Key Laboratory of Enhanced Oil Recovery in Carbonate Fractured-vuggy Reservoirs, CNPC, Urumchi, Xinjiang, China

Jialiang Wang
Mud service company, BHDC, CNPC, Tianjing, China

Wu Long
SINOPEC Northwest Company of China Petroleum and Chemical Corporation, Urumchi, Xinjiang, China
Key Laboratory of Enhanced Oil Recovery in Carbonate Fractured-vuggy Reservoirs, CNPC, Urumchi, Xinjiang, China

Jiaxue Li*
Engineering Institute, China University of Petroleum (Beijing) @Keramay, Keramay, Xinjiang, China

Zhengli Qin
College of Petroleum Engineering, China University of Petroleum (Beijing), Beijing, China

ABSTRACT: The buried depth of oil and gas reservoirs in Tarim Basin is 7000–8000m, and the length of the above salt formation is 4500–6000 m. The drilling period of this section accounts for more than 50% of the total drilling period. It is the critical factor for speeding up the drilling of ultra-deep wells in this block. Actual drilling experience shows that there is a great relationship between drilling efficiency and drilling fluid performance control. However, at present, the determination of drilling fluid performance parameters in this block mainly depends on experience for adjustment and final determination, lacking theoretical guidance. In order to solve the above problems, the author puts forward an optimization method based on the combination of theoretical calculation, software simulation, and drilling big data analysis. Combined with the formation characteristics of the block and the requirements of engineering, the author puts forward the key points of drilling fluid rheology control in each upper salt formation. This study enhances the theoretical and pertinence of drilling fluid fluidity design and provides a basic guarantee for drilling speed increase in salt formation in the block.

1 INTRODUCTION

The exploration and development in the foreland area of Tarim Basin are mainly concentrated in the Kuqa depression in the north of the basin. Kelasu thrust belt, serving as the main natural gas producing area of Kuqa depression, is located in the middle of Kuqa depression. It is composed of two secondary structural units: The northern monoclinal belt and Kelasu structural belt. The block has a scale of trillion cubic meters of natural gas reserves (Hu 2020; Wang & Zhou 2019). From 2017 to 2021, more than 70 wells were drilled in this area, with an average well depth of > 7000m and an average thickness of upper salt formation > 6000m. The drilling period accounts

*Corresponding Author: liji3211979@163.com

DOI 10.1201/9781003319412-22

for more than 50% of the total drilling and completion period. The average production time of salt formation is more than 95%, while the complex accident time is less than 3%. Therefore, the complex accident has little impact on the drilling period of upper salt formation, and the drilling speed increase of upper salt formation mainly lies in improving the mechanical ROP, which also puts forward new requirements for the rheology of drilling fluid (Li et al. 2013; Yin et al. 2012).

At present, the design principle of drilling fluid rheology is mainly derived from the drilling fluid design specification, and the design of drilling fluid rheological performance parameters is mainly determined by the designer's experience (Shen et al. 2016; Yin et al. 2012; Wang et al. 2016). A specific drilling fluid system is generally designed with general drilling fluid performance parameters, which does not fully consider the influence of formation on drilling fluid. Therefore, according to the formation characteristics and drilling fluid construction experience, the author uses the big data processing method to establish the rheological property design method of drilling fluid, so as to improve the pertinence of drilling fluid design to the formation (Rong et al. 2019; Zhang et al. 2016).

2 APPLICATION OF DRILLING FLUID IN UPPER SALT FORMATION

2.1 *Physicochemical properties of ground layer upper salt*

The composition of each layer of ultra-deep wells in the Tarim Basin changes little. The content of nonclay minerals accounts for 74–82%, including quartz (36–40%), feldspar (16–19%) and calcite (13–21%). The content of clay minerals is high (14–23%), the content of illite in clay minerals is high (7–12%), followed by illite/Montmorillonite Mixed minerals (5–6%), and the content of kaolinite is low without montmorillonite. The mineral content of illite/Montmorillonite Mixed minerals in suweiyi formation (E2–3s) and Kangcun formation (N1–2k) is higher, followed by Kuqa formation (N2k), and jidick formation (N1j) is the lowest, as shown in Tables 1 and 2.

Table 1. Rock analysis results of Suprasalt strata in Tarim Basin.

Formation	Non clay minerals %	Quartz %	Feldspar %	Calcite %	Dolomite %	Other %
N2k	81.91	36.52	16.92	21.2	1.77	5.5
N1-2k	79.92	37.3	19.22	19.81	2.41	1.18
N1j	82.68	42.51	19.12	20.25	0.63	0.17
E2-3s	74.13	41.57	17.04	13.81	1.22	0.49

Table 2. Analysis results of clay minerals in salt strata of Tarim Basin.

Formation	Clay minerals %	Illite/Montmorillonite %	Illite %	kaolinite %	Chlorite %	Other %
N2k	16.09	5.24	7.74	0.52	2.12	0.47
N1–2k	17.64	6.13	8.18	0.19	1.67	1.47
N1j	14.87	4.44	7.58	0.58	1.47	0.8
E2–3s	23.29	6.46	12.6	0.22	3.24	0.77

2.2 *Application of drilling fluid in upper salt formation*

Ultra-deep wells in the Tarim basin generally adopt the four-spud in structure, including the first spud to expose the surface stratum, the second spud to the Kuqa stratum, the third spud to the salt top, the fourth spud to especially drilling the salt gypsum stratum, and the fifth spud to drill the target stratum. Therefore, all the upper salt strata are mainly the second spud and the third spud to expose the stratum.

Basically, the mature drilling fluid system commonly used in Tarim Basin is adopted for all salt formations: bentonite polymer system is adopted for the first spud, polymer system is adopted for the upper part of the second spud, and potassium polysulfonate drilling fluid system is adopted for the lower part of the second spud and the third spud. A few wells use the organic salt system, high-performance water-based drilling fluid system, and oil-based drilling fluid system. Therefore, this paper only studies the rheological optimization of common drilling fluid systems.

2.3 *Problems of drilling fluid in upper salt formation*

At present, the rheological parameters in drilling fluid design are first calculated by the unified empirical formula of the design software, as shown in Table 3. For ultra-deep wells in Tarim Basin, there are great differences in geological conditions, so the requirements for drilling fluid fluidity should be different, and the unified empirical formula cannot solve this problem. This requires adjustment and final determination based on the experience of drilling fluid designers. On the one hand, this puts forward high requirements for the experience of drilling fluid designers. On the other hand, there are differences in the final determination of performance range due to the lack of theoretical guidance.

Table 3. Rheological property design table of drilling fluid.

Density g/cm^3	Funnel viscosity s	Yield Point Pa	plastic viscosity mPa·s	gel strength Pa (10s)	gel strength Pa (10min)
1.3–1.72	40–75	3–14	9–45	1–5	3–15

The monitoring and inspection results of the drilling fluid performance of the ongoing drilling show that the nonconformities between the rheology of the on-site drilling fluid and the design performance of the drilling engineering are mainly concentrated in the funnel viscosity, plastic viscosity, and dynamic shear force of the drilling fluid.

3 RHEOLOGICAL OPTIMIZATION DESIGN OF DRILLING FLUID IN UPPER SALT FORMATION

In order to maintain high borehole cleaning efficiency and ensure the carrying and suspension of rock cuttings like gravel, appropriate drilling fluid shear and viscosity are required. At the same time, in order to increase the drilling speed in salt formation, it is necessary to strengthen the hydraulic parameters, so that the hydraulic power of drilling fluid can be used to assist rock breaking, so higher drilling fluid displacement is required. In order to meet the requirements of displacement and pump pressure, lower flow resistance and circulating pressure consumption are required, that is, lower plastic viscosity is required.

The upper salt formation in Tarim Basin is drilled in three spuds, of which the second and third spuds are long, so it is subdivided into upper and lower well sections, that is, five well sections such as the first spud well section, the second upper well section, the second lower well section, the third upper well section, and the third lower well section are used for segmented fluidity optimization.

3.1 *Optimum design of funnel viscosity*

Funnel viscosity comprehensively judges the tendency of viscosity change in each stage during the drilling operation in combination with other parameters to jointly characterize the rheology of drilling fluid. Therefore, funnel viscosity belongs to the key performance of each drilling fluid.

(1) First spud section
The formation of the first spud section is Xiyu formation(Q). Due to the shallow well depth and short drilling time, the project focus is mainly to prevent the surface leakage of drilling fluid from damaging the rig foundation. Therefore, it is necessary to improve the flow resistance and wall

protection capacity of drilling fluid. According to the modified Ergun equation, the flow resistance is related to the flow pattern index(n) and kinematic viscosity (μ) of drilling fluid, which can be comprehensively characterized by funnel viscosity(FV). According to practical experience, the funnel viscosity is generally greater than the 80s, which has a good protective effect. Therefore, the recommended range is funnel viscosity, which is greater than the 80s.

(2) Second and third spud sections
The funnel viscosity (FV) of the upper section of the second spud, the lower section of the second spud, the upper section of the third spud, and the lower section of the third spud is optimized by using the method of real drilling big data analysis. That is, the frequency of funnel viscosity data in the same well section in the block is calculated, the cumulative distribution map of funnel viscosity is drawn, and the funnel viscosity optimization chart is established, as shown in Figure 1. The performance parameters cumulatively distributed in the range of 10%–90% in Figure 1 are used as the optimization performance parameters. This performance parameter represents the optimized performance range for the successful application of drilling fluid in the well section of the block and reflects the requirements of Geology and drilling technology for drilling fluid. Therefore, the performance optimization design of funnel viscosity is finally determined, as shown in Table 4.

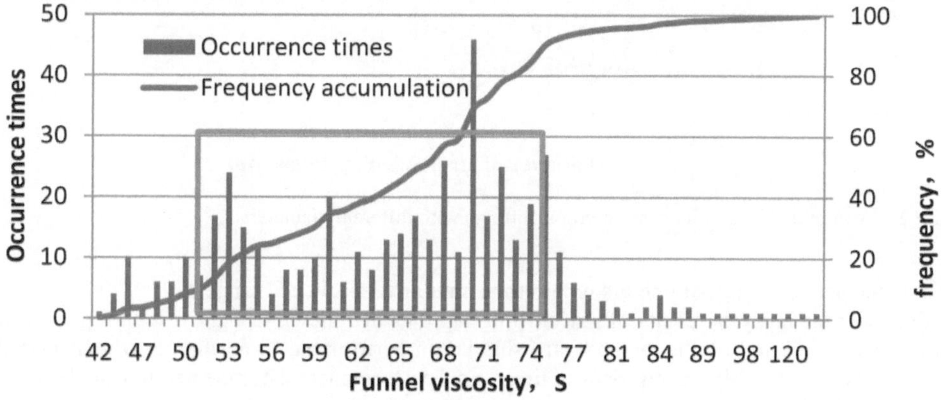

Figure 1. Funnel viscosity performance of the third spud Section.

Table 4. Funnel viscosity performance optimization design.

Section	The upper part of the second spud	The lower part of the second spud	The upper part of the third spud	The upper part the third spud
Funnel viscosity/s	41–68	45–75	55–75	50–80

3.2 *Optimum design of Yield Point*

Yield Point reflects the interaction force between clay particles and polymer molecules when drilling fluid flows in laminar flow, that is, the ability to form spatial grid structure. It is the main parameter affecting rock carrying of drilling fluid.

Firstly, the software is used for simulation calculation to determine the lower limit value of Yield Point that meets the requirements of cuttings carrying and the upper limit value that meets the requirements of pump pressure. Then, combined with the analysis results of real drilling big data, the performance parameters are within the cumulative frequency range of 10%–90%. Finally, judging whether the dynamic shear force is the key to monitoring the performance of each opening and giving the performance optimization table.

The upper well section of the second spud is taken as an example for optimization. Firstly, it is needed to take the particle diameter of 5mm and design the minimum displacement of 50L/s according to the simulation calculation of the software. The minimum Yield Point meeting the rock cuttings carrying requirements is 4.5pa. When the Yield Point is 10 and 20Pa, the pump pressure is 19.4mpa and 20.9mpa respectively, which has little impact on the pump pressure, as shown in Figure 2. Therefore, the lower limit and upper limit of Yield Point are 4.5pa and 20Pa respectively.

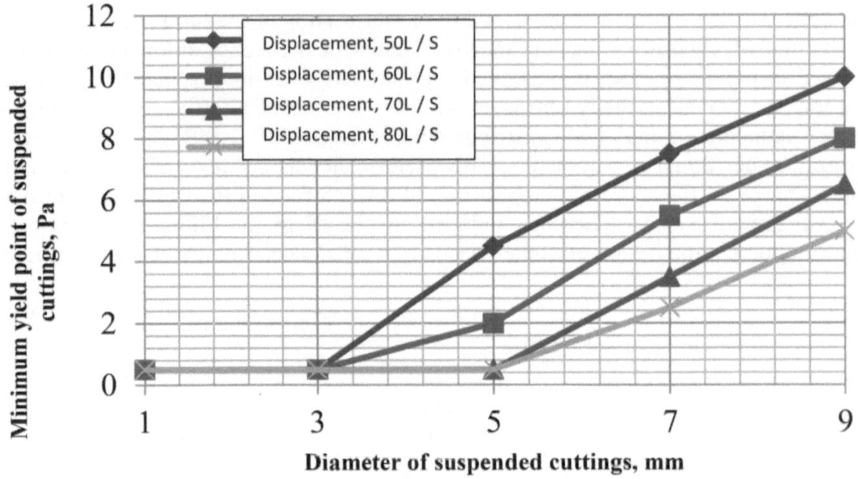

Figure 2. Minimum Yield Point carrying rock cuttings with different diameters.

Then, the performance parameters within the cumulative frequency range of 10%–90% in combination with the analysis results of real drilling big data are shown in Figure 3. In Figure 3, the performance in the red box is the software simulation performance, and the performance in the green box is the real drilling big data optimization performance. The intersection of the two is taken as the final optimization performance of Yield Point. The same optimization idea is adopted to optimize the lower well section of the second spud, the upper well section of the third spud, and the lower well section of the third spud. Since the drilling fluid density is more than 1.6g/cm from the upper well section of the third spud, the final dynamic shear force performance optimization is shown in Table 5.

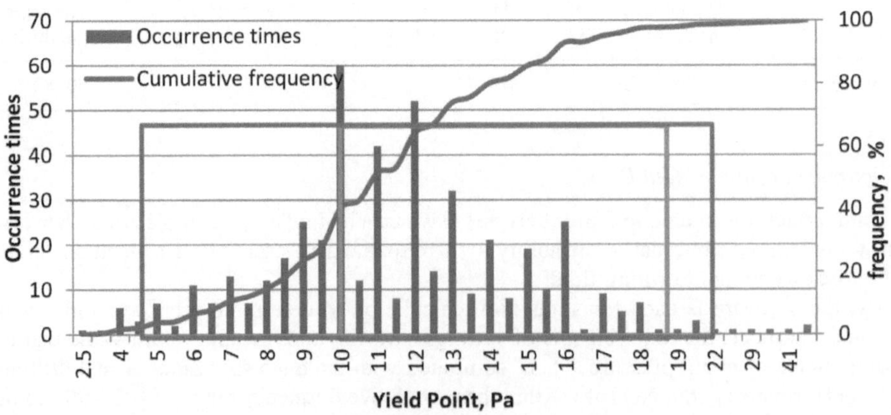

Figure 3. Yield Point performance of the third spud Section.

Table 5. Yield Point performance optimization design.

Section	The upper part of the second spud	The lower part of the second spud	The upper part of the third spud	The upper part of the third spud
Yield Point/Pa	4.5–10	5.5–14	7–14	6–16

3.3 *Optimum design of plastic viscosity*

Plastic viscosity reflects the strength of internal friction between suspended solid particles, between solid particles and liquid phase within continuous liquid phase, when the damage and recovery of grid structure in drilling fluid are in dynamic equilibrium under laminar flow. It is the main rheological parameter affecting pump pressure.

Firstly, well plan software is used for simulation calculation to determine the upper limit value meeting the pump pressure requirements. Then, combined with the performance parameters within the cumulative frequency of 10%–90% of the real drilling big data analysis results, it is finally judged whether the plastic viscosity is the key to monitoring the performance of each opening and giving the performance optimization table.

The upper well section of the second spud is taken as an example for optimization. The design minimum displacement is set as 50L/s and the working pump pressure is set as 25MPa. According to the simulation calculation of well plan software, the maximum plastic viscosity is 41mpa·s, as shown in Figure 4.

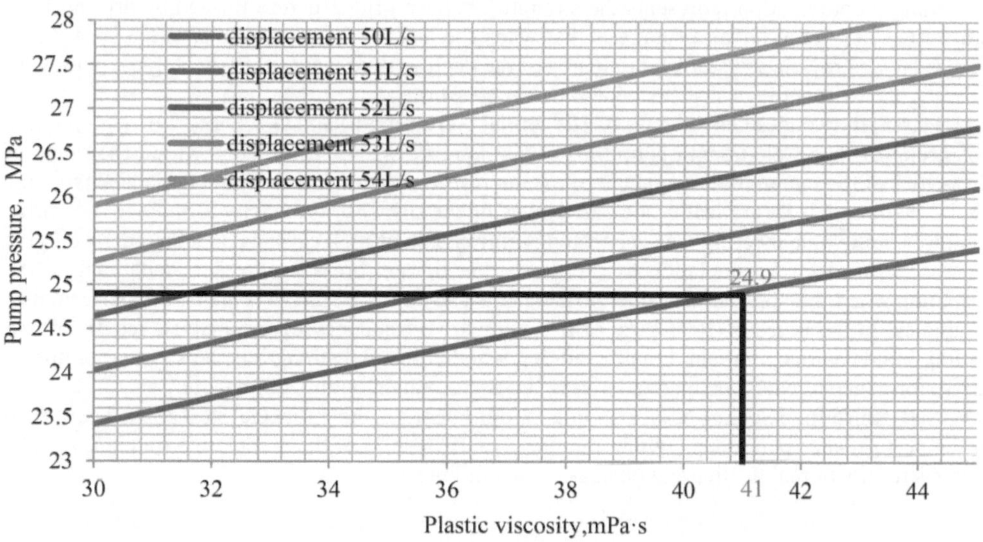

Figure 4. Relationship between plastic viscosity and pump pressure under different displacement.

Then, the performance parameters within the cumulative frequency range of 10%–90% in combination with the analysis results of real drilling big data are shown in Figure 5. The same optimization idea is adopted to optimize the lower well section of the second spud, the upper well section of the third spud, and the lower well section of the third spud. Since the drilling fluid density in the upper well section of the second spud is ≤ 1.3g/cm, the solid content in the drilling fluid is small and there is a certain space from the solid capacity limit of the drilling fluid. Therefore, the plastic viscosity does not belong to the key performance of the upper well section of the second spud. The final plastic viscosity performance optimization is shown in Table 6.

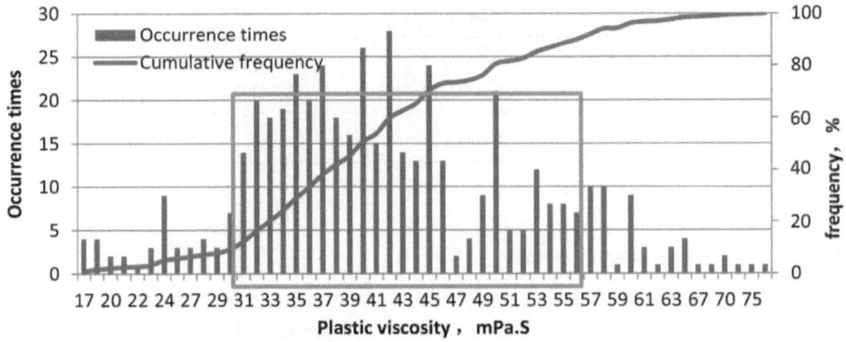

Figure 5. Plastic viscosity performance of the third spud Section.

Table 6. Plastic viscosity performance optimization design.

Section	The upper part of the second spud	The lower part of the second spud	The upper part of the third spud	The upper part of the third spud
Plastic viscosity/mPa·s	10–27	14–45	29–57	25–55

3.4 *Optimum design of gel strength optimization*

Gel strength optimization represents the strength of space grid structure formed by drilling fluid in a static state. Its physical meaning is that when the drilling fluid is stationary, the shear force required to destroy the structure per unit area of the drilling fluid is the main rheological parameter affecting the rock cuttings suspension and pump pressure.

Gel strength includes Gel strength of 10s (initial Gel strength) and Gel strength of 10min (final Gel strength). The Gel strength is optimized by the performance parameters within the range of 10%–90% of the cumulative frequency of the analysis results of real drilling big data. Firstly, the final Gel strength is theoretically calculated by using the suspension capacity formula to determine the lower limit value meeting the suspension requirements of rock cuttings, and then the well plan software is used for a calculation to verify whether the gel breaking pressure will fracture the formation when the upper limit value of the final shear force is 30MPa. Finally, the performance parameters in the range of 10%–90% of the cumulative frequency of the analysis results of real drilling big data are combined. Finally, judging whether the Gel strength is the key to monitoring the performance of each opening and giving the performance optimization table.

With the third spud Section as an example for optimization, the initial Gel strength is determined by the performance parameters within the range of 10%–90% of the cumulative frequency of the analysis results of real drilling big data, as shown in Figure 6.

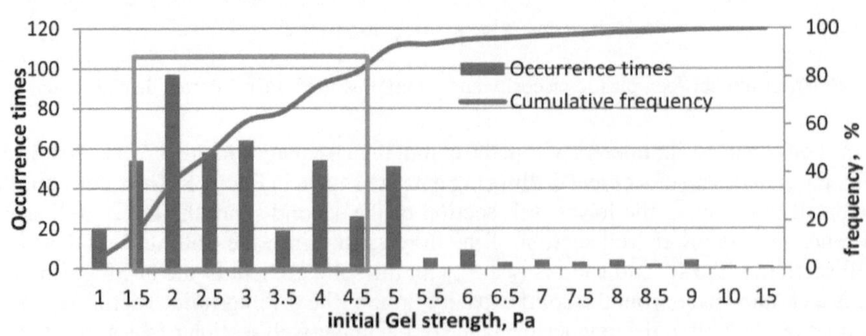

Figure 6. Initial Gel strength of the third spud Section.

For the final Gel strength, the lower limit of the Gel strength meeting the suspension requirements of rock cuttings with a diameter of 3mm is 6 Pa by using the suspension capacity formula (Formula 1), as shown in Figure 7.

$$\tau_s = \frac{d(\rho_{rock} - \rho)g}{6} \tag{1}$$

Where: τ_s*-Minimum Gel strength of suspended cuttings, Pa; d-Diameter of rock particles, m;* ρ_{rock}*-Density of rock, kg/m*^{3}*;* ρ*-Density of drilling fluid, kg/m*^{3}*; g-Gravitational acceleration, m/s*^{2}*.*

After that, the software is used for the calculation to verify that when the upper limit of final Gel strength is 30MPa, the gel breaking pressure will not fracture the formation, as shown in Table 7.

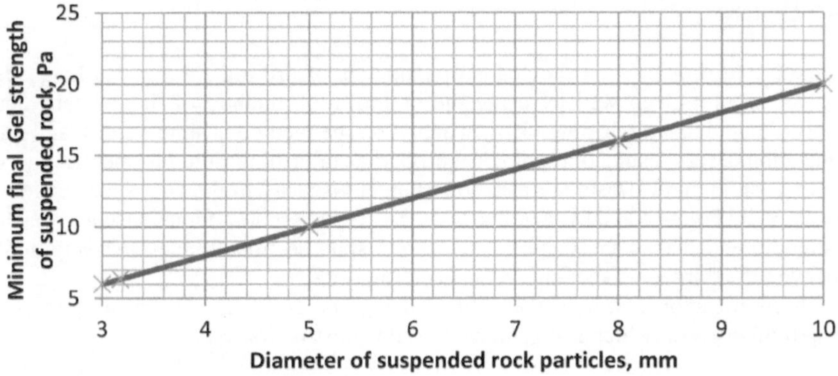

Figure 7. Minimum final Gel strength of suspended rocks with different diameters.

Table 7. Additional pressure required for fracturing formation in the upper well section of the second spud.

Well depth, m	Drilling fluid density, g/cm^3	Drilling fluid column pressure, MPa	Formation fracture pressure gradient	fracture pressure, MPa	Additional pressure, MPa
3040	1.30	38.8	1.80	53.7	14.9

Finally, the performance parameters within the cumulative frequency range of 10%–90% based on the analysis results of real drilling big data are shown in Figure 8. The same optimization idea is adopted to optimize the lower well section of the second spud, the upper well section of the third spud, and the lower well section of the third spud. The drilling fluid density is more than 1.6g/cm from the upper well section of the third spud. Therefore, the static shear force does not belong to

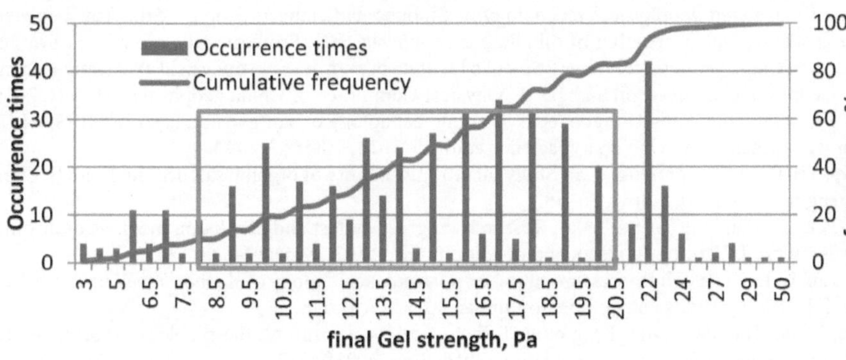

Figure 8. Final Gel strength of the third spud Section.

the key monitoring performance of the third spud in the section, and the final static shear force performance optimization is shown in Table 8.

Table 8. Gel strength performance optimization design.

Section	The upper part of the second spud	The lower part of the second spud	The upper part of the third spud	The upper part of the third spud
Initial Gel strength, Pa	1–4	1–5	1–3.5	1.5–5
Final Gel strength, Pa	6–20	6–20	5–21.5	5–21

4 CONCLUSIONS

(1) It is feasible to optimize the rheological parameters of drilling fluid in the long open hole section of an ultra-deep well by using the optimization method of theoretical calculation, software simulation, and drilling big data analysis. The method is simple, easy, and reliable.
(2) Optimization design results of rheological properties of drilling fluid in super deep well upper salt formation in Tarim Basin are as below: funnel viscosity in the first spud section, funnel viscosity, Yield Point and Gel strength in the second spud section, funnel viscosity, Yield Point, Gel strength and plastic viscosity in the second spud section, funnel viscosity and plastic viscosity in the third spud section are the key properties.
(3) Through optimization research, the optimal design range of rheological properties of each drilling fluid is given, which is conducive to field construction.

FUNDING

This work was supported by the National Major Science and Technology Projects of China titled "Key engineering technologies of ultra-deep oil and gas wells in marine carbonates" (NO. 2017ZX05005-005) and Sinopec Science and Technology Key Project titled "Study on quality and speed-up drilling and completion technology in No. 5 fault zone of Shunbei 1 Block" (NO. P2002). The authors would like to thank them.

REFERENCES

Hu W G. Development technology and research direction of fractured- vuggy carbonate reservoirs in Tahe Oilfield[J]. Reservoir Evaluation and Development, 2020, 10 (2): 1–10.

Jiaxin Wang, Yan Zhou. Study on remaining oil distribution of fractured vuggy carbonate reservoir in Tahe 4 area [J]. Petrochemical Technology, 2019, 26 (09): 302–303 + 356.

Junqing Rong, Maojin Tan. Effectiveness analysis of Ordovician carbonate reservoirs in zones 6 and 7 of Tahe Oilfield [a]. Professional Committee of oil and gas geophysics of China geophysical society, Research Institute of petroleum geophysical technology of Sinopec and Jiangsu geophysical society. Proceedings of 2019 academic annual meeting of oil and gas geophysics [C]. Professional Committee of oil and gas geophysics of China geophysical society Sinopec Institute of petroleum geophysical exploration technology, Jiangsu Geophysical Society: oil and gas geophysical Committee of China geophysical society, 2019:4.

Li Junwei, Zhao Jingfang, Yang Hongbo, et al. Drilling technology of salt gypsum layer in MISSAN Oilfield [J]. Journal of Changjiang University (science edition). 2013, 10(16): 92–94.

Shen Wenqi, Yin Da, Wang Zhilong, et al. Study on settling stability of organic salt drilling fluid[J]. Chemistry and Bioengineering. 2016, 33(5): 56–58.

Wang Jan, Peng Fanfan, Xu Tongtai, et al. Research Progress on test and prediction methods of drilling fluid settling stability[J]. Drilling fluids and completion fluids. 2016, 29(5):79–63.

Yin Da, Ye Yan, Li Lei, et al. High-pressure saltwater treatment technology of Keshen 7 well in Tarim Piedmont structure[J]. Drilling fluids and completion fluids. 2012, 29(5):6–8.

Zhang Yan, Xiong Hanqiao, Ding Feng et al. Effect of oil-water ratio on the rheology of oil-based drilling fluid[J]. Science technology and Engineering. 2016, 16(12): 238–242.

Advances in Mineral Resources, Geotechnology and Geological Exploration – Rashid & Zhang (eds)
© 2023 Copyright the Author(s), ISBN 978-1-032-33377-9

Study on diffusion behavior of water molecules in mineral and vegetable oil

Qingdong Zhu*, Ran Xu, Wenbing Zhu, Haozhe Wang, Zhaoliang Gu & Mengzhao Zhu
State Grid Shandong Electric Power Company Electric Power Research Institute, Jinan, China

ABSTRACT: Water content in insulating oil used for transformers is one of the crucial reasons accounting for its insulation deterioration. Lots of experiments were carried out to study the water influence on dielectric properties of oil insulation, but little is known about the mechanisms behind it, especially at the molecular level. To this end, the diffusion behaviors of water molecules in mineral and vegetable oil were studied by the molecular dynamics approach. The relation between different mass fractions of water content (1%, 3%, and 5%) in mineral and vegetable oil and its diffusion coefficients were studied. We found that the higher the mass fraction of water content is (up to 5%), the smaller the water molecules' diffusion coefficient is. In addition, this diffusion coefficient of water molecules has a greater value in mineral oil than that in vegetable oil when the mass fraction of water content is the same. These findings can be explained by our calculations of free volume fraction and the interaction energies.

1 INTRODUCTION

Vegetable and mineral oil are commonly used as insulating materials in oil-immersed transformers. Its insulating property is one of the most important factors influencing the safe operation of power transformers. In the long-term operation process, the insulating material will gradually age and degrade due to various factors such as partial discharge, space charge accumulation, oil temperature rises, and so on. The aging and degrading of insulating oil could finally lead to dielectric failure which would endanger the safe operation of transformers. Water content in insulation oil is one of the crucial reasons accounting for its insulation deterioration. Lots of experiments were done to investigate the influence of water content in oil on its dielectric properties, however, little is known about the mechanisms behind it, especially at the molecular level. With the rapid development of molecular dynamics algorithms and computational capabilities of the computer, the molecular simulation approach has been gradually used to investigate how water content influences the physical and chemical properties of insulating oil in recent years. SiO_2 nanoparticles were found to be a good filler for reducing the diffusion coefficient of H_2O molecules both in mineral and vegetable oil through molecular dynamics simulation (Tian et al. 2018; Wang et al. 2019; Zhang et al. 2021); The relationship between water content and furfural molecules diffusion were also studied both in mineral and vegetable oil (Liao et al. 2011; Pan et al. 2019). However, a comprehensive investigation of diffusion behaviors of water molecules themselves both in mineral and vegetable oil was not yet studied. In this paper, we use the molecular dynamics approach to simulate the diffusion behavior of water molecules with different mass fractions in mineral and vegetable oil. The relation between different mass fractions of water content and diffusion coefficient was in studied and the mechanisms behind it were also revealed by studying the free volume fraction and interaction energy between H_2O and oil molecules.

*Corresponding Author: 165671795@qq.com

DOI 10.1201/9781003319412-23

2 MOLECULAR MODELLING AND SIMULATION DETAILS

The vegetable oil modelled in this paper is based on the biological temperature insulating oil developed by ABB. Bio-Temp Oil is a high oleic vegetable oil containing more than 75% triglyceride oleate (also known as cis-9-octadecanoic acid), less than 10% triglyceride dienoate, less than 3% of acid triglycerides, less than 8% of saturated triglycerides, and 0.1–0.3% of antioxidants. The oleic acid triglyceride is used as the main component of our vegetable oil model as the content of oleic acid triglycerides in vegetable oil exceed 80%. The component molecule, oleic acid triglyceride, is shown in Figure 1(a).

Mineral oil is widely used in power transformer insulation systems due to its excellent insulating and thermal conductivity properties. Mineral oil is a mixture of various hydrocarbons. According to the different mass fractions of various hydrocarbon components in the oil, it can be mainly categorized into paraffin-based oil and naphthenic oil. Compared to paraffin-based oil, naphthenic oil has superior insulating performance and a simple production process, so our simulation here is based on naphthenic mineral oil. However, due to the complex composition of naphthenic oil, a simplified model consisting of a hydrocarbon chain ($C_{12}H_{26}$) and four cycloparaffins ($C_{14}H_{28}$, $C_{13}H_{24}$, $C_{16}H_{28}$, and $C_{16}H_{26}$) were modelled in our simulation and the component molecules are shown in Figure 1 (b). The mass fraction of each component, $C_{12}H_{26}$, $C_{14}H_{28}$, $C_{13}H_{24}$, $C_{16}H_{28}$ and $C_{16}H_{26}$, are 11.8%, 15.7%, 28.4%, 23.6% and 10.1%, respectively.

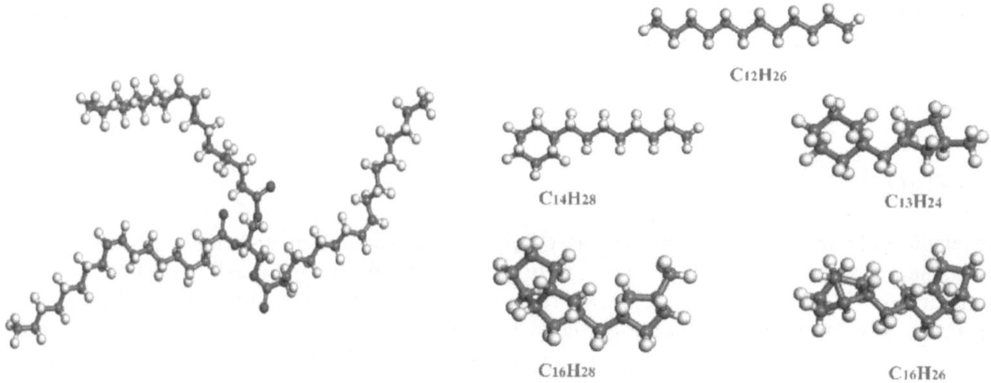

(a) Component molecules in vegetable oil (b) Component molecules in mineral oil

Figure 1. Component molecules in vegetable and mineral oil.

In order to simulate the influence of different mass fractions of water molecules in oil on their diffusion behavior, water molecules with a mass fraction of 1%, 3%, and 5% were added to the vegetable oil model and the mineral oil model, respectively. Both models were generated by Molecular Dynamics (MD) approach using OPLS forced field (Damm et al. 1997) in LAMMPS. Both configurations of mineral and vegetable oil were prepared by melting the initial sets of molecules in NPT simulations with an isotropic barostat holding the pressure at 1atm. During the first 5ns, the temperature was ramped up from 300 K to 500 K. Subsequently the simulation continued at a constant temperature of 500 K for an additional 5 ns during which the material melted. The melt was then quenched by cooling down from 500 K to 320 K (the operating temperature of insulating oil in power transformers) over 5 ns. The final densities of our mineral oil model and vegetable oil model were 0.84g/cm^3 and 0.83 g/cm^3, respectively, which are in good accordance with their actual densities. Finally, an additional 400ps of NVT simulations were employed for sampling and analyzing purposes with a constant temperature of 320 K. The resulting representative morphologies of vegetable oil with water (5%) and mineral oil with water (5%) were shown in Figure 2.

(a) Vegetable oil with water (b) Mineral oil with water

Figure 2. Vegetable and mineral oil model with 5% of water content.

3 RESULTS AND DISCUSSIONS

3.1 *Diffusion coefficient of water molecules in vegetable and mineral oil*

The diffusion coefficient D can be calculated by the mean square displacement (MSD), as shown in Eq. (1).

$$D = \frac{1}{6N} \lim_{t \to \infty} \frac{d}{dt} \sum_{i=1}^{n} (r_i(t) - r_i(0))^2 = \frac{1}{6}a \quad (1)$$

where $r_i(t)$ and $r_i(0)$ are the position vectors of the i_{th} atom at time t and time zero. N is the number of water molecules in the model and a represents the slope of the linear fitting of the MSD curve.

Figure 3 shows the mean square displacement curves of water molecules in the vegetable oil and mineral oil models with different mass fractions of water contents. The linear fitting curves of these six models are presented as well in order to calculate the diffusion coefficient of water molecules in different systems. The calculated diffusion coefficient of water molecules in vegetable and mineral oil with different mass fraction are shown in Table 1.

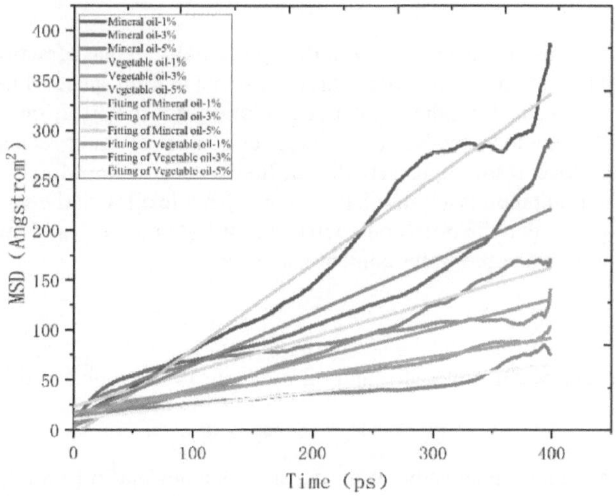

Figure 3. Linear fitting of different MSD curves for water molecules in mineral and vegetable oil.

The results in Table 1 show that with the increase in mass fraction of water molecules, the diffusion coefficient of water molecules decreases both in vegetable and mineral oil. In addition, the diffusion coefficient of water molecules in mineral oil is greater than that in vegetable oil with the same mass fraction of water. This implies that with the same mass fraction of water molecules, it is more difficult for vegetable oil to form a "water-bridge" than mineral oil, indicating a better breakdown strength of vegetable oil than mineral oil. This is well consistent with the experiment. The reason for the differences in diffusion behaviors of water molecules in mineral and vegetable oil could be analyzed by the calculation of free volume for water molecules and the interaction energy between water molecules and oil molecules.

Table 1. Slope a and diffusion coefficient D in mineral and vegetable oil.

	Vagetable-1%	Vegetable-3%	Vegetable-5%	Mineral-1%	Mineral-3%	Mineral-5%
a	0.30101	0.18701	0.13663	0.85805	0.52341	0.34903
D(Å^2/s)	0.05	0.031	0.022	0.143	0.087	0.058

3.2 *Free volume fraction (FFV)*

The free volume is an important factor affecting the diffusion behavior of different molecules. It can provide the necessary space for small particles to move, and the size and shape of the free volume play an important role in the diffusion behavior of small particles in the medium. We use the hardball probe method to calculate the free volume with the van der Waals radius of H_2O, which is 1.5Å. The calculated free volume (V_f) and its free volume fractions (FFV) for water molecules in vegetable and mineral oil with a mass fraction of 1%, 3%, and 5% are shown in Table 2.

Table 2. FFV for water molecules in vegetable and mineral oil.

	Vagetable-1%	Vegetable-3%	Vegetable-5%	Mineral-1%	Mineral-3%	Mineral-5%
Vo (Å^3)	14816.61	14568.59	15189.13	16017.23	15808.14	16676.07
V_f (Å^3)	1706.83	1628.62	1693.15	2046.78	1877.18	1766.58
FFV (%)	0.103297498	0.100549416	0.100291548	0.113307068	0.106143400	0.0957877528

As shown in Table 2, the FFV decreases with the increase of the mass fraction of water content both in vegetable and mineral oil, indicating a larger possible motion space while the water content in an oil is lower. This is well consistent with our calculations for the diffusion coefficient of water molecules both in mineral and vegetable oil. The higher content of water reduces the free volume for H_2O molecules to move, resulting in a smaller diffusion coefficient. Moreover, the free volume of water molecules in mineral oil is greater than that in vegetable oil with the same mass fraction of water, which also explains that the diffusion coefficient of H_2O molecules in mineral oil is greater than that in vegetable oil when the water content is the same.

3.3 *Interaction energy*

The interactions between water molecules and oil molecules can be obtained by Equation (1), as shown below:

$$E_{inter} = E_{total} - (E_w + E_o) \tag{2}$$

where E_{inter} is the interaction energy between water molecules and oil molecules in our model; E_{total}, E_w and E_0 represent the total potential energy of the oil-water model, the potential energy of water, and the potential energy of oil, respectively. The calculated interaction energy between water molecules and vegetable/mineral oil molecules is presented in Table 3. Positive interaction energy indicates a repulsive force between molecules; While negative interaction energy indicates an attractive force between molecules. The absolute value of interaction energy reflects the force strength between molecules.

Table 3. Interaction energy between water and mineral $E_{inter(W-M)}$/ vegetable oil $E_{inter(W-V)}$.

	Mass fraction of water molecules		
	1%	3%	5%
$E_{Mineral}$(Kcal/mol)	−126.661	−122.285	−96.516
E_{water}(Kcal/mol)	−5.787	−55.389	−116.115
$E_{inter(W-M)}$ (Kcal/mol)	−23.944	−43.076	−70.242
$E_{Vegetable}$(Kcal/mol)	−581.901	−567.062	−562.046
E_{water}(Kcal/mol)	−8.980	−78.517	−146.813
$E_{inter(W-V)}$ (Kcal/mol)	−8.489	−22.296	−32.796

As shown in Table 3, a stronger attractive force exists between water molecules and oil molecules when the mass fraction of water is higher, both in vegetable and mineral oil. As the interaction energy is more negative when the mass fraction of water in oil in higher. The higher attractive force between water and oil restricts the motion of water molecules, leading to a reduction of its diffusion coefficient. This is also in accordance with our previous findings on the relation between the diffusion coefficient of water molecules and their mass fraction. Additionally, the interaction energy between water molecules and mineral oil molecules is more negative than that in vegetable oil with the same mass fraction of water, indicating a smaller attractive force between water molecules and oil molecules could be found in mineral oil than that in vegetable oil, leading to an easier motion behavior of water molecules in mineral oil than that in vegetable oil. This also explains our previous findings of the differences in the diffusion coefficient of water molecules in mineral and vegetable oil.

3.4 *Centroid trajectory*

The centroid trajectories of water molecules in mineral and vegetable oil are also studied in order to visualize the motion region of water molecules. The centroid coordinates of water molecules were calculated by the coordinates of each oxygen atom on water molecules for simplification and the trajectories are presented in Figure 4.

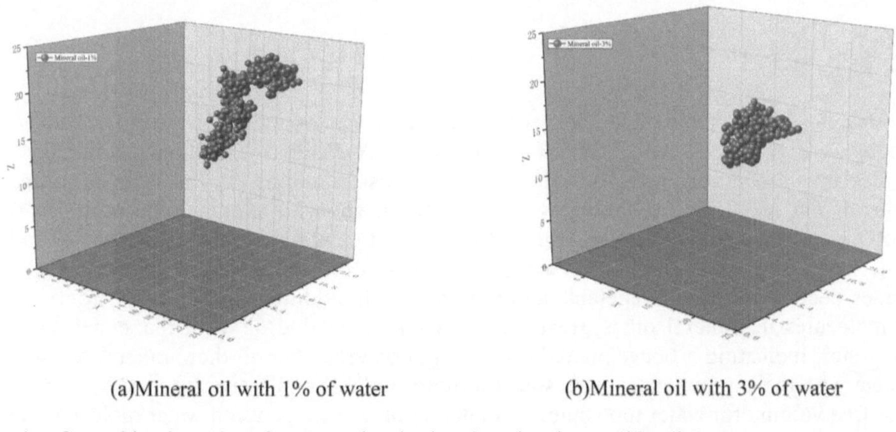

(a)Mineral oil with 1% of water (b)Mineral oil with 3% of water

Figure 4. Centroid trajectories of water molecules in mineral and vegetable oil.

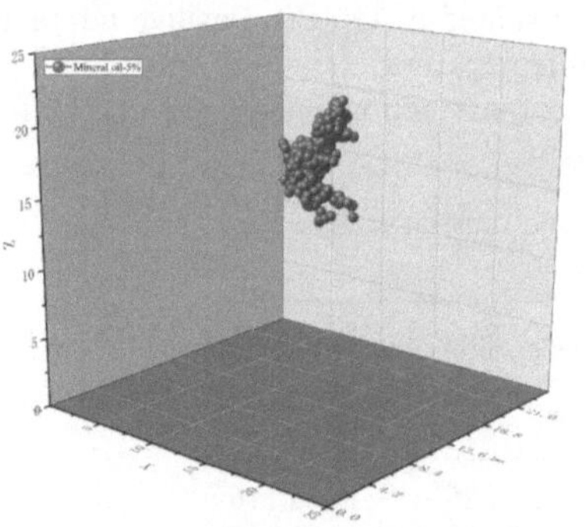

(c)Mineral oil with 5% of water

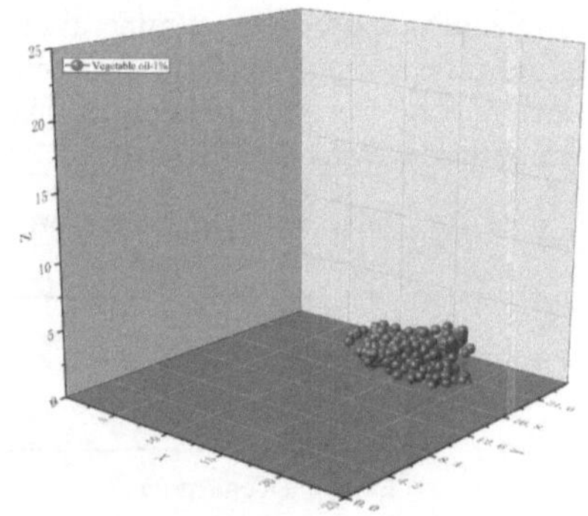

(d)Vegetable oil with 1% of water

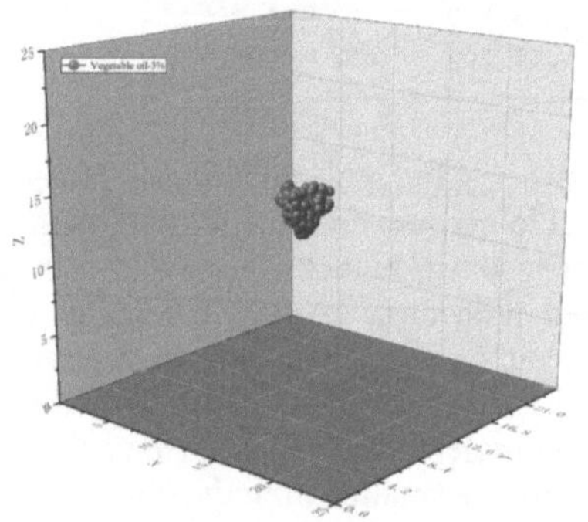

(e) Vegetable oil with 3% of water

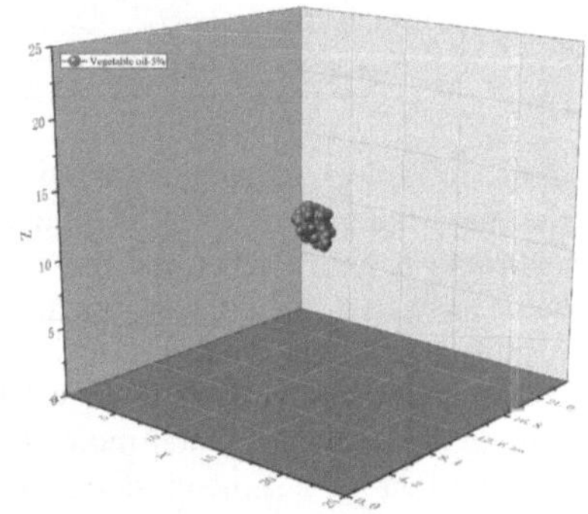

(f)Vegetable oil with 5% of water

Figure 4. Continued.

Figure 4 visualizes the motion region of water molecules in mineral and vegetable oil by plotting the centroid trajectories of water molecules. It could be seen that the higher the mass fraction of water is, the smaller the motion region of water molecules is. And the centroid trajectories of water molecules in mineral oil account for a larger space than that in vegetable oil when the mass fraction of water is the same. Both findings are well consistent with our calculations for the diffusion coefficient of water molecules in Section 3.1.

4 CONCLUSIONS

Both mineral and vegetable oil models with a different mass fraction of water molecules (1%, 3%, 5%) were generated with OPLS force filed in LAMMPS by melting the initial sets of molecules in NPT simulations with an isotropic barostat holding the pressure at 1atm. The densities of our resulting morphologies are well consistent with experimental data. By fitting the MSD curve of water molecules, the diffusion coefficients were calculated. We found that with the increase in mass fraction of water molecules in oil, the diffusion coefficient of water molecules decreases both in vegetable and mineral oil. In addition, the diffusion coefficient of water molecules in mineral oil is greater than that in vegetable oil with the same mass fraction of water, indicating a better breakdown strength of vegetable oil than mineral oil, which is consistent with the experiment. The reasons for these findings could be supported by our calculations of free volume for water molecules and interaction energy between water molecules and oil molecules.

ACKNOWLEDGMENTS

This work was financially supported by the Science and Technology Project of State Grid Shandong Electric Power Company (520626200070).

REFERENCES

Damm, W., Frontera, A., TiradoâRives, J., Jorgensen, W.L., (1997) Opls all-atom force field for carbohydrates. J. Comput. Chem., 18:1955–1970.

Liao, R.J., Zhou, X., Yang, L.J., Zhu, M.Z., Gong, C.R., Li, W., (2011) Impacts of Water Within Transformer Oil on the Diffusion and Distribution Characteristics of Furfural by Molecular Dynamics. High Volt. Eng., 37: 1321–1328.

Pan, Z., Zhu, M.Z., Ye., W.Y., Wang, W.L., Ma, X.Z., Wang, S.L., (2019) Molecular Dynamics Vegetable Simulation on Oil on Furfural Effect of Water in Diffusion. Insul. Mater., 52:55–62.

Tian, W.X., Tang, C., Wang, Q., Zhang, S.L., Yang, Y.L. (2018) The Effect and Associate Mechanism of Nano SiO_2 Particles on the Diffusion Behavior of Water in Insulating Oil. Mater., 11: 2373–2473.

Wang, Q., Zhang, S.L., Tian, W.X., Liu, X., Du, D.Y., Tang, C., (2019) Effect of different contents of nano-SiO_2 particles on water diffusion behavior in insulating oil. AIP Adv., 9:105022.

Zhang, Z.X., Zhou, H.B., Liu, D.S., Zhao, F., Li, W.T., Tang, C., (2021) The Effect of Nano-Silica Modified with Silane Coupling Agents on the Diffusion Behavior of Water Molecules in Palm Oil Based on Molecular Simulation. IEEE Access, 9:82213–82220.

Advances in Mineral Resources, Geotechnology and Geological Exploration – Rashid & Zhang (eds)

Classification and safety precautions of blasting technology

Xiyong Qian*
Henan University of Technology, Zhengzhou, Henan Province, China

ABSTRACT: Mine blasting belongs to the application of engineering blasting technology in mining. All kinds of blasting techniques are used in mine engineering, whether in the process of tunneling, open-pit mining, or underground stope mining. This paper mainly introduces the classification of engineering blasting, the main methods used in mine blasting, several kinds of blasting technology usually used in the underground stope, and introduces the relevant control blasting technology and basic mine blasting safety technology in detail. Blasting technology can be classified according to the form of medicine package, the charging mode, and charging space shape. Besides, workers must have learned safety precautions of blasting technology before they start working. Classification and safety precautions of blasting technology allow workers have a better understanding of their projects so that the blasting work can be safe and reliable and then their working efficiency can be enhanced.

1 INTRODUCTION

Engineering blasting methods can be divided into two categories according to the shape of the charge bag, the charge way, and the charge space form.

1.1 *Classification according to the form of medicine package*

According to the shape of the charge bag, namely, the explosive action and characteristics of the charge bag are classified. According to this method, it can be divided into four types:

1.1.1 *Concentrated drug package method*

Theoretically speaking, the shape of the package is spherical, the initiation point starts from the center of the sphere, and the detonation wave radiates outward in the form of a spherical expansion, that is, the explosion effect is evenly distributed to the surrounding medium. However, in engineering practice, it is almost impossible to process the drug bag into this shape. Usually, the drug bag is made into a cube or rectangular shape, and the longest side of a cuboid is no more than 6 times the shortest side. At this point, the charge can be called the centralized charge, usually called the charge chamber method and the charge pot method.

1.1.2 *Lengthening cartridge method - cylindrical cartridge*

The package is made into a long strip, which can be cylindrical or square, but in general, it is cylindrical. Once the explosive is excited and exploded, the wavefront form of the generated detonation wave is cylindrical, that is, the detonation wave formed after the explosion spreads around in the form of a cylindrical wave and acts on the surrounding medium. Bags with a length greater than 6 times the shortest edge or diameter are usually called extended bags. In the practical

*Corresponding Author: qxy001024@163.com

 DOI 10.1201/9781003319412-24

engineering application, the strip charge blasting method in deep hole blasting, borehole blasting, and charge chamber blasting belongs to the extended charge blasting method.

1.1.3 *Flat drug package method*

The blasting of this cartridge is different from the first two methods, it does not need to drill nor dig chambers, but directly lay explosives on the surface of the medium, so the explosion effect is only on the surface of the medium contact cartridge, most of the energy is lost to the air, the detonation wave generated should be regarded as a plane wave. For example, in the processing of mechanical parts, the use of a round cake-shaped charge package, blasting coated on the surface of the medium, is the processing of mechanical parts explosive processing method. However, the planar charge bag method in chamber blasting is different from this method. It is a charged plane with an equivalent concentrated charge bag or strip charge bag arranged at a certain distance. The detonation wave produced during blasting is also similar to the plane wave.

1.1.4 *Shape bag method*

This is a particular shape of the charge used to achieve a particular explosive effect. The most widely used is the shaped explosive method, which is to process one end of the cartridge shell into conical or parabolic pits, so that the detonation wave generated by the explosive explosion gathers on its focal point or axis according to the surface of the conical or parabolic pits, forming high-energy jet and penetrating a part of the medium in contact with it. This cartridge is used militarily as an armor-piercing round to penetrate tank decks and other military targets; in practical engineering, it is used to cut metal sheets (for example, the cutting of old hulls, the demolition of steel-structural buildings and structures), the secondary crushing of rock chunks and the perforation in frozen soil, etc.

1.2 *According to the charging mode and charging space shape of different classification*

According to the different charge modes and charge space shapes, it can be divided into four blasting methods (Kusmaul 1987).

1.2.1 *Medicine chamber method*

This is a commonly used blasting method in a large number of earthwork excavation projects. Its advantage is that the construction machinery and tools are relatively simple, not limited by geographical conditions and climate, the greater the number of projects, the higher the efficiency. Generally speaking, the medicine chamber blasting method according to the size of the powder chamber volume within the rock mass excavation, also can be divided into the big room, small room, and bar type. The three kinds of powder chamber methods have each room of an explosive loading capacity, small to hundreds of kilograms, big to hundreds of tons. The charge blasting chamber capacity can reach thousands of tons, and our country had many times and one thousand tons and tons of big blasts.

1.2.2 *Medicine pot method*

In order to achieve the blasting method of loading more charge, a small amount of explosive is put into the bottom of the ordinary hole for non-blocking blasting, so that the bottom of the hole is expanded into a round pot after repeated blasting. The blasting method of the charge pot belongs to the centralized charge package, which is suitable for the blasting of medium and hard rock. It can obtain more earth and stone volume by less hole blasting under the construction condition of small engineering quantities and insufficient drilling machines and tools. At present, this blasting method is only used under individual conditions.

1.2.3 *Perforation method*

Usually, according to the different hole diameter and hole depth, the hole depth is not less than 5m, and the hole diameter is not less than 50mm called deep hole blasting, otherwise known as shallow hole blasting and hole blasting. In terms of charge structure, it belongs to a kind of extended charge, which is the most widely used and the largest blasting method in engineering blasting.

1.2.4 *Naked medicine bag method*

This is the simplest and most convenient blasting construction method. It is not necessary to drill holes, but to lay the explosive directly on the surface of the blasted object and simply cover it. This blasting method is often used to remove dangerous objects, traffic obstacles, secondary crushing of large rocks, dangerous rock treatment, and so on (Kachanov 1992).

1.3 *Classification of dust*

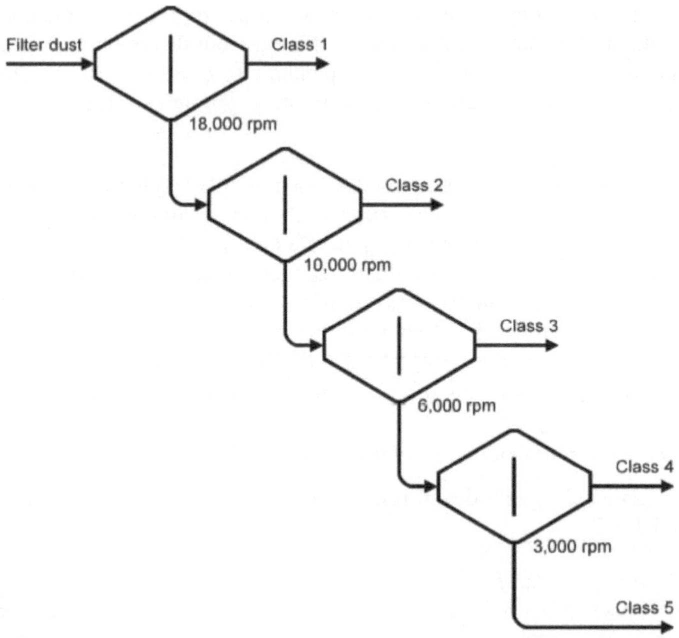

Figure 1. Classification procedure [3].

A laboratory classifier 100MZR from Hosokawa Alpine was used for the dry classification of dust. The procedure for the classification is illustrated in Figure 1. In the first classification step, the finest dust fraction was separated from the mixture and collected in the cyclone at the outlet of the classifier as Particle Class 1. The remaining coarse fraction was used as feed material in the next classification step, where the classifier was operated at a reduced speed. Thus, the cut size diameter of the classification was shifted to coarser particle size. In this second classification step, the material was split into particle Class 2 and a new coarse fraction. This procedure was repeated twice so that the dust mixture was separated into five dust particle classes. The speed of the classifier in the four classification runs was 18000rpm, 10000rpm, 6000rpm, and 3000rpm, and the airflow through the classifier was constant at 50 m^3/h. In the first classification step, it was found that some fine dust material is lost at the cyclone outlet because the collection efficiency of the cyclone for very fine material is limited. The amount of the fine dust lost was calculated for the first classification run by a mass balance.

2 ROADWAY DRIVING AND BLASTING

Roadway driving blasting includes the blasting of various underground passages such as drift, shaft, inclined shaft, patio, and tunnel. Its common characteristic is that under the condition of the single free surface, restricted by the excavation section, the blasting footage is generally only 1 ~ 3m per time. In order to form a certain shape of tunnel section, different types of boreholes must be arranged on the working surface. The specification and direction of the roadway should be strictly guaranteed, and the requirements of concentration of blasting pile and uniformity of fragmentation should be met. The utilization rate of the blasting hole should be high, the surrounding area should be smooth, and the material consumption should be less. The parameters of roadway driving include hole diameter, unit explosive consumption, hole spacing, hole depth, hole number, charge quantity, filling length, etc. (Budiansky et al. 1997).

In particular, it is a blasting technology developed in the 1950s that the one-hole and one-hole blasting is used in deep hole blasting driving. Its characteristic is that the one along the patio with drilling cavern full high drilling deep holes, blasting can be divided into a number of blasting, from bottom to top is presented. The rock under blasting by gravity whereabouts, gun smoke roadway discharge through deep hole leads to the upper level, the charging, stuffing, initiating such as homework at the upper-level tunnel or cavern rock drill. Compared with common tunneling methods, deep hole blasting tunneling has the advantages of high efficiency, safe operation, and good working conditions. It is suitable for vertical or steeply inclined roadway excavation such as patio, pass, and filling well. The blasting parameters of the open-hole blasting scheme mainly include a number of deep holes, sectional height, and so on (Huang et al. 1994).

3 CONTROLLED BLASTING

Definition of controlled blasting: According to the engineering requirement and blasting environment, scale, object, such as specific conditions, through careful design, construction and protection, and other technical measures, strictly control and medium crushing explosion energy release process, both to achieve the expected blasting effect, and the scope of blasting, direction and must be, air shock wave and noise of the blasting seismic wave and broken objects flying hazard control within the prescribed limit, the double control of blasting effect and blasting harm is called controlled blasting.

A risk analysis of blast damage to building infrastructure can be represented, for convenience, as having three levels, each level progressively requiring more useful probabilistic measures of reliability and thus resulting in more useful probabilistic measures of reliability and risk (see Figure 2) (Mark et al. 2008). These are summarized as:

Level 1: Fragility (or vulnerability) curves.

Level 2: Probability of failure conditional on the occurrence of a specific threat scenario. BRCs can be generated from this information.

Level 3: Probability of failure obtained from the aggregation of conditional risks if the relative or absolute threat probabilities are known or inferred by export opinion.

3.1 *Differential blasting*

Millisecond blasting is a kind of blasting technology that arranges the blasting order and reasonable time difference of each hole ingeniously. Because the time interval of blasting is a millisecond, millisecond blasting is also called millisecond blasting (Xu et al. 2003). The correct application of differential blasting can reduce the bulk rate after blasting, the intensity of seismic wave and air shock wave, the flying distance of fragments, get a good blasting effect, and easy to clear the accumulation.

The key to differential blasting technology is the choice of the time interval. A reasonable time difference can ensure a good blasting effect, otherwise, it will cause bad consequences, which cannot reach the design purpose, and even the explosion rejection, increase the harm of seismic waves and other accidents. The emergence of multi-stage high-precision non-electric millisecond detonators, electronic detonators, and magnetoelectric detonators provides good conditions for the wide application of this blasting technology. Differential blasting technology is widely used in

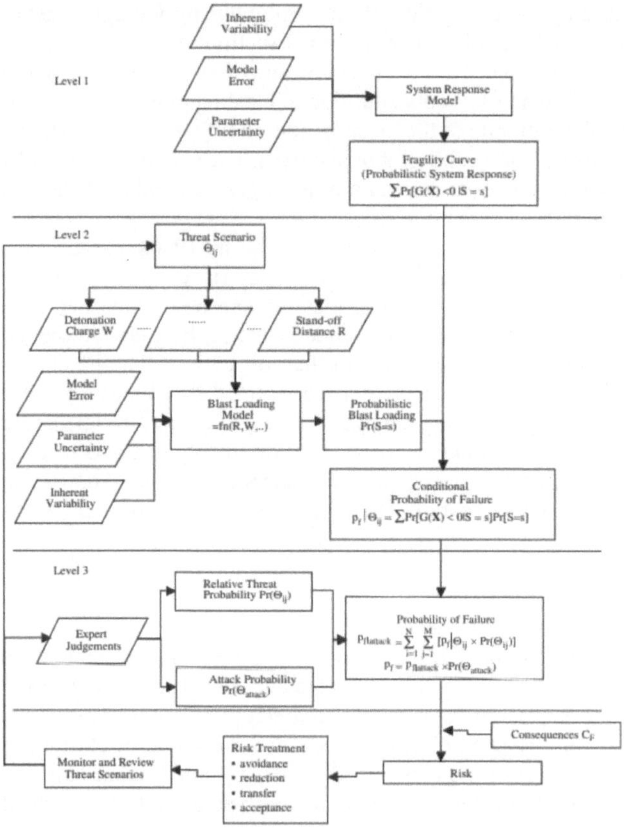

Figure 2. Illustration of models and probabilistic data required for risk analysis of blast damage to building infrastructure (Mark et al. 2008).

open-pit and underground excavation and urban controlled blasting, and directional blasting for dam construction by large chamber blasting has also been applied.

3.2 *Extrusion blasting (ballast-blasting)*

Extrusion blasting (ballast blasting) refers to the blasting under the condition of covering loose rock blocks in front of the free so that the rock is further crushed by extrusion. This method is commonly used in open-pit and deep-hole underground blasting. The extrusion effect is limited by the number of blasting rows: if the number of blasting rows is large, the loose rock blocks in the front will be more and more compacted, and the gap will be smaller and smaller, so the extrusion effect of deep-hole blasting in the rear row is not available. When the loose coefficient reaches 1.1, it is the limit value. If the number of blasting rows is adjusted well, it can be regarded as an effective method to reduce the bulk rate (Xiao et al. 2002).

Squeeze blasting (ballast blasting) is the blasting in which the compensating space is not enough for the free fragmentation of the exploded ore. According to the way of compensation space, squeeze blasting can be divided into small compensation space squeeze blasting and loose rock squeeze blasting. Squeeze blasting compensation in a small compensation space. The volume of space accounts for only 12%-20% of the volume of the ore exploded. The mining roadway and cutting groove between ore sections to be exploded are used as compensation space. Compressional blasting of loose ore and rock mainly compact the loose ore and rock on the mining side by the kinetic energy obtained from ore caving during blasting. This kind of blasting is also called lateral extrusion blasting.

Under suitable conditions, the use of extrusion blasting (ballast-blasting) will promote the development of mining science and technology, reduce consumption and increase efficiency, and bring remarkable economic benefits to mine production and management.

3.3 *Pre-split blasting and smooth blasting*

Pre-split blasting and smooth blasting are two kinds of blasting techniques. Smooth blasting is a blasting operation in which dense holes are arranged along the excavation boundary, uncoupled charge or low-power explosive is adopted and detonated after the main blasting area to form a smooth contour surface. Pre-splitting blasting is a blasting operation in which dense holes are arranged along the excavation boundary, uncoupled charge or low-power explosive is adopted to initiate detonation before the main blasting area, so as to form pre-cracks between the blasting area and the reserved area, so as to weaken the damage of the reserved rock mass by the main blasting and form a smooth contour (Li et al. 1998). The ultimate purpose and effect of the two blasting technologies are the same, both are to make the rock face smooth and stable after blasting, so as to ensure that the surrounding rock is not damaged, and the blasting mechanism of the two technologies is also very similar. The difference between both of the pre-splitting blasting is within the intact rock mass, in front of the blasting excavation blasting in advance, so along the excavation section and do not need to keep some boundaries opened a crack, is used to partition the blasting effect of reserved rock mass damage, and in the project after the completion of keep some new smooth surface; Smooth blasting is to reserve a protective layer of a certain thickness (smooth blasting layer) when blasting is close to the excavation boundary, and then conduct intensive drilling for this protective layer. These holes are continuously detonated after blasting other holes, so as to obtain a smooth slope and contour surface after blasting (Zheng et al. 2003).

4 CONCLUSIONS

Engineering blasting technology can be classified into different categories according to the form of medicine package and the charging mode and the charging space shape. Roadway driving blasting includes the blasting of various underground passages and the one-hole, and one-hole blasting is used in deep hole blasting driving. Controlled blasting can be divided into differential blasting: extrusion blasting, pre-split blasting, and smooth blasting.

Blasting safety has been accompanied by the development of blasting technology. A successful blast must, first of all, be safe. Blasting work is often risky, but blasting technology is not a risky technology, it must be safe and reliable, so the blasting workers of all countries attach great importance to the study of blasting safety technology.

Overall grasp of various blasting technology should not only have certain mathematics, mechanics, physics, chemistry, and engineering geological knowledge but also has certain construction technology and experience accumulation, as blasting engineering technical personnel should be familiar with all kinds of physical and mechanical properties of the medium, blasting action principle, method of blasting, blasting method and blasting parameters calculation principle, construction technology, and knowledge. At the same time, it is necessary to be familiar with the rules of blasting action, such as seismic wave, air shock wave, fragmentation scattering, and destruction range, as well as the corresponding safety protection knowledge.

Classification and safety precautions of blasting technology allow people who are working on blasting projects to have a better understanding of engineering blasting methods and precautions of blasting technology. It can help reduce the risk of hurt and death in blasting projects so that they can be safe and reliable. Once the safety of workers has been guaranteed, their motivation and enthusiasm will be enhanced and the work efficiency will be improved.

ACKNOWLEDGMENTS

I would like to show my appreciation to my professor and teacher, as for my professor, who helped me to tackle many problems related to my major, and gave me some directions about my paper. As for my teacher, she gave assistance with the techniques of writing an article and gave me some very specific advice. Without their help, I believe I could not have finished this paper so smoothly.

REFERENCES

Budiansky B. and O'Connell R. J, Elastic Moduli of cracked solid, Int. J.Solids, 1997, 12:81–87.
Christ of Lanzerstorfer, Michaela Kroppl, Air classification of blast furnace dust collected in a fabric filter for recycling to the sintering process, 2014, A-4600 Wels, Austria.
Huang Y et al. Generalized Self-consistent Mechanics method for Microcracked Solids, J. Meeh. Phys. Solids, 1994, 42:1273–1291.
Kachanov M., Effective Elastic properties of Crack solids, Critical Review of Some Basic Concepts, App. Mech. Review, 1992, 45(7): 304–335.
Kusmaul J.S., A New Constitutive Model of Frasmeniation of Rock under Dynamic Loading, in 2nd Int. symp. On Rock Rrag by Blasting, Keystone, 1987: 412–424.
Li Shou-ju, Liu Ying-xi. Analysis of crack propagation process under blasting load [J] Journal of Rock Mechanics and Engineering, 1998, 17: 888~891.
Mark G. Stewart, Michael D. Netherton, Security risks and probabilistic risk assessment of glazing subject to explosive blast loading, 93(2008): 627–638.
Xiao Zheng-xue, Zhang Zhi-cheng, Guo Xue-bin. Research on crack developing law of rock fracture controlled blasting [J] Chinese Journal of RockMechanics and Engineering, 2002, 21(4): 546–549.
Xu Hong-tao, Lu Wen-bo. Study on dynamic unloading effect during the process of fragmentizing rock by blasting [J] Rock and Soil Mechanics, 2003, 24 (6): 969–973.
Zheng Chang-qing, Pan Jin, Lu Yun-ming, et al. Discussion on rock slipping blasting [J] Blasting, 2003, 20(3): 14–16.

Advances in Mineral Resources, Geotechnology and Geological Exploration – Rashid & Zhang (eds)

Research on swelling and deformation characteristics of surrounding rock of Lijiaxia diversion tunnel, NW China

Xiaofan An* & Guanghong Ju
Northwest Engineering Corporation Limited, Power China, Xi'an, China

ABSTRACT: Swelling rock is a kind of rock mass with significant expansion and softening properties after water absorption. Due to the complex deformation and mechanical mechanism, the underground engineering constructed in swelling grounds is prone to the instability of surrounding rocks and supporting structures. Taking the Lijiaxia diversion tunnel as an object, this study analyzes the geological and disaster conditions in detail. On this basis, the physical and expansion tests of relevant clay rock are carried out. Furthermore, a finite element model is established for the section with prominent deformation, and the expansion and stress characteristics of surrounding rocks during tunnel excavation are revealed. The research has a guiding significance for the construction and deformation control of swelling rock tunnels.

1 INTRODUCTION

Swelling rock refers to a type of rock whose water content increases with time and volume expands due to physical and chemical reactions with water. When rock mass is disturbed, especially humidity changes, the properties of swelling rock usually change dramatically. The resulting expansion exerts great pressure on structures, which seriously affects the stability of underground engineering (Barla 2008; Hawlader et al. 2005).

Swelling rock is widely distributed around the world and has been found in more than 40 countries. Grob (1975) reported on the destruction of floors due to expansion in some road tunnels in Switzerland, with a maximum floor heave of 90 cm within a few months. In the United States, damage to buildings, roads, and pipelines caused by rock and soil expansion cost $2.3 billion annually, twice the combined economic losses from typhoons, floods, and earthquakes (Huang et al. 1986). The swelling rock and soil areas in Sudan cover more than one-third of the entire land, and the water systems, buildings, and roads in these areas are often severely damaged (Tang et al. 2014). In the tunnels in operation in Norway, 75% of additional reinforcement costs are related to expansive clay (Selmer-Olsen et al. 1989). Related disasters have also occurred in mines, railways, and water conservancy projects in China. For example, the Yunling Tunnel has been repeatedly deformed and damaged due to swelling rocks, resulting in disasters such as floor heaving and lining cracking (Wang et al. 2010). Since main roadways in the Shenbei Mining are located in expansive weak rocks, almost all of them are destroyed within 5 months after supporting (Jin 2003). The lining of the Dugongling Tunnel in Changping Expressway was cracked half a year after the operation, and the cumulative length of the left tunnel disease reached 1580m (Liu et al. 2020). Similar problems also occur in the Beijing Metro Station, Nanjing Meishan Iron Mine, Shannxi Yuntaishan Tunnel, etc. (Fan 2008).

To sum up, building caverns and maintaining safety in expansive formations has become one of the critical topics in rock engineering (He 2014; Tang 2012; Yu et al. 2020). However, the existing theory of swelling rock is relatively weak, coupled with its complex mechanism and properties, there is less research on tunnel engineering for this type of rock. In this study, the Lijiaxia diversion tunnel is taken as an object, starting from the analysis of geological conditions and tunnel diseases,

*Corresponding Author: anxiaof@nwh.cn

DOI 10.1201/9781003319412-25

the physical and expansion tests of relevant rock samples are carried out. Furthermore, a typical section of the excavated tunnel is selected to establish a numerical model, and the expansion and stress characteristics of surrounding rock in the excavation are analyzed. This study provides references for constructing tunnels in swelling rock and mastering the stability of surrounding rocks and supporting structures.

2 ENGINEERING BACKGROUND

There are 12 tunnels in the north canal of the Lijiaxia water diversion project (NW China), with a total length of 17.149km, accounting for 56.2% of the north canal, all of which are free-flow tunnels. Among them, tunnel 10# has a length of 10.725km and is constructed in three sections (see Figure 1). The designed excavation is a U-shaped section of 3.0m × 3.0m. Temporary supports adopt the form of mat reinforcement, spraying anchor, and steel centering, and permanent supports adopt reinforced concrete lining with a thickness of 0.30m. After construction, the clearance section is horseshoe-shaped at 2.3m×2.3m (see Figure 2).

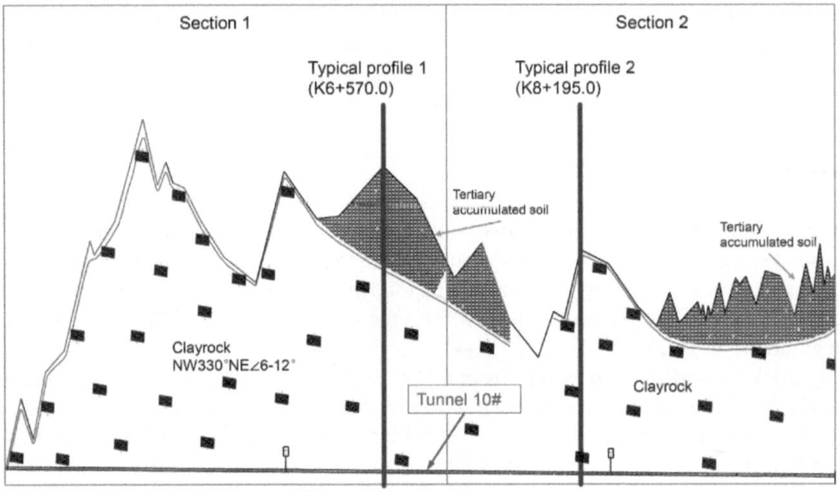

Figure 1. Stratum of the excavated section of tunnel 10#.

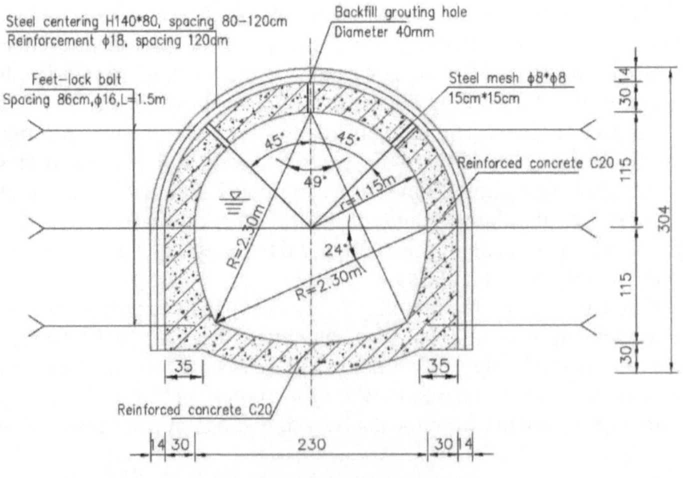

Figure 2. Design parameters of tunnel cross-section (K8+195.0).

The lithology of the stratum traversed by the tunnel is the clay rock, silt, landslide mass, and loose gravel. Among them, the length of the Tertiary clay rock tunnel is 14.462km, accounting for 84.33% of the total length; the length of the Quaternary silt tunnel is 1.584km, accounting for 9.24% of the total length. Detailed geological features are described as follows:

(1) Clayrock section. The brick-red clay rock mixes with glutenite, with a medium-thick layered structure and gentle dip. Rock stratum is dominated by layer fissures, with an occurrence of NW330° NE$\angle 6 - 12°$. Strongly weathered rocks are 3-5m thick, fragmented-blocky, and have poor integrity, and weakly weathered rocks are mostly massive-layered. Although groundwater activity in this section is slight, the surrounding rock of the tunnel is soft-extremely soft rock as a whole, with poor self-stability, and the arch crown and walls are easy to collapse.
(2) Quaternary loose layer. The stratum is located near the entrance and exit of the tunnel. Walls and arch crowns are easily deformed due to the poor stability of the surrounding rock. During construction, the phenomenon of collapse while excavating may occur, or even roof fall. Therefore, excavation needs to be supported immediately or in advance, and the entire section should be lined.

Most of the tunnel 10# pass through the clay rock area. The rock is rich in montmorillonite, illite, kaolinite, and other hydrophilic substances, which are easy to expand and soften after absorbing water. During the period of excavation and initial supports, the maximum deformation of surrounding rock reached 50–60 cm, resulting in serious damage to the excavated section. The main manifestations are the uplift of the arch bottom and the excessive lateral displacement of walls, as well as the severe deformation of steel centering, cracking of the spray layer, water seepage, and collapse. The large deformation occupied the design clearance of the tunnel, affected the construction of permanent lining, and posed a great threat to construction safety. Therefore, it is of great significance for the tunnel to understand the expansion properties of surrounding rock and its influence on the stress of supporting structures.

3 PHYSICAL TEST OF EXPANSION PROPERTIES

The following laboratory tests are designed, using a lever-type compressometer to measure the expansion properties of the clay rock.

(1) Free expansion test: measure the free expansion rate of rock samples.
(2) No-load expansion test: measure the expansion ratio of rock samples under the condition of no-load confinement.
(3) Loaded expansion test: for three rock samples with different initial water contents, measure the expansion ratios under four vertical pressures (100kPa, 200kPa, 400kPa, 800kPa) and confinement respectively.
(4) Expansion force test: for three rock samples with different initial water contents, measure the expansion force under four vertical pressures (100kPa, 200kPa, 400kPa, 800kPa) and confinement respectively.

In addition, several sets of geotechnical tests are carried out to measure the water content, dry-wet density, saturation degree, and liquid-plastic limits of the samples, the results are shown in Table 1. The average free expansion rate of rock samples is determined to be 10% by free expansion tests. According to the expansion tests with and without load, the relationship between the expansion increment Δs and time T of rock samples with different water contents under various pressures are shown in Figure 3 (only the curves with w_0 = 18.5% are shown here).

From the expansion force tests with and without load, the following rules can be obtained: (1) Due to the differences between samples, the expansion force (rate) does not correspond well with the pressure in the same set of tests, and even subsidence by water immersion (expansion force is zero). (2) The expansion force (rate) of rock samples with different water contents under external load is smaller. (3) The expansion force (rate) under no-load is relatively large, and the effect of load on the expansion force (rate) is more sensitive than that of water content.

Table 1. Statistics of geotechnical test results.

Sample	Water content w_0 (%)	Wet/ Dry density ρ_w/ρ_d(g/cm^3)	Void ratio e_0	Saturation degree S_r(%)	Pressure P (kPa)	Expansion rate 1 δ_{ep1}(%)	Expansion rate 2 δ_{ep2}(%)	Expansion force P_e (kPa)	Free expansion rate δ_{ef} (%)	Liquid/ Plastic limit w_L/w_P
1	5.6	2.30/2.18	0.325	50	100	−0.91	−0.35	0	10.0	41.0/ 23.8
2		2.24/2.12	0.364	45	200	−1.09	0.06	37.5		
3		2.42/2.30	0.259	63	400	−1.75	−0.24	0		
8		2.10/1.99	0.453	36	800	−5.62	−0.60	0		
7	9.3	2.28/2.09	0.385	70	100	−0.47	0.34	75.0		
11		2.19/2.00	0.442	61	200	−2.18	−0.06	0		
12		2.29/2.10	0.378	71	400	−2.55	0.17	40.6		
13		2.25/2.06	0.401	67	800	−3.22	−0.16	0		
4	18.5	2.11/1.78	0.624	86	100	−0.27	1.14	256.3		
5		2.22/1.87	0.542	98	200	−3.18	−0.45	0		
6		2.18/1.84	0.571	93	400	−2.47	0.41	150.0		
27		2.21/1.86	0.551	97	800	−3.56	0.36	100.0		
14	7.9	2.19/2.03	0.424	54	0	8.85	8.85	951.6		

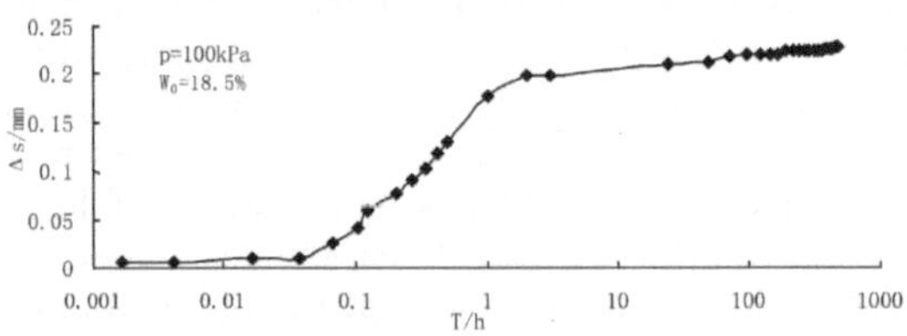

(a) p = 100kpa (sample 4)

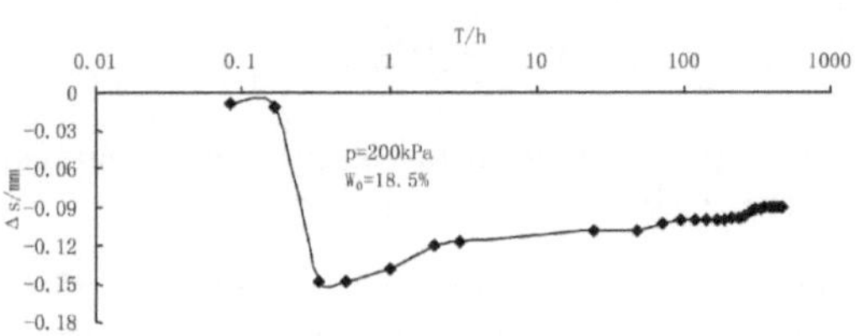

(b) p = 200kpa (sample 5)

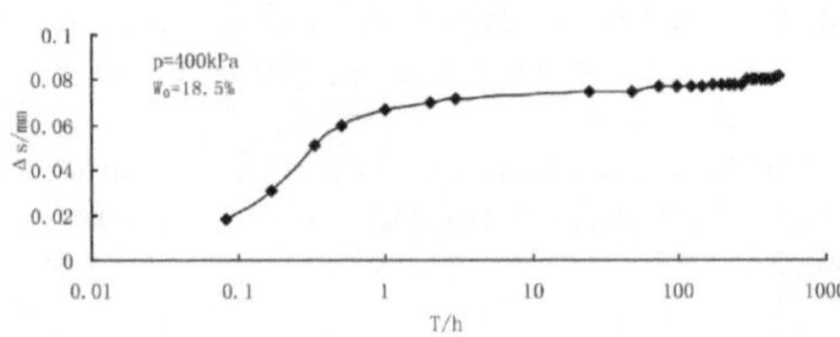

(c) p = 100kpa (sample 6)

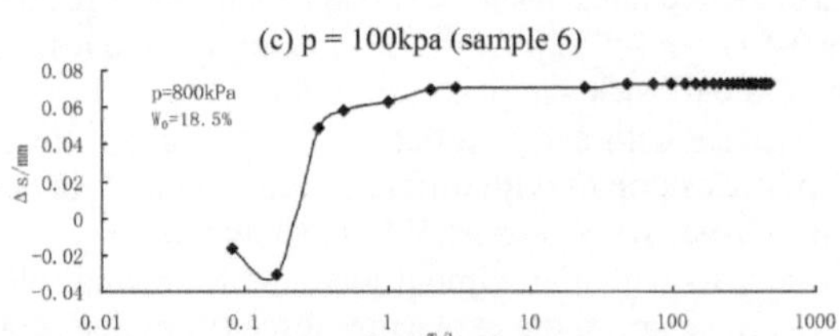

(d) p = 200kpa (sample 27)

Figure 3. Relationship between expansion increment and time under different pressure when w0 = 18.5%.

4 NUMERICAL ANALYSIS OF STRESS-STRAIN CHARACTERISTICS

Based on the engineering geological data of the Lijiaxia area, the test of expansion properties and the results of parameter inversion, physical and mechanical parameters of the clay rock are determined as shown in Table 2. The K8+195.0 section in Figure 1 (buried depth of 230 m) is selected to establish a finite element model, as shown in Figure 4 left. Initial supports are shotcrete with a thickness of 14cm (including steel centering) and six $\varphi 16$ bolts with a length of 1.5m and a spacing of 0.86m. Support structures and monitoring points are shown in Figure 4 right.

Table 2. Physical and mechanical parameters of surrounding rock.

Deformation modulus (GPa)	Poisson's ratio	Cohesion (MPa)	Friction angle (°)	Density (kN/m^3)	Saturated water content (%)	Free expansion rate (%)	Expansion force (kPa)	Softening coefficient
0.5	0.33	0.45	32	22.6	15.0	10.0	600.0	0.12

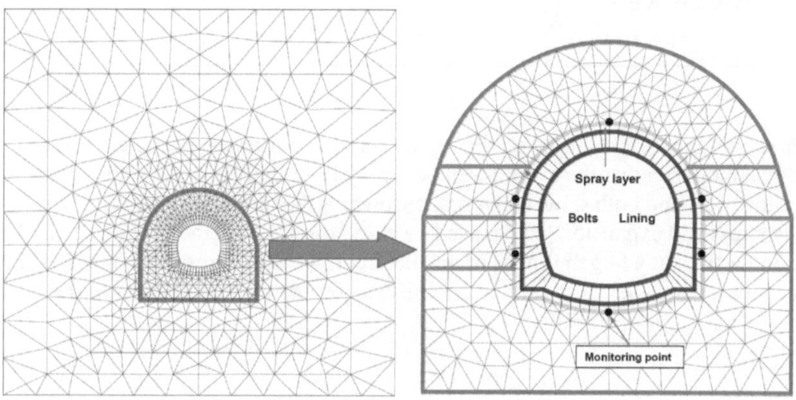

Figure 4. Finite element model and its details.

Under the combined action of excavation and expansion-softening, the stress, displacement, and deformation states of surrounding rock are shown in Figures 5(a)~5(c), respectively. It can be seen that the main axis of stress deflects approximately parallel to the tunnel outline, and tensile stress appears at the arch bottom (Figure 5(a)). The deformation of surrounding rock in this section generally converges to the interior of the tunnel, that is, the subsidence of the arch crown, the uplift of the arch bottom, and the opposing deformation of sidewalls (Figure 5 (c)).

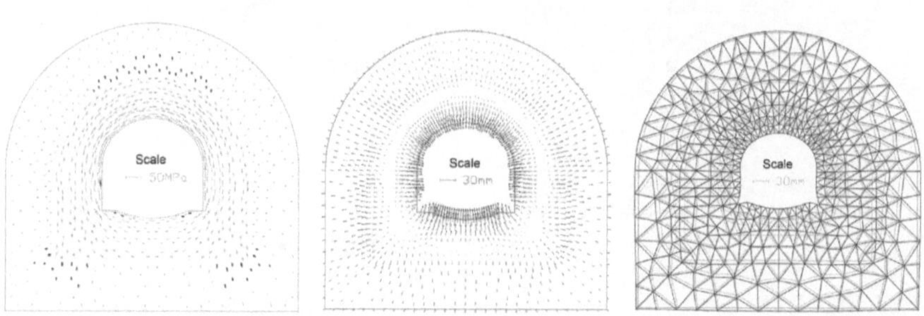

(a) (b) (c)

Figure 5. Stress, displacement, and deformation of surrounding rock after excavation.

Monitoring results of key points in the numerical model show that the maximum deformation of surrounding rock is about 40mm, which is distributed at the top and bottom of the arch (see Table 3). The deformations of the spandrel and the wall are basically the same, about 25mm. In addition, the major principle stresses at key points are compressive stresses and roughly point inside the tunnel, the maximum value is about 0.5MPa at the arch bottom, and the minimum value is about 0.3MPa at the wall. The minor principal stresses are approximately parallel to the excavation profile, the maximum value is about 11.0MPa at the spandrel, and the minimum value is about 1.0MPa at the arch bottom.

Table 3. Displacement and stress of key points.

Location		Displacement (mm)		Stress (MPa)	
		Horizontal	Vertical	σ_{max}	σ_{min}
Arch crown		−0.01	−37.93	−0.41	−3.64
Spandrel	Left	24.45	−8.52	−0.50	−10.73
	Right	−24.44	−8.55	−0.50	−10.76
Middle of wall	Left	24.42	−4.00	−0.28	−9.38
	Right	−24.42	−3.99	−0.30	−9.42
Arch bottom		−0.03	32.90	−0.53	−0.93

Figure 6 shows the deformation of the initial support structure (concrete spray layer) after excavation. It can be seen that the overall trend of spray layer deformation is consistent with that of the surrounding rock, and both show convergence towards the tunnel. However, under the influence of secondary stress and expansion load caused by excavation, local cracking occurs in the concrete spray layer. In particular, a large number of tensile cracks (tensile stress of 2.38MPa) appear at the wall on both sides, which seriously affect the construction and safe operation of the tunnel.

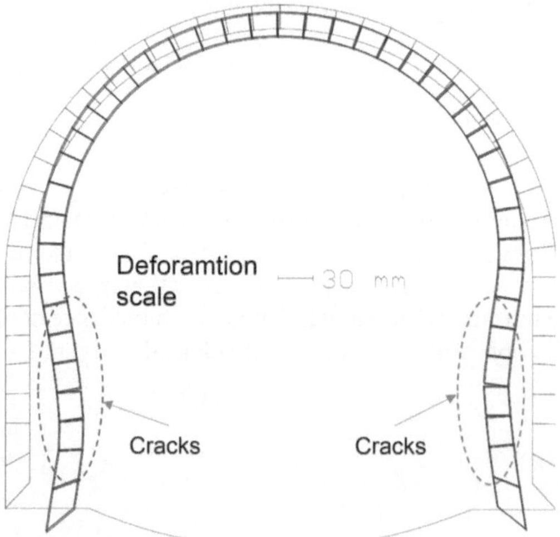

Figure 6. Deformation and failure characters of spray layer.

5 CONCLUSION

The Lijiaxia tunnel passes through the water-rich expansive stratum, causing serious damage to the excavated section. And the deformation occupies the design clearance of the tunnel and affects the construction of linings. Based on laboratory tests of relevant rocks, the expansion, deformation, and stress characteristics of the surrounding rock are analyzed using numerical simulation. The conclusions are as follows:

(1) The clay rock is rich in montmorillonite, illite, and other highly hydrophilic substances, which are prone to expand and soften after absorbing water. Its saturated free expansion rate is between 0.40 and 0.585, and its softening coefficient is between 0.015 and 0.096. According to the results of engineering analogy, this type of surrounding rock has poor self-stability and is prone to large deformation.
(2) Numerical modelling shows that under the action of expansion load and water pressure, the surrounding rock deforms into the tunnel as a whole after excavation, especially if the bottom heave exceeds 30mm. A large number of cracks appear in concrete spray layers, and the tensile stress of the side walls is about 2.38MPa. These disasters seriously affect the construction and safe operation of the tunnel.
(3) The main reason for local damage to the tunnel (manifested as steel centering twisting, spray layer cracking, and floor heave) is the delay in lining construction. As a result, the deformation caused by stress adjustment and rheology of surrounding rock has not been suppressed in a timely and effective manner. Meanwhile, the initial support cannot withstand excessive pressures, which leads to cracking and large deformation.
(4) Under the condition of ensuring the timely construction of support structures, the future work is to study the cooperative stress and deformation of the surrounding rock, initial support, and secondary lining of the tunnel for the unexcavated section.

ACKNOWLEDGMENTS

Financial support for this research is provided by the Major Science and Technology Project of Northwest Engineering Corporation Limited, Power China.

REFERENCES

Barla, M. (2008) Numerical simulation of the swelling behavior around tunnels based on special triaxial tests. Tunn. Undergr. Sp. Tech., 23(5): 508–521.
Fan, Q. (2008) Swelling rock and Engineering. Science Press, Beijing.
Grob, H. (1975) Swelling and heave in Swiss tunnels. B. Int. Assoc. Eng. Geol., 14: 55–60.
Hawlader, B.C., Lo, K.Y., Moore, I.D. (2005) Analysis of tunnels in shaly rock considering three-dimensional stress effects on swelling. Can. Geotech. J., 42(1): 1–12.
He, M. (2014) Latest progress of soft rock mechanics and engineering in China. J. Rock Mech. Geotech., 6(3): 165–179.
Huang, S.L., Aughenbaugh, N.B., Rockaway, J.D. (1986) Characterization of the swelling potential of shale strata. Proceedings of the 27th U.S. Symposium on Rock Mechanics, Alabama.
Jin, Y. (2003) Cause distortion and study of repair measure in Xiling Xueshan tunnel of Dashuang road. Southwest Jiaotong University.
Liu, N., Li, N., Xu, C. et al. (2020) Mechanism of secondary lining cracking and its simulation for the Dugongling tunnel. Rock Mech. Rock Eng., 53: 4539–4558.
Selmer-Olsen, R., Palmstrom, A. (1989) Tunnel collapses in swelling clay zone. Tunnel & Tunnelling, 21(11): 49–51.
Tang, S., Tang, G. (2014) Engineering properties and mitigation methods of expansive soil in Munga, Sudan. Soil Eng. Found., 27(4): 37–40.
Tang, S.B., Tang C.A. (2012) Numerical studies on tunnel floor heave in swelling ground under humid conditions. Int. J. Rock Mech. Min., 55: 139–150.
Wang, Y., Tang, X., Tan, X., et al. (2010) Mechanism analysis of floor heave in Yunling Tunnel. Rock Soil Mech., 31(8): 2530–2534.
Yu, J., Liu, G., Cai, Y., et al. (2020) Time-Dependent deformation mechanism for swelling soft-rock tunnels in coal mines and its mathematical deduction. Int. J Geomech., 20(3): 04019186.

Advances in Mineral Resources, Geotechnology and Geological Exploration – Rashid & Zhang (eds)
© 2023 Copyright the Author(s), ISBN 978-1-032-33377-9

Study on design method of super long and super deep retaining wall structure with prestressed anchor cable

Yinan Wang & Guangxiu Fang*
Department of Civil Engineering, College of Engineering, Yanbian University, Yanji, Jilin, China

Dingguo Wang
Jilin Shuangli Construction Project Management Co., Ltd, Yanji, Jilin, China

Wei Jang
Jilin Province Transportation Construction Group Co., Ltd, Yanji, Jilin, China

Huiquan Wang
Yanji Construction Project Quality Supervision Station, Yanji, Jilin, China

ABSTRACT: Taking the design of prestressed anchor cable type extra-long and extra-deep retaining wall structure of Guantang Hot Spring Resort Hotel Project in Yanji City, Jilin Province as an example, the geological and hydrological conditions were analyzed, the reliability and applicability of geotechnical parameters and indicators, frost depth and frost swelling, and site stability were evaluated, and recommendations for foundation design and construction were put forward, along with the design method. Through support monitoring and verification, it shows that the selection, support design scheme, and design measures of the slope retaining wall support project are practical and feasible, and meet the requirements of the current technical specification for retaining and protecting building foundation excavations (JGJ-120-2012), and shorten the construction period and reduce the project cost, and provide a reference for the design of similar retaining wall structures.

1 INTRODUCTION

Although the deep foundation pit support structure is a temporary support structure during construction, its correct and reasonable selection, calculation, and design have a significant impact on the safety, duration, and economic efficiency of the project (China Academy of Building Sciences JGJ-120-2012). Row pile support (Ministry of Housing and Urban-rural Development of the People's Republic of China GB-50497-2019) is usually used in the case of excavation depths of 6 to 14 m, where the row pile consists of a series of piles in close proximity to each other and serves to resist earth pressure and prevent slope instability (Ying et al. 2019). At present, it is important to carry out research on the calculation method of the support method for the extra-long and extra-deep retaining walls of specific structures, when the design uses prestressed anchor cable type retaining walls (Xu et al. 2019), which is not uniform, and there is little research on the optimization of the design method. This paper analyzes and researches the practical design method of prestressed anchor retaining walls with engineering cases for the application of similar projects.

*Corresponding Author: fangjian888888@163.com

 DOI 10.1201/9781003319412-26

2 PROJECT OVERVIEW AND SURVEY VOLUME

2.1 *Project overview*

The Guantang Hot Spring Resort & Spa project is located in the south of Yanji City, Yanbian Korean Autonomous Prefecture, Jilin Province, east of the intersection of Yanan Road and Yanlong Road, near Yanji Dinosaur Museum and Huanershan National Forest Park. The supported object is the west side soil of the Guantang Hot Spring Resort Hotel project, and the project plan is shown in Figure 1.

The maximum excavation depth of the retaining wall is 32 meters in zone VI. The slope height of this project belongs to the partial and sub-section project with a certain scale and high risk. It is necessary to draw up a special safety technical design scheme and organize experts to demonstrate it.

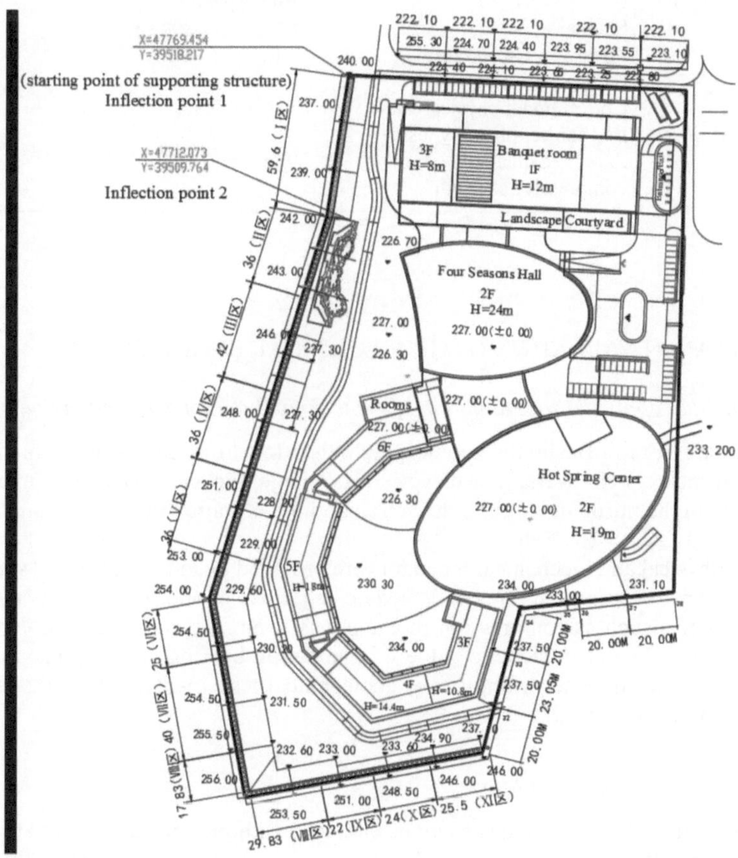

Figure 1. Project plan.

2.2 *Survey work volume*

According to section 3.1 of Code for investigation of geotechnical engineering GB50021-2001 (2009 edition), the importance level of the project is selected as Grade 2, the complexity level of the site is Grade 2, the complexity level of the foundation is Grade 2, and the comprehensive determination of geotechnical engineering survey-grade is set as Grade B. According to the technical requirements of the tender documents, a combination of drilling sampling, indoor geotechnical test, standard penetration test, and wave velocity test is used as the exploration method. A total of

54 exploration holes were completed, including 52 soil penetration holes and 2 wave velocity test holes. The total depth of exploration was 1482.70 m. The final workload is summarized in Table 1.

Table 1. List of survey workload.

Job description			Workload
Measurement	Hole coordinates and elevation		54 points
	Drill hole	Number of holes	54 Holes
		Hole depth	22.0~30.0m
		Total footage	1482.70m
Drilling	Sampling (pieces)	Class II slightly disturbed soil sample	119
		Rock specimens	15
	Standard penetration tests	52 groups, 204 sections	
	Shear wave speed test (single-hole method)	2 holes, 57.0m	
Indoor geotechnical tests	Conventional physical and mechanical property tests	119 groups	
Rock tests	Saturated uniaxial compressive strength	9 groups	

3 ANALYSIS AND EVALUATION OF GEOTECHNICAL ENGINEERING

3.1 *Reliability and applicability evaluation of geotechnical parameters and indicators*

In the eastern part of Yanji Basin, the geomorphic unit is low hills, consisting of muddy siltstone forming a basal rock system overlain by clayey soils of large thickness. The topography of the site is generally low in the north and high in the south, and the elevation of the measured exploration hole opening is 237.22m~263.49m.

Most of the physical and mechanical indicators are provided by indoor tests, and standard penetration tests and wave velocity tests are conducted in the field to provide empirical or recommended values of parameters such as compression modulus based on relevant codes and combined with regional experience. The characteristic value of foundation bearing capacity fak for each layer of foundation is determined according to the geotechnical test, in-situ test results, and regional experience, as shown in Table 2.

3.2 *Analysis of hydrological conditions*

The clayey soil layer in the proposed site can be endowed with groundwater in the abundant water year or season, and the groundwater type is upper layer stagnant water. During this survey, no groundwater was seen within the depth of the exploration holes.

3.3 *Site stability and suitability evaluation*

(1) According to the exploration and site investigation data, there are no geological hazards such as karst, landslide, debris flow, dangerous rock, and collapse at the proposed site.
(2) No large groundwater extraction wells and void areas are found in the vicinity of the proposed site, and no geotechnical environmental geology and adverse geological effects such as ground subsidence are found in the site area.
(3) According to the investigation and data analysis of the fracture structure in this area, there is no active fracture structure in the near field area.

Table 2. Characteristic values of bearing capacity of each layer of foundation f_{ak}(kPa).

Name of the soil layer	Natural Gravity (KN/m^3)	Compression Modulus Es (MPa)	Standard values of uniaxial compressive strength f_{rk}(kPa)	Load-bearing force characteristic value f_{ak}(kPa)	Penetration Coefficient	Shear strength indicators Cohesion C (kPa)	The angle of internal friction (°)
Tilling the soil①	18.8	/	/	100	3.5×10^{-5}	20	18
Powdered clay with sand②	19.2	5.3	/	180	2.0×10^{-6}	29	15
Fully weathered muddy siltstone③	24.0	23.0	/	240	4.2×10^{-5}	25	22
Strongly weathered muddy siltstone④	24.7	38.0	/	400	2.5×10^{-6}	45	25
Medium-weathered muddy siltstone⑤	25.0	/	768	750	1.0×10^{-6}	70	30

3.4 *Seismic intensity*

According to "the Code for Seismic Design of Buildings" (GB 50011-2010) (2016 version), the seismic intensity of the proposed site is 6 degrees, the design basic seismic acceleration value is 0.05g, the design seismic grouping is Group I, and the characteristic period value is 0.35s.

3.5 *Frost depth and freeze swell evaluation*

According to the "Standard Freezing Depth Line Map of Chinese Seasonal Frozen Soil", the standard freezing depth of the site is 1.60m.

Frost swelling evaluation: the ① layer of plain fill is IV strong frost swelling soil, the ② layer of powder clay III grade frost swelling soil.

3.6 *Site stability assessment*

The rocks of the proposed site are monoclinic, with a stable distribution of rock layers and a simple geological structure. There is no new active fracture structure passing through the site, landslide, debris flow, karst cave, and other adverse geological phenomena are not developed, the surface is 1.0~2.0m thick cultivated soil, covered with trees and weeds, the site is stable. The site is a general lot where construction can take place.

3.7 *Suggestions for foundation design and construction*

The slope is upper soil and lower rock slope, higher, according to the drilling results, combined with the local experience, need to pile anchor support system form excavation foundation. The support pile can use a screw bore pressurized pile, anchor pull structure can use anchor cable, slope surface spray concrete to protect, and set a certain number of drainage holes. The slope support parameters are shown in Table 3.

The design should consider the influence of ground load on the support system such as mechanical equipment construction and vehicle operation, minimize the pile load on the edge of the berm, and reduce the vehicle operation during the construction process. Excavated soil should be removed in time to avoid piling up on the edge of the berm and causing adverse effects on the system.

Table 3. List of calculation parameters of slope support.

Geotechnical layer serial number and name	The characteristic value of bond strength between geotechnical layer and anchor solid f_{rb} (kPa)	The friction coefficient of the geotechnical layer on the base of the retaining wall μ
Powdered clay sandwich②	25	0.28
Fully weathered muddy siltstone ③	80	0.42
Strongly weathered muddy siltstone ④	160	0.51
Medium-weathered muddy siltstone ⑤	300	0.58

4 STRUCTURAL CALCULATION AND DESIGN METHODS

4.1 *Design basic information and parameters*

In this paper, the row pile anchor retaining wall is designed and verified, and the sketch is shown in Figure 2. According to the geological report of the construction site and the national specification "Technical specification for retaining and protection of building foundation excavations" JGJ120-2012, the information related to the columns is calculated as shown in Table 4, and the information related to the anchor rods is shown in Table 5.

Table 4. Information about the column.

Column height above the embedment point (m)	3.500	Column capacity (kN/m^3)	25.00
Column embedment depth (m)	16.500	Column concrete strength grade	C30
Column section shape	Round	Column reinforcement method	Uniform reinforcement
L Diameter D(m)	1.100	Column longitudinal reinforcement level	HRB400
Column center distance (m)	2.000	Column hoop level	HRB400
Column calculation method	Elastic pivot point method	Diameter of column longitudinal reinforcement (mm)	25
Column bottom support condition	Fixed	Column as(mm)	80
Automatic calculation of initial elasticity coefficient	√		
Initial elasticity coefficient A(MN/m^3)	–		

Basic information on anchor rods (ropes), as shown in Table 5.

4.2 *Design and calculation of structures*

4.2.1 *Geotechnical pressure calculation*

Calculation of active earth pressure according to the formula (6.2.3) of the Slope Code: E_a = 46.585(kN) E_x = 43.776(kN) E_y = 15.933(kN) Action point height Z_y = 1.167(m), Since the Cullen earth pressure can only calculate the earth pressure of a single layer of soil, the indicators of multiple

Table 5. Basic information on anchor rods (ropes).

Anchor number	Anchor rods Type	Vertical Spacing (m)	Incidence angle (°)	Free section Length (m)	Length of anchorage section (m)	Anchor solids Diameter (mm)	Horizontal preload (kN)	Level Stiffness (MN/m)	Tendon slurry Strength f_b (kPa)	Over-excavation Depth (m)
1	Anchor cable	3.000	15.00	4.00	5.00	200	100.000	10.32	2950.00	–

layers of soil within the wall height range are considered as a single layer of soil after weighted average. The indicators are as follows: γ = 18.600(kN/m^3), C = 19.800(kPa), φ = 14.800(°)

4.2.2 *Column internal force calculation*

The excavation depth of the current calculation condition is 3.500, and the stratigraphic calculation method of the 1st, 2nd, and 3rd embedded sections are all adopted as the m method. The backside is the retaining side and the face side is the non-retaining side. The calculation considers the maximum bending moment on the backside, the maximum bending moment on the face side, and the maximum shear force. When the anchor rod (cable) is designed to be under pressure, i.e., the horizontal tension is negative, the horizontal tension is taken as 0.

4.2.3 *Column reinforcement calculation*

The internal force design value is used for reinforcement calculation, and the internal force standard value is used for crack calculation.

4.2.4 *Overall stability test*

The Swedish strip method is chosen, and the overall stability of the most unfavorable sliding surface needs to be verified. The center of the circle is (15.64091, 35.00000); the radius is 36.39684(m); The safety factor is 5.668; the total decline force = 114.373(kN); the total slip resistance=648.261(kN); the partial sliding force of soil body = 114.373(kN); the partial slip resistance of soil body=648.261(kN); the anti-slip force of ribbons = 0.000(kN).

In summary, the overall stability test satisfies: Minimum safety factor = 5.668 ≥1.25

4.2.5 *Design solutions*

The design sketch of the row pile anchor retaining wall is shown in Figure 2.

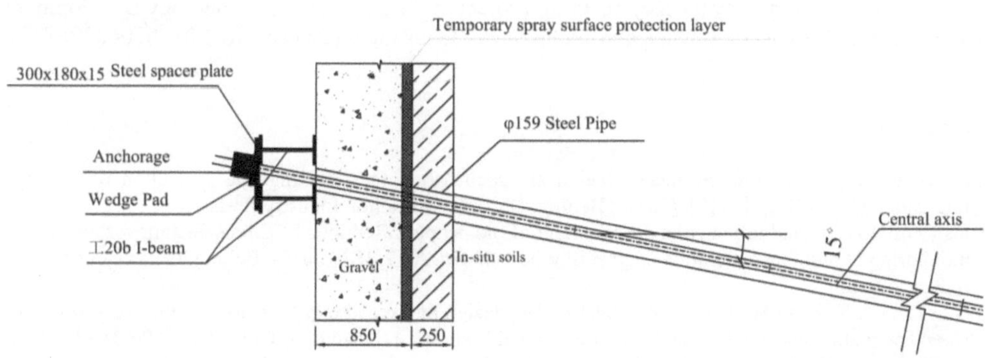

Figure 2. Design sketch of row pile type anchor retaining wall.

The elevation design of the row pile-type anchor retaining wall is shown in Figure 3.

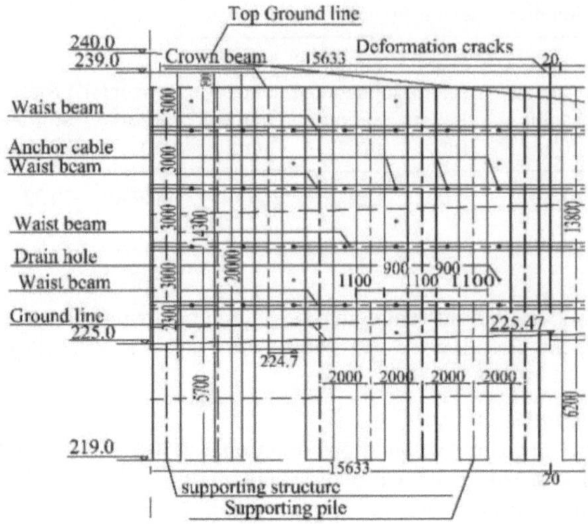

Figure 3. Elevation design of row pile type anchor retaining wall.

5 CONCLUSIONS

Taking the design of extra-long and extra-deep row pile + prestressed anchor cable support structure of Guantang Hot Spring Resort Hotel project in Yanji City, Jilin Province as an example, the reliability and applicability of geotechnical parameters and indicators, frost depth and frost swelling, and site stability were evaluated based on geological and hydrological conditions, and recommendations for foundation design and construction were put forward. The final row of piles used auger bored compression piles with the fifth layer of medium weathered muddy siltstone as the base bearing layer. At the same time, the design method for specific structure retaining walls was proposed. Through support monitoring and verification, it shows that the selection, design method, design scheme, and measures of the slope retaining wall support project are practical and feasible, which provides a reference for the design of similar projects.

ACKNOWLEDGMENT

This paper was supported by the Key Tackling Project of the Science and Technology Development Plan of the Jilin Provincial Science and Technology Department (Project No. 20170204032SF).

REFERENCES

China Academy of Building Sciences. Technical specification for retaining and protection of building foundation (JGJ-120-2012) [S]. (2012). Beijing: China Construction Industry Press.

Ministry of Housing and urban-rural development of the people's Republic of China. Technical standard for monitoring of building excavation engineering (GB-50497-2019) [S]. (2019). Beijing: China Construction Industry Press.

Xu J.L, Fang G.X, Zhao Sh. F, Han W.Q. (2019). The design of Excavating Piles Bored with Long Auger for Deep Foundation and Anchor Cable and Reinforced Methods. [J]. Building Technology, 50(03):335–337.

Ying R.P, Fang G.X, Zhao Sh. F, Han W.Q. (2019). Design and deeper construct to artificial digging hole pile as supporting structure of deep foundation pit [J]. Building Technology, 50(01):88–91.

Advances in Mineral Resources, Geotechnology and Geological Exploration – Rashid & Zhang (eds)
© 2023 Copyright the Author(s), ISBN 978-1-032-33377-9

Study on internal structure and interface development characteristics of thick oil layer in block D of L oilfield

Shuangshuang Gong*
The Sixth Oil Production Plant of Daqing Oilfield, Daqing, China

ABSTRACT: The reserve ratio of a thick oil layer above 2.0m in an L oilfield is more than 62.5%. Although it has entered the late stage of high water cut development, the thick oil layer is still the main object of oilfield development. With the deepening of oilfield development, the comprehensive water cut of the oilfield is gradually rising, resulting in the increasing difficulty of water polymer two drive production. Especially, the serious impact of structural heterogeneity and physical heterogeneity in thick oil layers has plagued the more efficient development of thick oil layers. In order to further excavate the remaining oil in the thick oil layer, this paper focuses on the three types of thick oil layers of class A, class B, and class C above 1.0m. Based on the genetic analysis and type division of the internal structure interface of the thick layer, the development characteristics of different types of structure interface are studied, which provides a geological basis for tapping the potential of water drive measures in the future.

1 INTRODUCTION

The study area is area D, focusing on the anatomical study of the thick layer of the water drive layer of SAIIgroup ~ SAIII layer in area D. using the method of "cycle comparison, hierarchical control, genetic analysis, and closed tracking" under the control of standard layer, the two-level structural interface between the sedimentary unit and the structural unit is traced on the skeleton section.

Based on anatomical research and according to the types and development characteristics of thick layers in different sedimentary environments, taking the rock outcrop profile and modern sedimentary profile as the theoretical basis, according to the lithological characteristics, the internal interface of thick oil layer is divided into five types: mudstone interlayer, silty mudstone, calcareous siltstone, argillaceous siltstone, and calcareous rock.

2 INTERLAYER TYPE DIVISION BASED ON SAND BODY GENETIC MECHANISM

There are three main forms of genetic mechanism: first, in the normal water period, when the carrying capacity of the river is significantly weakened, the argillaceous silt is mainly deposited, which is obviously different from the lithology of the channel sand body, forming the interlayer inside the sand body; Second, the argillaceous components and aeolian fine-grained sediments are deposited interactively in the dry season with intermittent flooding, all of which are deposited directly on the bottom of the exposed sandy riverbed and the top of the sandy core beach dam to form internal interlayer, and the core is represented by mudstone and silty mudstone; Third, the argillaceous sandstone or siltstone deposited in the dry season forms a calcareous interlayer under the oxidizing environmental conditions (Dong 1992) of pedogenesis, and the core is shown as a calcium layer. However, in the riverbed with water flow, there is still continuous deposition of sand and no interlayer is developed. Therefore, according to the distribution position and genetic mechanism of interlayer in a thick layer, it is determined that the types of interlayer mainly include I, II, III, and IV.

*Corresponding Author: gongshuang1petrochina.com

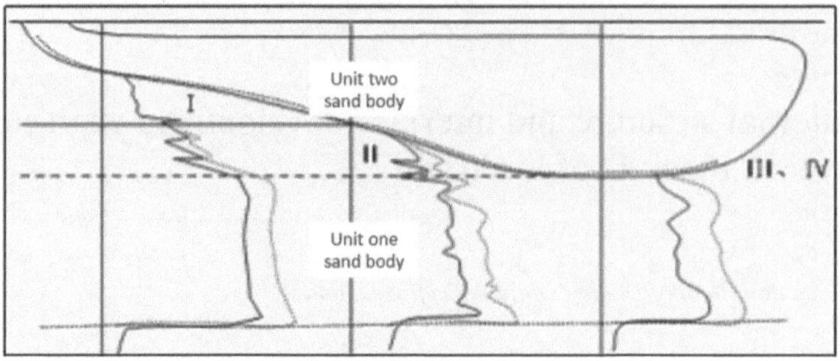

Figure 1. Cross-section of interlayer type.

3 RESEARCH ON STRUCTURAL INTERFACE CHARACTERISTICS

Class I structural interface: it refers to the mudstone or calcareous layer between the interpretation sand layers. For argillaceous rocks with a thickness greater than 0.4m, the air permeability is generally less than $20 \times 10^{-3}\mu m^2$, the return range is large and obvious on the logging curve, and the sand body is divided into upper and lower natural interpretation sand layers; Class II structure interface: calcareous layer or physical interlayer between two effective thickness sections in a well logging interpretation sand layer, with thickness generally greater than 0.4m and permeability of $20 \sim 120 \times 10^{-3}\mu m^2$; Class III structure interface: the interlayer in the effective thickness interpretation section, the thickness is generally less than 0.4m, and the permeability is $20 \sim 120 \times 10^{-3}\mu m^2$; Class IV structural interface: the permeability grading interface existing in the superposition of two stages of sandstone, with a thickness of less than 0.1m, that is, ultra-thin interlayer.

Based on the existing classification standards of different types of thick oil layers (Zeng & Jzcson 2009), the development quantity and thickness of internal structural interfaces of three different types of thick layers in the remaining intervals of water drive in 20 core wells, as well as the development position in the sand body are statistically analyzed. The internal structure and interface type characteristics of different types of thick reservoirs are determined:

First, the internal structure and interface development characteristics of class a thick oil layer. The distributary sandstone is mainly composed of a straight channel and curved channel with a thickness of more than 2.0m, and the distributary sandstone is mainly distributed in the rhythmic channel with a thickness of more than 2.0m. It can be observed that the distributary sandstone is mainly distributed in the channel with a thickness of 2.0m, and the distributary sandstone is mainly distributed in the channel with a thickness of more than 2.0m. Some are of class II and IV. Among them, the class III structural interface is mainly argillaceous siltstone, argillaceous siltstone, and siltstone; Class II structural interfaces are mainly silty mudstone, argillaceous siltstone, and siltstone. In the positive rhythm type of oil layer, the structural interface is mostly distributed at the top of the rhythm, mainly in the embankment surfaces and river overtopping surfaces with weak hydrodynamic conditions; Some class IV structural interfaces are mainly distributed in the middle of the rhythm and formed by the multi-stage river cutting and overlapping. In the class, an oil layer with complex rhythm, the class III and IV structural interfaces are distributed in all parts of the rhythm (usually more than 0.5m from the bottom limit of effective thickness), which is mainly formed by multi-stage channel superposition and cutting.

The second is the development characteristics of the internal structure interface of the class B thick oil layer. Class B thick oil layer is a reservoir with an effective thickness of more than 1.5m and a positive rhythmic sandstone thickness of more than 2.0 m. Through core observation, it can be determined that it is mainly river sedimentation, which is mainly developed in low bend distributary channel, low bend straight distributary channel, and dendritic delta sedimentary model. Its internal structure interface type is mainly class III, and a few contain class II and class IV interfaces. Among them, the class III structural interface is mainly distributed in the middle and upper part of the rhythm, and the lithology is argillaceous siltstone, argillaceous siltstone, and siltstone; the Class II interface is mainly distributed at the top of the positive rhythm, which is silty mudstone or argillaceous siltstone; A few class IV interfaces are mainly distributed in the middle of the rhythm. The reason for the formation of the interface is similar to the positive rhythm in class a thick layer, which is mainly formed under the condition of weak hydrodynamic force in the continuous sedimentation of river and delta, which is usually distributed in the bank surfaces and river overtopping surfaces with weak hydrodynamic conditions.

Table 1. Statistics of internal structure and interface development characteristics of different types of thick reservoirs on core wells.

	Interlayer in thick oil layer							Thick oil layer (Sand body)		
				Position					Average thickness (m)	
Thick layer	Type	Quantity (PCs.)	Thickness (m)	Upper part	Middle part	Under part	Component	Depositional model	Offsurface thickness	Effective thickness
A	II	8	7.2	8	/	/	Silty mudstone, Pelitic siltstone, Siltstone	Highly curved distributary channel; Low bend distributary channel; Low bend straight distributary channel	6.5	5.9
	III	308	55.2	132	132	44	Pelitic siltstone, Argillaceous siltstone, Siltstone		8.2	6.8
	IV	32	0	12	20	/	/		5.6	5.4
B	II	3	1.2	3	/	/	Silty mudstone, Pelitic siltstone, Siltstone	Low bend distributary channel; Low bend straight distributary channel; Dendritic delta	2.6	1.6
	III	69	13.2	45	27	/	Pelitic siltstone, Argillaceous siltstone, Siltstone		2.7	1.6
	IV	6	/	/	6	/			2.8	1.7
C	I	25	9.1	25	/	/	Mudstone, Silty mudstone, Calcareous sandstone	Low bend distributary channel; Dendritic delta Zhituo transitional delta	3.1	1.2
	II	5	2.8	3	/	2	Pelitic siltstone, Silty mudstone, Siltstone		2.8	1.0
	III	15	3.0	5	10	/	Argillaceous siltstone, Pelitic siltstone, Siltstone		1.5	1.1

The third is the development characteristics of the internal structure interface of the class C thick oil layer. Class C thick oil layer is a combination of low water flooded thin difference layer or surface layer with a continuous effective thickness of more than 1.0m, sandstone thickness of more than 2.5m, and interbedded cumulative sandstone thickness of more than 2.5m. It is mainly developed in the sedimentary models of low bend distributary channel, branched Delta, and branch Tuo transition Delta (Jiang 2011). Through core observation, it can be determined that the internal structural interface type of thick layer is mainly class I interface, followed by class III interface, and class II is the least developed. Among them, the class I interface is mainly composed of mudstone, silty mudstone, silty mudstone, and calcareous rock between the thin differential layer or surface layer, which are mainly distributed in the upper part of the composite sand body; the Class III interface is mainly developed in the middle and upper part of the continuous effective thickness, and a few are developed in the lower part of the effective thickness. This is mainly because the class C thick layer is mostly developed in the underwater environment of the Delta and formed intermittently under the influence of Lake wave action.

4 STATISTICAL RESULTS OF STRUCTURAL INTERFACE DISTRIBUTION CHARACTERISTICS IN ZONE D

Using the subdivision results, the drilling conditions of interface types between structural units are counted. According to the statistical results, according to the type of interlayer, in the SA II 2+3 section, it is mainly class a thick layer, its structural interface is mainly class III and IV, and the sand body is in superposition and shear superposition contact; In the section of SAII 4 ~ SAIII 3, according to the interlayer type, the interface of class B thick layer structure is mainly class I and II, and the sand body is the mainly layered interval (Table 2).

Table 2. Development of interlayer between structural units of SII ~ SIII 3 in area DUnit.

		Class I interlayer			Class II interlayer		Class III interlayer		Class IV interlayer	
Unit name	Total number of wells (mouth)	Number of wells (mouth)	thickness (m)	proportion (%)	Number of wells (mouth)	proportion (%)	Number of wells (mouth)	proportion (%)	Number of wells (mouth)	proportion (%)
S II1+2_1	1887	1810	1.53	95.9	28	1.4	28	1.4	21	1.1
S II1+2_2	1887	1692	1.07	89.6	38	2.0	106	5.6	51	2.7
S II2+3_{1a}	1887	969	0.41	51.3	174	9.2	413	21.8	331	17.5
S II2+3_{1b}	1887	970	0.5	51.4	171	9.0	478	25.3	268	14.2
S II2+3_{2a}	1887	692	0.39	36.6	174	9.2	436	23.1	585	31.0
S II2+3_{2b}	1887	1301	0.74	68.9	99	5.2	307	16.2	180	9.5
S II4	1887	1627	0.52	86.2	58	3.0	143	7.5	58	3.0
S II5+6_a	1887	1293	0.48	68.5	115	6.0	272	14.4	206	10.9
S II5+6_b	1887	1704	1.06	90.3	30	1.5	116	6.1	36	1.9
S II7+8_a	1887	1393	0.91	73.8	66	3.5	247	13.0	180	9.5
S II7+8_b	1887	1785	1.03	94.5	15	0.7	49	2.6	37	1.9
S II9_a	1887	1620	0.24	85.8	63	3.3	113	5.9	90	4.7
S II9_b	1887	1722	0.96	91.2	43	2.2	80	4.2	41	2.1
S II10+11_a	1887	1298	0.91	68.7	79	4.1	284	15.0	225	11.9
S II10+11_b	1887	1758	1.37	93.1	12	0.6	59	3.1	57	3.0
S II12	1887	1710	1.06	90.6	31	1.6	76	4.0	69	3.6
S II13+14_a	1887	1398	0.89	74.0	95	5.0	214	11.3	179	9.4
S II13+14_b	1887	1676	1.07	88.8	44	2.3	93	4.9	73	3.8
S II15+16_a	1887	1385	0.8	73.4	78	4.1	240	12.7	183	9.7
S II15+16_b	1887	1717	1.06	90.9	32	1.7	80	4.2	57	3.0
S III 1+2_a	1887	1350	0.97	71.5	64	3.3	260	13.7	212	11.2
S III 1+2_b	1887	1644	1.02	87.1	30	1.5	139	7.3	73	3.8
S III 3_a	1887	1392	0.6	73.7	78	4.1	202	10.7	214	11.3
S III 3_b	1887	1278	0.3	67.7	76	4.0	235	12.4	297	15.7
S III 3_c	1887	1525	0.9	80.8	70	3.7	146	7.7	145	7.6

5 STUDY ON THE PREDICTION METHOD OF DETERMINING DIFFERENT TYPES OF STRUCTURAL INTERFACES IN THE SEDIMENTARY UNIT OF SA II FORMATION

The prediction of different types of structural interfaces in sedimentary units is mainly based on the sedimentary model as the constraint condition, and the classification, prediction, and tracking are carried out according to the development law of the internal interfaces of different river channels sand bodies (Li 1999). Guided by the distribution mode of the four-level interface inside the channel sand body of different river types, the method of "hierarchical control and rhythm comparison" is adopted to analyze the rhythm (Zhang & Chen 1983), interlayer and permeability difference characteristics in the thick sedimentary unit, and trace the layer correspondence of the four-level interface inside the channel sand body within the well cluster. The thickness of each structural unit meets conditions of more than 1.0m.

The first is the research on the prediction and tracking method of class a thick layer, that is, the fourth level interface of braided channel sand body. In the braided river sedimentary model, the channel sand body is strongly lenticular, the sand bodies are cut each other, the sand body is thick and presents multiple rhythmic sections, the interlayer between rhythms is usually thin and rough, mainly class III and IV interlayer, which is argillaceous siltstone with horizontal occurrence distribution. The four-level structure interface in the channel sand body can be traced and compared according to the distribution range of class III and IV interlayer and its corresponding relationship on the logging curve.

The second is the research on the prediction and tracking method of class a thick layer, that is, the internal four-level interface of a highly curved channel sand body. The river channel is large in scale and curvature, and there are a certain number of abandoned river channels at the edge and inside of the river channel, which traps the meandering zone into several point bar sand bodies. The interior of the point dam sand body is in the accumulation mode of lateral accretion, which is formed by a series of crescent-shaped lateral accretion bodies inclined to the concave bank. The lateral accretion bodies are often separated by a thin lateral accretion interlayer, mainly in the upper part of the point dam sand body. The interlayer types are mainly Classes II, III, and IV, which are inclined to the concave bank or oxbow lake, with an inclination of about 15°. The structural interface should be traced and compared according to the interlayer inclination direction.

The third is the research on the prediction and tracking method of the four-level interface in the thick oil layer dominated by class A and B thick layers, that is, the low bend straight channel sand body. Due to the narrow scale of the river channel and the weak flow intensity and cutting ability, the lateral migration ability is weak. It is dominated by the vertical aggradation of the cutting filling type. The interlayer in the unit is of class I, II, and III. the interlayer lithology becomes thinner and all of them are horizontal occurrences. In the unit, it can be traced and compared according to the interlayer type and its corresponding relationship on the logging curve.

The fourth is the research on the prediction and tracking method of the internal four-level interface of underwater distributary channel sand body in the delta front environment dominated by class B and C thick layers. Due to entering the underwater environment and under the influence of wave action, the scale of the river channel is narrow and the flow intensity and cutting ability are weak, so the lateral migration ability is weak. It is mainly vertical aggradation. The Intercalations in the unit are class I, II, and III, and all of them are horizontal occurrences. In the unit, it can be traced and compared according to the intercalation type and its corresponding relationship on the logging curve.

6 CONCLUSION

1. The sand body interlayer in the thick oil layer in Block D of the L oilfield is divided into three categories based on the genetic mechanism. The first is the argillaceous silt deposited in the normal water period; the second is the internal interlayer formed by argillaceous components

and aeolian fine-grained sediments in the dry season of intermittent flooding; the third is the calcareous intercalation deposited in the dry season.

2. According to the distribution position and genetic mechanism of interlayer in a thick layer, the main types of interlayer are class I, II, III, and IV.
3. It describes the structural interface distribution characteristics of class A, class B, and class C thick layers under different sedimentary environments.
4. The prediction methods of structural interfaces in different types of thick sedimentary units are determined.

REFERENCES

Dong An Li. (1992). Three-dimensional fault combination Petroleum geophysical exploration [J], 9 (3): 24–26.

Hongliu zeng and Alan Jzcson. (2009). How thin is a thin bed? An alternative perspective[J]. The ldading dege,(28):1192–1197.

JiangYan. (2011). Integrated well-to-seismic fault interpretation technology and its application in LMD oilfield: A case study on Northern SongLIiao basin, Northeast China[C]. Shen Zhen: SPG/SEG 2011 International Geophysical Conference,1592–1595.

Li Wenqing. (1999). Geologicalbasis of oil and gas field development Beijing: Petroleum Industry Press.

Zhang Baozheng, Chen Qi. (1983). At al Fundamentals of geology Beijing: Geological Publishing House.

Advances in Mineral Resources, Geotechnology and Geological Exploration – Rashid & Zhang (eds)
© 2023 Copyright the Author(s), ISBN 978-1-032-33377-9

Comprehensive evaluation of source rocks characteristics in Dehui Area, Songliao Basin

Fan Yang*
Liaohe Oilfield Exploration and Development Research Institute, Panjin, Liaoning, China

ABSTRACT: The Songliao Basin in north-eastern China is a large basin rich in hydrocarbon resources, of which the Dehui area has made an important breakthrough at depth. However, the current evaluation and understanding of hydrocarbon source rocks have not been studied scientifically and systematically. The lack of oil and gas source comparisons has resulted in a lack of clear and reasonable direction for the next step in finding favorable exploration zones and targets in the Dehui area. A comprehensive evaluation of its hydrocarbon source rocks was carried out by combining rock sample analysis tests, logging, and drilling and logging data. The results show that the source rocks in the Dehui area are widely distributed, with a large thickness of up to 600m or more, with type II_2 and III kerogens, a high total organic carbon content (TOC), a mature-high maturity stage of thermal evolution and medium hydrocarbon generation potential. The main development of moderate-good hydrocarbon source rocks shows a promising future.

1 INTRODUCTION

The comprehensive evaluation of hydrocarbon source rocks mainly includes the evaluation of both the distribution and geochemical characteristics of hydrocarbon source rocks (Zhao et al. 2006; Chen et al. 1997; Li et al. 2006). Among them, the distribution of hydrocarbon source rocks includes the thickness of hydrocarbon source rock development and the area of distribution; the geochemical characteristics of organic matter include three aspects of organic matter abundance, organic matter type, and organic matter maturity.

The Songliao Basin in northeastern China is a large territory, which is rich in oil and gas resources, of which the Dehui area has made important breakthroughs in deeper layers and has successively discovered three oil and gas reservoirs, including Nong'an, Buhai, and Xiaohelong, showing great promise (Figure 1). At the same time, some geological problems need to be solved urgently. The evaluation and understanding of hydrocarbon source rocks have not been studied scientifically and systematically. The lack of oil and gas source comparison has led to the lack of a clear and reasonable direction for the next step in finding favorable exploration zones and targets in the Dehui area. Therefore, it is urgent to start with hydrocarbon source rock evaluation, to clarify the distribution of hydrocarbon source rock formations and areas, to understand the hydrocarbon generation capacity of hydrocarbon source rock, to identify the cause of natural gas, to analyze the source of natural gas, and to lay the foundation for the next step of analysis of reservoir formation process and reservoir formation law.

In response to the main problems in the Dehui area, this paper will start from the collected geological data, analyze and assay the collected rock samples, collect and collate data to evaluate the hydrocarbon source rocks in the Deyhui area in terms of development distribution, organic matter abundance, type, and maturity, and assess their resource potential. It is expected to provide assistance to hydrocarbon exploration in the study area.

*Corresponding Author: yangf02@petrochina.com.cn

DOI 10.1201/9781003319412-28

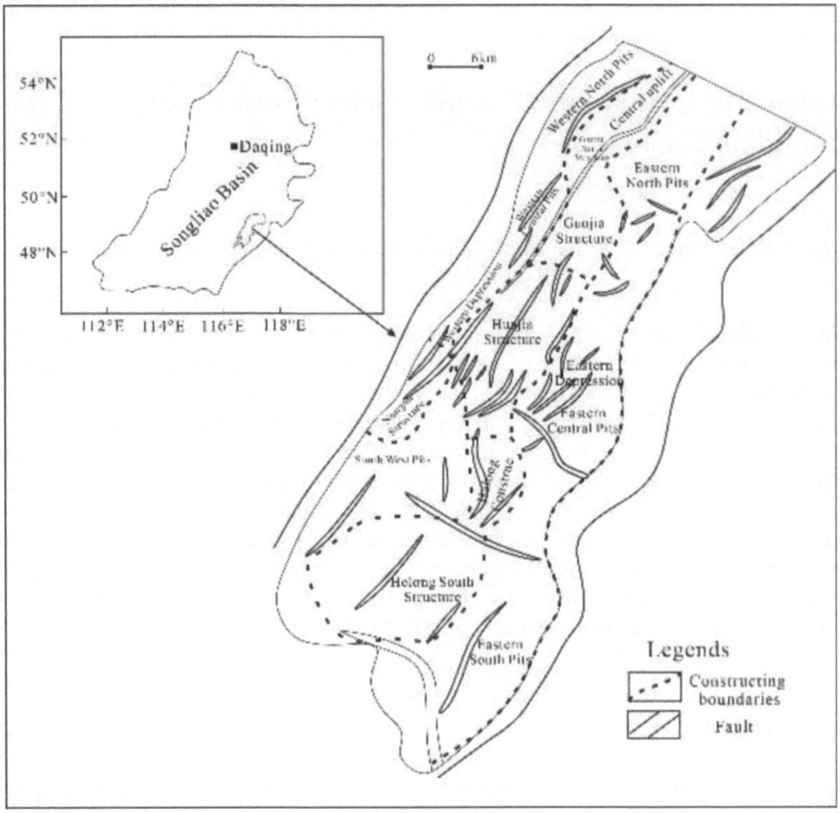

Figure 1. Location and structural characteristics of the study area.

2 METHODS

The collected samples are used for analysis and testing. For the set study, Rock-Eval pyrolysis data, total organic carbon content, specular reflectance, micro fraction, biomarker compounds, and major elements are obtained to analyze the geochemical characteristics of hydrocarbon source rocks in terms of organic matter type, abundance, and maturity, thereby evaluating their resource potential.

3 RESULTS AND DISCUSSION

The rock samples, logging letters, drilling, and logging data were integrated to comprehensively analyze and evaluate the deep hydrocarbon source rocks of the Dehui area in terms of their spreading characteristics, organic matter abundance, organic matter type, and organic matter maturity.

3.1 *Distribution characteristics*

Three sets of hydrocarbon source rocks, the Yingcheng, Shahezi, and Huoshi Ling Formations, are developed in the Dehui area. The source rocks of the Huosiling Formation are less widespread and mainly develop in the central part of the area, with a maximum thickness of 600 m. The source rocks of the Yingcheng Formation are distributed in the north and south. The Shahezi Formation is the most widely distributed, with high-value areas mainly in the north, forming a north-south strip, and the average cumulative thickness of the mudstone can reach over 200m.

3.2 *Geochemical characteristics*

According to microfracture, elemental analysis, Rock-Eval pyrolysis data, and biomarker compound characteristics (Fu et al. 1991; Huang et al. 1984; Zhang et al. 2006), most the deep hydrocarbon source rocks in the Dehui area belong to Type III and Type II_2, while a small amount of hydrocarbon source rocks in the Shahezi and Huosiling Formations belong to Type I and Type II_1 (Figure 2).

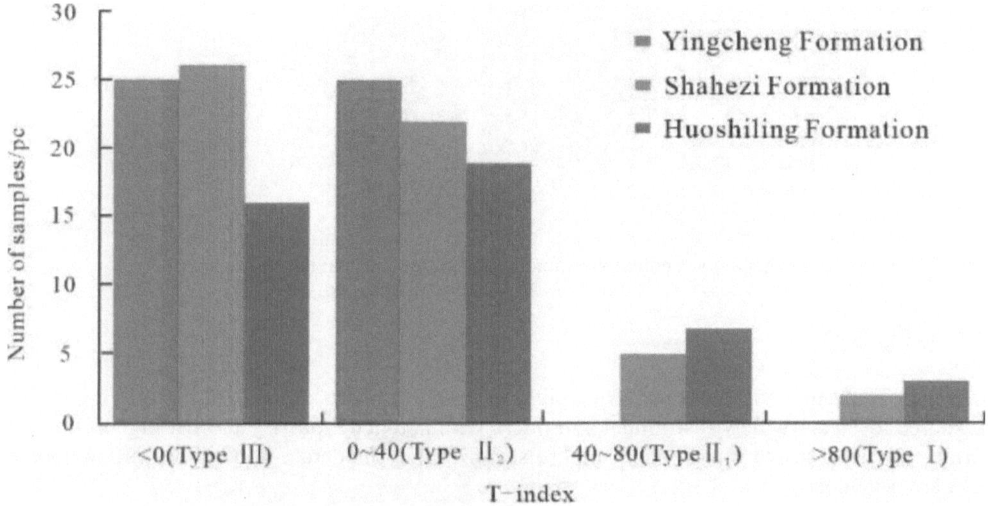

Figure 2. Distribution map of kerogen type in the Dehui area.

The organic carbon content of the deep hydrocarbon source rocks in the Deyhui area is generally high and the organic matter abundance is moderate to good. The TOC content of the hydrocarbon source rocks of the Yingcheng, Shahezi, and Huoshi Ling Formations is mainly distributed between 0.6% and 4.0%, with the Shahezi Formation being the very good hydrocarbon source rock, followed by the Yingcheng and Huoshi Ling Formations.

According to the results of Ro analysis and testing, the overall maturity of deep hydrocarbon source rocks in the Dehui area is high and increases with depth, with Ro values mainly distributed in the range of 0.5% to 2.0%. The source rocks are mainly in the mature-high maturity stage of evolution.

The overall distribution of soluble organic matter content in the deep hydrocarbon source rocks of the Dehui area is wide, with the medium hydrocarbon source rocks accounting for the highest proportion, ranging from 0.05% to 0.1%. The total hydrocarbon content is generally moderate, with slightly poorer hydrocarbon source rocks of the Huoshi Ling Formation.

3.3 *Hydrocarbon generation potential*

Hydrocarbon adsorption (S_1) and hydrocarbon pyrolysis (S_2) can be obtained from source rock analysis, and hydrocarbon generation potential (S_1+S_2) is also one of the important parameters for characterizing organic matter abundance, which can more directly reflect the hydrocarbon generation capacity of a group of source rocks (Cheng et al. 1995). The hydrocarbon generation potential of hydrocarbon source rocks in the Dehui area is poor to moderate overall, with slightly poorer hydrocarbon source rocks of the Huolsling Formation. The hydrocarbon source rocks of the Yingcheng Formation, Shahezi Formation, and Huosiling Formation have moderate hydrocarbon potential between 2 mg/g and 6 mg/g, accounting for about 25%, 35%, and 20% of the samples, respectively. Good source rocks with hydrocarbon potential greater than 6mg/g account for about 15%, 6.1%, and 7.2% of the samples, respectively (Figure 3).

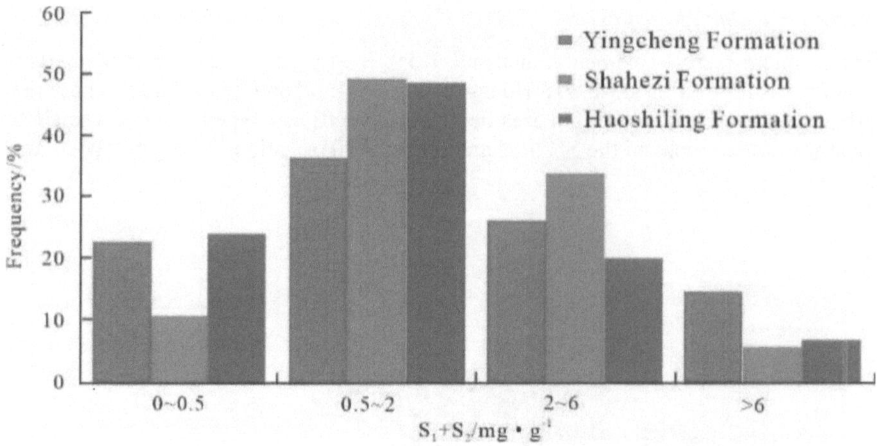

Figure 3. Hydrocarbon generation potential evaluation of source rocks in the Dehui area.

4 CONCLUSION

The source rocks in the Dehui area have a high potential for hydrocarbon exploration. In general, the source rocks are widely distributed and thick, with kerogen types II_2 and III, high TOC, and thermal evolution in the mature-high mature stage. Mainly moderate-good hydrocarbon source rocks are developed.

ACKNOWLEDGMENTS

This work was financially supported by the National Natural Science Foundation of China.

REFERENCES

Chen, J.P., Zhao, C.Y., He, Z.H. (1997) Exploration on the evaluation criteria of hydrocarbon generation potential of coal-based organic matter. Petroleum Exploration and Development, (1): 1–5, 91.

Cheng, K.M., Wang, T.G., Zhong N.N., et al. (1995) Geochemistry of hydrocarbon rocks. Science Press.Beijing. pp. 135–138.

Fu, J.M., Sheng, G.Y., Xu, J.Y., et al. (1991) Application of biomarker compound parameters to discern paleosedimentary environments. Geochimica, (1): 1–12.

Huang, D.F., Li, J.C., Zhang, D.J. (1984) The validity, limitations, and relevance of types of cheese roots and their classification parameters. Acta Sedimentologica Sinica, (3): 18–33, 135–136.

Li, J.K., Liu, W., Song, L.B. (2006) Study on hydrocarbon generation conditions of deep hydrocarbon source rocks in the Xujiaweizi Fault. Natural Gas Industry, (6): 21–24, 155.

Zhang, W.Z., Yang, H., Li, J.F., et al. (2006) On the dominant role of high-quality oil source rocks in low permeability hydrocarbon reservoir enrichment in the long-7 section of Ordos Basin: Analysis of the characteristics and mechanism of strong hydrocarbon production and discharge. Petroleum Exploration and Development, (3): 289–293.

Zhao, Z., Zhong, N.N., Huang, Z.L. (2006) Hydrocarbon drainage conditions in carbonates and their influence on the lower limit of organic matter abundance in hydrocarbon source rocks. Geochimica, (2): 167–173.

Advances in Mineral Resources, Geotechnology and Geological Exploration – Rashid & Zhang (eds)
© 2023 Copyright the Author(s), ISBN 978-1-032-33377-9

Treatment for highway tunnel crossing super-large karst cave: A case study

Xubin Liu
Department of Civil and Environmental Engineering, Shantou University, Shantou, Guangdong, P.R. China; Third Engineering Co., Lth of CCCC Fourth Highway Engineering Co., Lth, Beijing, P.R. China

Weiqiang Lin*
Department of Civil and Environmental Engineering, Shantou University, Shantou, Guangdong, P.R. China

Yanning Wang
Department of Civil and Environmental Engineering, Shantou University, Shantou, Guangdong, P.R. China; State Key Laboratory for Geomechanics and Deep Underground Engineering, China University of Mining and Technology, Xuzhou, Jiangsu, China

Yu Jin
Fujian Tunnel and Urban Underground Space Engineering Technology Research Center, Huaqiao University, Xiamen, Fujian, China

ABSTRACT: This paper presented a case study of super large Karst cave in highway tunnel construction, based on the generalized analysis of engineering geology and hydrogeology in the Taiping Tunnel project of Xuan'en- Hefeng Highway in Hubei Province, a variety of technical schemes for crossing the karst tunnel are compared and selected, so comprehensive approach with double-arch open-cut tunnel combing backfilling and underground excavation was proposed. This technique can reduce the backfilling volume and meet the requirements of safety and economy. Meanwhile, monitoring measurements are carried out for the project, and the results show that the tunnel settlement had experienced two deformation stages, the deformation in the rapid settlement period can account for more than 82.6% of the total deformation, and the deformation in convergent settlement period accounts for less than 20% of the total deformation. This technology can significantly improve construction efficiency and safety and effectively prevent major risks.

1 INTRODUCTION

Tunneling in karst terrain always faced difficulties including tunnel water inrush (Li 2013; Lin 2020; Wang 2019), mud inrush, tunnel roof collapse (Park 2020; Wang 2016), ground collapse[6] and other disasters (Lv 2020) which will cause serious economic losses. Therefore, it is necessary to further study the disposal scheme. In this paper, relevant research work is carried out and a new scheme of karst cave backfilling treatment is put forward. The practice proves that the scheme is feasible and provides examples for similar projects.

1.1 *Project description*

The Taiping tunnel's maximum buried depth of the left line is 137m, the total length is 655m, and the starting and ending pile number is ZK56+840~ZK57+495. The maximum buried depth of the right line is 145m, the total length is 658m, and the number of starting and ending piles is YK56+812~YK57+470.

*Corresponding Author: 21wqlin@stu.edu.cn

DOI 10.1201/9781003319412-29

1.2 *Karst cave geological conditions*

Karst cave of Taiping Tunnel: Left line ZK57+020~ZK57+168, length of 148m. Right line YK57+002~YK57+129, length of 127m. The plane shape is shown in Figure 1.

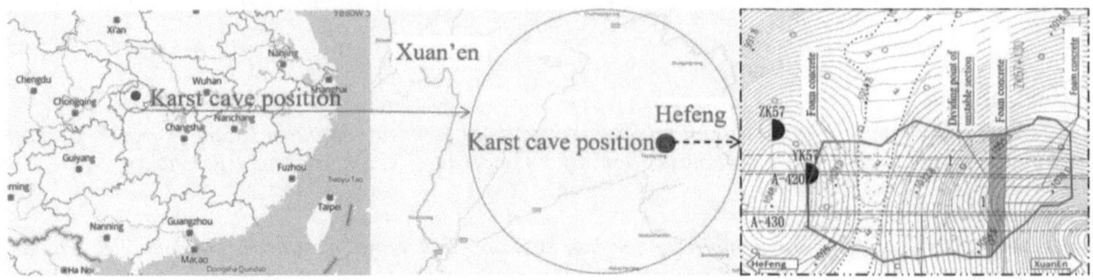

Figure 1. Plane relationship between Taiping.

Stalactites are distributed linearly along the joint fissure plane in the partial dome shown in Figure 2.

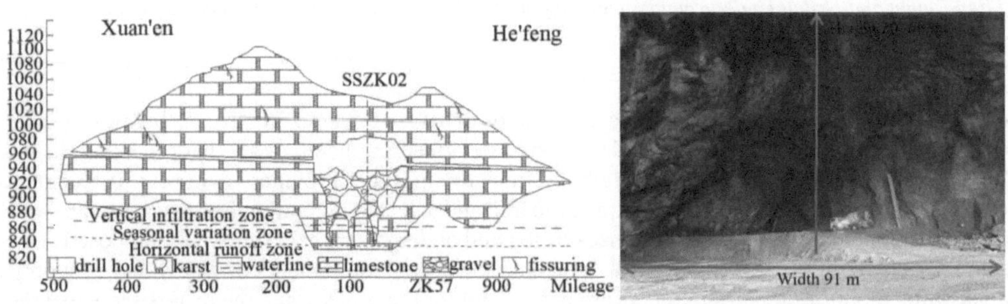

Figure 2. Karst cave of Taiping tunnel.

1.3 *Textual hydrogeology*

According to the hydrogeological research report, the tunnel is located in the underground river recharge and runoff area, and the catchment area is small. The water level during the flood period is about 860m, which will not cause large water flow through the tunnel body (elevation is about 940m).

2 METHODS SOLUTION

2.1 *Advance geological prediction*

The detection results of TRT6000 in the left line tunnel are shown in Figure 3.

2.2 *Treatment design*

After comparing several schemes from the angles of economy, technical feasibility, and construction difficulty, the scheme of double arch open tunnel + backfill underground excavation is finally determined.

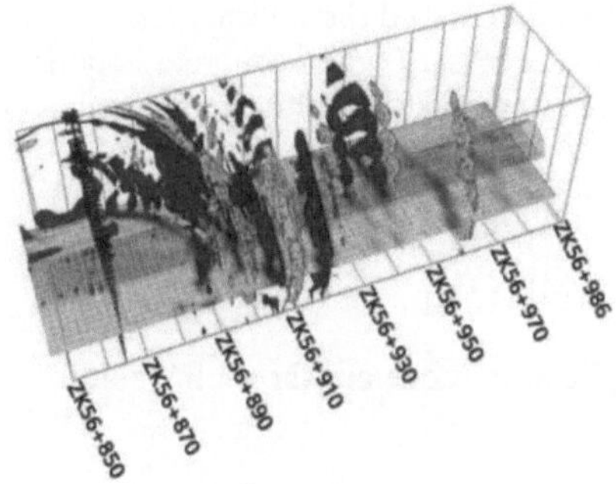

Figure 3. Three-dimensional view of ZK56+864~ZK56+986 section of the left line.

2.3 *Construction plan*

The construction is carried out in the sequence of backfilling of dark cave section, construction of dark cave section and construction of open cave section, backfill section as shown in Figures 4(a) and 4(b).

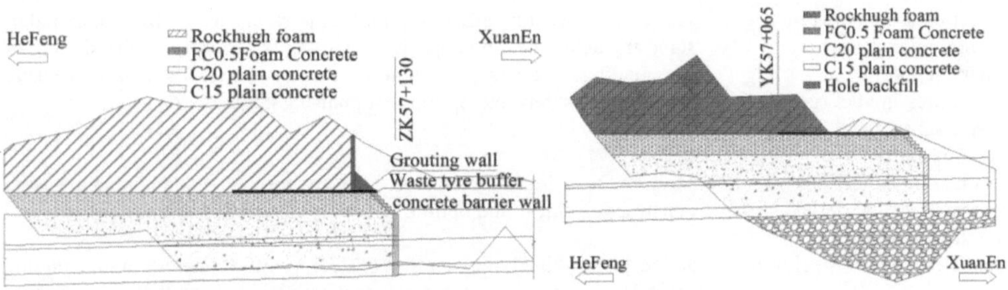

Figure 4. Profile of backfilling: (a) left line and (b) right line.

After the backfilling of FC0.5 foamed concrete is completed, a grouting retaining wall is built above it to prevent the leakage of foam. The grouting retaining wall is equipped with sensors to monitor the internal pressure and ensure the filling effect of the polymer foamed resin layer. As shown in Figure 5(a) safety shed shall be set in the opening area to ensure construction safety, as shown in Figure 5(b).

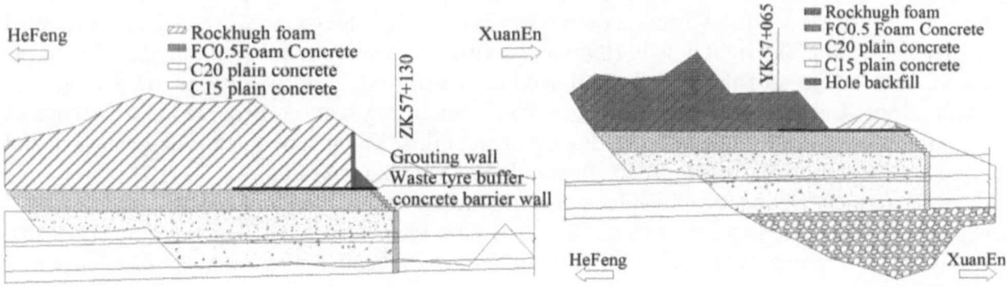

Figure 5. (a) Profile of retaining wall. (b) Profile of safety protection shed frame.

Double-layer secondary lining: (1) 30cm thick sand cushion is applied in the open tunnel section, and the last gravel layer must be levelled. (2) 150cm foundation is adopted in the open tunnel section: the reserved position of deformation joints is consistent with the foundation, and the steel bars are embedded in the open tunnel foundation and the internal and external arches of the open tunnel. (3) Vibration and compaction are necessary when concrete is poured inside the arch of the open cave. (4) For pouring the tires hanging outside of the open tunnel, a 20T crane is used to manually coordinate the tire installation.

3 TUNNEL SAFETY ASSESSMENT

Monitoring items and their application scope are shown in Table 1.

Table 1. Monitoring item form.

Monitoring project	Scope of application
Geological observation of palm surface	Preliminary judgment of the surrounding rock ahead and the stability of the tunnel structure.
Displacement of cavity top	To judge the stability of cavity top and tunnel structure; provide reasonable supporting time for secondary lining; judge the rationality of the initial support design and construction method to guide the design and construction.
Convergent displacement of lining	
Vault sink	
Arch feet sank	
Contact pressure between a surrounding rock and primary support	Judge the supporting effect of initial support on surrounding rock; understand the actual bearing capacity of two-layer support and the sharing of surrounding rock pressure; monitor the stress of tunnel structure to ensure construction safety and optimize supporting parameters.
Contact pressure between primary support and secondary lining	
The stress of shotcrete in primary anchorage	Monitor the change of internal stress of support and second lining to ensure construction safety and optimize support parameters.
Secondary lining stress	
The internal force of steel arch	Judge the combined support effect of steel arch and concrete on surrounding rock; The actual working state of steel arch depends on the specific situation to decide whether to take reinforcement measures; Judge the bearing capacity of initial support, ensure construction safety and optimize support parameters.

Dark hole pile: ZK57+020~ ZK57+ 130 to YK57+002~YK57+111, with a total length of 148m. open tunnel pile: ZK57+130~ZK57+168 to YK57+111~YK57+129, with a total length of 127m. According to the specification, a monitoring section should be arranged every 3-5m in the karst section of the dark cave. The left section of the dark cave is 110m long, and 23 sections are arranged with a spacing of 5m. The right hole is 109m long, with each section spacing of 5m and 23 sections arranged. The left hole of the open tunnel section is 38m long, and 9 sections are arranged at intervals of 5m. The right hole is 18m long, with 5 monitoring sections arranged at a distance of 5m. The monitoring items include dome displacement, initial lining convergence deformation, and arch foot settlement displacement. A schematic diagram is shown in Figure 6.

The measurement frequency of displacement around tunnel lining: 1 to 15 days every day, 16 to 1 month every two days, 1 to 2 times a week in 1 to 3 months, 1 to 3 times a month after 3 months.

As shown in Figures 7(a) and 7(b), the settlement of vault and ground surface appeared respectively.

It can be seen from Figure 7(a) that settlement can be divided into two stages: the first stage is the rapid settlement stage, mainly 0-60d after excavation, which has a high settlement rate and rapid deformation development, accounting for more than 82.6% of the total deformation. The second stage is the settlement convergence stage, which occurs 60 days after excavation, and the settlement in this stage accounts for less than 20% of the total deformation. The monitoring results

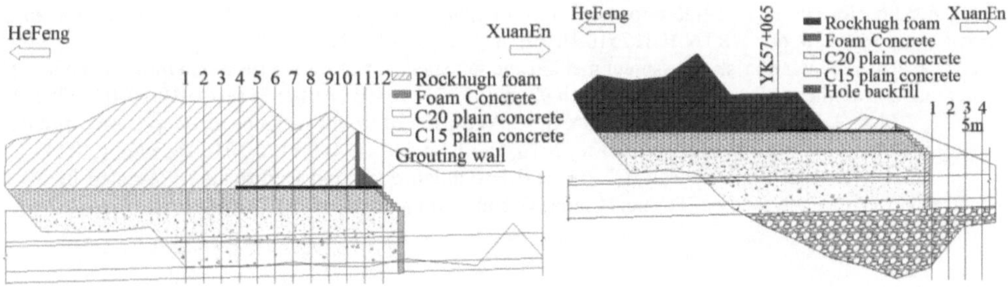

Figure 6. Layout profile of monitoring section: (a) Dark cave segment (b) Open tunnel segment.

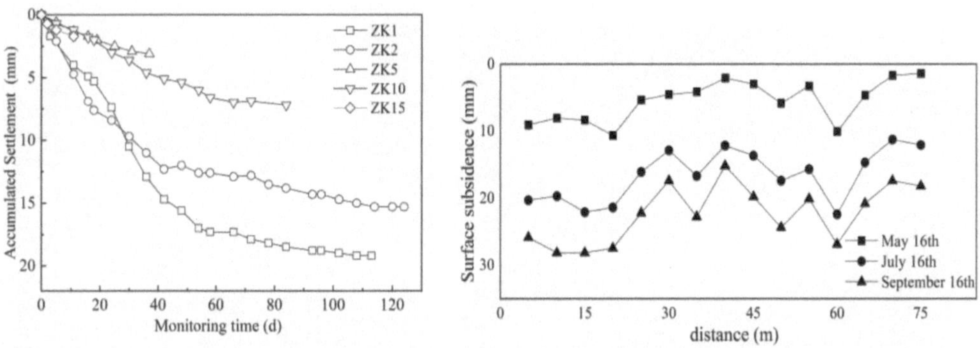

Figure 7. (a) Cumulative settlement at different measuring points. (b) Surface subsidence at different times.

show that the underground excavation scheme of double arch open cave + super-large karst cave backfilling is economical, feasible, and safe.

4 CONCLUSIONS

(1) The construction scheme of a double-arch open tunnel + backfill underground tunnel can reduce safety risks during construction while having less backfill and lower cost. At the same time, the tire filling as a buffer layer can mitigate the impact of possible rock falls, greatly reducing the risk during operation. (2) The tunnel settlement goes through two deformation stages, the deformation in the rapid settlement stage accounts for more than 82.6% of the total deformation, and the deformation in the convergent settlement stage accounts for less than 20% of the total deformation. (3) Safety risk assessment can control the safety risks of highway construction, strictly prevent the occurrence of major production safety accidents, reduce casualties and economic losses, and ensure the safety of highway tunnel construction.

REFERENCES

Li, S. C. et al. Risk assessment of water inrush in karst tunnels based on attribute synthetic evaluation system. Tunn Undergr Sp Tech 38, 50–58, doi:10.1016/j.tust.2013.05.001 (2013).

Lin, C. J. et al. A new quantitative method for risk assessment of water inrush in karst tunnels based on variable weight function and improved cloud model. Tunn Undergr Sp Tech 95 (2020).

Lv, Y. X., Jiang, Y. J., Hu, W., Cao, M. & Mao, Y. A review of the effects of tunnel excavation on the hydrology, ecology, and environment in karst areas: Current status, challenges, and perspectives. J Hydrol 586 (2020).

Park, D. & Michalowski, R. L. Three-dimensional roof collapse analysis in circular tunnels in the rock. Int J Rock Mech Min 128, doi: ARTN 10427510.1016/j.ijrmms.2020.104275 (2020).
Wang, X. T. et al. An interval risk assessment method and management of water inflow and inrush in course of karst tunnel excavation. Tunn Undergr Sp Tech 92, doi:ARTN 10303310.1016/j.tust.2019.103033 (2019).
Wang, Y. C., Jing, H. W., Zhang, Q., Luo, N. & Yin, X. Prediction of Collapse Scope of Deep-Buried Tunnels Using Pressure Arch Theory. Math Probl Eng 2016, DOI: Art 262817410.1155/2016/2628174 (2016).
Zhu, W., Zhang, Y., Liu, Z. K. & Zhu, Q. Pre- and Postcollapse Ground Deformation Revealed by SAR Interferometry: A Case Study of Foshan (China) Ground Collapse. J Sensors 2020 (2020).

Advances in Mineral Resources, Geotechnology and Geological Exploration – Rashid & Zhang (eds)
© 2023 Copyright the Author(s), ISBN 978-1-032-33377-9

Simulation study on the tensile performance of connection joints between self-locking and unlocking steel structure modules

Xinpeng Tian
Urban and Rural Construction College, Agriculture University of Hebei, Baoding, China

Xingwang Liu*
Urban and Rural Construction College, Agriculture University of Hebei, Baoding, China
Department of Civil Engineering, Tianjin University, Tianjin, China

Yan Liu, Yang Chen & Yuntong Liu
Urban and Rural Construction College, Agriculture University of Hebei, Baoding, China

ABSTRACT: For the connection joints between self-locking and unlocking steel structure module elements, the tensile model of the joint is established by ABAQUS finite element software. Through parametric analysis, the stress characteristics of corner parts and plug-ins when the joint is in tension and the influence of plug-in neck width on the tensile bearing capacity of the joint are studied. The results show that the connection joints between self-locking and unlocking steel structure module elements have good bearing capacity; the important factor affecting the bearing capacity of the joint is the plug neck width. Reducing the plug neck width is conducive to improving the yield-bearing capacity of the joint.

1 INTRODUCTION

Modular steel structure building is a modern architectural form. The conventional architectural form is divided into prefabricated and fully decorated in the factory, transported to the site, and assembled into an integral 3D modular unit (Goodier 2007; Lawson 2008). As a high-end product of prefabricated buildings, MSB has outstanding advantages such as fast construction speed, excellent project quality, saving manpower and reducing environmental impact, and has broad development prospects (Ferdous 2019; Kamali 2016; Lopez 2016). Modular steel structure buildings are mainly used for projects with small construction site space, no impact on the surrounding environment, and high construction speed requirements. Examples include urban apartments and hotels, dormitories, and hospitals for epidemic prevention (Annan 2009). Modular steel buildings are widely used in Japan, North America, and Sweden, and are gaining acceptance in China and Australia (Lawson 2012). Modular buildings have "multi-beam and multi-column" connection features at various locations, which is different from the "one-column and two-beam" connection features in traditional buildings (William 2019). The structural model, mechanical efficiency, and performance level of the connection between modules play a key role in the convenience of construction site assembly and the safety of the overall structure, and the connection between modules has become the focus of researchers (Chen 2017; William 2018; Srisangeerthanan 2020). At present, the connection between modules of MSB mostly adopts welding and bolt connection, and a few adopt prestressed connection. The welding method has a large workload on-site, so it is difficult to realize the four-sided welding of the upper and lower column base plates, and it is difficult to ensure the welding quality. Bolt connection has strict requirements for installation accuracy, and holes need to be made on beams or columns, resulting in the conflict between the installation process and the architectural

*Corresponding Author: liuxw@hebau.edu.cn

decoration of the module. Although the prestressed connection avoids on-site welding and reserved operation holes, the prestressed connection construction is complex and requires high construction requirements. In addition, at present, most studies have focused on semi-structured modular unit systems (specimens with structural member length, i.e., beams and columns considering inflection points) (Chen 2017; Deng 2018; Lee 2017; Sanches 2018). The failure mainly occurs at the beam edge or column end section. There is still obvious excess bearing capacity in the connection area, so the bearing capacity and calculation method of the connection between modules cannot be obtained.

To overcome the above limitations, an innovative self-locking and unlocking connection between steel structure module elements is introduced in this study. As shown in Figure 1, its detailed structure is shown in Figure 2. This kind of joint can solve the problem of corner connection of module units in modular buildings. The stress of the joint is clear, and it can resist horizontal shear and vertical tension. The structure is simple and reasonable, the installation is fast and convenient, and it can be recycled, disassembled, and replaced. However, there is little research on this type of joint at present.

In this paper, the finite element analysis of the connection joints between the self-locking and unlocking steel structure module elements is mainly carried out to investigate the stress distribution, bearing capacity mechanism, and failure form of the joint under the action of tension, as well as the influence of different parameters on the tensile bearing capacity and stress state of the joint, and the structural suggestions for the joint are put forward.

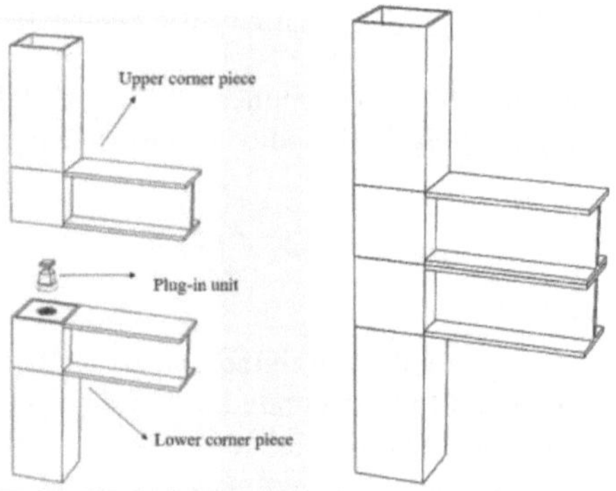

(a) Disassemble (b) The whole

Figure 1. Connection nodes between units of self-locking and unlocking steel structure module.

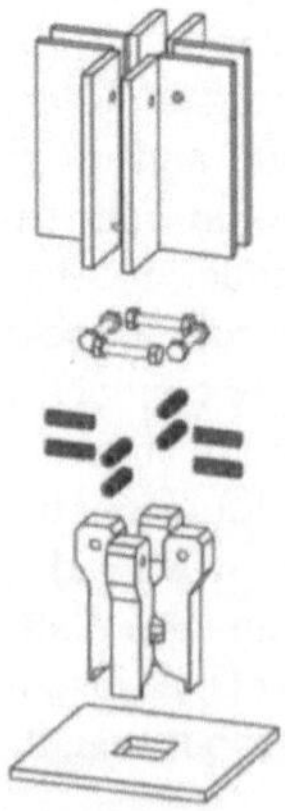

Figure 2. Internal detail of upper corner.

2 FINITE ELEMENT ANALYSIS

2.1 *Model establishment*

The finite element model is established by using the finite element analysis software ABAQUS. The model numbers are T-1~T-5. The components of the model mainly include a module column, lower corner piece, upper corner piece, and plug. Each component is simulated by c3d8r solid element and structured grid division technology. The steel adopts a linear strengthened elastic-plastic constitutive model, and the welding parts such as corner piece and module column adopt binding contact. The model is shown in Figure 3.

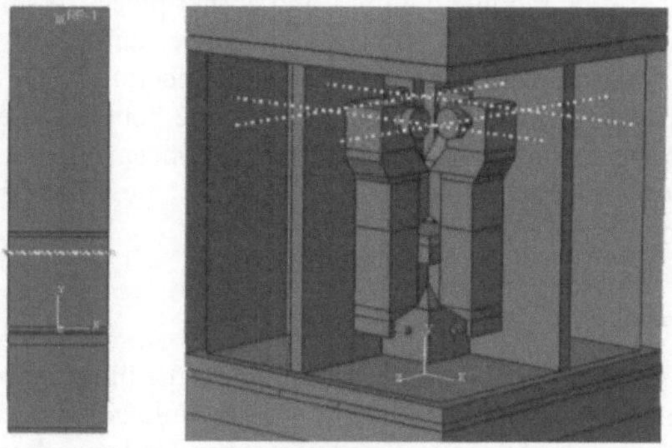

Figure 3. Finite element model.

The mesh division of the node connection core area of the node finite element model is dense, and the mesh division of other areas is sparse. The mesh division of the node model is shown in Figure 4.

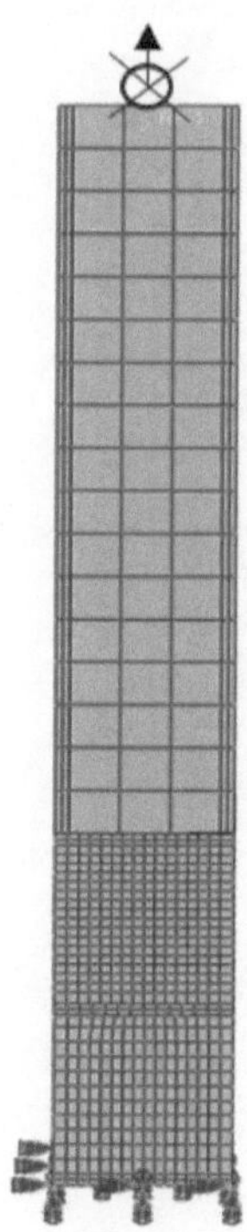

Figure 4. Grid division, boundary conditions, and load application.

2.2 *Boundary conditions and load application*

As shown in Figure 4, in the tensile finite element model of the connecting node between the self-locking and unlocking steel structure module elements, the reference point RP1 is set at the upper module column end, and the reference point RP1 is coupled with the module column section, and the lower module column end restricts three-way translation and three-way rotation. The load is applied in the form of axial displacement through the reference point RP1.

2.3 *Component assembly and interaction*

Tie constraints are used between module columns and corner pieces to simulate the welding connection between upper and lower module columns and upper and lower corner pieces. The top plate of the lower corner and the bottom surface of the connecting plate, the bottom plate of the upper corner and the top surface of the connecting plate, the bottom plate of the plug-in and the connecting plate, and the surfaces of all parts in the corner adopt surface contact. The normal action is hard to contact and allows separation after contact, and the tangential action is a penalty function with a friction coefficient of 0.3.

2.4 *Constitutive relation*

The constitutive relation of steel studied in this paper adopts the linear strengthened elastic-plastic model. According to the constitutive relationship of the module column, the yield strength is fy= 354MPa, the tensile strength is fu = 508MPa, the plastic strain corresponding to the ultimate strength is 0.00208, and the elastic modulus is 202906.99MPa; According to the constitutive relationship of corner pieces, the yield strength is fy = 344MPa, the tensile strength is fu = 492MPa, the plastic strain corresponding to the ultimate strength is 0.00192, and the elastic modulus is 174065.15MPa.

2.5 *Finite element results*

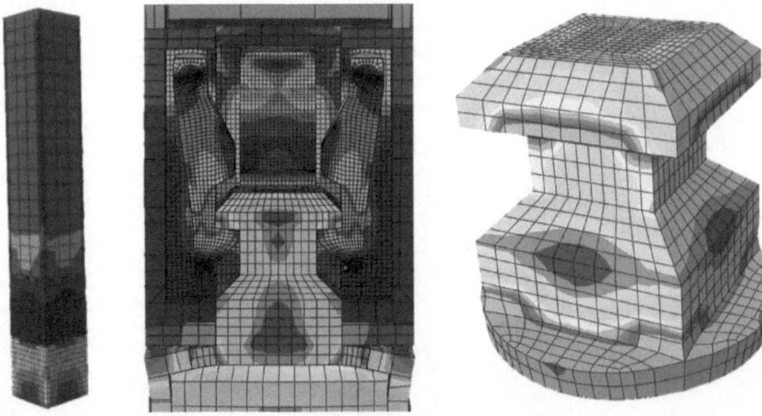

(a)External integral (b) internal integral section (c) plug

Figure 5. Finite element stress nephogram.

2.6 *Parametric analysis*

2.6.1 *Influence of plug neck width*

Through the finite element analysis, the ultimate bearing capacity and yield bearing capacity of the connection node model between the corner self-locking and unlocking steel structure module elements are obtained. The yield-bearing capacity is determined by the overall yield method. According to the node load-displacement curve, the intersection of the extension line of the initial linear segment and the tangent of the nonlinear segment whose slope is approximately constant is defined as the yield-bearing capacity of the model. The comparison results are shown in Figure 6 and listed in Table 1.

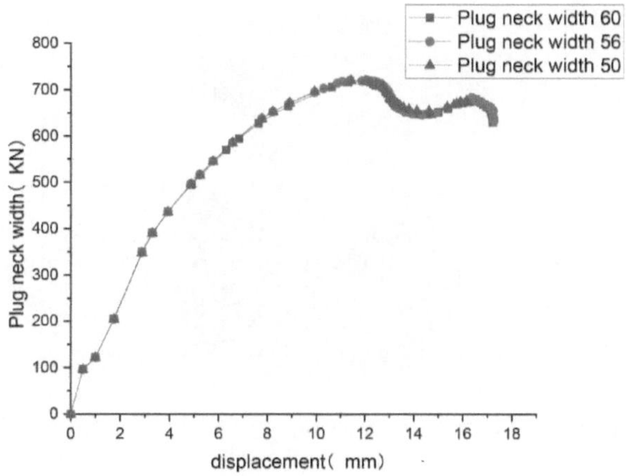

Figure 6. Comparison of the load-displacement curve.

Table 1. Summary of model parameters and bearing capacity.

Model number	Plug neck width (mm)	Bolt height (mm)	Yield load (KN)	Ultimate bearing capacity (KN)
T-1	60	16	570.7827	716.712
T-2	56	16	576.9924	720.615
T-3	50	16	577.0834	720.062

With the increase of the plug neck width, the tensile ultimate bearing capacity of the joint decreases, and with the increase of the plug neck width, the effect of reducing the bearing capacity becomes less and less significant. Considering the economy, it is recommended to select the neck width of 56mm. Under monotonic static load, the variation law of the stress nephogram of each model is basically the same. Take model T-2 as an example. Figure 7 shows the stress nephogram

Figure 7. Cloud diagram of model T-2 stress.

of model T-2 under static load. The maximum stress of the model occurs in the snapshot of the upper corner piece, the plug bottom plate, and the top plate of the lower corner piece. The stress of the connecting plate and square steel pipe is relatively small. Under the state of ultimate bearing capacity, it is shown as the tensile yield of the top plate of the lower corner piece and the tensile yield of the bottom plate of the rotating piece.

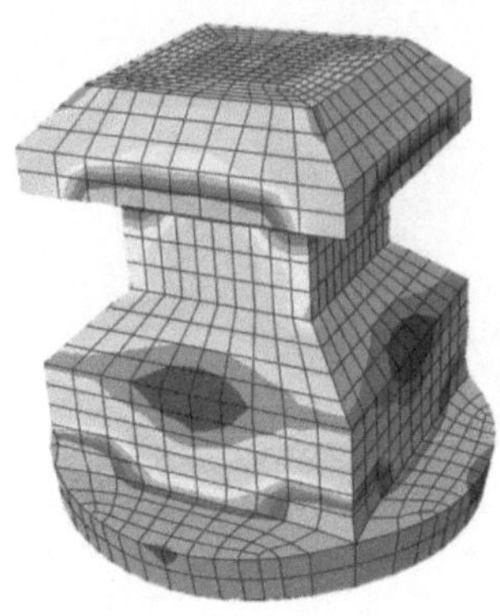

Figure 8. Cloud diagram of model T-2 plug deformation.

Extract the plug deformation cloud diagram as shown in Figure 8. The bulge deformation occurs around the plug and the bottom plate under the action of axial tensile load. The reason for the bulge deformation is that the tensile force is transmitted to the plug through the upper corner piece, resulting in the bulge around the plug and the bottom plate. In order to deeply study the deformation law of the plug, the deformation law on the central axis parallel to the y-axis along one side of the surface is extracted, as shown in Figure 9. The key points are the two points of the ends on both sides of the corresponding central axis and the edge of the hole of the corresponding central axis. Under the same other conditions, the plug thickness increases. Compared with model T-1, the deformation of the two points around the edge of the corresponding central axis plug decreases by 12.5%. With the increase of the plug neck width, the vertical deformation around the plug neck width decreases. The reason is that the plug neck width increases and the joint stiffness increases. Under the same load, the deformation of the joint decreases relatively.

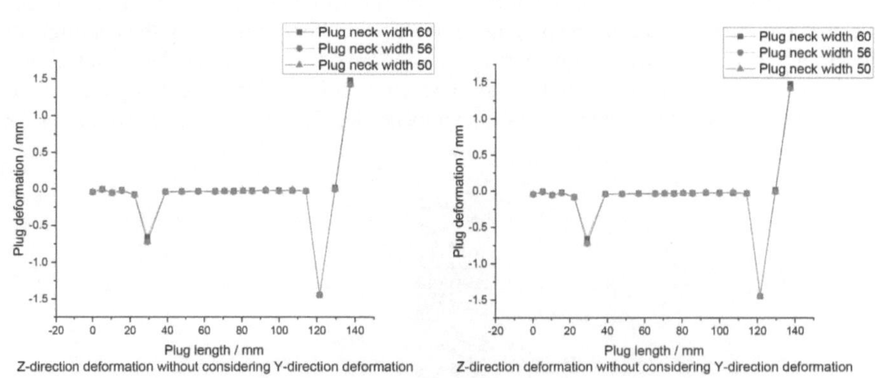

Figure 9. Deformation curve of the plug when the width and thickness of the plug neck change.

2.6.2 *Influence of the thickness of the top plate of the lower corner*

Through the finite element analysis, the yield-bearing capacity and ultimate bearing capacity of the connection node between the self-locking and unlocking steel structure module elements under different top plate thicknesses of the upper corner when the top plate thickness of the lower corner is 20mm are obtained, as shown in Table 2. The analysis shows that the ultimate bearing capacity of the joint increases with the increase of the top plate thickness of the lower corner, but the increase range is less than that when the plug neck width increases.

As the top of the lower corner piece bears the tensile force transmitted from the upper module column and transmits the tensile force to the plug, the top plate swells due to tensile deformation. The analysis shows that with the decrease of the top plate thickness of the upper corner, the deformation of the upper corner increases, the plate never yields to nearly yield, and the section plastic development area gradually expands, so as to make full use of the material. Therefore, considering the economy and the bearing capacity of the joint, it is suggested that the thickness of the top plate of the upper corner piece should be 16mm.

Table 2. Model parameters and bearing capacity (roof thickness of different lower corners).

Model number	Plug neck width (mm)	Top plate thickness of lower corner (mm)	Yield load (KN)	Ultimate bearing capacity (KN)
T-1	60	16	570.7827	716.712
T-4	60	20	611.3328	718.788

2.6.3 *Influence of shaft diameter*

Table 3 shows the tensile bearing capacity of joints under different shaft diameters when the thickness of the top plate of the lower corner is 20mm and the thickness of the bottom plate of the upper corner is 16mm. The analysis shows that with the increase of the shaft diameter, the tensile bearing capacity of the joint increases, and the deformation of the shaft decreases. Due to the tensile force transmitted by the upper corner, the middle part of the rotating shaft diameter is stressed, resulting in bending at both ends. Considering the economy and bearing performance of the joint, it is suggested that the diameter of the rotating shaft should be 20mm

Table 3. Model parameters and bearing capacity (different shaft diameters).

Model number	Plug neck width (mm)	Shaft diameter (mm)	Yield load (KN)	Ultimate bearing capacity (KN)
T-1	60	20	570.7827	716.712
T-5	60	25	583.4622	725.624

2.6.4 *Comparison of finite element results*

Through the comparison with the finite element simulation of the tensile joint of the corner rotating module element, the stress nephogram in Figure 10 is obtained. By quoting the data in the author's article (Wang 2021), the bearing capacity value is in good agreement with the author's simulation results. Table 4 shows the comparison of bearing capacity values. The error of ultimate bearing capacity between the model and the test is about 2.3%.

Table 4. Comparison of bearing capacity.

Category	Yield load / KN	Ultimate bearing capacity / KN
Original data	405.1250	528.261
Contrast	417.2953	540.305

Figure 10. Comparison of stress nephogram.

3 CONCLUSIONS AND PROSPECT

This paper mainly focuses on the numerical simulation of a structural form of spliced joints between modules of modular buildings applied to a column load-bearing modular structural system. However, for a new structural form of joints, there are still deficiencies in this paper, which need to be supplemented and developed in the follow-up research work. The following conclusions and Prospects are as follows:

(1) The numerical simulation lacks theoretical derivation and test verification. In the next step, it is planned to prepare experimental research with theoretical support.
(2) In this paper, in the self-locking and unlocking steel structure module element connection joint simulation research, the specimen only selects the module corner connection part, and the constraint of the node module beam is slightly different from the actual situation. In the subsequent experimental research, the overall structure test of the column-bearing modular structure system can be considered to obtain a more accurate and comprehensive performance of splicing joints between modules.
(3) The tensile bearing capacity of spliced joints of traditional modular buildings is about 400KN, the bearing capacity of connecting joints between self-locking and unlocking steel structure modules is increased by about 79%, and the final plug top plate reaches the yield state. The connecting joints between self-locking and unlocking steel structure modules have a high tensile bearing capacity and good deformation performance, which meet the requirements of connecting joints of modular buildings, It is proved that the connection joints between self-locking and unlocking steel structure modules have good static performance.
(4) The tensile bearing capacity of connecting joints between self-locking and unlocking steel structure modules is related to the plug neck width, which is greatly affected and decreases with the increase of plug neck width.
(5) The stress concentration at the connection between the card shoot and the plug is obvious, which can be considered to be thickened.

REFERENCES

Annan, C. D. MA Youssef, & Naggar, M. E. (2009). Experimental evaluation of the seismic performance of modular steel-braced frames. ENGINEERING STRUCTURES.

Annan, C. D., M. A. Youssef & M.H. El Naggar. (2009). Seismic Vulnerability Assessment of Modular Steel Buildings. Journal of Earthquake Engineering (8)

AW Lacey, Chen, W., Hao, H., & Bi, K. (2019). Review of bolted inter-module connections in modular steel buildings. Journal of Building Engineering.

Bi, J. X., Wang, H. P., Li, Y. M., Ding, Y., & Dai, X. M. (2020). State-of-the-art on the connection in modular steel construction. Jianzhu Jiegou Xuebao/Journal of Building Structures, 30(3), 33–40.

Chen Zhihua, Liu Jiadi, Yu Yujie, & Rengjing Yan. (2017). Experimental study of an innovative modular steel building connection – sciencedirect. Journal of Constructional Steel Research, 139, 69–82.

Chen, Z., Liu, J., & Yu, Y. (2017). Experimental study on interior connections in modular steel buildings. Engineering Structures, 147(Sept. 15), 625–638.

D Lopez, & Froese, T. M. (2016). Analysis of costs and benefits of panelized and modular prefabricated homes. Procedia Engineering, 145, 1291–1297.

Deng, E. F., Liang, Z., Yang, D., & Luo, Y. B. (2018). Seismic behavior and design of cruciform bolted module-to-module connection with various reinforcing details. Thin-Walled Structures, 133, 106–119.

Deng, E. F., Zong, L., Ding, Y., Dai, X. M., Lou, N., & Chen, Y. (2018). Monotonic and cyclic response of bolted connections with welded cover plate for modular steel construction. Engineering Structures, 167(JUL.15), 407–419.

Goodier, C., & Gibb, A. (2007). Future opportunities for offsite in the UK. Construction Management & Economics, 25(6), 585–595.

Kamali, M., & Hewage, K. (2016). Life cycle performance of modular buildings: a critical review. Renewable & Sustainable Energy Reviews, 62, 1171–1183.

Lacey, A. W., Chen, W., Hao, H., & Bi, K. (2017). Structural response of modular buildings – an overview. Journal of Building Engineering, 45–56.

Lawson, R. M., Asce, M., Ogden, R. G., & Bergin, R. (2012). Application of modular construction in high-rise buildings. Journal of Architectural Engineering.

Lee, S., Park, J., Shon, S., & Kang, C. (2018). Seismic performance evaluation of the ceiling-bracket-type modular joint with various bracket parameters. Journal of Constructional Steel Research, 150(NOV.), 298–325.

R. M. Lawson, M. P. Byfield, S. O. Popo-Ola & P. J. Grubb. (2009). Robustness Of Light Steel Frames And Modular Construction. Construction materials (CM1), DOI:

Rafaela, S., Oya, M., & Brent, R. (2018). Experimental investigations of vertical post-tensioned connection for modular steel structures. Engineering Structures, 175, 776–789.

Seungjae, L., Jaeseong, P., Euishin, K., Sudeok, S., Kang, C., & Hosoon, C. (2017). Verification of the seismic performance of a rigidly connected modular system depending on the shape and size of the ceiling bracket. Materials, 10(3), 263.

Srisangeerthanan, S., Hashemi, M. J., Rajeev, P., Gad, E., & Fernando, S. Review of performance requirements for inter-module connections in multi-story modular buildings. Journal of Building Engineering, 28.

Wahid Ferdous, Yu Bai, Tuan Duc Ngo & Priyan Mendis. (2019). New advancements, challenges, and opportunities of multi-story modular buildings – A state-of-the-art review. Engineering Structures, 183, 883–893.

Wang Xiaodun, Cong Ziyi, Liu Jiadi (2021). Study on the tensile behavior of rotary inter-module connection joint Building Structure 51(13): 93–101.

Advances in Mineral Resources, Geotechnology and Geological Exploration – Rashid & Zhang (eds)
© 2023 Copyright the Author(s), ISBN 978-1-032-33377-9

Study on fine combination method of isolated breakpoints in $N1^{\#}$ fault area of oilfield A

Guangyi Li*
The Sixth Oil Production Plant of Daqing Oilfield, Daqing, China

ABSTRACT: The study of scattered small faults represented by isolated breakpoints is an important research direction in the middle and later stages of oilfield development. Although these faults have a small fault distance and short extension, they control the distribution of oil and water and the connection between oil and water wells and play an extremely important role in oilfield development. The occurrence and combination of scattered small faults are difficult to determine because there are few wells drilled and the three-dimensional seismic reflection signal is weak. In this paper, the distribution characteristics and formation principle of the remaining isolated breakpoints after the reorganization of large faults are systematically analyzed. Through the plane and section display of breakpoint data in two-dimensional drawing software and three-dimensional visual and structural modeling in modeling software, the research on small faults is strengthened, and the best combination method of isolated breakpoints is discussed.

1 INTRODUCTION

Since the infilling adjustment of oilfield A, the recognition of faults has been gradually deepened and the recognition results have been refined. At present, the occurrence and shape of large-scale faults have been basically determined, and the combination rate of breakpoints in the whole oilfield is only 74.7%. The insufficient utilization of scattered fault data represented by isolated breakpoints not only affects the fineness of structural research but also hinders the guidance of its achievements to oilfield production. Due to the irregularity of isolated breakpoints, relatively large faults will be more complex and multiple solutions in the combination process. There are some limitations in using the combination method of the original large faults. Taking the $N1^{\#}$ fault area as an example, this paper combines the uncombined isolated breakpoints in this area. In the combination process, the method of exploration and combination is adopted, and various data are comprehensively used to complete the identification of scattered small faults and improve the combination rate of breakpoints.

2 BLOCK OVERVIEW

Oilfield A is located at the north end of Daqing placanticline, with flat terrain. Structurally, it belongs to a tertiary structural belt of Daqing placanticline, which is an asymmetric short-axis anticline gas cap reservoir controlled by structure. The whole structure is cut by two groups of large faults ($N2^{\#}$and $N1^{\#}$) extending in the northwest direction and divided into three large blocks in the south, middle and North with different areas. The high point of the trap structure is −614.7m, the closing contour is −700m, the closing height is 85.0m and the closing area is 25.9km^2. The fault is mainly formed in the sedimentary period from the Mingshui formation to the Sifangtai formation in the late Yanshan movement. It is mainly a tensile fault. The shallow stratum fault is

*Corresponding Author: liguangyi1@petrochina.com

 DOI 10.1201/9781003319412-31

more developed than the oil layer fault, and the fault inheritance between each layer is good (Dong 1992). The strike is mainly NW, NNW, and NE; The extension length is 0.189 ~ 4.29km, and the fault dip angle is 45°~60°. The plane distribution of faults has obvious zoning, and the isolated breakpoints are concentrated. The fault distance is between 0.6m and 192.5m. The fault distance of the same fault is large in the middle and small at both ends; On the section, the upper fault distance is large and the lower fault distance is small.

N1# fault area is located in the north of the south block of oilfield A, with 4 large faults (N1#, N2#, N4#, N3#) developed. The west end of N1# fault intersects with N4# fault in the south, forming a local graben structure; The central to Eastern ends of N1# fault and N5# fault in the South form a graben structure; N3# fault and N5# fault are arranged in steps. The east end of the N1# fault and N11# fault in the north are arranged in a ladder shape. The control area of the fault area is 6.13km^2, and the combination rate of breakpoints in this area is 85.7%.

3 STUDY ON COMBINATION METHOD OF SCATTERED SMALL FAULTS

3.1 *Formation mechanism and characteristics of scattered small faults*

Scattered small faults can be divided into two types according to their formation mode:

One is the associated small fault produced during the formation of the Sartu anticline. In the process of anticline formation, a series of faults with different sizes and occurrences are associated, among which some faults with relatively small scale are mostly distributed between the two wings or some large faults in the core of the anticline, which are similar to the occurrence and shape of large faults, but the extension range and fault distance are small, most of them do not intersect with large faults and are limited by large faults. The analysis shows that the formation period of these small faults is later than that of large faults.

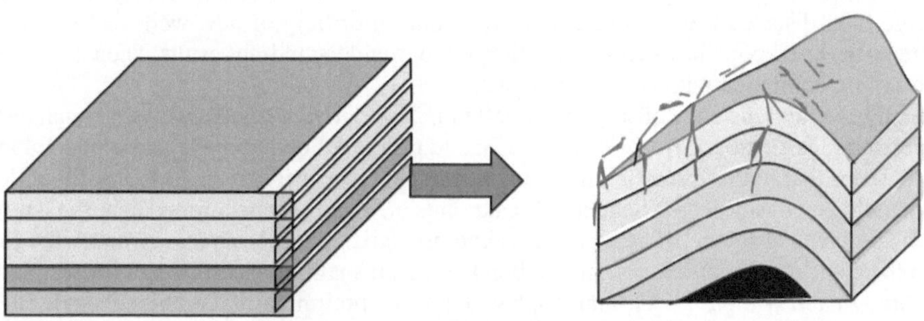

Figure 1. Schematic diagram of fault formation of the anticline.

The other is derived from small faults produced in the formation of main (large) faults. In the same period or later stage of the formation of large faults, due to the up and down dislocation of the two walls of the fault, the internal stress field in a certain area changes, resulting in a series of derived small faults and small folds. For example, due to the large fracture bandwidth of the main fault and the loose material structure of the rock stratum, the hanging wall rock stratum close to the fault plane drops faster under the action of gravity, while the rock stratum far away from the fault plane drops slower. Therefore, tension is generated locally in the hanging wall rock stratum of the main fault. Under this tension, the rock stratum develops into a derived small fault after generating small cracks, such small faults are mostly located in shallow strata. On the plane, they are mostly concentrated near the main fault, sometimes appear in groups, and the scale is smaller than that of the previous type. Most of the sections are connected with the main fault to form a zigzag shape (Sun 1995).

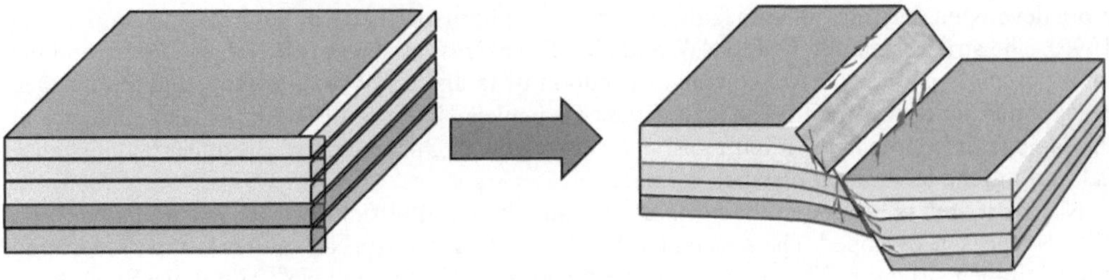

Figure 2. Schematic diagram of the main fault and derived fault.

3.2 *Thinking and theoretical basis of scattered small fault combination*

At present, there are three main breakpoint combination methods: using logging data to combine breakpoints, using seismic data to combine breakpoints, and using three-dimensional modeling software to combine breakpoints, but each method has some limitations. When combining large faults, due to their regularity, the three methods can well determine the scale and occurrence of large faults, but they have their own advantages and disadvantages in fault interpretation accuracy. However, most of the breakpoint fault distances encountered by scattered small faults are relatively small and the resolution of seismic data is low. Therefore, the seismic data interpretation method is no longer suitable for identifying small faults, while the three-dimensional modeling technology is more suitable for faults with concentrated distribution and obvious regularity and some correction work after the completion of fault combination. Therefore, for scattered small faults. This paper mainly focuses on combining faults based on logging data.

All breakpoint data encountered during drilling in a block are mixed with large and small fault distances. Firstly, large fault distance breakpoints with regional distribution characteristics should be selected. Generally speaking, their combination law is obvious and it is easy to form faults. Even large fault distance breakpoints encountered during drilling of new wells can be classified according to the breakpoint law according to the previous understanding results, Then the remaining breakpoints are combined with secondary faults.

Generally, the fault distance of scattered small faults is relatively small and the regional distribution characteristics are not obvious. Therefore, in addition to following the breakpoint combination principle, according to the formation mechanism and characteristics of small faults, the following aspects should be considered: ① Scattered breakpoints are mostly concentrated near the large fault layer; ② The break distance of scattered breakpoints varies slightly and sometimes irregularly; ③ When the plane combination of scattered breakpoints, it shall not pass through the well location drilled through and without breakpoint display; ④ For the preliminarily combined small fault, the positional relationship with the large fault should be verified, that is, the large fault limits the small fault, and the small fault can only connect with the large fault rather than stagger the large fault (An 2001).

4 CAUSES AND COMBINATION RESULTS OF ISOLATED BREAKPOINTS IN OILFIELD A

4.1 *Causes of isolated breakpoints*

Isolated breakpoints refer to scattered breakpoints that cannot be combined into a fault by using the technical means at that time. Breakpoint combination needs to meet two conditions: section and plane can be combined. There are at least two continuous sections on the section. According to the principle of "three points forming a plane", there must be more than three points on the plane to determine that the breakpoints can be combined. Through the preliminary analysis of all isolated breakpoints in the N1$^{\#}$ fault area, the reasons for the isolation of breakpoints are as follows:

① because the fault does not extend to the oil layer, this part of breakpoints are not combined to form isolated breakpoints according to the needs of production; ② Due to the technical restrictions on the formation of isolated breakpoints and combination of sections, it can not meet the technical constraints on the plane; ③ Incomplete well data can not be combined to form an isolated breakpoint; ④ Failure to combine breakpoints to form isolated breakpoints due to incorrect interpretation of breakpoint data; ⑤ Breakpoints that cannot be combined using existing technologies (Zeng & Hentz 2001).

Table 1. Breakpoint isolation reason questionnaire.

Serial number	project	Number (PCs.)	Cumulative number (PCs.)	Proportion (%)	Cumulative proportion (%)
A	Not extended to reservoir combination	63	63	35.2	35.2
B	Technical method restrictions not combined	48	111	26.8	62.0
C	Incomplete data not combined	22	133	12.3	74.3
D	Breakpoint interpretation error	2	135	1.1	75.4
E	other	44	179	24.6	100

4.2 *Combination methods and steps of isolated breakpoints*

Before the combination of isolated breakpoints, the preliminary statistics of these isolated breakpoints are carried out. From the statistical table, it is found that among the 179 isolated breakpoints, there are 139 isolated breakpoints developed in the upper stratum and 40 isolated breakpoints developed in the oil layer. There are 26 isolated breakpoints below 3M (including 3M) in the upper stratum, accounting for 18.7%; There are 41 breakpoints with a fault distance of 3m to 8m (including 8m), accounting for 29.5%; There are 72 breakpoints with a fault distance of more than 8m, accounting for N1 8%. There are 40 isolated breakpoints in the oil layer, 22 of which are below 3M (including 3M), accounting for 55%; There are 12 breakpoints with a fault distance of 3m to 8m (including 8m), accounting for 30%; There are 6 breakpoints with a fault distance of more than 8m, accounting for 15%.

Table 2. Breakpoint isolated distance range questionnaire.

Fault distance	The isolated breakpoint of the upper stratum		The isolated breakpoint of the reserve	
	Quantity (PCs.)	Percentage (%)	Quantity (PCs.)	Percentage (%)
≤3M	26	18.7	22	55
3~8M	41	29.5	12	30
>8M	72	51.8	6	15
Total	139	100	40	100

It can be seen from the above data that the isolated breakpoints with large fault distances are mostly distributed in the upper stratum, and the isolated breakpoints and fault distances of the oil layer are relatively small, which proves that the small fault represented by the isolated breakpoints of the oil layer is smaller, and the combination method for large faults will have great limitations.

The breakpoint combination of oilfield A has gone through five stages: in the 1960s, the structural form and macro fracture characteristics of oilfield A were preliminarily found out by using light spot seismic records combined with exploration and well data; in the 1980s, the structure of oilfield A was clarified through basic well pattern data, and the structural map was published; in the 1990s, after the well pattern was densified, the top structural map of nine oil reservoir groups in oilfield A was published; in 2003, through three-dimensional geological modeling, the top structure map of the small layer was formed; by 2008, oilfield a has achieved full coverage of 3D3C earthquake. At the same time, with the continuous densification of the well pattern, the fault combination method of oilfield a has been deepened, and a set of fault combination methods of well seismic combination under the condition of dense well pattern has emerged. At the same time, through the application of various mapping software, the fault combination has entered a new stage (Zhang & Chen 1983).

Due to the poor accuracy of seismic data in fault identification of small fault distance, the identification of small fault should be based on drilling data and refer to the general idea of seismic data. For faults with fault distance greater than 8m, the seismic results should be the main idea. For breakpoints below 8m, a two-dimensional logging combination is mainly used. In the two-dimensional combination of the original large fault, the step of section combination first and then plane verification is adopted. Due to the scattered distribution of scattered breakpoints and no obvious regularity, it is difficult to determine the occurrence of the fault on the section map. Therefore, for the combination of scattered breakpoints, the method of two-dimensional plane combination and then section verification and the drawing method of three-dimensional modeling are adopted.

The specific steps are as follows: ① Import all well numbers in this area into Shuanghu software; ② Create two new layers in the software, and import the combined and noncombined breakpoint data into the two layers respectively; ③ Make a preliminary plane combination of isolated breakpoints according to the principle of "three points into a plane"; ④ Verify the combined fault in the fault profile; ⑤ Determine the fault cutting depth in combination with seismic data. If it cannot be determined by the earthquake, it shall be determined according to the original manual experience; ⑥ In the modeling software, the breakpoint combination results are re-displayed and modeled in three dimensions to further identify and correct the combination results.

4.3 *Isolated breakpoint combination results*

Through the verification of two-dimensional plane combination and profile of all isolated breakpoints in the N1$^{\#}$ fault area, the repeated research of three-dimensional visualization and structural modeling analysis, the attribution of 179 scattered breakpoint data was re-identified and recombined, and 15 small faults were identified in this consensus. 128 breakpoints were combined, and the breakpoint combination rate was 95.9%, an increase of 10.2 percentage points.

5 CONCLUSION

1. The isolated breakpoint is not a single point, but a small fault with a smaller scale;
2. Combining isolated breakpoints by plane and then the section is more conducive to determining the occurrence of micro faults;
3. The micro fault after the oil layer combination can be further verified by using the dynamic data of oil and water wells.

REFERENCES

An P, Moon W M, Kalantzis F. Reservoir characterization using seismic waveform and feed forward neural net-works [J]. Geophysics, 2001, 66(5):1450–1456.

Dong An Li Three-dimensional fault combination Petroleum geophysical exploration [J], 1992, 9 (3): 24–26.
Sun Yanju Discussion on fault interpretation and plane combination method of breakpoints Petroleum geophysical exploration [J], 1995, 30 (4): 108–113.
Zeng Hongliu, Hentz T F, Wood L J.Stratal slicing of Miocence-Pliocene sediments in Vermilion Block 502, Tiger Shoal area, offshore Louisiana [J].The Leading Edge, 2001, 20 (4):408–418.
Zhang Baozheng, Chen Qi, et al Fundamentals of geology Beijing: Geological Publishing House, 1983.

Advances in Mineral Resources, Geotechnology and Geological Exploration – Rashid & Zhang (eds)
© 2023 Copyright the Author(s), ISBN 978-1-032-33377-9

Study on rock mechanics characteristics and its influence on hydraulic fracture propagation in BZ27 oilfield

Kechao Gao, Suogui Shang, Jinxin Ma & Qiangyong Gao
Tianjin Branch of CNOOC (China) Ltd. Company, Binhai, Tianjin, P.R. China

R.H. Hu
Southwest Petroleum University, Chengdu, Sichuan, P.R. China

ABSTRACT: The low permeability reservoir in the BZ27 oilfield needs hydraulic fracturing stimulation. In this paper, the dynamic and static Young's modulus and Poisson's ratio are tested in the laboratory and the corresponding regression relationship is established. The cohesion force and internal friction angles of rock are obtained by empirical relation. Based on the logging information of BZ27 well, the in-situ stress characteristics were interpreted. Furthermore, it is found that hydraulic fractures can easily cross through the rock interface and lead to high fracture growth, which is favourable to stimulation treatment in multi-layers. Lastly, hydraulic fracture widths under different pumping rates are calculated according to the hydraulic fracturing volume (about 400 m^3) of adjacent similar oilfields, and the maximum allowable fracture proppant particle size is selected as 20/40 mesh (0.425–0.85 mm) for BZ27 oilfield.

1 INTRODUCTION

Hydraulic fracturing is essential for the economic development of low permeability reservoirs. The BZ27 low permeability reservoir is deeply buried (more than 4000m), with many small layers and severe heterogeneity, which makes the design of a hydraulic fracturing plan very challenging. The key difficulties that must be solved are rock mechanical properties, distribution characteristics of in-situ stress, hydraulic fracture height extension mode at rock interface, and selection of allowable proppant size.

Daneshy (1978) believes that the shear strength of the rock interface is the core factor influencing the hydraulic fracture penetration through the rock interface. Ben-Naceur (1990) established a theoretical model to judge the propagation of hydraulic fractures through interlayer interfaces in multilayer media. Warpiniski (1999) investigated the hydraulic fracture height growth from the in-situ stress contrast. Hu (1996) made a comprehensive analysis of the influence of geological and engineering factors on the height propagation of hydraulic fractures and gave the order of the main factors controlling the height extension of hydraulic fractures on the base of theoretical simulation.

There are two main methods to obtain rock static mechanical properties (Lu 2005): one is a core experiment in the laboratory; the other is to obtain dynamic mechanical parameters based on the logging interpretation and then convert them into static data. The former has formed standard experimental specifications and methods. The latter is widely used because it can obtain the continuous distribution of mechanical parameters that change continuously with well depth (Guo 2019; He 2008; Zhao 2015).

In-situ stress is usually characterized as vertical principal stress and two horizontal principal stress. Logging interpretation is one of the most important methods to obtain in-situ stress (Yin

 DOI 10.1201/9781003319412-32

2018). Although there are some differences between the in-situ stress obtained from logging interpretation and the actual in-situ stress of rock strata, it is still widely used in engineering (Song 2012; Zhao 2015).

In this paper, the dynamic and static mechanical properties and in-situ stress of the BZ27 oilfield were obtained by experimental testing and logging interpretation. It has been proven that the multilayer in well BZ27 has been hydraulically fractured based on the established criterion of hydraulic fracture extension at the rock interface, and a reasonable proppant particle size was selected based on the simulation of hydraulic fracture width. The hydraulic fracturing plan is more targeted and has practical guiding significance.

2 EXPERIMENTAL TESTING OF ROCK MECHANICS PROPERTIES

All the core samples used in the experiments were from the target formation of the BZ27 oilfield. The samples were made into Φ25 mm×50 mm cores and then tested according to the Chinese national standard (GB/T 50266–2013: Standard for experimental methods of engineering rock mass).

2.1 *Dynamic rock mechanic properties*

(1) Dynamic elastic modulus and Poisson's ratio

Firstly, the compressional & shear wave velocities of six rock samples were measured by SCMS-E multi-parameter instruments. The results are displayed in Figure 1. Then the dynamic Young's modulus (YM) and Poisson's ratio (PR) are calculated according to Formulas (1) and (2), the experimental results of YM and PR of samples are shown in Figure 2.

$$E_d = \frac{\rho V_s^2 (3V_p^2 - 4V_s^2)}{V_p^2 - V_s^2} \tag{1}$$

$$\mu_d = \frac{V_p^2 - 2V_s^2}{2(V_p^2 - V_s^2)} \tag{2}$$

where v_p = compressional wave velocity, m/s; v_s = shear wave velocity, m/s; ρ = rock density, Kg/m^3; E_d = dynamic Young's modulus, MPa; μ_d = dynamic Poisson's ratio.

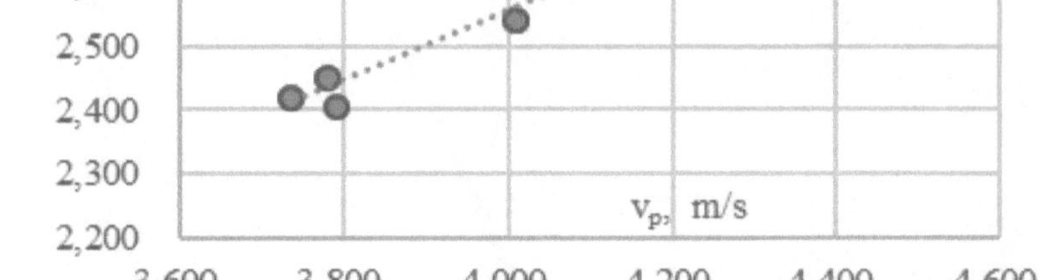

Figure 1. Wave velocity of rock samples.

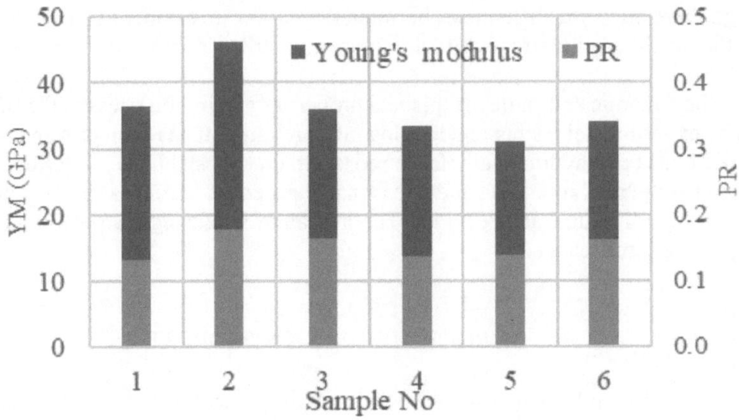

Figure 2. Results of dynamic YM and PR of rock samples.

(1) Cohesion and internal friction angle

Rock cohesion and internal friction angle are calculated by following empirical formulae (Chen, 2011).

$$C = A(1 - 2\nu_d)\left(\frac{1+\mu_d}{1-\mu_d}\right)^2 \rho^2 V_p^4 (1 + 0.78V_{cl}) \tag{3}$$

$$\varphi = 36.545 - 0.4952C \tag{4}$$

where C = cohesion force, MPa; A = experience parameters; V_{cl} = mud content of sandstone, decimal; φ = internal friction angle.

The logging information of well BZ27 was used to interpret the mud content, and the cohesion of rock samples was 9.85–36.3 MPa with an average of about 15 MPa. The internal friction angle was 18.56°–31.6°, with an average of 29.1°.

2.2 *Static mechanical properties*

Four core samples were tested according to the Chinese national standard by RTR-1000 HTHP triaxial rock mechanics testing system, and the stress-strain curve of rock samples was shown in Figure 3. Static mechanical parameters obtained from experimental tests are shown in Figure 4.

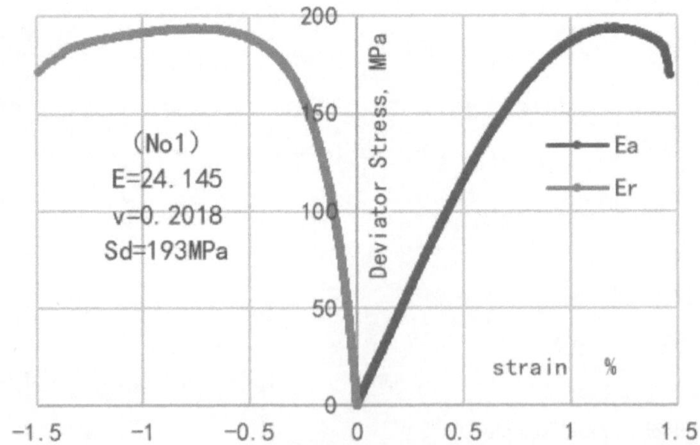

Figure 3. Stress-strain curves of triaxial compression tests.

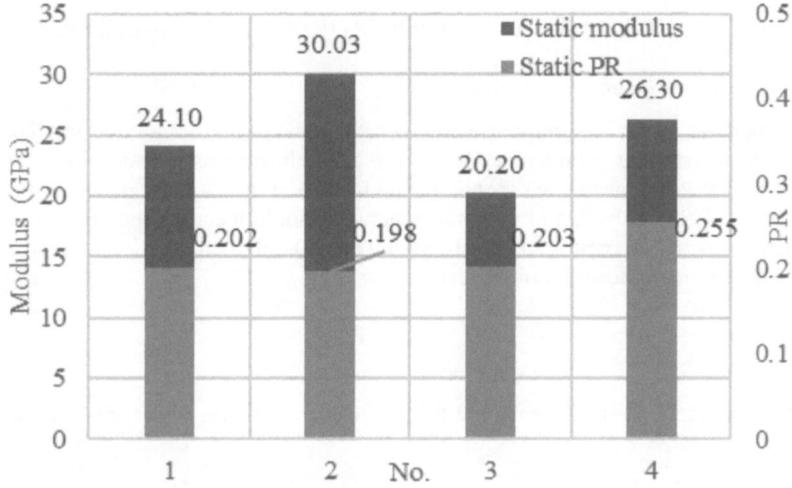

Figure 4. Results of static mechanical properties.

2.3 *Relationship of dynamic and static mechanical parameters of rock*

Based on the experimental results of dynamic and static mechanical parameters of rock, the dynamic and static parameter regression relationship is established as shown below.

$$E = 0.5338E_d + 4.6938 \quad \nu = 0.0809\mu_d + 0.0192$$

where E = Static elastic modulus of rock, GPa; ν = static Poisson's ratio, dimensionless.

3 ANALYSIS OF IN-SITU STRESS PROFILE CHARACTERISTICS

Generally, in-situ stress consists of gravity stress and tectonic stress. The vertical component of in-situ stress can be calculated by density data, and the horizontal in-situ stress can be calculated by the Newberry model. The principal stress distribution profile is interpreted based on the relationship between rock dynamic and static mechanical parameters and logging information from the well, the results of well BZ27 are demonstrated in Figure 5. The vertical stress is about 92MPa, the horizontal maximum stress is 88-93.6MPa, and the horizontal minimum principal stress is 65–71 MPa.

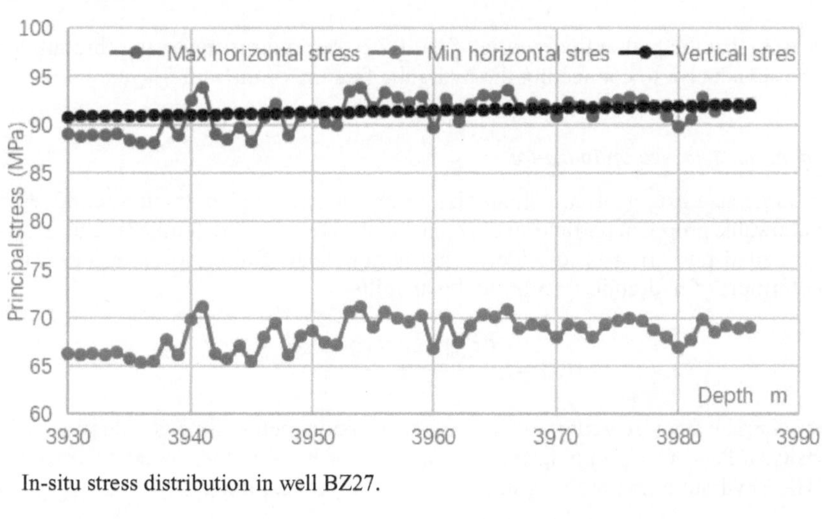

Figure 5. In-situ stress distribution in well BZ27.

4 INFLUENCE OF ROCK MECHANICS PROPERTIES ON HYDRAULIC FRACTURING

4.1 *Hydraulic fracture height propagation mode*

It is assumed that the vertical hydraulic fracture is orthogonal to the strata interface. The rocks on both sides of the interface are homogeneous, isotropic, and linear elastic materials.

According to the elastic mechanic's theory, the total stress at the crack tip under the action of the far-field stress could be derived, and hydraulic fractures should meet the maximum tensile stress criterion when crossing the rock interface. Renshaw (1995) obtained the discriminant formula of hydraulic fracture crossing to strata interface.

$$\frac{C/f+\sigma_y'}{T+\sigma_x'} > \frac{1+f}{3f} \tag{5}$$

where f = friction coefficient, dimensionless; T = tensile strength, MPa; $\sigma i'$ = effective stress (i=x,y), MPa.

Make $\lambda_{cr} = \frac{1+f}{3f}$, $\lambda = \frac{C/f+\sigma_y'}{T+\sigma_x'}$

When $\lambda > \lambda_{cr}$, the hydraulic fracture will cross the rock interface;

When $\lambda < \lambda_{cr}$, the hydraulic fracture will slip at the interface.

The calculation results of rock mechanics parameters and in-situ stress based on the in-situ stress and rock mechanics parameters of the BZ27 oilfield are shown in Figure 6.

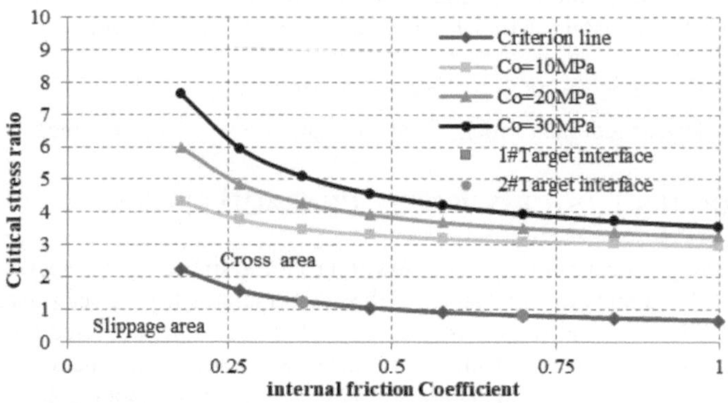

Figure 6. Discrimination results from hydraulic fracture interface.

It can be seen that the hydraulic fracture of the BZ27 reservoir is traversing through the matrix rock, which is conducive to connecting the hydraulic fracture of the multi-layer.

4.2 *Proppant particle size optimization*

In general, large-size proppant and small-size proppant have higher fracture conductivity. The maximum allowable proppant particle size is controlled by dynamic hydraulic fractures. At present, water-based gelled fracturing fluid widely used is power-law fluid. Nordgren (1972) gives the calculation formula of hydraulic fracture width as follows:

$$w(t) = f_2\left[\frac{2\mu(1-v^2)Q^2}{cEH_f}\right]^{1/4} t^{1/8} \tag{6}$$

where w = hydraulic fracture width, m; f_2 = unit conversion factor, dimensionless; μ = fracturing fluid viscosity, mPa·s; Q = pumping rate, m^3/min; c = filtration coefficient of fracturing fluid, m/min$^{0.5}$; H_f = hydraulic fracture height, m; t = pumping time, min.

Figure 7 shows the dynamic fracture width at different pumping rates according to the fracturing fluid volume (400 m^3) of adjacent block BZ27. According to the bridge plugging principle, sand plugging will not occur as long as the hydraulic fracture width is greater than 6 times the proppant diameter. Therefore, the widely used 20/40 mesh (0.425–0.85 mm) proppant size can be used in the BZ27 field.

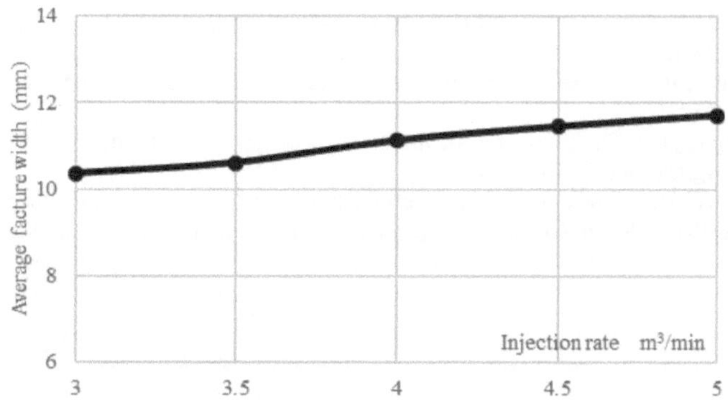

Figure 7. Relationship between fracture width and pumping rate.

5 CONCLUSIONS

- The transformation relationship of dynamic and static mechanical parameters of rock is obtained from laboratory tests: $E=0.5338E_d+4.6938$, $\nu = 0.0809\nu_d + 0.0192$. The static Young's modulus of the rock is 20–30 GPa, with an average of 25.1GPa. The static Poisson's ratio is 0.198–0.255, with an average of 0.2145. The cohesion force is 9.85–36.3 MPa, with an average of 15MPa, and the internal friction angle is 16.56–31.6° with an average of 29.1°.
- The principal stress profile is interpreted based on the relationship between rock dynamics from well BZ27. The vertical stress is about 92MPa, the horizontal maximum stress is 88-93.6MPa, and the horizontal minimum principal stress gradient is 65-71MPa.
- Whether under average or minimum stress gradient, the hydraulic fracture encounters the rock layer interface, the fracture propagation model is always passing mode, so the hydraulic fracture easily enters the bottom shielding layer and communicates with the bottom water.
- Because the hydraulic fracture easily passes through the interface of rock formation, it is beneficial to implement the combined fracture treatment.

REFERENCES

Ben Naceur, K. & T. Eric (1990). Mechanisms controlling fracture height growth in layered media. SPE16433

Chen, M. & Y. Jin (2011): Fundamentals of rock mechanics in petroleum engineering. Beijing. Petroleum Industry Press. 109.

Daneshy, A.A. (1978). Hydraulic fracture propagation in layered formations. Society of Petroleum Engineers Journal. (2), 33–41.

Guo, P. F. & H. C. Deng (2019). Rock mechanics characteristics and influence factors analysis of Chang 8 Reservoir in the Southern Margin of Ordos Basin. Science Technology and Engineering. 19(18), 189–198.

He, S. Y. & Y. M. Shi (2008). The method for the acquirement of conventional logging response-based lithomechanical Parameters. Xinjiang Petroleum Geology. 5, 662–664.

Hu, Y. Q. & S. Q. Ren (1996). Control analysis of fracture height in hydraulic fracturing. Petroleum Geology & Oil Field Development in Daqing. 2, 55–58.

Lu, B. P. & H. Z. Bao (2005). Advances in calculation methods for rock mechanics parameters. Petroleum Drilling Techniques. 5, 47–50.

Nordgren R P (1972). Propagation of a vertical hydraulic fracture. SPE Journal. 8, 306–314.
Reshaw, P (1995). An experimentally verified criterion for propagation across unbounded frictional interfaces in brittle, linear elastic materials. International Journal of Rock Mechanics Mining Science and Geomechanics, 32(5): 237–249.
Song, L. & G. Hareland (2012). Minimum horizontal stress profile from logging data for Montney formation of Northeast British Columbia. SPE Canadian Unconventional Resources Conference. Society of Petroleum Engineers.
Warpinski, N, R. & R. A. Schmidt (1999). In-situ stresses: the predominant influence on hydraulic fracture containment. SPE 8932.
Yin, X. Y. & N. Ma (2018). Review of in-situ stress prediction technology. Geophysical Prospecting for Petroleum. 57(4), 488–504.
Zhao, J. & F. L. Yang (2015). The Logging Calculation and Calibration Methods for Crustal Stress. Science Technology and Engineering. 5(17), 42–46.
Zhao, J. L. & Z. D. Cai (2015). Study on method for establishing rock mechanics parameter profiles of Chang-8 reservoir in Ordos Basin. Journal of Xi'an Shiyou University (Natural Science Edition). 30(03), 47–52.

Advances in Mineral Resources, Geotechnology and Geological Exploration – Rashid & Zhang (eds)
© 2023 Copyright the Author(s), ISBN 978-1-032-33377-9

Analysis of the influence of soft foundation treatment measures on the engineering characteristics of adjacent pile foundations and treatment effect under surcharge

Jingbin He*
Power China Northwest Engineering Corporation Limited, Shaanxi, Xian, China

Yunxiu Dong
School of Civil Engineering, Longdong University, Qingyang, Gansu, China

ABSTRACT: To deal with the impact of soft foundation on the characteristics of pile foundation lateral deformation under surcharge, this paper analyzes the treatment effect of different treatment measures and different treatment zone distances on soft foundation under surcharge based on on-site monitoring and numerical simulation analysis. Studies show that the weak interlayer has the highest impact on the piles closest to the outer side of the pile, and the pile foundation exhibits the largest deformation. The design should improve the bearing capacity of the side piles; the pile body displacement and the rigid long and short piles should be treated with cement mixing piles and rigid long and short piles. The law of soil resistance that changes on the side of the pile is basically same. The treatment effect of rigid long and short piles is better than that of cement mixing piles; there are certain differences in the treatment effect when the distance of the treatment area is different. The treatment area should neither be very far from the pile foundation nor close. The treatment effect is found to be best when the treatment area is 8 m away.

1 INTRODUCTION

When building a bridge pile foundation in a river valley area, due to high roadbed filling behind the abutment, the bridge pile foundation not only is affected by the vertical load and horizontal load from the superstructure but also by the side pile load. In terms of lateral displacement, especially when there is a weak interlayer under the subgrade fill, the bridge pile foundation is is at risk of large lateral displacement (Feng 2008; Handbook of Foundation Treatment (Third Edition) 2008; Zheng et al. 2012). The lateral deformation is an important indicator in evaluating the stability of bridge pile foundations, and researchers at home and abroad have carried out extensive research on the influence of soft foundation on the lateral deformation of adjacent pile foundations under the action of surcharge.

When there is a weak interlayer under the heap load, according to its force and deformation law and its influence on adjacent structures, Li (Li & Zhu 2007a, 2007b) analyzed the lateral displacement mode, heap load size, and heap load underground overload conditions. The influence of distance and pile spacing on the mechanical properties of adjacent pile foundations was determined. Feng (Feng et al. 2018) analyzed the lateral displacement mode, degeneration law, and the influence on adjacent pile foundations of soft soil foundations under the condition of heap loading and even overload. Based on the investigation of the geological conditions of high-speed railway bridge and on-site stacking, Dong (Dong et al. 2015) analyzed the stress and deformation characteristics of the bridge piers and abutments under the action of single-sided stacking. Ma (Ma et al. 2014) used the finite difference method to analyze the influence of the heap load on the lateral displacement

*Corresponding Author: 503940792@qq.com

DOI 10.1201/9781003319412-33

of the passive pile. Side stacking seriously affects the force and deformation characteristics of adjacent pile foundations, thereby affecting the safety of the project. For this reason, it needs to be treated to ensure safe operation of the bridge. Relying on the observation data of soft foundation deformation of Suizi Expressway, Chen (Chen et al. 2013) discussed the treatment effects of gravel pile treatment and plastic drainage slabs on the lateral deformation of soft soil foundation. Yang (Yang & Zhu 2003, 2004) analyzed the effects of reinforcement width, reinforcement depth, and reinforcement elastic modulus on the lateral deformation of adjacent piles in overloaded soft soil foundations.

At present, when the pile foundation is piled on a large area on one side, the pretreatment of the soft foundation is often ignored in the project; however, after the completion of the bridge pile foundation construction, anti-slide piles, and cutting squares are set on the filling side of the bridge abutment. The load reduction method reduces the stress on the bridge pile foundation. However, the lateral displacement of the pile foundation occurs at this time. Larger displacement affects the bearing characteristics and stability of the pile foundation. Therefore, this paper is based on the engineering problem of large lateral displacement of pile foundation under heap load in a certain bridge project, combined with the actual situation of the site. Midas finite element software analyzes the treatment measures of pretreatment, and the treatment distance is close to the heaping side. The influence of pile foundation engineering characteristics was determined.

2 PROJECT OVERVIEW

The lower structure of the bridge was originally designed as column piers. Among them, the pile diameter of the 1# pier on the side of the heap load is 2.0 m and the pile length is 35 m. During the construction process, the top of the 1# bridge pier was offset from the pile load, and the offset reached 12 cm. After supplementary survey, it was found that there is a large area of weak interlayer under the side of the 1# bridge pier. During the stacking process, the soft soil layer is squeezed to produce lateral squeezing, which leads to large displacement of the bridge pier.

When the engineering disease occurs, it leads to delay in the construction progress and increases construction cost. Therefore, this article proposes the construction idea of handling first and then stacking after the detailed survey, using different treatment measures and treatment distances to treat the weak interlayer foundation and evaluate it. The treatment effect ensures safe operation of the bridge project.

3 ESTABLISHMENT AND VERIFICATION OF NUMERICAL SIMULATION MODEL

3.1 *Model establishment and parameter selection*

This paper uses Midas finite element software to establish a numerical simulation model to analyze the deformation laws and mechanical characteristics of the bridge pier column after composite foundation treatment. The calculation model is shown in Figure 1. The bridge foundation and rigid long and short piles are made of concrete materials, the ideal elastic constitutive model is used for analysis, and the ideal elastic-plastic model linear Mohr-Coulomb yield criterion is used for the analysis of foundation soil and cement mixing piles. We fixed the displacement in x and y directions around the model. The bottom surface of the model is fixed in x, y, and z directions. Behind the pile is a pile carrier with a height of 30 m. From top to bottom, the soil layers around the pile comprise soft soil layer, strongly weathered rock layer, and medium weathered rock layer, with layer thicknesses of 16 m, 6 m, and 42 m, respectively. The parameters of each soil material, cement mixing pile, and rigid long and short piles are shown in Table 1.

3.2 *Model verification*

According to the on-site monitoring results, the top offsets of 1#-1, 1#-2, and 1#-3 are 64.6 mm, 66.3 mm, and 69.9 mm, respectively. Compared with the data simulated by Midas software, which

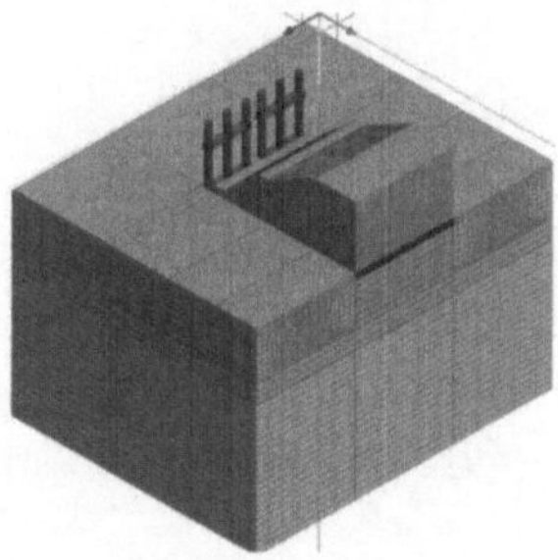

Figure 1. Calculation model.

Table 1. Model calculation parameters.

Material	Elastic Modulus (Pa)	Poisson's ratio	Cohesion (kPa)	Internal friction angle (°)	Bulk density (kN/m^3)
Piers, pile foundations	3.15×10^{10}	0.2	—	—	25
Heap carrier	2.30×10^{6}	0.35	0	25	22
Soft soil	3.50×10^{6}	0.35	28	11	17.6
Strongly weathered mudstone	1.20×10^{9}	0.33	25	15	22
Moderately weathered mudstone	2.50×10^{9}	0.25	100	25	27
Rigid long and short pile	2.00×10^{10}	0.2	—	—	25
Cement mixing pile	4.50×10^{8}	0.23	40	35	25

is 62.62 mm, 62.37 mm, and 62.31 mm, the errors are 3.1%, 5.9%, and 10.8%, respectively. That is, the numerical simulation results are consistent with the on-site monitoring results. Therefore, it can be considered that the model and parameter selection established in this paper are reliable.

3.3 *Calculation conditions*

1) Treatment measures
The influence of cement mixing piles and rigid long and short piles on the lateral deformation of adjacent pile foundations under surcharge is analyzed. Two types of treatment measures adopt plum blossom pile arrangement form. The pile foundation parameters of two types of treatment measures are: (1) Cement mixing pile: The cement mixing piles are 18 m long and 0.5 m in diameter. There are 28 rows of 16 in each row. The distance between the piles along the bridge is 1 m and the distance between the piles in the transverse direction is 2 m. (2) Rigid long and short piles: These are long piles 22 m in length, 14 m in short piles, and a pile diameter of 0.5 m. There are 14 rows of long and short piles each, 16 in each row. The distance between piles in the bridge direction is 1 m and the distance between piles in the transverse direction is 2 m.
2) Distance of treatment area
When the distance between the treatment area and the pile is different, its influence on the adjacent pile foundation is also different. Therefore, taking the treatment area distance l as the influencing factor, analyze the influence of cement mixing piles and rigid long-short pile composite foundation on the lateral deformation of adjacent pile foundations when the treatment area distances are 2 m, 4 m, 6 m, 8 m, 10 m, 12 m, 14 m. The distance diagram of the treatment area is shown in Figure 2.

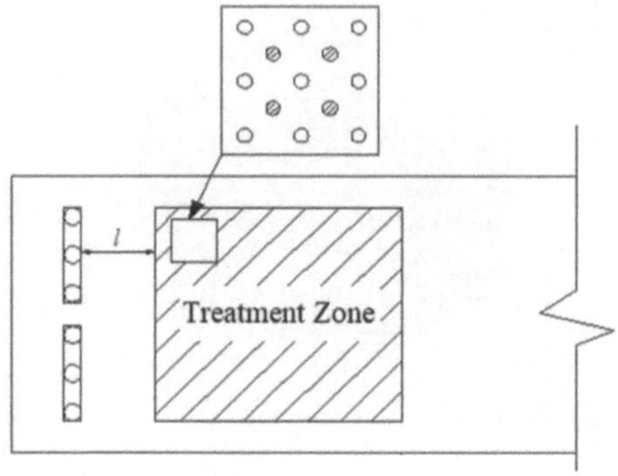

Figure 2. Schematic diagram of the treatment area distance.

4 ANALYSIS OF NUMERICAL SIMULATION RESULTS

4.1 *The effect of treatment measures on the force and deformation characteristics of adjacent pile foundations*

Taking the treatment area distance l = 8 m as an example, the influence of cement mixing pile and rigid long and short pile treatment measures on the force and deformation characteristics of adjacent pile foundations are analyzed.

(1) Analysis of lateral deformation of pile foundation
Figure 3 shows the displacement of piles 1#-1, 1#-2, and 1#-3 when different treatment measures are adopted.
It can be seen from Figure 3 that when the composite foundation is used for treatment, the lateral displacement of the pile foundation is significantly reduced compared with the untreated case, and the displacement of the outer pile (1#-1) is greater than the displacement of the inner pile (1#-3); When the cement mixing pile is used, the pile body produces a positive displacement. When the rigid long-short pile is used, the pile foundation will have a turning point after entering the bearing layer. The pile top displacement reduction of the cement mixing pile and the rigid long and short pile is 66.35%–66.71% and 85.48%–93.03%, respectively. It can be seen that the effect of rigid long and short piles in treating weak interlayers under heap loads is better than that of cement mixing piles.

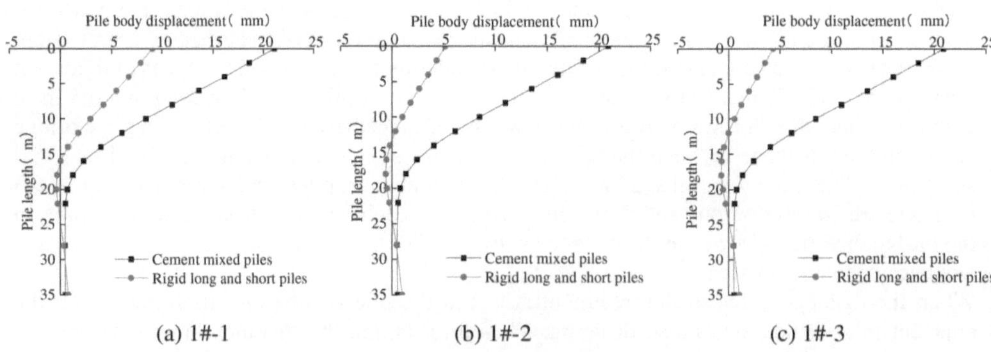

Figure 3. Pile body displacement under two treatment measures.

The main reasons are: 1) When the pile foundation is closer to the outside, because of the smaller area of the treated soil around the outside soil, there is still a weak interlayer around, so a larger lateral squeezing force is generated and the displacement increases. Therefore, in the foundation treatment, the displacement and deformation of the outer piles should be considered emphatically. 2) When the cement mixing pile is used for treatment, since the cement mixing pile is flexible, its strength is low, and the restraint effect on the soft soil after treatment is small, the pile body produces a large displacement, and the deformation into the bearing layer is relatively reduced. However, when rigid long and short piles are used, their strength is high, the soil restraint is strong, and the displacement of the pile body is small. Therefore, it is better to use rigid long and short piles to deal with weak interlayers under surcharge, which also shows that the flexible pile composite foundation is not suitable for dealing with this kind of engineering problem.

(2) Analysis of soil resistance of pile side

The soil resistance in front of piles 1#-1, 1#-2, and 1#-3 under the two treatment measures is shown in Figure 4.

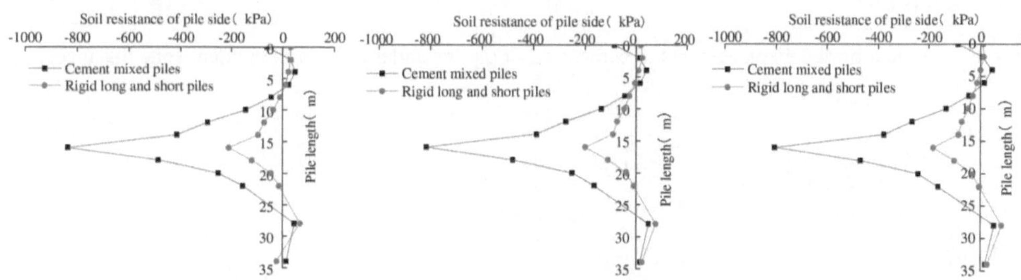

(a) 1#-1 (b) 1#-2 (c) 1#-3

Figure 4. The resistance of the soil in front of the pile foundation with two treatment methods.

It can be seen from Figure 4 that the soil resistance of the pile side under the two treatment measures has a soil resistance peak at the rock–soil interface (16 m). When cement mixing piles are used, the maximum pile side soil resistance is 836.32 kPa, 815.81 kPa, and 806.08 kPa. When rigid long-short piles are used, the maximum pile side soil resistance is 210.20 kPa, 200.08 kPa, and 183.20 kPa. Compared with cement mixing piles, the resistance of rigid long and short piles is reduced by 74.87%, 75.47%, and 77.23%, respectively.

This is because: 1) Due to the low strength of the soft soil layer, lateral deformation occurs under the action of the pile load, which causes the pile body to be laterally compressed, and the weak soil layer on the side of the pile cannot provide the corresponding soil resistance. The pile body is considerably deformed. Due to large deformation of the pile body at the interface of the rock-soil layer, the soil resistance generated by the rock layer is large, so the soil resistance reaches the peak value at the interface of the rock-soil layer. After entering the rock formation, the soil resistance is reduced due to the strong restraint effect of the rock formation on the pile body and small displacement.

2) When the pile load is the same, due to high strength of the rigid long and short piles, the strength of the treated composite foundation increases and can bear the effect of overburdening, so that the lateral squeezing force acting on the pile foundation is reduced, and the soil resistance is decreased. However, the cement mixing pile has low strength and can only bear part of the overburden load, the lateral squeezing force acting on the pile foundation increases, and the soil resistance is relatively large.

4.2 *The influence of the treatment zone distance on the force and deformation characteristics of adjacent pile foundations*

In the two treatment measures, the vertical bridge displacements of piles 1#-1, 1#-2, and 1#-3 at different treatment area distances are shown in Figures 5 and 6, respectively. The pile top displacements of piles 1#-1, 1#-2, and 1#-3 under different treatment area distances are shown in Figure 7.

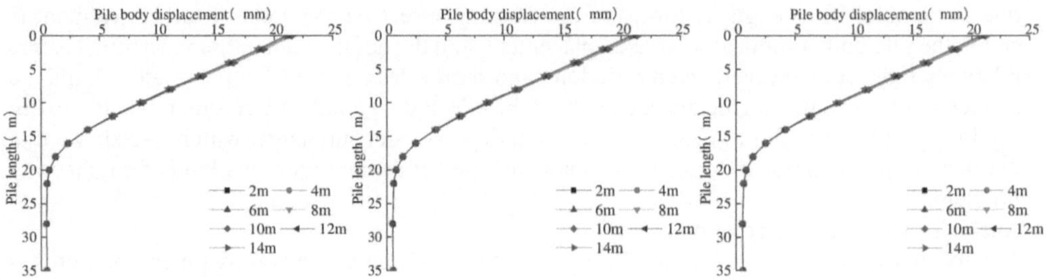

(a) 1#-1 (b) 1#-2 (c) 1#-3

Figure 5. Vertical bridge displacements of cement mixing piles under different treatment zone distances.

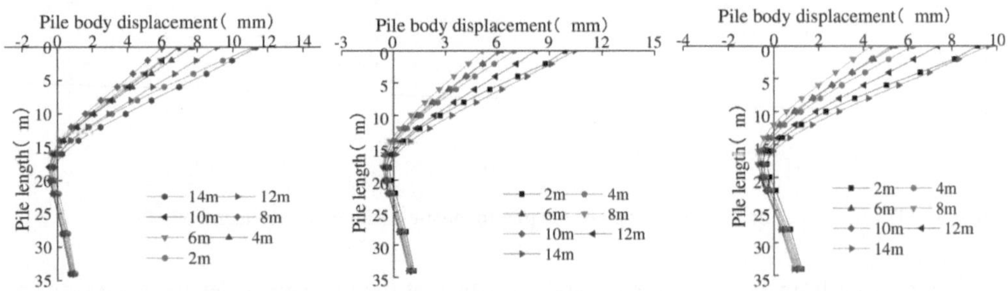

(a) 1#-1 (b) 1#-2 (c) 1#-3

Figure 6. Vertical bridge displacements of rigid long and short piles under different treatment zone distances.

It can be seen from Figures 5–7 that the displacement of the pile body along the bridge adjacent the pile body is consistent with the distance from different treatment areas, and large lateral displacement occurs in the soft soil layer. As the distance from the treatment area gradually increases, the overall displacement of the pile body first decreases and then increases. When the treatment area is 8 m away, the pile displacement is the smallest overall. Under the two treatment measures, when the treatment area is 14 m away from the pile body, the displacement is the largest, followed by 12 m, 2 m, 4 m, 6 m, and 10 m. In addition, the displacement of the bottom pile top from

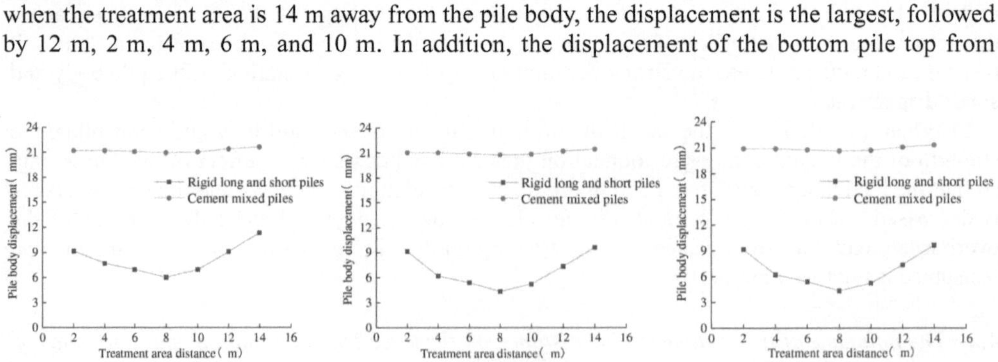

(a) 1#-1 (b) 1#-2 (c) 1#-3

Figure 7. Pile top displacement under different treatment zone distances.

different treatment areas is very small when the cement mixing pile is processed, while the displacement of the bottom pile top from different treatment areas is significantly different when the rigid long-short pile is processed.

The reasons are as follows: 1) Due to low strength of cement mixing piles, the ability to resist the lateral squeezing force of the soil is weak, so the change in the treatment area has less impact on the treatment effect than rigid long and short piles. 2) When the distance between the treatment area is small, because the treatment area is close to the pile foundation as a whole, and most of the soft soil foundation remains untreated after the treatment area, the untreated soft soil is squeezed by the overburden to produce a lateral squeezing force. Due to the large range of untreated soft soil, the resulting squeezing force causes the overall displacement of the treatment area to act on the pile foundation, making the resulting displacement larger. When the treatment area is relatively far away, it can bear the overlying load and resist the soft foundation behind the treatment area. The soft soil area in front of the treatment area is relatively large, and the lateral squeezing force of the soft soil directly acts on the pile foundation under the action of the overburden, thereby causing a large displacement. When the distance of the treatment area is 8 m, the treatment area can resist the overburden of the treatment area and the soft squeezing force behind it, while reducing the soft soil in front of the treatment area and having less influence on the lateral displacement of the pile foundation.

Thus, it can be seen that the distance of the treatment area has a significant impact on the treatment effect of the lateral displacement of the pile foundation. It should neither be very far nor close. A better treatment effect can be achieved at 8 m distance.

5 CONCLUSIONS

Based on on-site monitoring and numerical simulation analysis, this paper establishes a finite element model that conforms to the actual project, and comprehensively compares the treatment effects of two treatment measures for pile foundations on a single side of a large area pile in soft soil areas. The following conclusions are obtained:

1) When there is a weak interlayer under the surcharge, it produces large lateral deformation to squeeze the pile foundation, and the pile closest to the outer side of the pile is in the most unfavorable position. The design can appropriately improve the bearing performance of the side pile foundation.
2) When cement mixing piles and rigid long and short piles are used to treat the overlying soft foundation, the displacement of the pile body and the change in soil resistance of the pile side are basically the same. However, the treatment effect is quite different, and the treatment effect of rigid long and short piles is better than that of cement mixing piles. It shows that compared with flexible pile composite foundation, rigid pile composite foundation can effectively reduce the lateral displacement of pile foundation on one side of large area pile in soft soil area, while flexible pile composite foundation is not suitable for solving this kind of engineering problems.
3) Under different treatment measures, the treatment effect will be different when the treatment area is at different distances. The treatment distance should neither be very far from the pile foundation nor close. The treatment effect is best when the treatment area is 8 m away.
4) This article only considers the impact of the treatment zone distance and treatment measures on the piles adjacent to the pile. The influence of the pile strength and pile size in the composite foundation is not clear, and further research can be carried out.

REFERENCES

Chen Jibin, Zhao Qihua, Peng Sheqin, Chen Ziyang. The lateral deformation of soft soil foundation treated by different methods. J. Rock and Soil Mechanics, 2013, 34(S1): 340–347. DOI: 10.16285/j.rsm.2013.s1.052.

Dong Liang, Niu Bin, Gu Mu, Li Dongsheng. Numerical analysis of the influence of large-area single-side heap load on high-speed railway bridge piers and abutments. J. Railway Construction, 2015(01): 39–42.
Feng Zhongju, Li Xiaoxiong, Su Hangzhou, Wang Xiqing, Wang Fuchun, Jian Xinlong, Xu Wangui, Ye Fangcai. Research on transverse and axial bearing characteristics of bridge pile foundations in deep soft ground area. J. China and Foreign Highway, 2018, 38(01): 123–127. DOI:10.14048/j.issn.1671-2579.2018.01.027.
Feng Zhongju. Special area foundation engineering. M. People's Communications Press, 2008.
Handbook of Foundation Treatment (Third Edition). M. Beijing: China Construction Industry Press, 2008.
Li Zhongcheng, Zhu Xiaojun. Analysis of soil lateral displacement mode and influence on adjacent pile foundation under ground overload condition. C.//Proceedings of the 9th National Symposium on Numerical Analysis and Analytical Methods of Rock and Soil Mechanics., 2007a:819–824.
Li Zhongcheng , Zhu Xiaojun. Analysis of soil lateral displacement mode and its influence on adjacent pile foundation under ground overload conditions. J. Rock and Soil Mechanics, 2007b, 28(S1): 809–814. DOI:10.16285/j.rsm .2007.s1.168.
Ma Yuangang, Wang Yanfen, Chen Chen. Stress analysis and treatment measures of bridge passive pile deflection under supercharged load. J. Bridge Construction, 2014, 44(04): 22–26.
Yang Min, Zhu Bitang. Analysis of active reinforcement of overloaded soft soil foundation to control the lateral deformation of adjacent pile foundation. J. Journal of Building Structures, 2003(04): 76–84.
Yang Min, Zhu Bitang. Analysis of passive reinforcement of overloaded soft soil foundation to control the lateral deformation of adjacent pile foundation. J. Chinese Journal of Rock Mechanics and Engineering, 2004(11):1912–1918.
Zheng Gang, Gong Xiaonan, Xie Yongli, Li Guangxin. Summary of the development of foundation treatment technology. J. China Civil Engineering Journal, 2012, 45(02): 127–146. DOI:10.15951/j.tmgcxb.2012.02.023.

Advances in Mineral Resources, Geotechnology and Geological Exploration – Rashid & Zhang (eds)
© 2023 Copyright the Author(s), ISBN 978-1-032-33377-9

Research on indoor test method for shear strength of rock-concrete interface in transmission line engineering

Zhaoqing Zhu*, Shijun Ding, Yin Man & Wenzhi Yang
China Electric Power Research Institute, Beijing, China

ABSTRACT: In order to solve the technical problems faced in the application of transmission line foundation engineering, the indoor test method for shear strength of rock-concrete interface has been established. The shear strength test of the rock-concrete interface has been completed, and the field conditions are simulated for the indoor test. The calculation method of shear strength of rock-concrete interface is proposed, and the calculation results can provide a reference for the design of transmission line foundations. The indoor test method for shear strength of rock-concrete interface can provide technical services for the design of bearing capacity of relevant foundation in the field construction of scientific research project experiment and can provide technical support for optimizing the investigation scheme and construction design scheme of similar foundations in transmission lines.

1 INTRODUCTION

In recent years, with the large-scale construction of power grid, especially DC UHV projects, the survey, design, construction, and management of rock-socketed digging piles, rock anchors and other types of foundations are facing some technical problems, especially the determination of shear strength of rock-concrete interface and how to design such foundations (Cheng et al. 2002; Lu et al. 2008).

At present, there is no mature indoor test method for the shear strength of the rock-concrete interface (Wang et al. 2015; Xu & Gong 2020). The value of design parameters based on shear performance mainly depends on experience, resulting in large randomness and conservatism. For the in-site foundation, the static load test used to determine the shear strength of the rock-concrete interface is not only expensive, difficult, and long-term, but also has no obvious reference function when the geological conditions are complex.

Using indoor test equipment to determine the shear strength of the rock-concrete interface has the advantages of low cost, convenience, and high efficiency. Therefore, on the basis of the existing test equipment, it is of great practical and theoretical significance to research indoor test methods and specific supporting instruments to test the shear strength of the rock-concrete interface.

2 TEST PROCESS

2.1 *Preparation of indoor test specimens*

The test specimens were obtained from the in-situ rock on-site by a drilling machine. The test specimens are annular, as shown in Figure 1. Generally, the outer diameter is 100mm to 300mm, the inner diameter is 70mm to 250mm, and the height is 300mm to 1000mm.

Both ends of the test specimen shall be ground flat in the laboratory and then placed in the test device. The aperture gap between the outer side of the test specimen and the test device shall be filled, the corrugated steel bar shall be placed inside, and concrete shall be poured.

*Corresponding Author: clementfox@163.com

DOI 10.1201/9781003319412-34

Figure 1. Test specimens.

2.2 *Test loading method*

The maintenance load method is adopted and the load is applied in stages. Generally, the load increment of each level is the same. And the loading amount of each level is estimated by the product of the shearing force of the rock-concrete interface and their contact area. The shearing force can be taken as 50kPa to 100kPa.

2.3 *Test data collection method*

The maintenance time of each level of loading shall not be less than 5 minutes. The load, the displacement of the top or bottom of the test specimen and concrete, and the relative displacement of the contact surface between the test specimen and concrete shall be recorded every 1 minute. The automatic system is adopted for test loading and test data collection.

2.4 *Test status determination rules*

When the relative displacement between the concrete and the top of the test specimen is not greater than 0.1mm within 2 minutes, it indicates that the deformation reaches a stable state under this level of load, and the next level of load can be carried out. When the ratio of the displacement at this level to that at the previous level is not less than 10 on the load-displacement curve (inflection point) or the specimen is damaged, the test can be terminated.

2.5 *Shear strength calculation method*

When the displacement corresponding to the inflection point is not greater than 4mm, the load corresponding to the inflection point is the calculated load (Qt). When the inflection point displacement is greater than 4mm or there is no inflection point, the load at this level corresponding to the displacement of 4mm shall be taken as the calculation load (Qt).

The shear strength of the rock-concrete interface (τ) is calculated according to the following formula:

$$\tau = \frac{Q_t}{\pi dh} \quad (1)$$

In the formula: d represents the inner diameter of the specimen, and h represents the height of the specimen.

3 TEST RESULTS AND DISCUSSIONS

3.1 *Test situation*

The indoor tests of the shear strength of the rock-concrete interface were conducted in the geotechnical engineering laboratory of China Electric Power Research Institute, as shown in Figure 2.

The specimen parameters of the test are shown in Table 1.

Table 1. Parameters table of specimens.

Serial number	Category	Length/ cm	Internal diameter/ cm	Degree of weathering	Integrity
1	Limestone	10	8	Micro-Medium	Complete
2	Limestone	30	8.5	Micro-Medium	Complete

Figure 2. Test in the geotechnical engineering laboratory.

3.2 *Test results*

The load-displacement curves of the tests are shown in Figure 3.

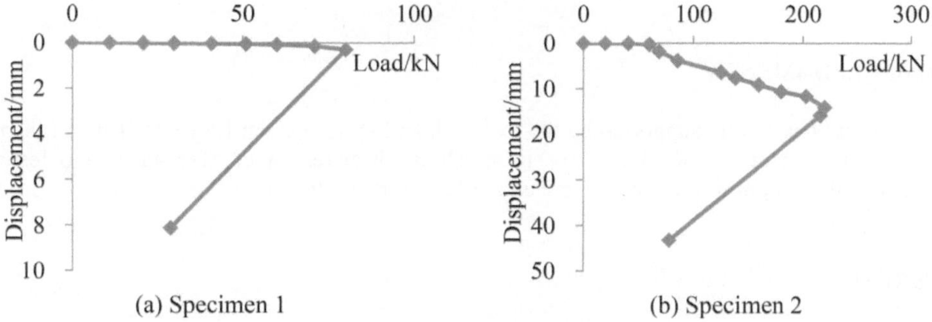

Figure 3. The load-displacement curve.

During the test of specimen 1, the displacement increases slowly in the early stage of loading. When it is loaded to 70kN, the displacement growth accelerates, and the displacement at this level is 0.06mm, while the previous cumulative displacement is 0.08mm. Continue loading, when the 9th level load is applied, it cannot be loaded according to the predetermined load. The actual load is only 29kN, the displacement increases obviously, the concrete is separated from the core, and the test is finished.

During the test of specimen 2, the displacement increases slowly in the early stage of loading. When the actual load reaches 70kN, the displacement growth accelerates, and the displacement at this level is 1.55mm, while the previous cumulative displacement is 0.28mm. When the 12th level load is applied, it cannot be loaded according to the predetermined load, and the actual load is 216kN. When the 13th level load is applied, it cannot be loaded according to the predetermined load. During the loading process, the concrete is separated from the rock, and then the actual load is reduced to 78kN.

The calculation results are shown in Table 2 by formula (1).

Table 2. Shear strength calculation results.

Serial number	Q_t/kN	d/mm	h/mm	τ/kPa
1	80	80	100	3183
2	220	160	300	1459

4 CONCLUSION

Based on the results and discussions presented above, the conclusions are obtained as below:

(1) The shear strength test of the rock-concrete interface has been successfully completed, and the measured values of displacement and load were obtained. The test results are true and reasonable, which can provide an important basis for the research of the shear mechanism of the rock-concrete interface.
(2) The shear strength test device of the rock-concrete interface is designed reasonably, realizes the effective simulation of the field conditions, and has the function of an indoor test.
(3) The indoor test method used for the investigation and detection of shear strength of the rock-concrete interface can provide technical services for the design of bearing capacity of relevant foundation in the field construction, and can provide technical support for optimizing the investigation scheme and construction design scheme of similar foundations such as rock-socketed digging pile.

ACKNOWLEDGMENTS

This work was financially supported by the Research and Development Project of China Electric Power Research Institute (GC83-19-013): Research on Shear-resistance Mechanism and Testing Technology of Strength of the rock-concrete interface under multi-scale.

REFERENCES

Cheng Y.F., Shao X.Y., Zhu Q.J.: (2002) Current Situation of Foundation Works and Existing Problems for Transmission Lines in China. Electric Power Construction 23(3), 32–34.

Lu XL, Cheng Y.F.: (2008) Review and new development on transmission lines tower foundation in China. CIGRE 2008 Session: B2-215. Paris.
Wang W.D., Wu J.B., Nie S.B.: (2015) Field Loading Tests on Large-diameter Rock-socketed Bored Piles of Wuhan Greenland Center Tower. Chinese Journal of Geotechnical Engineering 37 (11) 1945–1954.
Xu J, Gong W.M.: (2020) A New Method for Predicting the Ultimate Shaft Resistance of Rock-socketed Drilled Shafts. Proceedings of the Institution of Civil Engineers–Geotechnical Engineering 173(2) 169–186.

Advances in Mineral Resources, Geotechnology and Geological Exploration – Rashid & Zhang (eds)
 ISBN 978-1-032-33377-9

Research on finite element method for non-aqueous contaminant transport in unsaturated soil

Hongyun Yin, Hong Li* & Longtan Shao
Faculty of Infrastructure Engineering, Dalian University of Technology, Dalian, China

ABSTRACT: Non-aqueous pollutants are one of the important sources of soil and groundwater pollution. Due to the complexity of geological conditions and pollutant properties, it is difficult to obtain an accurate solution through analytical methods, so a numerical model needs to be established. Based on the interaction mechanism of water, air, and NAPL in porous media, a finite element numerical model was established to describe the migration of NAPL pollutants in the soil unsaturated zone, and the corresponding Fortran program Mig-3PF was written. Based on the mathematical model of groundwater seepage, the partial differential equations for the control of immiscible pollutants in the ground are deduced. The governing equations take into account the compressibility of air and adopt a mixed expression form of saturation and pressure. The numerical difficulty brought about by the distinct non-linearity of the temporal evolution of saturation-dependent variables is overcome by the mixed-form formulation. The process of discrete solutions using the Galerkin finite element method is described in detail. Finally, the NAPL Simulator and the self-written code Mig-3PF are used to simulate and compare the same example, which preliminarily verifies the correctness of the self-developed code.

1 INTRODUCTION

In recent years, the pollution of soil and water environment has been paid more attention in China. Hydrocarbons wasted in the production process of the petroleum industry, leakage of oil storage tanks in gas stations, and organic garbage leachate discarded by the food industry, are important sources of soil groundwater pollution (Dong & Fu 2009; Ma 2021; Yang et al. 2013). Such pollutants have minimal solubility in water and are called non-aqueous phase liquid (NAPL). When NAPL seeps into the ground, it tends to be a long-term source of pollution because it is extremely difficult to degrade. It can cause persistent pollution of soil, groundwater, and air and pose a threat to biological communities (Liu et al. 2010).

In the 1860s, the movement of NAPL in porous media first received attention in the field of oil and gas reservoir engineering. Then the multiphase flow problem in petroleum engineering is adjusted according to typical groundwater pollution conditions so that it can be applied to the field of groundwater pollution (Abriola & Pinder 1985a, 1985b). In this field, most of the numerical simulation studies by Chinese scholars are based on foreign source code or commercial software, such as FEFLOW, MOFAT, NAPL simulator, COMSOL, etc. based on finite element method, STOMP, TOUGH2, etc. based on finite difference method (Kamaruddin et al. 2011). However, starting from the basic theory and mathematical model, establishing a numerical discrete format, independently developing the source code, and completing the verification, there are not many such studies.

The process of non-aqueous fluid infiltrating into the subsurface and flowing through the soil unsaturated zone has a strong nonlinear flow mechanism driven by interfacial tension. At this stage,

*Corresponding Author: hong.li@dlut.edu.cn

 DOI 10.1201/9781003319412-35

the three-phase fluids of NAPL, water, and air interact, and the physical mechanism is very complex. In this paper, a mathematical model of three-phase transport in porous media is established for the whole process of infiltration of non-aqueous pollutants into the soil.

2 FUNDAMENTAL THEORY OF THREE-PHASE FLOW IN POROUS MEDIA

2.1 *Governing equations*

In the research domain, a tiny cuboid unit is taken as the representation of elemental volume (REV), it can represent the average physical properties near the point, as shown in Figure 1.

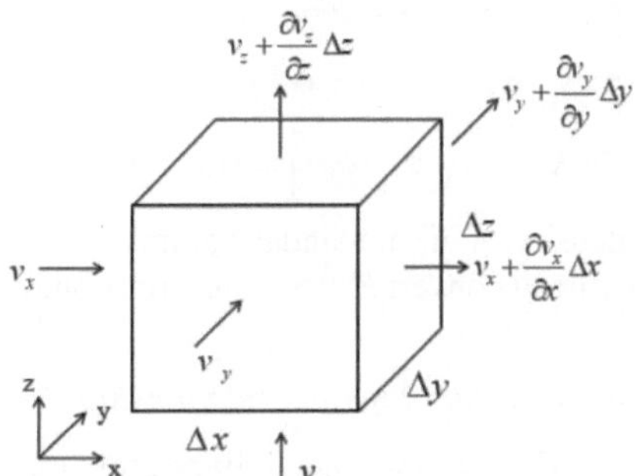

Figure 1. Mass conservation of a fluid flowing through a unit domain.

After a unit time, the mass (Q_i) of the fluid flowing into the unit can be expressed as:

$$Q_i = \rho v_x \Delta y \Delta z + \rho v_y \Delta x \Delta z + \rho v_z \Delta x \Delta y \tag{1}$$

where, ρ(kg/m^3) is the density; v_x, v_y, v_z (m/sec) represent the fluid velocity in three directions.

The fluid mass (Q_o) flowing out of the cuboid element can be expressed as:

$$Q_o = \left[\rho v_x + \frac{\partial(\rho v_x)}{\partial x}\Delta x\right]\Delta y \Delta z + \left[\rho v_y + \frac{\partial(\rho v_y)}{\partial y}\Delta y\right]\cdot \Delta x \Delta z + \left[\rho v_z + \frac{\partial(\rho v_z)}{\partial z}\Delta z\right]\Delta x \Delta y \tag{2}$$

The cumulative storage mass (Q_r) of the fluid in the cuboid unit can be expressed as:

$$Q_r = \frac{\partial(\rho \phi S)}{\partial t}\Delta x \Delta y \Delta z \tag{3}$$

Where, ϕ (−) is the porosity of the medium, and S (−) is the saturation of the fluid.

According to the law of conservation of mass ($Q_i - Q_o = Q_r$) , the fluid continuity equation in the control volume can be expressed as:

$$-\frac{\partial}{\partial x_i}(\rho v_i) = \frac{\partial}{\partial t}(\rho \phi S) \tag{4}$$

The motion equation of the fluid in the porous medium is expressed by Darcy's law:

$$v_f = -\frac{k_{rf}}{\mu_{rf}} K_w \frac{\partial}{\partial x_j}(\psi_f + \rho_{rf} z)(f = w, n, a) \tag{5}$$

Where, w, n, a represent the water, NAPL, and gas; k_{rf} (=k_f/k_s) is the relative permeability coefficient of the fluid f, k_f is the fluid unsaturated permeability coefficient, k_s is the soil inherent saturated permeability coefficient; μ_{rf} (=μ_f/μ_w) is the viscosity coefficient of fluid f relative to water, μ_f and μ_w are the viscosity coefficients of fluid f and water; ρ_{rf} (=ρ_f/ρ_w) is the ratio of the density of the fluid f to the density of water; K_w (m^2) is the saturated permeability coefficient tensor, ψ_f(m) is the pressure head of fluid f.

Substitute Darcy's law (5) into the mass conservation Equation (4), and further simplify the equation:

$$\nabla\cdot\left[\rho_f\frac{k_{rf}}{\mu_{rf}}K_w\nabla\left(\rho_{rf}z+\psi_f\right)\right]=\phi\frac{\partial\left(\rho_fS_f\right)}{\partial t}\quad f=w,n,a \tag{6}$$

Since the compressibility of water and NAPL is extremely small, it is not considered. Then, the governing equation of the water-NAPL-air three-phase flow considering the air compressibility is:

$$\begin{cases}\nabla\cdot k_{rw}K_w\nabla\left(z+\psi_w\right)=\phi\frac{\partial S_w}{\partial t}\\ \nabla\cdot\frac{k_{rn}}{\mu_{rn}}K_w\nabla\left(\rho_{rn}z+\psi_n\right)=\phi\frac{\partial S_n}{\partial t}\\ \nabla\cdot\left[\rho_a\frac{k_{ra}}{\mu_{ra}}K_w\nabla\left(\rho_{ra}z+\psi_a\right)\right]=\phi S_a\frac{\rho_{atm}}{\psi_{atm}}\frac{\partial\psi_a}{\partial t}+\phi\rho_a\frac{\partial S_a}{\partial t}\end{cases} \tag{7}$$

Where, ρ_{atm} (kg/m^3) is the density of air at standard atmospheric pressure, and ψ_{atm}(m) is the corresponding value of pressure head converted from one atmospheric pressure.

2.2 *Three-phase flow saturation-capillary pressure relationship*

This paper adopts the three-phase flow saturation-capillary pressure relationship model proposed by Parker and Lenhard (Parker & Lenhard 1990). The establishment of this model is based on the initial assumption proposed by Leverett (Leverett 1941) regardless of the air storage in water and NAPL storage and NAPL storage water. As shown in Figure 2, in the three-phase system, the wettability relationship is water>NAPL>air, and the pore fluid with higher wettability tends to reside near the contact point of soil particles.

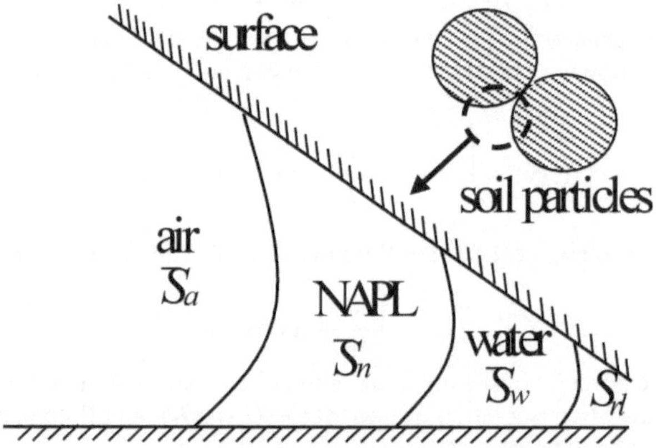

Figure 2. Distribution of three-phase fluid in pores.

S_{rw}, S_{rn}, and $S_{r\ell}$ represent the residual saturation of water, NAPL, and liquid phase, respectively, and $S_{r\ell}=S_{rw}+S_{rn}$. While $\overline{S}_w$, $\overline{S}_n$, and $\overline{S}_\ell$ represent the effective saturation of water, NAPL, and liquid phase, this paper does not consider the residual air saturation, the following relationship holds:

$$S_w+S_n+S_a=S_\ell+S_a=1 \tag{8}$$

$$\overline{S}_f=\left(S_f-S_{rf}\right)/\left(1-S_{r\ell}\right)f=w,n,\ell \tag{9}$$

$$\overline{S}_a=S_a/\left(1-S_{r\ell}\right) \tag{10}$$

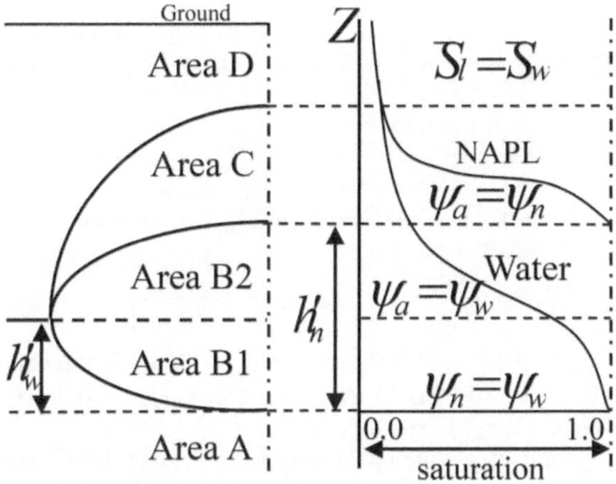

Figure 3. Steady-state saturation and pressure distribution.

When seepage reaches a steady-state, the relationship between capillary pressure head and saturation is shown in Figure 3. The basin can be divided into the following areas: Area A is a saturated zone, $\psi_w \geq 0$ and $\psi_w \geq \psi_n$; Areas B1 and B2 are NAPL and water two-phase regions, $\psi_n > \psi_w \geq 0$, $\psi_w \leq 0$ and $\psi_n \geq 0$; Area C is the three-phase area of air, NAPL, and water, $\psi_w < \beta_{an}\psi_n < 0$; Area D is the two-phase region of air and water, $\beta_{an}\psi_n \leq \psi_w < 0$, where $\beta_{an} = \sigma_{aw}/\sigma_{an}$, $\beta_{nw} = \sigma_{aw}/\sigma_{nw}$. $\beta(-)$ is the scaling parameter, σ_{aw}, σ_{an}, and σ_{nw} are the interfacial tension between air and water, air and NAPL, water, and NAPL.

The effective saturation in each region, using the van Genuchten (Van Genuchten 1980) model, is expressed by the relational expression of the capillary head as follows:

$$\begin{cases} \overline{S}_w = 1 & \text{Area A} \\ \overline{S}_w = (1 + (\alpha\beta_{nw}\psi_{cnw})^l)^{-y} & \text{Area B1, B2, C} \\ \overline{S}_w = (1 + (\alpha\psi_{caw})^l)^{-y} & \text{Area D} \end{cases} \quad \begin{cases} \bar{S}_\ell = 1 & \text{Area A, B1, B2} \\ \bar{S}_l = \left(1 + (\alpha\beta_{an}\psi_{can})^l\right)^{-y} & \text{Area C} \\ \bar{S}_\ell = \bar{S}_w & \text{Area D} \end{cases} \tag{11}$$

Where, α (m^{-1}) and $l(-)$ are both van Genuchten model constants, and dimensionless constant $\gamma = 1 - 1/l$ is also constants.

2.3 *Three-phase flow relative permeability coefficient-saturation relationship*

Using the Mualem (Mualem 1976) model, the relative permeability coefficient of three-phase fluids is related to the effective saturation as follows:

$$\begin{cases} k_{rw} = \bar{S}_w^{1/2}\{1 - (1 - \bar{S}_w^{1/\gamma})^\gamma\}^2 \\ k_{rn} = (\bar{S}_l - \bar{S}_w)^{1/2}\left\{(1 - \bar{S}_w^{1/\gamma})^\gamma - \left(1 - \bar{S}_l^{1/\gamma}\right)^\gamma\right\}^2 \\ k_{ra} = (1 - \bar{S}_l)^{1/2}\left(1 - \bar{S}_l^{1/\gamma}\right)^{2\gamma} \end{cases} \tag{12}$$

3 NUMERICAL DISCRETIZATION FORMULATIONS USING THE FINITE ELEMENT METHOD

Considering Equation (11), Taylor series expansion of $S_w^{n+1,m+1}$, $S_n^{n+1,m+1}$, and $S_a^{n+1,m+1}$ can be expressed:

$$S_w^{n+1,m+1} = S_w^{n+1,m} + \tau_1 \frac{\partial S_w}{\partial \psi_{cnw}}\bigg|^{n+1,m} \Delta\psi_{cnw} + \tau_2 \frac{\partial S_w}{\partial \psi_{caw}}\bigg|^{n+1,m} \Delta\psi_{caw} + O\left(\delta^2\right) \tag{13}$$

$$\begin{aligned} S_n^{n+1,m+1} = S_n^{n+1,m} &- \tau_1 \frac{\partial S_w}{\partial \psi_{cnw}}\bigg|^{n+1,m} \Delta\psi_{cnw} - \tau_2 \frac{\partial S_w}{\partial \psi_{caw}}\bigg|^{n+1,m} \Delta\psi_{caw} \\ &- \tau_3 \frac{\partial S_a}{\partial \psi_{can}}\bigg|^{n+1,m} \Delta\psi_{can} - \tau_4 \frac{\partial S_a}{\partial \psi_{caw}}\bigg|^{n+1,m} \Delta\psi_{caw} + O\left(\delta^2\right) \end{aligned} \tag{14}$$

$$S_a^{n+1,m+1} = S_a^{n+1,m} + \tau_3 \frac{\partial S_a}{\partial \psi_{can}}\bigg|^{n+1,m} \Delta\psi_{can} + \tau_4 \frac{\partial S_a}{\partial \psi_{caw}}\bigg|^{n+1,m} \Delta\psi_{caw} + O\left(\delta^2\right) \tag{15}$$

In the formula, $\Delta\psi = \psi^{n+1,m+1} - \psi^{n+1,m}$. According to Figure 3, the values of τ_1, τ_2, τ_3, τ_4 are as follows: Area A: τ_1=0, τ_2=0, τ_3=0, τ_4=0; Area B1, B2: τ_1=1, τ_2=0, τ_3=0, τ_4=0; Area C: τ_1=1, τ_2=0, τ_3=0, τ_4=0; Area D: τ_1=1, τ_2=1, τ_3=0, τ_4=1.

In addition, C_{aw}, C_{nw}, and C_{an} are the specific moisture capacity of water and air, NAPL, and water, air and NAPL, respectively, which can be obtained from the following formulas:

$$C_{aw} = \phi \frac{\partial S_w}{\partial \psi_{caw}}\bigg|^{n+1,m} = \phi\left(1 - S_{r\ell}\right) \frac{\partial \bar{S}_w}{\partial \psi_{caw}} \tag{16}$$

$$C_{nw} = \phi \frac{\partial S_w}{\partial \psi_{cnw}}\bigg|^{n+1,m} = \phi\left(1 - S_{r\ell}\right) \frac{\partial \bar{S}_w}{\partial \psi_{cnw}} \tag{17}$$

$$C_{an} = -\phi \frac{\partial S_a}{\partial \psi_{can}}\bigg|^{n+1,m} = \phi\left(1 - S_{r\ell}\right) \frac{\partial \bar{S}_\ell}{\partial \psi_{can}} \tag{18}$$

Putting Equation (11) into the above Equations (16), (17), and (18), we can obtain:

$$C_{aw} = \left[-\phi(1 - S_{rw})\alpha\gamma l(\alpha\psi_{caw})^{l-1}\right] / \left[1 + (\alpha\psi_{caw})^l\right]^{\gamma+1} \tag{19}$$

$$C_{nw} = \left[-\phi(1 - S_{rw})\alpha\beta_{nw}\gamma l(\alpha\beta_{nw}\psi_{cnw})^{l-1}\right] / \left[1 + (\alpha\beta_{nw}\psi_{cnw})^l\right]^{\gamma+1} \tag{20}$$

$$C_{an} = \left[-\phi(1 - S_{rw})\alpha\beta_{an}\gamma l(\alpha\beta_{an}\psi_{can})^{l-1}\right] / \left[1 + (\alpha\beta_{an}\psi_{can})^l\right]^{\gamma+1} \tag{21}$$

Taking the water-phase governing equation as an example, the higher-order terms are omitted from Equation (13), and Equations (16) and (17) are brought in to get:

$$\Delta S = S_w^{n+1,m+1} - S_w^{n+1,m} = \frac{\tau_1}{\phi} C_{nw}^{n+1,m}\left(\delta_n^{n+1,m} - \delta_w^{n+1,m}\right) + \frac{\tau_2}{\phi} C_{aw}^{n+1,m}\left(\delta_a^{n+1,m} - \delta_w^{n+1,m}\right) \tag{22}$$

Where $\delta^{n+1,m} = \psi^{n+1,m+1} - \psi^{n+1,m}$. However, Equation (22) cannot consistently satisfy the requirement of mass conservation. As shown in Figure 4, the soil water content characteristic curve in the process of drainage or infiltration has strong nonlinear characteristics. When the saturation value is relatively small, corresponding to a small saturation increment, the actual capillary pressure head value changes greatly. If Equation (22) is substituted into Equation (6) to construct an equation system in which the saturation term is eliminated and the pressure head is used as the unknown, extremely high convergence accuracy is required to ensure that it gradually approaches the correct one during the linear selection process. In reality, this approach is extremely inefficient or even hopeless.

This paper selects and implements the numerical method of the hybrid formulation advocated by M.A. Celia et al (Celia, Bouloutas et al. 1990). For the processing of time steps, the method proposed by Milly (Milly 1985) is adopted. The problem of the conservation of mass is well solved. The space is discretized by the finite element method of the Galerkin scheme, and finally, the governing Equations (23), (24), and (25) are obtained.

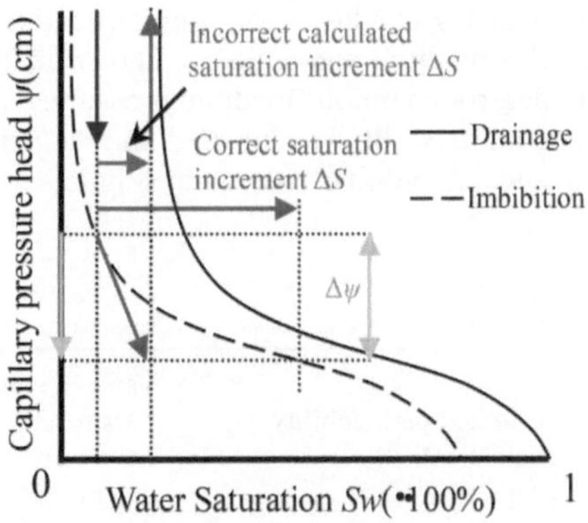

Figure 4. Schematic diagram of numerical difficulty caused by strong nonlinearity.

$$F\delta_{ai}^{n+1,m} + E\delta_{ni}^{n+1,m} - (E+F)\,\delta_{wi}^{n+1,m} - \sum_{j=1}^{N} \overline{k}_{rw}^{n+1,m} K_w \int_V \nabla N_i \nabla N_j dV \delta_{wj}^{n+1,m} = M_w - I_w \tag{23}$$

$$\begin{aligned} &-(F-H-G)\,\delta_{ai}^{n+1,m} - (E+G)\delta_{ni}^{n+1,m} + (E+F-H)\delta_{wi}^{n+1,m} \\ &-\sum_{j=1}^{N} \frac{\overline{k}_{rn}^{n+1,m}}{\mu_{rn}} K_w \int_V \nabla N_i \nabla N_j dV \delta_{nj}^{n+1,m} = M_n - I_n \end{aligned} \tag{24}$$

$$\begin{aligned} &\phi \frac{\rho_{atm}}{\psi_{atm}} S_a^{n+1,m} \frac{1}{\Delta t} \int_V N_i dV \delta_{ai}^{n+1,m} - \rho_a^{n+1,m} \sum_{j=1}^{N} \frac{\overline{k}_{ra}^{n+1,m}}{\mu_{ra}} K_w \int_V \nabla N_i \nabla N_j dV \delta_{aj}^{n+1,m} - \rho_a^{n+1,m}((G+H)\delta_{ai}^{n+1,m} \\ &+ G\delta_{ni}^{n+1,m} + H\delta_{wi}^{n+1,m}) = \rho_a^{n+1,m}(M_a - I_a) - \phi \int_V N_i dV \frac{\rho_{atm}}{\psi_{atm}} S_a^{n+1,m} \frac{\psi_a^{n+1,m} - \psi_a^n}{\Delta t} \end{aligned} \tag{25}$$

In the formula,

$$E = \frac{\tau_1}{\Delta t} \overline{C}_{nw}^{n+1,m} \int_V N_i dV; F = \frac{\tau_2}{\Delta t} \overline{C}_{aw}^{n+1,m} \int_V N_i dV; G = \frac{\tau_3}{\Delta t} \overline{C}_{an}^{n+1,m} \int_V N_i dV; H = \frac{\tau_4}{\Delta t} \overline{C}_{aw}^{n+1,m} \int_V N_i dV$$

$$I_f = \phi \int_V N_i dV \frac{S_f^{n+1,m} - S_f^n}{\Delta t}; M_f = \sum_{j=1}^{N} \frac{\overline{k}_{rf}^{n+1,m}}{\mu_{rf}} K_w \int_V \nabla N_i \nabla N_j dV \cdot \left(\psi_{fj}^{n+1,m} + \rho_{rf}^{n+1,m} z_j\right)$$

Where, n represents the time step, and m represents the number of iterations under the same time step. Where, $\overline{C}$ is the average value of C in the unit, and δ is the unknown variable.

4 VALIDATION OF SELF-DEVELOPED SOURCE PROGRAM MIG-3PF

According to the above principles and formats, the author developed the independent source code Mig-3PF after writing and debugging it in Fortran language. To verify its correctness, the author used the NAPL Simulator (Guarnaccia et al. 1997) and a self-compiled program to simulate the same model.

As shown in Figure 5, the boundary conditions and initial conditions have been marked and given. The property parameters of medium and fluid are shown in Table 1 and Table 2. It is a 20cm×30cm 2D model. The underground porous medium is assumed to be uniform and isotropic, and the total simulation time is 80min. NAPL was injected into the upper left corner of the model with a flow rate of 0.9 cm^3/min and a duration of 10 min to simulate the infiltration and migration process of NAPL.

Table 1. Physical properties of soil media.

Medium	Porosity	Saturated permeability coefficient (cm·s^{-1})	Van Genuchten parameters		NAPL residual saturation (100%)
			α (cm^{-1})	l (−)	
Sand	0.50	0.08	0.27	5.9	0.08

Table 2. Physical property of three-phase fluids.

Fluid	Density ($g \cdot cm^{-3}$)	Viscosity ($Pa \cdot s$)	Scaling parameters	β (−)
water	1.0	100×10^{-3}	water-air	1.00
NAPL	0.80	180×10^{-3}	water-NAPL	1.80
air	125×10^{-3}	1.0×10^{-5}	air-NAPL	2.25

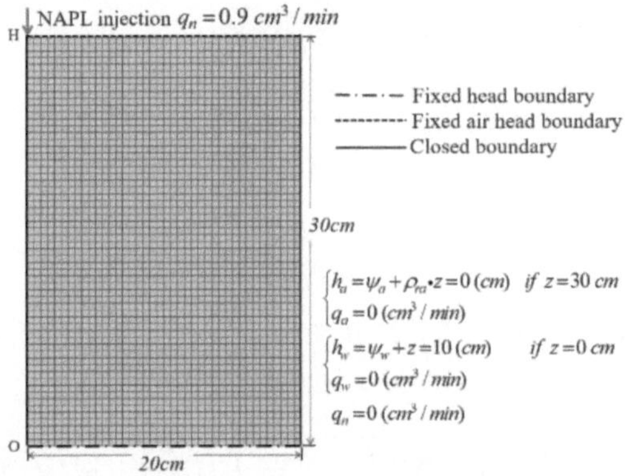

Figure 5. Schematic illustration of the model.

The initial distribution of water saturation before NAPL injection is shown in Figure 7. Among them, the calculation process of the NAPL Simulator is: at 0–30000s, the water head is given at the bottom of the model, the water moves upward, simulates the natural state water saturation distribution, and continuously injects NAPL during the period of 30000s–30600s. The self-programming program directly calculates the water saturation distribution in a steady state according to the given initial conditions, and continuously injects NAPL in the period of 0-600s.

At the initial moment, the relationship curve between water saturation S and Y coordinate values is shown in Figure 7. The curve is basically consistent, indicating that the autonomous program can correctly simulate the initial distribution of water.

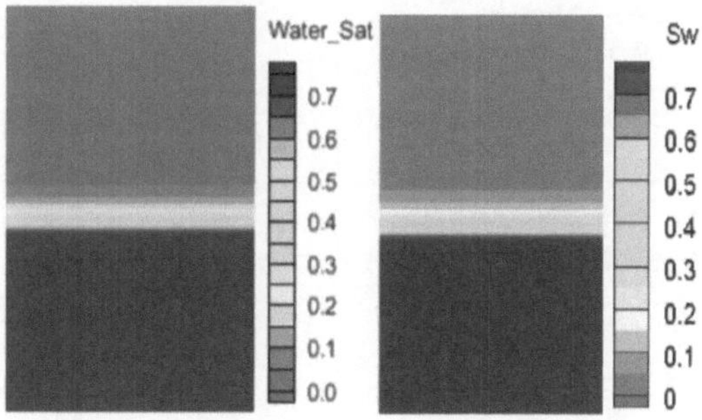

Figure 6. Initial distribution of water saturation.

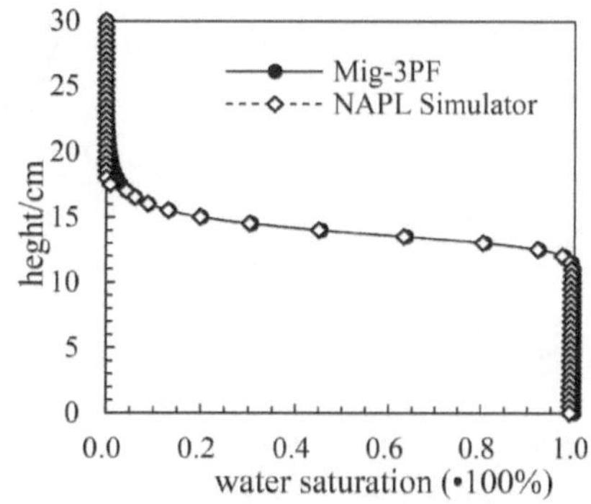

Figure 7. Distribution of Sw at the initial time (Left) NAPL Simulator; (Right) Mig-3PF.

The simulation results of the spatiotemporal distribution of NAPL saturation are shown in Figure 8. Since the injected NAPL is less dense than water, it floats above the water table.

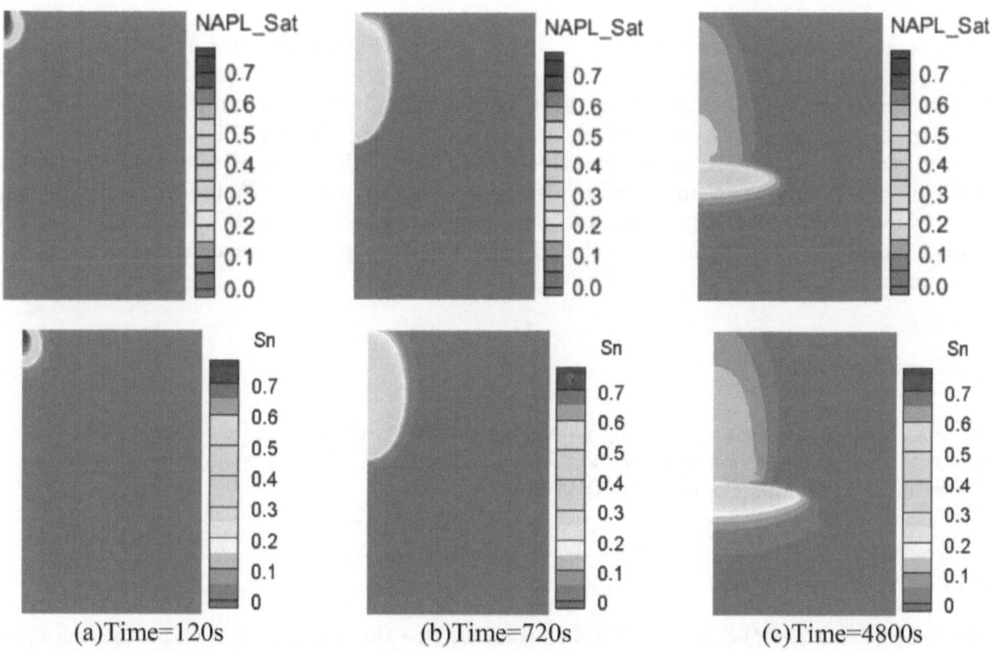

Figure 8. Distribution of NAPL saturation at each moment: (Up) NAPL Simulator; (Down) Mig-3PF.

At 4800s, the saturation of NAPL in the model follows the distribution curve of the OH line in the model, as shown in Figure 9. The NAPL saturation changes simulated by the two programs are basically the same, but there are still some differences. There are many reasons for the difference: the hysteresis models of the two are different; This section also points out that in setting the water saturation distribution, the two adopt different methods, and the initial state formed is not the same. In addition, the discrete meshes near the water surface are not refined, and the two finite element integration methods are different. In short, comparing the running results of the two programs, the simulated state process is roughly the same, and the difference has a cause and an acceptable range, which preliminarily proves the correctness of the development program.

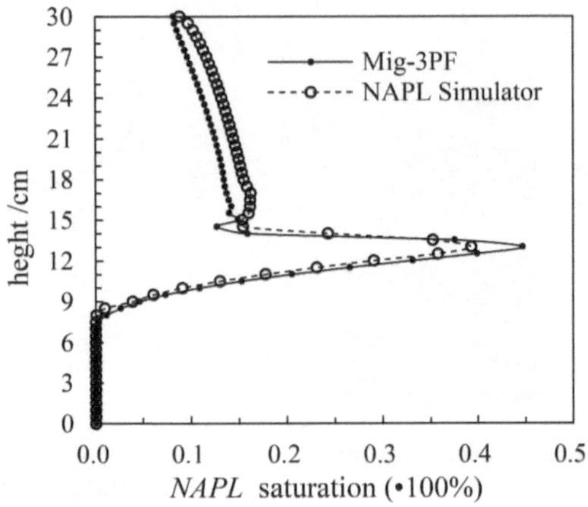

Figure 9. NAPL saturation distributing along model left side at time 4800 sec.

5 CONCLUSION

In this paper, the interaction mechanism of NAPL, water, and air in the process of NAPL passing through the soil unsaturated zone is described in detail. A hybrid method of equation solution and the time step is used to solve the numerical problems caused by strong nonlinearity. The discrete format of the governing equations is given, and the autonomous code Mig-3PF is written according to the format. The correctness of the autonomous code is preliminarily verified through the NAPL simulator. This program can simulate the infiltration of immiscible fluids into the subsurface and has preliminary quantitative guidance for predicting the distribution and changes of pollution. However, it still needs to be further developed and verified according to the actual problems in the laboratory and field.

ACKNOWLEDGMENTS

This work was financially supported by Wuhan University of Science and Technology, Open Foundation of State Environmental Protection Key Laboratory of Mineral Metallurgical Resources Utilization and Pollution Control (HB201904).

REFERENCES

Abriola L M, Pinder G F. (1985a) A Multiphase Approach to the Modeling of Porous Media Contamination by Organic Compounds: 2. Numerical Simulation[J]. Water Resources Research. 21(1): 19–26.

Abriola L M, Pinder G F. (1985b) A multiphase approach to the modeling of porous media contamination by organic compounds; 1, Equation development[J]. Water resources research. 21(1): 11–18.

Celia M A, Bouloutas E T, et al. (1990) A general mass-conservative numerical solution for the unsaturated flow equation[J]. Water Resources Research. 26(7): 1483–1496.

Dong W F, Fu D Q. (2009) Review on Environmental Pollution Accidents in Recent Years in China[J]. Environmental Science & Technology, 32(07): 75–77.

Guarnaccia, J., Pinder, G., Fishman, M. (1997) NAPL: Simulator documentation. USA: National Risk Management Research Laboratory, US Environmental Protection Agency.

Kamaruddin S A, Sulaiman W N A, Rahman N A, et al. (2011) A Review of Laboratory and Numerical Simulations of Hydrocarbons Migration in Subsurface Environments[J]. Journal of Environmental Science and Technology. 4(3): 191–214.

Leverett M C. (1941) Capillary Behavior in Porous Solids[J]. Transactions of the AIME. 142(01): 152–169.

Liu Y, Cheng L, Ding A, et al. (2010) Quick Assessment of Contamination Threat to Groundwater after NAPL Spill[C]. IEEE.

Ma J. (2021) Problems and suggestion for investigation and risk assessment of sites impacted by volatile organic compounds in China[J]. Chinese Journal of Environmental Engineering, 15(1): 3–7.

Milly P C D. (1985) A mass-conservative procedure for time-stepping in models of unsaturated flow[J]. Advances in Water Resources. 8(1): 32–36.

Mualem Y. (1976) A new model for predicting the hydraulic conductivity of unsaturated porous media[J]. Water Resources Research. 12(3): 513–522.

Parker J C, Lenhard R J. (1990) Determining three-phase permeability-saturation-pressure relations from two-phase system measurements[J]. Journal of petroleum science & engineering. 4(1): 57–65.

van Genuchten M T. (1980) A Closed-form Equation for Predicting the Hydraulic Conductivity of Unsaturated Soils[J]. Soil Science Society of America Journal. 44(5): 892–898.

Yang M X, Yang Y S, Du X Q. (2013) Organic fractions and their environmental implications of petroleum contaminated groundwater[J]. China Environmental Science, 33(06): 1025–1032.

Advances in Mineral Resources, Geotechnology and Geological Exploration – Rashid & Zhang (eds)

Determination of thickness of the reserved protective layer of temporary filling body

Yang Liu*, Weihua Wang & Yiwen Dai
School of Resources and Security Engineering, Central South University, Changsha, Hunan, China

ABSTRACT: With gradual depletion of shallow resources, deep mining has become an inevitable trend in acquiring resources in the 21st century. However, the stress environment and engineering geological conditions of deep rock mass are more complex than those in shallow ones, and the ground pressure management is the main problem faced, while backfill mining can effectively solve the ground pressure problem in deep mining. Therefore, backfill mining is being widely used. The key technology in backfill mining is to control blasting and protect backfill. Currently, most of the mines that adopt approach filling mining method face problems such as over-excavation and under-excavation, which are serious issues, the amount of approach support works is large, and the maintenance cost is high. The filling body is seriously damaged during mining, which leads to serious tailing mixing and abnormal dilution and loss index. The construction cost is high. The root cause lies in the lack of effective protection and control of the filling body. In this paper, starting from the practice of route mining in the Jiaojia Gold Mine, through theoretical analysis and numerical simulation, the thickness of the reserved protective layer for blasting the temporary filling body is determined, the blasting scheme is optimized, the field industrial test is carried out, and good results obtained.

1 INTRODUCTION

During deep mining, ground pressure management is the main challenge, and filling mining can effectively alleviate the ground pressure problem in deep mining (Dai et al. 2018). When the ore body enters the deep portion, the application of backfill mining becomes more extensive, and the key problem in backfill mining is the protection of backfill. Hu and Kemeny (1994) studied the relationship between dynamic blasting load and the stability of the filling body by using the principle of fracture mechanics and obtained the damage of the filling body under different loads. Ylmaz et al. (2014) obtained the relationship between the blasting stress wave and the strength of the filling body through ultrasonic testing. Bary et al. (2009) analyzed the effective elastic characteristics of cemented fillings by PFC. Liu and Li (2004) explored the reflection and refraction law of explosion stress wave at the interface of filling body through theoretical analysis and blasting test and analyzed the influence of blasting a dynamic load on filling body. Wen et al. (2021), based on the field blasting vibration test in the mining area, analyzed and demonstrated the influence of stope blasting vibration on the stability of filling on both sides. Huang et al. (2021) combined engineering examples and using the stress wave theory, analyzed the propagation of explosion stress wave in full tailings cemented backfill and its interaction with the medium interface. Based on the different failure mechanisms and failure criteria of the backfill, such as impact compression shear, shear, and reflection tension, the particle vibration velocity threshold of the backfill was studied. Zhu et al. (2018) studied the damage mechanism of the backfill under blasting dynamic load and put forward the calculation method of stress generated by blasting stress waves passing through the interface of the backfill. The numerical simulation carried out provides a certain theoretical basis for the optimization of blasting parameters and the stability analysis of the backfill in the two-step

*Corresponding Author: 1137510136@qq.com

DOI 10.1201/9781003319412-36

stopping of the backfill mining method. Hu et al. (2011) analyzed the action mechanism of pre-crack on explosion stress wave in blasting near filling body by ANSYS/LS-Dyna and concluded that pre-crack has good vibration reduction effect, and the critical interface of the filling body also has certain vibration reduction effect. Near the explosion source, the energy accumulation effect appears at the pre-crack and filling critical interface, and the closer to the pre-crack and the interface, the stronger is the energy-gathering effect. Jiang et al. (2020), based on the results of numerical simulations and field monitoring, studied the dynamic response mechanism of layered cemented backfill pillar under far-field blasting disturbance. Yang et al. (2018) studied the damage law of filling body caused by blasting load, discussed the relationship between dynamic compressive strength and the distance between blasting sources, and put forward several new measures to prevent the filling body from being damaged.

Because the blasting load will be smaller with the increase in distance from the blasting source, studying the damaged area of rock and backfill after blasting can effectively reduce the damage to surrounding rock and backfill. In addition, when smooth blasting is used, the charge of the second ring hole near the filling body is much larger than that of the smooth blasting hole, which may also damage the filling body. Therefore, it is very important to consider the influence of the explosion stress wave generated during the second hole blasting on the filling body. Therefore, to achieve controlled blasting effect, the optical key is to optimize the thickness of the reserved protective layer and the reasonable distance between the second circle of holes and the filling body.

2 THEORETICAL CALCULATION OF RESERVED PROTECTIVE LAYER THICKNESS

2.1 *Basic mechanical parameters*

The blasting performance of rock and numerical simulation results are closely related to rock mechanics parameters. The optimization of blasting performance of rock and reserved protective layer thickness cannot be separated from the rock mechanics test results and some basic mechanical parameters of existing fillings in mining areas. See Table 1 for the basic mechanical parameters of rock and fillings.

Table 1. Basic mechanical parameters of ore rocks and fillings.

Material type	Density/m^3	Compressive strength/ MPa	Tensile strength/ MPa	Modulus of elasticity/ GPa	Poisson's ratio	Bonding force/ MPa	Internal friction angle/ °
Ore rock	2811	68.6	8.85	26.2	0.24	17.86	38.3
1:4 filling body	1788	3.91	0.43	0.58	0.29	0.30	38.9
1:10 filling body	1819	1.75	0.16	0.18	0.32	0.16	44.4

2.2 *Theoretical calculation model*

The transflection of explosion stress wave at the interface B–B between ore body and filling body is approximated as the transflective propagation of one-dimensional stress wave at the interface of dissimilar media (Li 2014), as shown in Figure 1.

According to the one-dimensional stress wave theory:

$$\begin{cases} P_i + P_r = P_t \\ v_i + v_r = v_t \\ P = \rho D v \end{cases} \tag{1}$$

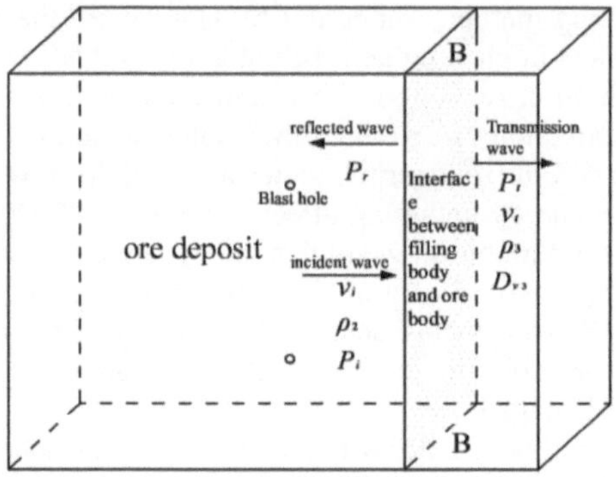

Figure 1. Transflection of stress wave at the interface between rock and filling body.

Can launch:

$$\begin{cases} P_r = \dfrac{\rho_3 D_{v3} - \rho_r D_{v2}}{\rho_r D_{v2} + \rho_3 D_{v3}} P_i \\ P_t = \dfrac{2\rho_3 D_{v3}}{\rho_r D_{v2} + \rho_3 D_{v3}} P_i \end{cases} \tag{2}$$

Type: ρ_3 — Filling density; ρ_r — Rock density; D_{v2} — rock wave velocity; D_{v3} — wave velocity of filling body; P_i — Incident stress; P_r — reflected stress; P_t — transmission stress.

When the filling body explodes, the shock wave generated by the explosion will be transmitted into the ore body and will be immediately attenuated into a stress wave. When the stress wave is transmitted into the interface between the ore body and the filling body, the stress wave transmitted from the ore body will be transmitted and reflected at the interface. According to the properties of the ore body and the filling body, the stress wave passes through the interface, that is, the transmitted wave is a compression wave. However, the stress wave reflected from the contact surface is a tensile wave, and the interface between the ore body and the filling body can only basically transmit a compressive wave. When the compressive stress generated by the transmitted compressive wave is greater than the dynamic compressive strength of the filling body, the filling body will be damaged or even collapsed. Therefore, to protect the filling body it must be ensured that transmission stress generated by blasting is less than the dynamic compressive strength of the filling body.

2.3 *Blasting load in the filling body*

For cylindrical charge, when air interval charge is used, the impact pressure on the hole wall is:

$$P_2 = \frac{1}{8}\rho_0 D_{v1}^2 \left(\frac{d_c}{d_b}\right)^6 \left(\frac{l_c}{l_c + l_a}\right) n \tag{3}$$

Type: d_c – explosive diameter; d_b – Borehole diameter; l_c – Charge length; l_a – Air interval length.

Near the explosion source, the rock is approximated as a fluid. After the explosive explodes, its blasting load is transmitted into the rock, forming a transmitted shock wave, and the shock wave will quickly spread outward and immediately decay into a stress wave (Dai 2001). The attenuation law of transmitted shock wave in rock can be expressed by formula (4):

$$P = P_2 \bar{r}^{-\alpha} \tag{4}$$

Type: P – peak pressure at the calculation point, P_2 – Impact pressure on the hole wall, $\bar{r}$ – specific distance, $\bar{r} = r/r_b$; r – Calculate the distance from the point to the charge center, m; r_b – Borehole radius, m; α – Load propagation attenuation index, $\alpha = 2 \pm \mu_d/1 - \mu_d = 2 \pm b$, b – Lateral stress coefficient; μ_d – Dynamic Poisson's ratio, within the loading rate range of engineering blasting, includes: $\mu_d = 0.8\ \mu$, μ – Static Poisson's ratio of rock; in the shock wave zone of rock, that is, the rock crushing zone, $\alpha = 2 + \mu_d/1 - \mu_d = 2 + b$. In is the stress wave area of rock, that is, the blasting crack area of rock is $\alpha = 2 - \mu_d/1 - \mu_d = 2 - b$.

Combined with formulas (2), (3), and (4), the transmission and reflection stress between ore body and filling body under the condition of uncoupled charging are as follows:

$$\begin{cases} P_r = \dfrac{\rho_3 D_{v3} - \rho_r D_{v2}}{8(\rho_r D_{v2} + \rho_3 D_{v3})} \rho_0 D_{v1}^2 \left(\dfrac{d_c}{d_b}\right)^6 \left(\dfrac{l_c}{l_c + l_a}\right) n\bar{r}^{-\alpha} \\ P_t = \dfrac{\rho_3 D_{v3}}{4(\rho_r D_{v2} + \rho_3 D_{v3})} \rho_0 D_{v1}^2 \left(\dfrac{d_c}{d_b}\right)^6 \left(\dfrac{l_c}{l_c + l_a}\right) n\bar{r}^{-\alpha} \end{cases} \tag{5}$$

According to the existing blasting mining conditions in the mining area, an uncoupled charge is adopted, which is a columnar charge. For smooth blasting hole, the length of hole charge $l_c = 0.44$ m, air interval length $l_a = 0.7$ m; choose no. 2 rock emulsion explosive with cartridge diameter $d_c = 32$ mm, length of single cartridge $l_c' = 22$ cm, single cartridge weight 0.2 kg, explosive density $\rho_0 = 1000$ kg/m^3, detonation velocity $D_{v1} = 3600$ m/s, density of rock $\rho_r = 2811$ kg/m^3, wave velocity $D_{v2} = 4000$ m/s. Filling density $\rho_3 = 1818.5$ kg/m^3, wave velocity $D_{v3} = 1800$ m/s, and the statistical results of transmission stress are shown in Table 2.

Table 2. Transmission stress at filling the body with different protective layer thickness.

Spacing/m	0.2	0.3	0.35	0.4	0.5
Filling internal load/MPa	10.39	5.08	3.87	3.06	2.07

According to the mechanical test of filling materials, the uniaxial compressive strength of filling materials with the cement-sand ratio 1:4 is 3.91 MPa, and that of filling materials with the cement-sand ratio 1:10 is 1.75 MPa. For the filling of road, the filling materials with the cement-sand ratio 1:10 are used in most areas of the road, and the filling materials with the cement-sand ratio 1:4 are used only near the top, and the filling materials with the cement-sand ratio of 1:4 are more than the cement-sand ratio of 1. The uniaxial compressive strength of the filling body of 10 is much larger, so the damage of the filling body of 1:10 lime-sand ratio is mainly considered. The research (Dai, 2002) shows that, under the blasting load, the dynamic compressive strength of rock will increase with the increase of loading rate, and the dynamic compressive strength of the filling body will also increase with the increase of loading rate. To ensure stability of the filling body, the cement-sand ratio is 1:10, and the critical value of dynamic compressive strength of the filling body is 3.50 MPa, that is, when the stress introduced into the filling body by the blasting load is greater than the dynamic compressive strength of the filling body of 3.50 MPa, it is considered that the filling body will be damaged at this time.

From Table 2, it can be seen that when the reserved protective layer thickness is greater than or equal to 0.4 m, the filling body will be in a safe state when blasting mining is carried out on the route; hence, the theoretical value of the reserved protective layer thickness is 0.4 m.

3 NUMERICAL SIMULATION OF BLASTING PARAMETERS OF TEMPORARY FILLING BODY

3.1 *Numerical calculation of reserved protective layer thickness*

(1) Numerical calculation model

In the model of reserved protective layer thickness, the hole spacing of smooth blasting holes is 60 cm, the hole diameter is 42 mm, the hole length is 2.2 m, the cartridge diameter is 32 mm, the cartridge length is 220 mm, the smooth blasting holes are filled with two charges, and the holes are blocked by 30 cm with air interval charging. Except for the free surface, all other interfaces are set as non-reflective boundary surfaces. Establish calculation models of the reserved protective layer thickness of 30 cm, 40 cm, and 50 cm respectively. The three-dimensional model with a thickness of 30 cm protective layer reserved is shown in Figure 2. The G-cm-S unit system is adopted, and the numerical calculation time is 1 ms.

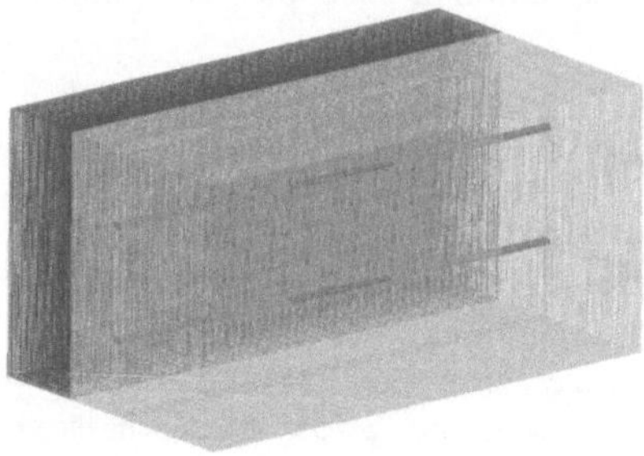

Figure 2. Stereo model with 30 cm protective layer thickness reserved.

(2) Model calculation results and analysis

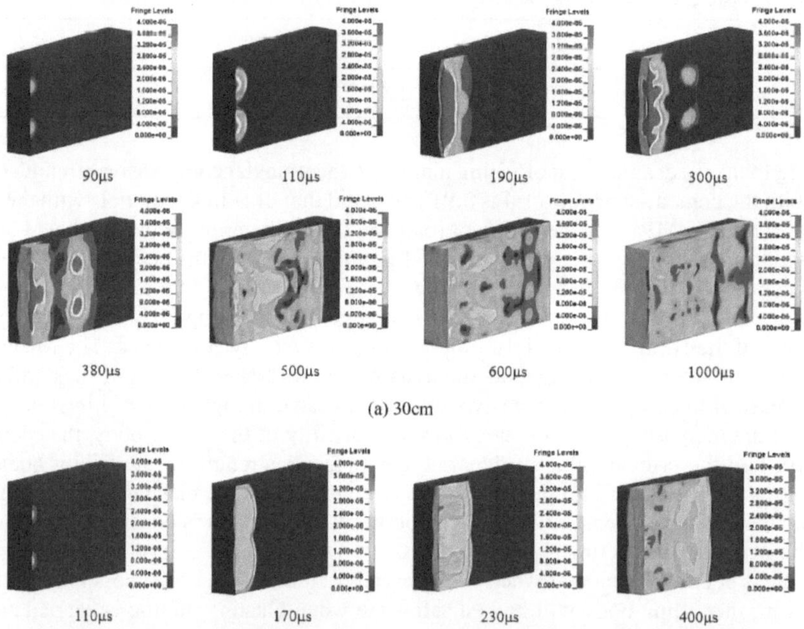

(a) 30cm

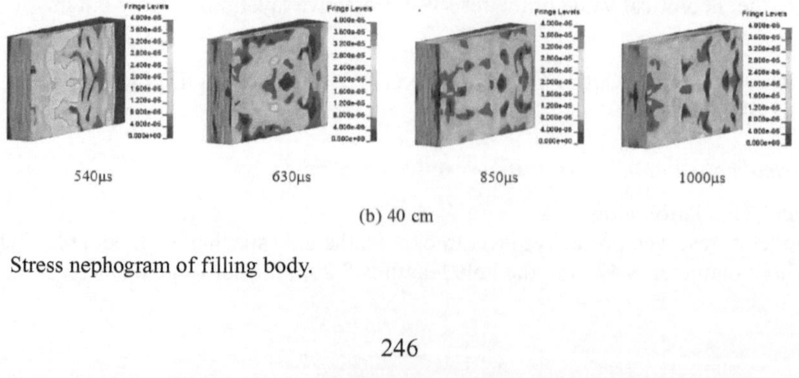

(b) 40 cm

Figure 3. Stress nephogram of filling body.

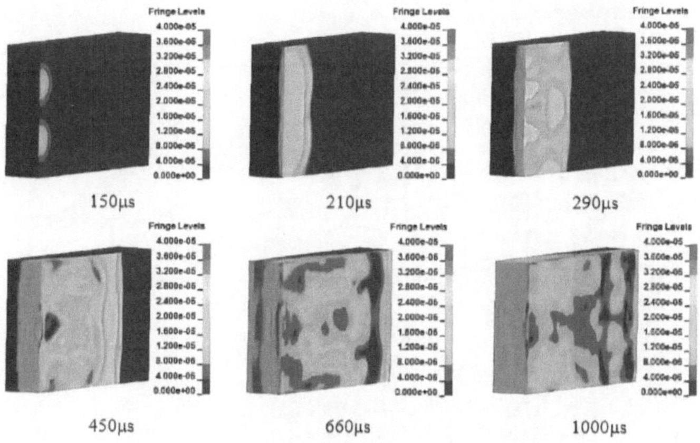

(c) 50 cm

Figure 3. Continued.

It can be seen from the table that when the protective layer thickness of 40 cm and 50 cm is reserved, there is no obvious stress danger area in the filling body. However, the stress of each unit point is not intuitive to determine from the stress cloud map, and only the rough danger area can be obtained from the stress cloud map, which is mainly concentrated near the charge at the bottom of the blast hole. Therefore, it is only necessary to analyze the stress in the filling body near the bottom of the charge to determine whether the filling body will be damaged or not. To ensure reliability of the analysis results, on the contact surface of the filling body near the bottom charge, a unit is taken as a recording point every 20 cm, as shown in Figure 4, the stress-time curve of each unit point is extracted, and the stress peak value of each unit point is recorded.

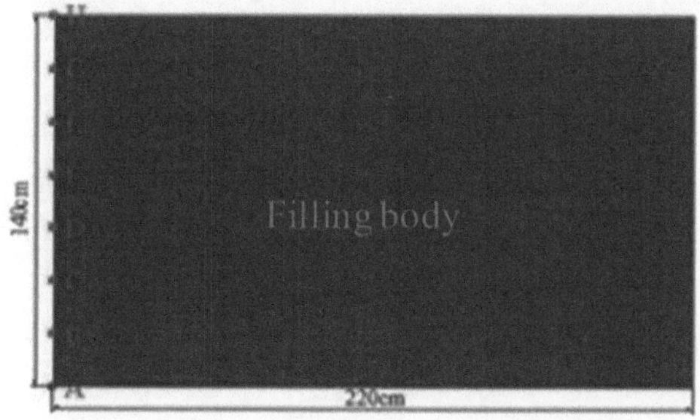

Figure 4. Schematic diagram of taking points at the interface of filling body.

The stress-time curve of each recorded point on the contact surface of the filling body near the bottom charge is shown in Figure 5.

From Figure 5, we can determine the specific stress of each recording point at each moment. As can be seen from the above figure, when a 30 cm protective layer is reserved, the stress of the extracted unit recording point is obviously higher than that when the distance between the smooth blasting hole and the filling body is 40 cm and 50 cm. According to Figure 4, in three simulation schemes with point A as the coordinate origin and 30 cm, 40 cm, and 50 cm protective layers

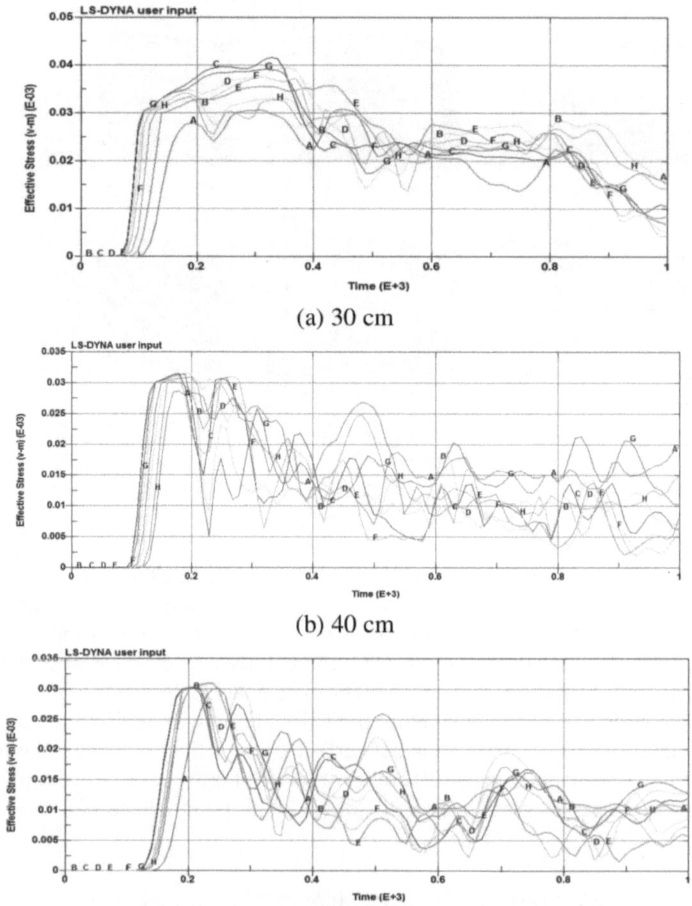

Figure 5. Stress-time curve of the unit point at the bottom of the interface.

reserved, Figure 6 shows the change of the stress peak value of each recording point on the contact surface of the filling body near the bottom charge with distance.

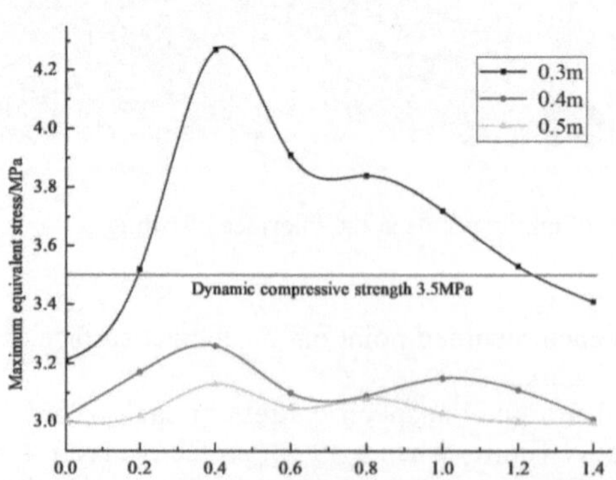

Figure 6. Equivalent stress peak of the unit point at the bottom of the interface.

It can be seen from the above figure that when a protective layer of 30 cm is reserved, there are 8 recording unit points selected on the contact surface of the filling body close to the bottom charge, among which the peak value of equivalent stress of 6 unit points is higher than 3.5 MPa, which indicates that when the smooth blasting hole is 30 cm away from the filling body, there is considerable risk of damage to the filling body, which will lead to the collapse of the filling body and is not conducive to the safety and stability of the entry stope. At the same time, it will also increase the ore dilution rate and increase the mining cost. However, when the smooth blasting hole is 40 cm and 50 cm away from the filling body, there is no obvious danger area in the stress cloud map of the filling body, and the stress peak value of all selected unit recording points does not reach 3.5 MPa. Therefore, it can be considered that when 40 cm and 50 cm protective layers are reserved, the filling body will not be damaged basically, as the charge of smooth blasting hole is less. When the safety protective layer thickness of 50 cm is reserved, large blocks or under excavation of route contour may occur during route blasting, hence the reserved protective layer thickness is 40 cm.

3.2 *Numerical calculation of the distance between the second ring hole and the filling body*

In the process of route mining blasting, because the charge of the second ring hole adjacent to the filling body is much larger than that of the smooth blasting hole, if the second ring hole is close to the filling body, the adjacent filling body may be damaged during the second ring hole blasting. For the second ring of holes, the row spacing is generally 0.7 m–0.9 m. In numerical simulation, the row spacing is 0.8 m, and the distance between the second ring of holes and the filling body is 0.9 m, 1.0 m, and 1.1 m, respectively. Schemes 1, 2, and 3.

(1) Model building

In the process of establishing the model, consider the numerical calculation model with the distance of 0.9m from the second ring hole to the filling body as an example. The three-dimensional section of the second ring hole is shown in Figure 7. Except for the free surface, the periphery of the model is set as a non-reflective boundary. The row spacing of auxiliary holes is 0.8 m, and a single hole is filled with seven charges, with a total charge length of 1.54 mm. A continuous charge is adopted, with a single hole charge of 1.4 kg, a hole length of 2.2 m, and a hole blockage of 0.3 m mG-cm-s unit system is adopted in the modeling process, and the calculation time is 3 ms.

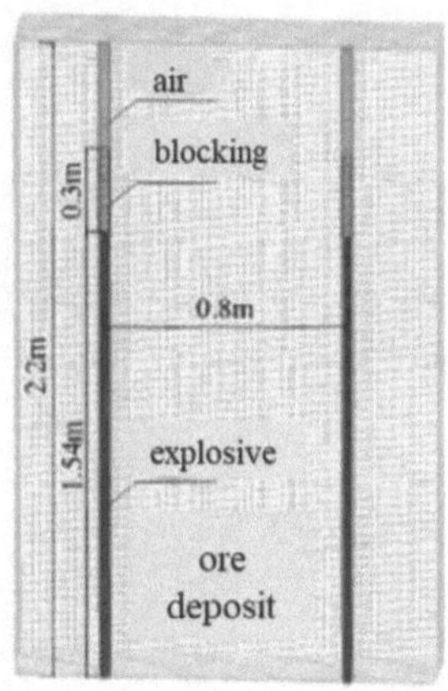

Figure 7. Three-dimensional section of the second circle auxiliary hole.

(2) Simulation results and analysis

From the stress nephogram of the filling body of the above three simulation schemes, it can be seen that there are obvious dangerous areas on the filling body in schemes 1 and 2. In the first scheme, at about 1500 μs, a dangerous area is seen at the bottom of the filling body; at about 780 μs, the dangerous area at the bottom of the filling body is the largest, and at 2450 μs, the dangerous

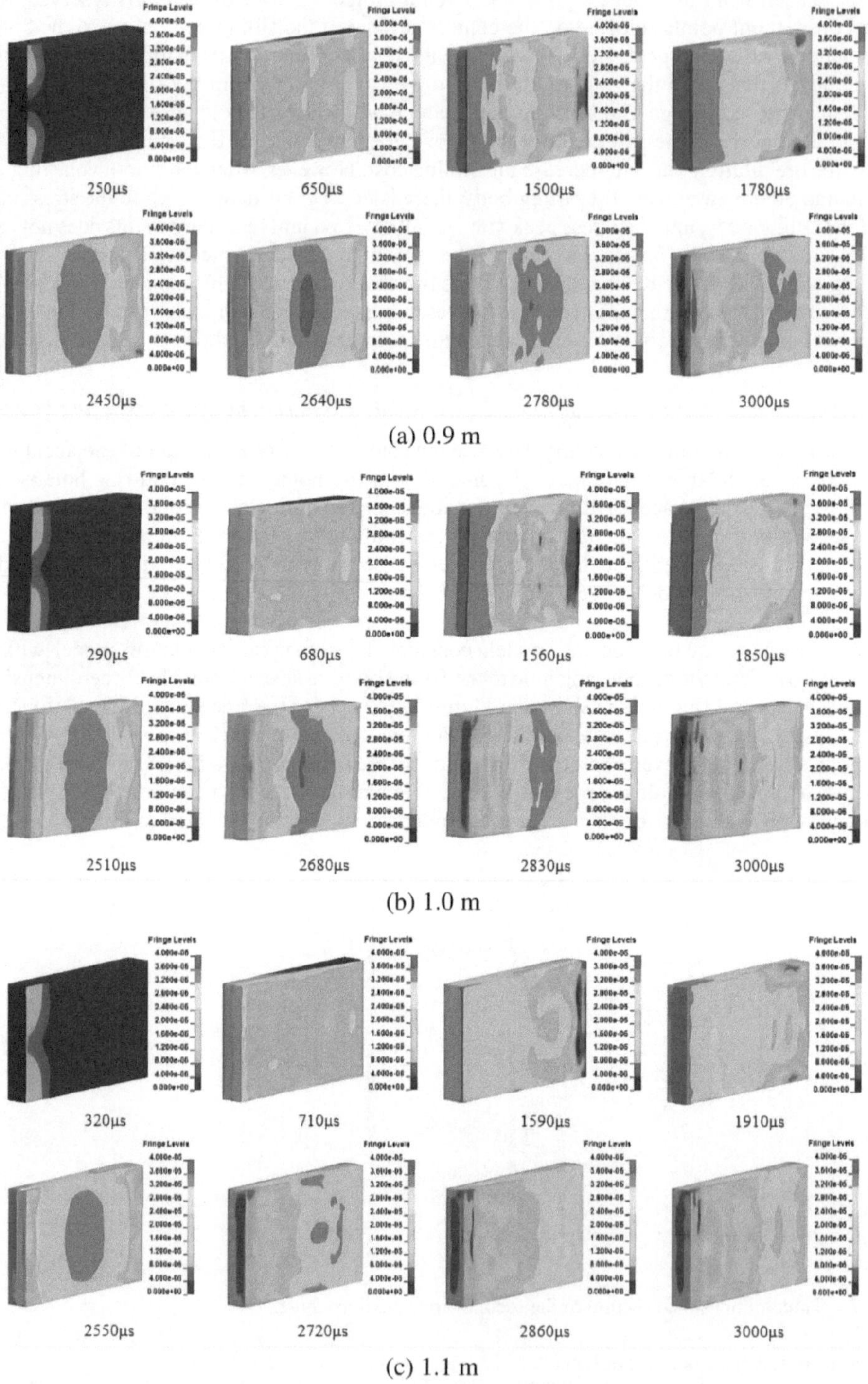

Figure 8. Stress nephogram of filling body.

area begins to appear in the middle of the interface between the ore body and the filling body. At about 640 μs. The danger reaches the maximum; at 3000 μs, the stress inside the filling body is still large. Option 2; at about 560 μs, a dangerous area appears at the bottom of the filling body; at about 850 μs, the dangerous area at the bottom of the filling body is the largest, and at 2510 μs, the dangerous area begins to appear in the middle of the interface between the ore body and the filling body. At about 680 μs, the dangerous area reaches the maximum; at 3000 μs, the stress inside the filling body is obvious in the stress in the first scheme. In the third scheme, around 1910 μs, the stress at the bottom of the filling body reached its peak, but there was no obvious dangerous area. At 2550 μs, a dark stress area appeared in the middle of the interface between the ore body and the filling body, and the filling body might be damaged. At 3000 μs, the stress inside the filling body is already small. To ensure reliability of the analysis results, a unit is still taken as a recording point every 20 cm on the contact surface of the filling body near the bottom charge, the stress-time curve of each unit point is extracted, and the stress peak value of each unit point is recorded. In the three schemes, the stress-time curve of the recording unit point on the contact surface of the filling body near the bottom charge is shown in Figure 9.

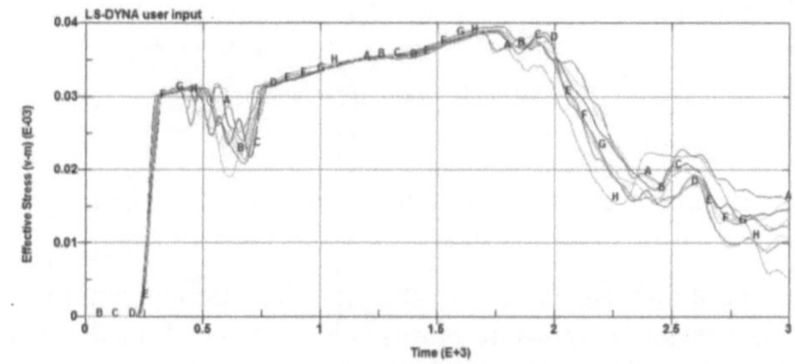

(a) 0.9 m

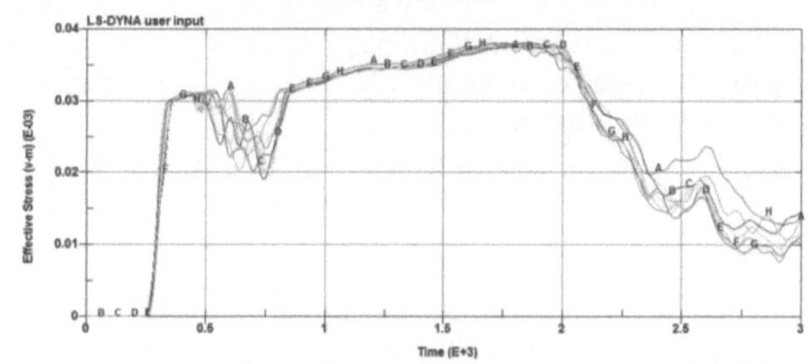

(b) 1.0 m

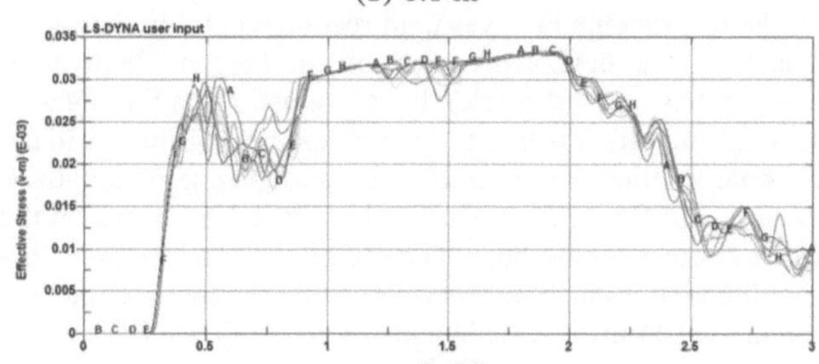

(c) 1.1 m

Figure 9. Stress-time curve of the unit point at the bottom of the interface.

As can be seen from the above figure, the stress peak at the bottom of the filling body in scheme 1 is around 1780 μs; in the second scheme, the peak stress at the bottom of the filling body is about

1850 μs; and in the third scheme, the peak stress at the bottom of the filling body is about 1910 μs, which is consistent with the stress cloud images of the three simulation schemes, thus ensuring the reliability of the analysis results. With point A as the coordinate origin, in the three simulation schemes, the change in the stress peak value of each recording point on the contact surface of the filling body near the bottom charge with distance is shown in Figure 10.

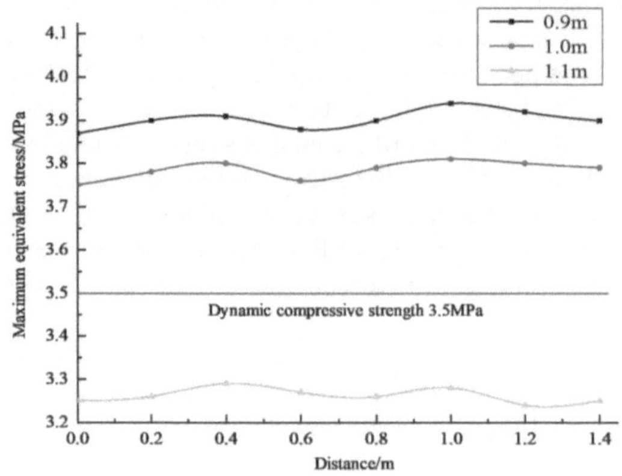

Figure 10. Equivalent stress peak of the unit point at the bottom of the interface.

It can be seen from the above figure that in the first and second schemes, the peak value of average effect force at the selected eight-unit points is higher than 3.5 MPa, and there is great risk of damage to the filling body at this time; however, in the third scheme, there is no dangerous area in the stress nephogram of the filling body, and the stress peak value of all selected unit recording points does not reach 3.5 MPa. Therefore, when the auxiliary hole is 1.1 m away from the filling body, the filling body will not be basically damaged. Therefore, the distance between the second auxiliary hole and the filling body is 1.1 m.

4 FIELD TEST

4.1 *Blasting test scheme design*

The cutting method of the test scheme involves unidirectional cutting, that is, the bottom hole is a cut hole. Field practice shows that downward extrusion cutting can effectively reduce the number of drilling holes in a single blasting and shorten the cycle operation time of a single blasting. The depth of the cut hole is 2.3 m, which is 0.1 m deeper than that of the auxiliary hole and smooth blasting hole. A single hole is filled with nine charges, and the charge length is 1.98 m. The hole depth of the auxiliary hole and the smooth blasting hole is 2.2 m, in which the auxiliary hole is filled with seven explosives in a single hole. The charge length is 1.54 m, the hole spacing of smooth blasting holes is 0.6 m, the single hole is filled with two charges, and the air interval charge is used. The charge length is 0.44 m, and the blocking length of each hole is 0.3 mm. The blasting design is shown in Figure 11.

The blasting experiment was carried out in the No. 5 approach of the J14011108 test stope in the middle section of 14. The bottom was filled with ore and rock on both sides. Capitalized Roman numerals in the figure represent the types of blast holes, that is, I mean cut holes (bottom holes), II means auxiliary holes, III means two smooth blasting holes, and IV means vault smooth blasting holes. Arabic numerals after capital Roman numerals represent the initiation sequence, that is, the initiation sequence of each hole is as follows: 2-3-4-5-6-7-8. The number of holes is as follows:

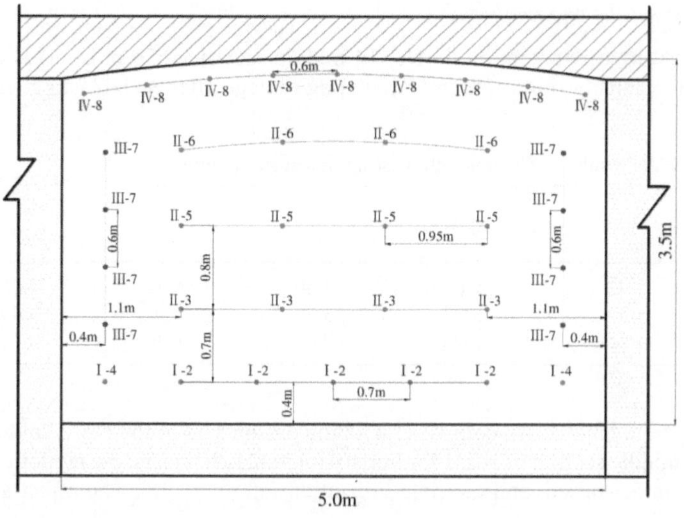

Figure 11. The bottom and both sides are filled with bodies.

Cut holes: The hole spacing is 0.7 m–0.8 m, and the number is 7; Smooth blasting holes: The hole spacing is 0.6 m, and the number is 17; Auxiliary holes: The hole spacing is 0.9 m–1.0 m, and the number is 12.

All blast holes are initiated at the bottom of the hole, and the charging structures of cut holes, auxiliary holes, and smooth blasting holes are shown in Figure 12.

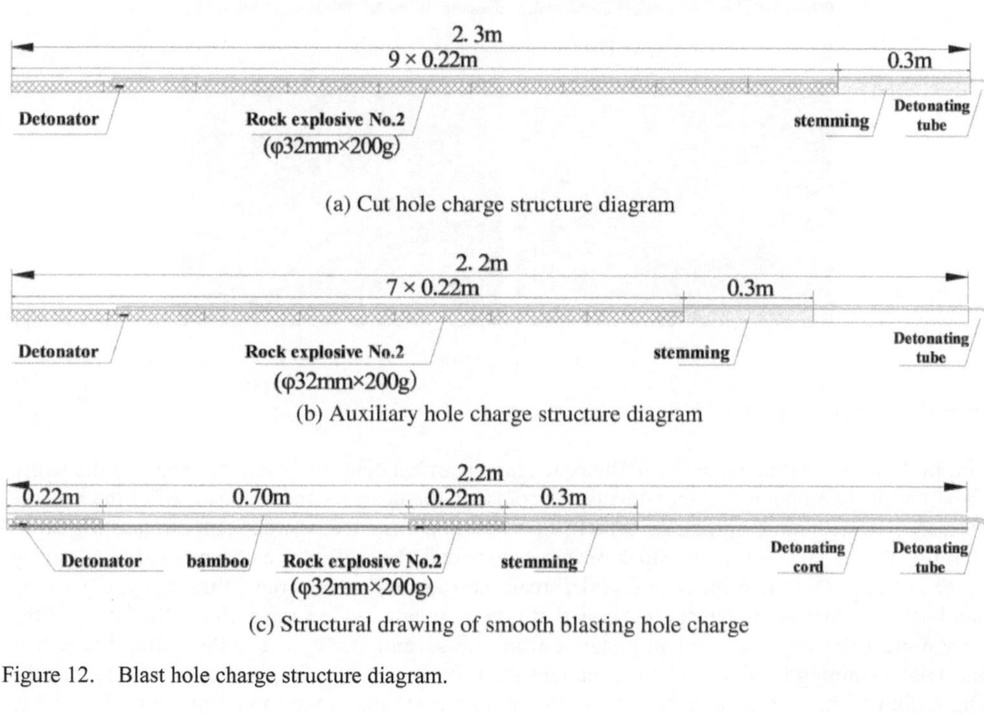

Figure 12. Blast hole charge structure diagram.

4.2 *Blasting test results and analysis*

Six blasting tests were carried out in the No. 14 middle section (-630 m level) J14011108 stope, and the statistical results of route size and circulating footage after each blasting test are shown in Table 3.

Table 3. Experimental results of 5# approach blasting in test stope (unit: m).

Number of experiments	1	2	3	4	5	6
Wide	4.93	5.05	4.80	4.95	4.85	4.96
Tall	3.43	3.52	3.45	3.40	3.49	3.47
Footage	2.01	1.98	2.04	1.99	2.05	2.03

It can be seen from table 3 that compared with the design size of the No. 5 route in J14011108 test stope in the middle section of No. 14, after six route blasting tests, the contour is 1.5% under-excavated, 1.1% under-raised, and the hole utilization rate is 91.7%. The effect after blasting is shown in Figure 13.

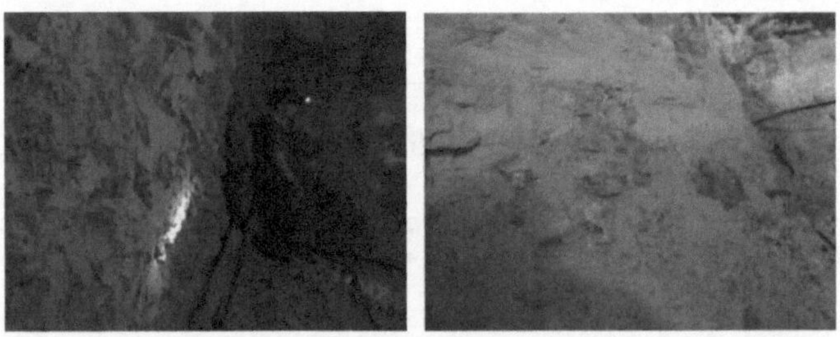

(a) The two gangs on the route

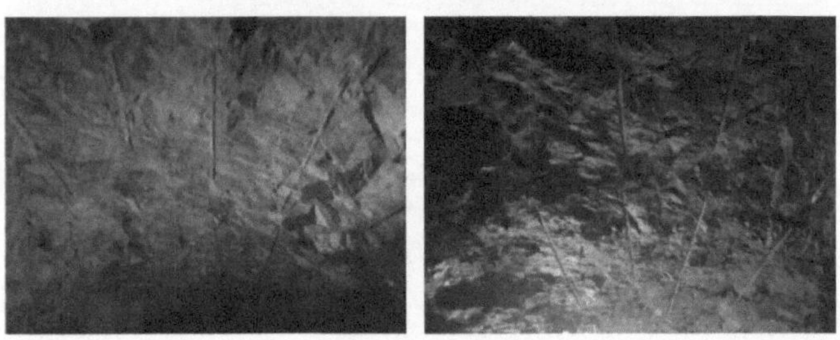

(b) Access roof

Figure 13. 108 Blasting test of 5# approach in test stope.

From the above statistical data and the post-blasting effect diagram, it can be seen that the width and height of the stope in the route test is very close to the design size, and the roof of the route is relatively flat, with more half holes, which shows that the distance between smooth blasting holes is 0.6 m. With regard to the two sides of the access road, the wall of the access road is relatively flat, basically without overbreak and underbreak, and for one side of the filling body, the filling body basically does not collapse. It shows that it is reasonable to reserve a 40 cm thickness of the safety protective layer and a 1.1 m distance from the second circle hole to the filling body. It is beneficial to mining by adjacent drift, reduces the mixing rate of tailings of backfill, reduces the mining dilution rate, considerably reduces the mining cost, and at the same time improves safety and stability of drift stope.

5 CONCLUSIONS

(1) The attenuation law of blasting load in rock was analyzed. According to the transflective law of one-dimensional stress wave at the interface of dissimilar media, the blasting load transmitted into the filling body was calculated, and the theoretical thickness of the protective layer reserved when blasting near the filling body was determined.
(2) According to the theoretical calculation and field practice, combined with the laboratory mechanical experiment of rock and filling body, the numerical calculation models of reserved 0.3 m, 0.4 m, and 0.5 m safety protective layer and 0.9 m, 1.0 m, and 1.1 m distance between the second circle hole and the filling body were established. By analyzing the overall stress nephogram of each simulation scheme, the stress nephogram of the filling body, and the stress-time curve of the unit recording point, the reasonable distance between the smooth blasting holes and the filling body was determined, that is, the reserved protective layer thickness was 0.4 m, and the distance between the two holes and the filling body was 1.1 m.
(3) The blasting engineering test was carried out in the J14011108 stope in the 14th middle section, and the results of the blasting test were statistically analyzed. It was found that the outline size of the route was basically consistent with the design, the outline of the roof and the two sides was flat, the overbreak and underbreak were controlled at about 1.5%, and the filling body was basically not damaged, which effectively reduced the supporting engineering quantity and the mixing rate of the filling body and the mining cost. The goal of safe and efficient mining in the Jiaojia Gold Mine was achieved, which provides some reference for the optimization of blasting parameters of the temporary filling body in drift filling mining at home and abroad.

REFERENCES

Bary. B & M. Ben Haha, et al. (2009). Numerical and analytical effective elastic properties of degraded cement pastes. J. Cement and ConcreteResearch. 39: 902–912.
Dai. J (2001). Calculation of crushing circle and fracture circle of rock in cylindrical charge blasting. J. Journal of Liaoning Technical University. 20(2):144–147.
Dai. J (2002). Rock dynamic characteristics and blasting theory. M. Beijing: Metallurgical Industry Press.
Dai. Y.W & Wang. W.H, et al. (2018). Roof stability analysis and support design of deep stope route mining. J. China Science and Technology of Safe Production. 14(02): 33–38.
Hu. J.H & T.Lei, et al. (2011). Dynamic Response Analysis of Pre-crack Blasting in Filling Environment. J. Journal of Central South University (Natural Science Edition), 06:1704–1709.
Hu. K.X & J. Kemeny (1994). A fracture mechanics analysis of the effect of backfill on the stability of cut and fill mine workings.J. International Journal of Rock Mechanics & Mining Sciences & Geomechanics Abstracts, 31(3): 231–241.
Huang. X.C & W.B. Lu, et al. (2021). Study on blasting vibration failure mechanism and vibration velocity threshold of cemented backfill with full tailings. J. Blasting, 2021, 38(01): 1–7.
Jiang. L.C & Y. Su Yong, et al. (2020) Dynamic response mechanism of layered cemented backfill pillar under the disturbance of horizontal stress wave of far-field blasting. J. Journal of rock mechanics and engineering, 39(01): 34–44.
Li, X.B. (2014). Foundation and application of rock dynamics. M. Science Press.
Liu. Z.X. & X.B. Li (2004). Study on the stability of high-stage backfill under blasting dynamic load. J. Mining and Metallurgy Engineering. 03: 21–24.
Wen. X & L. Zhao, et al (2021). Study on the propagation law of blasting vibration in filling. J. Mining and Metallurgy Engineering, 41(05): 32–35.
Yang.W & G.P. Li, et al. (2018). Failure law and prevention measures of cemented tailings backfill under blasting load. J. Mining Research and Development. 38(03):113–118.
Ylmaz.T & B Ercikd, et al. (2014). Assessment of strength properties of cemented paste backfill by ultrasonic pulse velocity test. J. Ultrasonics. 54: 1386–1394.
Zhu P.R. & W.D. Song, et al. (2018). Tensile mechanical response mechanism of cemented backfill under blasting dynamic load. J.Journal of Mining and Safety Engineering. 35(03):605–611.

Advances in Mineral Resources, Geotechnology and Geological Exploration – Rashid & Zhang (eds)
© 2023 Copyright the Author(s), ISBN 978-1-032-33377-9

Direct shear tests on three types of expansive soils at different water contents

Jianping Zhang, Sijiang Zhang, Zhao Liu & Fuping Zhang
State Grid Gansu Electric Power Company Construction Branch, Lanzhou, China
State Grid Gansu Electric Power Company, Lanzhou, China

ABSTRACT: Expansive soils are widely distributed in the foundation soil of water conveyance canals in Northern Xinjiang, and their strength characteristics are very obviously influenced by the water content. The increase of water content in the canal foundation soil caused by seepage and other reasons is very likely to cause canal slope collapse. The shear strength characteristics of three expansive canal foundation soils were studied using undrained direct shear tests, and the effect of soil body water content variations on the safety factor of simple homogeneous canal slopes was analyzed. The results show that with the increase in water content, the peak shear stress of the soil body decreases significantly. The peak shear stress of the green soil is low at 18 kPa at the water content of 30% and the normal stress of 300 kPa, with a decrease of more than 93%; the friction angles of the three soil bodies decrease significantly with the increase of water content, and there exists a critical water content value, and the internal friction angle decreases to almost 0 when the soil body water content is greater than this value. The critical water contents of the yellow, red, and green soil bodies were 21.9%, 22.8%, and 18.6%, respectively. The soil body cohesion decreases linearly with the increase of water content, and the decreased amplitudes of all three soil bodies were more than 80% after the water content increased from 10% to 30%. The safety coefficient of the canal slope decreases significantly with the increase of water content, and when the water content is greater than 22% at a volumetric weight of 19 kN m-3 and a slope height of 5 m, the safety coefficient of the green soil canal slope decreases to less than 1.

1 INTRODUCTION

Located in Altay Prefecture, the Northern Xinjiang Water Conveyance Canals have provided abundant water resources for Urumqi and the residents and factories along the route since its completion, which has promoted local economic and social development. However, the climatic environment and geological conditions along the canal are complex and variable, and there are a large number of expansive soils distributed in the canal base. Due to the canal leakage, the water content of expansive foundation soil will vary largely, and the long-term effect leads to damages to the canal slope such as sliding collapse and instability damage, which causes serious economic losses every year (Cai et al. 2019).

Expansive soil has the nature of expansion after absorbing water, and shrinkage after losing water. It is harder and stronger when the water content is low, but when touched with water, the strength is reduced due to the rapid water absorption, expansion, and softening, causing slope collapse and other disasters (Guo 2007; He et al. 2013; Wang 2011). For example, the north main canal of Naban Reservoir in Guangxi, the general main canal of the Irtysh–Karamay–Ürümqi Canal in Xinjiang, and many other places have been subjected to the expansive soil slope sliding collapse accidents (Liu 1997); several different types of expansive soil landslides have also occurred in the South-North Water Transfer Project (Gong et al. 2011; Ma et al. 2020; Zhang et al. 2014). The strength characteristics of expansive soils are closely related to the water content (Kong et al. 2000; Lu et

 DOI 10.1201/9781003319412-37

al. 1992). For unsaturated soils, some scholars have used water content instead of matric suction to reflect its effect on the shear strength of soils from a practical point of view (Ling et al. 2007; Ma et al. 2009). Linchang Miao et al (1999) conducted an experimental study on remodeled samples of expansive soils and concluded that the water content is the key factor affecting the strength, and strength decreases significantly with the increasing water content. Jiangzhu Shen et al. (2004) concluded that unsaturated expansive soils absorb water, expand, and soften during rainfall, leading to the loss of suction in the soil body and then causing increased deformation and reduced strength parameters, which is a key factor in inducing landslides. Bin Xu et al. (2011) studied the effects of water content, density, and fissure by direct shear test and triaxial test, showing that the water content and fissure have a large effect on the strength. Lianjie Zhang et al. (2015) showed that the water content and loads have a significant effect on the shear strength of remodeled expansive soils in their experimental study. Kassiff et al. (2011) showed that the landslide instability in expansive soils is caused by a large variation in water content resulting in a continuous decrease in strength. Lei Xu et al. (2015) tested the shear strength of some expansive soils along the Central Line Project of the South-North Water Transfer Project under the influence of freeze-thaw cycles, showing that the shear strength of the soil body decreases more obviously in the initial stage of the freeze-thaw cycle, and then the soil strength gradually increases due to the water content decreases. Junhua Wu et al. (2017) used direct shear apparatus to perform undrained fast shear of unsaturated expansive soil body under drying-wetting cycles, and the drying-wetting cycles significantly reduce the shear strength and weaken the cohesion more than weakening the friction force. Li Jingjing et al. (2017) showed that the shear stress-shear displacement curves obtained from the in-pore shear tests of expansive soils generally show a weak hardening deformation characteristic; water immersion has a significant effect on the shear strength, with a maximum reduction of 56.9%, mainly in terms of the reduction of cohesion, with a maximum value of 18.1 kPa. Jian Zhou et al. (2013) conducted direct shear tests on expansive soils which experienced the drying-wetting cycles in the slope model test, showing that the cohesion decreased significantly, the internal friction angle decreased less and analyzed the variation in the safety factor of the slope in ABAQUS. Leizhu Jia et al. (2014) tested and analyzed the shear strength of the expansive soils and the relationship between the strength index and water content, and they analyzed and calculated the retaining and protecting for the deep foundation excavation. Heping Yang et al. (2018) showed in their study that as the number of drying-wetting cycles increases, the cohesion value decays as an exponential function, which is the main cause of the shallow sliding collapse of slide slopes.

To sum up, the increase of the water content of expansive soils is one of the main factors causing the decay of soil strength, which in turn leads to the instability and damage of engineering structures. However, the mechanical properties of expansive soils under different geological conditions are very different. To better serve the prevention and control of water conveyance canals disasters in Northern Xinjiang, this paper selects three typical expansive foundation soils of the canal base, studies the shear strength and its parameter variations under different water content conditions, and analyzes the decay pattern of the slide slope safety factor caused by the increasing water content.

2 TEST MATERIALS AND TEST METHODS

2.1 *Test materials*

The soil samples used in the test were taken from Northern Xinjiang Water Conveyance Canals, and the collection area has a temperate continental climate with hot summers and cold winters. The three soil samples were all expansive soils, which are light yellow (Figure 1a), red (Figure 1b), and green (Figure 1c) in color, and are therefore numbered CH-Y, CH-R, and CL-G. The physical properties were tested according to the Test Methods of Soils for Highway Engineering, and the physical indexes are shown in Table 1, the particle size distribution of the test soil samples is shown in Figure 2.

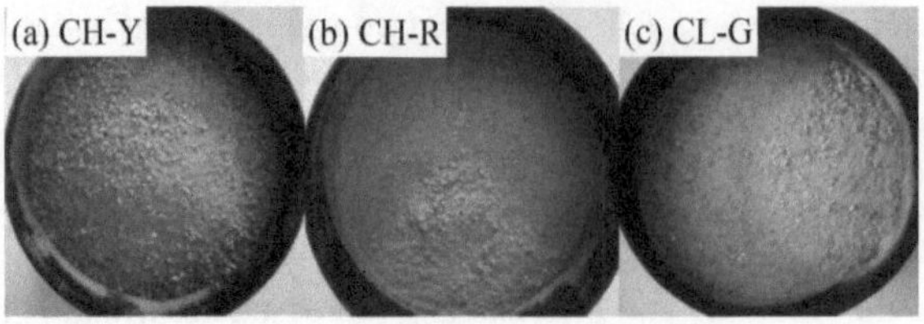

Figure 1. Pictures of three expensive soils, (a) yellow soil CH-Y, (b) red soil CH-R, and (c) Green soil CL-G.

Table 1. Basic properties of soils.

No.	Category	LL/%	PL/%	Proportion/	Optimal water content /%	Maximum density/ (g·cm^3)	Freedy expansion rate/%
CH-Y	High LL Clay	51.0	19.2	2.66	16	1.73	86
CL-G	Low LL Clay	46.9	23.5	2.69	19.1	1.71	76
CH-R	High LL Clay	57.6	26.8	2.7	21.8	1.67	90

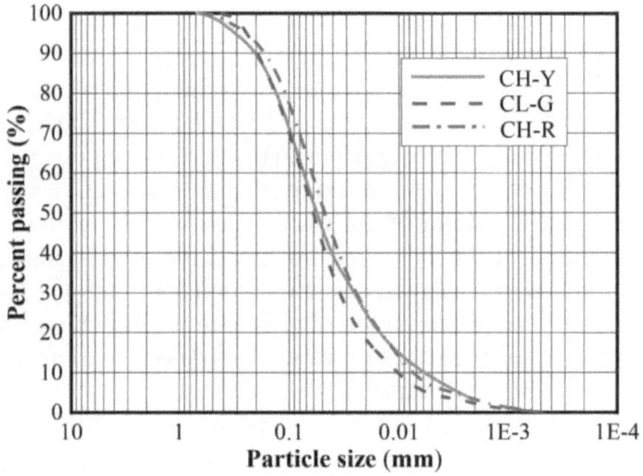

Figure 2. Distribution curves of soil particles.

2.2 *Test method*

The direct shear test specimens were prepared using 20 mm high, 61.8 mm diameter remodeled soil samples, and the preparation process consisted of the following steps: ① The test soil samples were collected in the field and dried naturally, mixed thoroughly, then crushed and filtered through a 2 mm aperture soil sieve. The initial water contents of the soil samples were measured. ② The distilled water is used to prepare the soil samples with the design-weight water contents of 10%, 14%, 18%, 22%, 26%, and 30%, and they are put in sealed bags to keep them moisturized for 24 hours after thorough mixing, and later, the drying method is used to measure water contents of wet soil samples to be used as actual water contents. Since the difference between the actual water content and the designed water content was within 0.3%, the designed water content was used to analyze the test data. ③ According to the designed dry density and water contents of soil materials, a certain mass of wet soil is weighed and loaded into the specimen bin evenly. Applied pressure to the soil sample loaded into the specimen bin slowly and uniformly using a standard sample press until the compactness was 95%. ④ The soil sample was ejected from the specimen bin using a jack, being cut with a ring cutter and the surface was corrected and leveled, and then the soil sample was wrapped with the cling film to prevent moisture dissipation and used as a test remodeled soil sample. Since the specimen bin was 150 mm in height, four sets of samples with the same water content can be made in one sample preparation process. The ZJ quadruple strain-controlled direct shear instrument manufactured by Nanjing Turang Instrument Factory Limited Company was used to carry out the undrained fast shear tests of the three expansive soils, and the test procedure was performed with reference to the Test Methods of Soils for Highway Engineering and the direct shear apparatus instructional manual (Specification of a soil test 1999). The shear mode was the displacement-controlled mode with the shear rate at 0.8 mm/min, the normal stress being set at 50, 100, 200, and 300 kPa, and the test temperature at 19°C at room temperature.

3 TEST RESULTS AND ANALYSIS

Figure 3 shows the relationship curves of shear displacement (δ) and shear stress (τ) for red soil (CH-R) at different water content and normal stress. It can be seen that the curve softening phenomenon is more obvious when the water content (wi) is low, while the curve shows a hardening characteristic when the water content is high. This is because when the initial water content of the expansive soil is low, the effect of matric suction makes the sample strength larger, and the curve shows a strain-softening type. When the water content is high, the matric suction decreases significantly, and the curve exhibits a strain-hardening type (Zhang et al., 2015a). At 10% water content (Figure 3a), the initial tangential shear stiffness and peak shear stress of the curve increase with the increasing normal stress (σN). However, as the water content increases, the variations of the above two parameters gradually decrease at different normal stress, especially after the water content is greater than 22%. This is because after the hydrophilic clay minerals in expansive soils are touched with water, the bound water film becomes thicker, and a significant expansive deformation occurs, making the interparticle distance increase. Thus, the interparticle interaction force is weakened and the effect of the normal stress gets weakened (Rao et al. 2001; Xu et al. 2018). The relationship curves between the shear stress and shear displacement of the yellow and green soil materials at different water contents are similar to those of red soil materials and will not be repeated here.

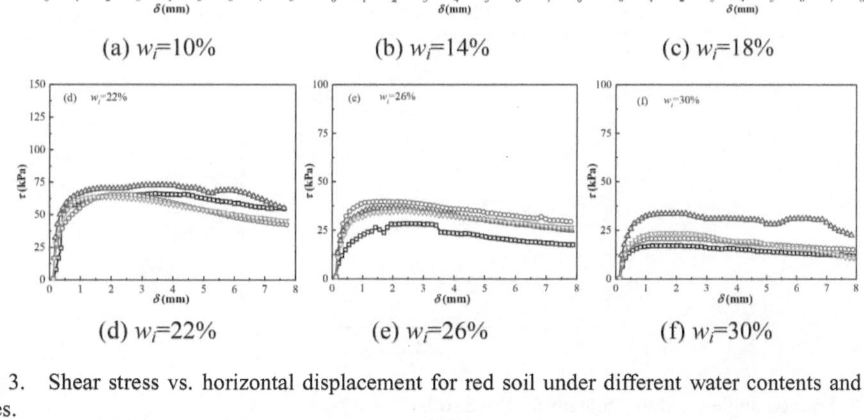

(a) w_i=10% (b) w_i=14% (c) w_i=18%

(d) w_i=22% (e) w_i=26% (f) w_i=30%

Figure 3. Shear stress vs. horizontal displacement for red soil under different water contents and normal stresses.

Figure 4 shows the relationship between the peak shear stress and the water content for three soil samples at different normal stress. In Figure 4a, it can be seen that at the same normal stress, the peak shear stress decreases significantly with increasing water content. This is because the shear strength of expansive soils mainly consists of the cohesion, friction angle, matric suction, and external restraint, while the cohesion, internal friction angle, and matric suction all decrease with the increase of water content of the soil (Zhang et al. 2015). For example, the peak shear stress decreases from 334 kPa at 10% water content to 20 kPa at 30% water content for yellow soils at 300 kPa. At lower normal stress, the decreasing trend becomes slower, for example, the peak shear stress decreases from 140 kPa at 10% water content to 9 kPa at 30% water content for yellow soils at a normal stress of 50 kPa. The red and green soils have similar variation patterns, only the values are different. Meanwhile, it can be found from Figure 6 that due to the water content increases, the interparticle interaction force weakens (Rao et al. 2001; Xu et al. 2018), and the effect of the normal stress on the peak shear stress gradually decreases, especially after the water content is greater than 18%.

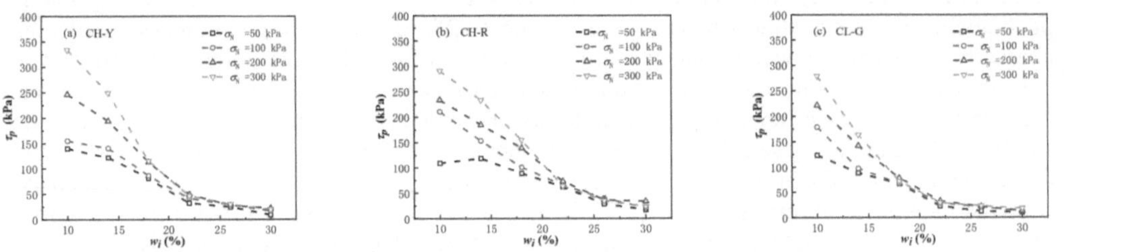

Figure 4. Peak shear stress vs. water content for three soils under different normal stresses.

The peak shear stress at four sets of normal stress was fitted using the Mohr-Coulomb criterion, and the soil strength parameters were analyzed. Figure 5 shows the variation pattern of the internal friction angle with water content for the three soil materials. It can be seen that the internal friction angle decreases significantly with increasing water content, which is similar to the findings of Linchang Miao et al. (1999), but differs from the results of Jingjing Li et al. (2017) and Jian Zhou et al. (2013), presumably due to the differences in physical properties of the soil materials. The internal friction angles of the yellow soil materials were 38.9°, 3.1°, and 2.3° at 10%, 22%, and 30% water contents, respectively; the internal friction angles of the red soil materials were 32.3°, 0.7°, and 2.0° at 10%, 22%, and 30% water contents, respectively; the internal friction angles of the green soil material were 30.4°, 1.2°, and 1.7° at 10%, 18% and 30% water contents, respectively. In Figure 5, it can be found that the friction angles of the three soils vary significantly below a certain critical water content (wc), while the internal friction angles are small and remain basically

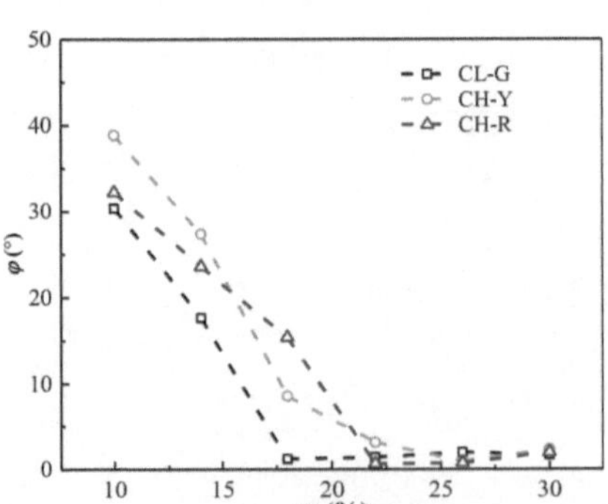

Figure 5. Friction angle vs. water content for three soils.

the same when the water contents are larger than the critical water content. It is inferred that the critical water content is related to the thickness of the interparticle water film. With the increase of the water content, the water film thickness around the particles increases, and when it reaches a certain degree, the particles could not contact effectively during the shearing process, and thus the friction cannot perform the effect. The critical water contents of the yellow, red, and green soil materials were 21.9%, 22.8%, and 18.6%, respectively.

Figure 6 shows the variation law of the cohesion with the water content for the three soil materials. It can be seen that as the water content increases, the cohesion basically shows a linearly decreasing trend (Zhang et al. 2015). After the water content increases from 10% to 30%, the cohesions of the yellow, red, and green soils decrease from 88, 109, and 106 kPa to 18, 10, and 9 kPa, respectively. Linear fitting of the relationship between the cohesion and water content of three soils gives the following relation.

$$c = m + nw_i \tag{1}$$

Where m and n are the fitting coefficients, the specific values are shown in Table 2.

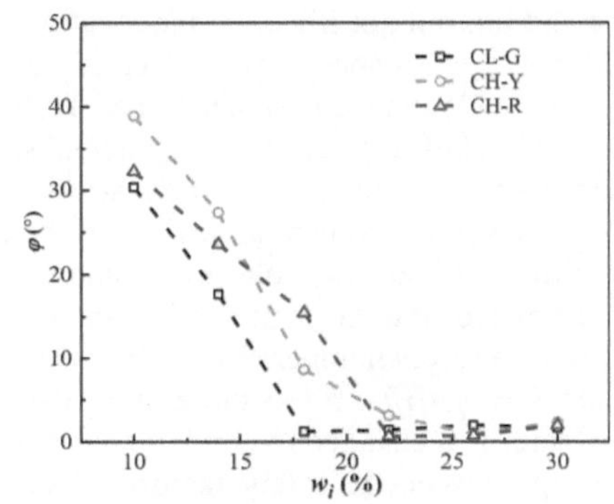

Figure 6. Cohesion vs. water content for three soils.

Table 2. Fitting coefficients of cohesion with water content

Soil Category	m	n	R^2
Yellow soils	114	–4.5	0.92
Red soils	163	–4.8	0.98
Green soils	144	–4.9	0.94

4 DISCUSSION

Figure 7 shows a typical cross-sectional diagram of the Northern Xinjiang Water Conveyance Canal. The canal's cross-section is trapezoidal, with a canal width of 5 m, a canal water depth of 4 m, and a slope ratio of 1:2 on both sides of the canal. The canal is in a water supply period from April to September every year, and the presence of unavoidable seepage makes the water content of the canal foundation soil increase. From the analysis of the above section, it can be seen that in the process of water content increases, the internal friction angle and cohesion of the three kinds of expansive foundation soils decline significantly, which will seriously affect the foundation soil carrying capacity, resulting in the slope being prone to have damage phenomena such as instability.

Therefore, the stability of typical soil quality slopes in the process of water content variation was analyzed.

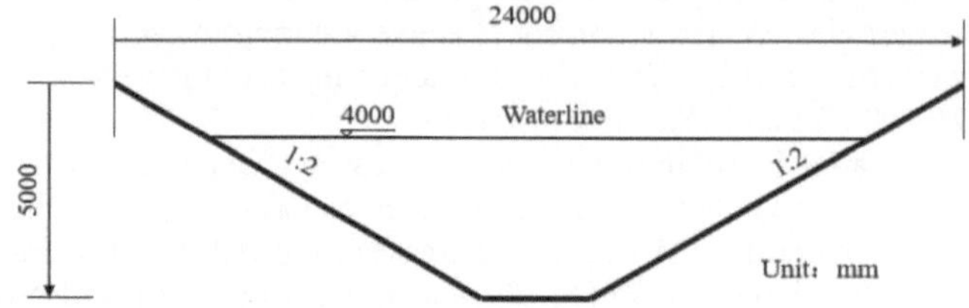

Figure 7. Cross-section of the canal.

Slope stability analysis initially was carried out using mainly the limit equilibrium method, and in recent years, the finite element method and finite difference method have been used more often, and much commercial software could analyze the stability of complex soils and structures. Taylor proposed a simple and fast diagram for calculating the stability of homogeneous slopes based on the friction circle method but did not illustrate the location of the sliding surface (Bao 2004; Michalowski 2002; Taylor 1937). To specify the effect of water content variation on the slope stability of the Northern Xinjiang Water Conveyance Canal, Taylor's diagram was used, other factors were neglected, only the effects of different water contents, soil volumetric weights, and slope heights on the safety factor were verified under the total stress conditions.

The calculation process is as follows: the slope ratio angle β of the canal slope, the volumetric weight γ, the friction angle φ of the soil body, the cohesion c, and the slope height H are known. firstly, the safety factor F was estimated, and the limit friction angle φ_d was calculated according to the formula $F = \tan\varphi/\varphi_d$. Then the safety number N was obtained by referring to the table, and Fs was calculated by the formula $N = c/\gamma HFs$. If F is close to Fs, the safety factor Fs is obtained, and if not, the estimated safety factor F is changed to continue the trial calculation.

Figure 8 shows the relationship between the safety factor and soil water content at different volumetric weights of the soil body for the green soil canal slope. It can be seen that the safety factor decreases significantly with the increase of soil water content. At the same water content, the safety coefficient decreases with the increase of the volumetric weight, but the change value decreases when the water content is high. Figure 9 shows the relationship between the safety coefficient and water content at different slope heights for the green soil material drainage slope. Again, it can be seen that the safety factor decreases significantly with increasing water content. At the same water content, the safety factor decreases significantly with the increasing slope height, for example, at the slope height of 15 m, the safety factor decreases to 1 after the water content of the soil body is more than 18%.

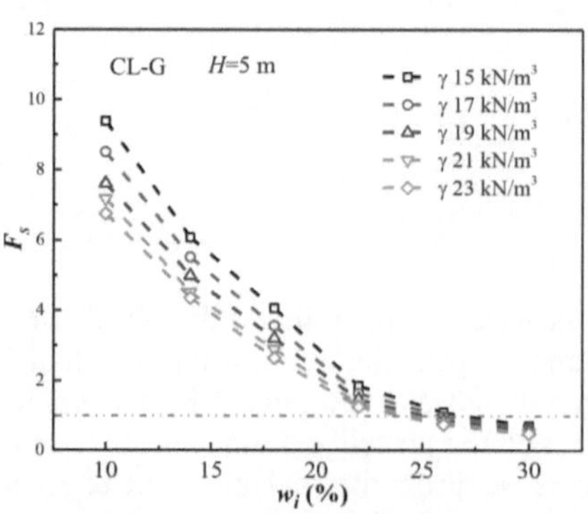

Figure 8. Safety factors vs. water content for uniform slope under different soil unit weight.

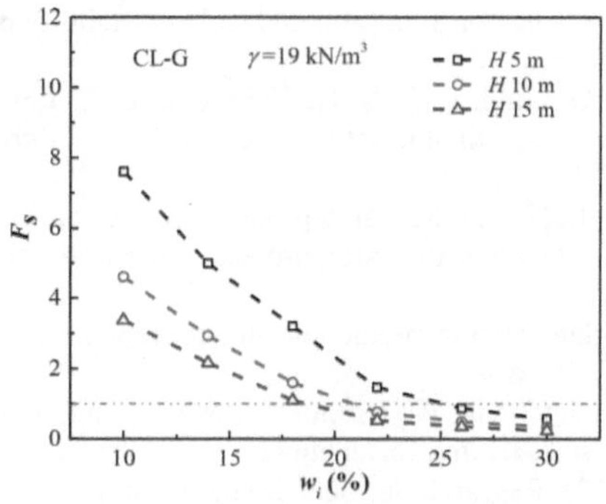

Figure 9. Safety factors vs. water content for uniform slope under different slope height.

5 CONCLUSION

Through fast shear tests on three types of the expansive soils, the effects of different water contents on the shear stress-shear displacement relationship, strength characteristics, and strength parameters of the foundation soil were studied, and the qualitative analysis of the effects of different water content conditions on the safety factor of simple homogeneous canal slopes was carried out. The following conclusions were obtained:

(1) Soil water content has a strong influence on the shear characteristics of three kinds of expansive soils. As the water content increases, the initial tangential stiffness of the shear stress-shear displacement curve decreases significantly; the peak shear stress of the soil body also decreases significantly, and the peak shear stress of the green soil material is as low as 18 kPa at the water content of 30% and normal stress of 300 kPa, decreasing by more than 93%.
(2) The friction angles of the three soils decrease significantly with the increase of water content, and there exists a critical water content, when the water content of the soil body is greater than the critical water content, the internal friction angle decreases to almost 0. The critical water contents of the yellow, red, and green soils were 21.9%, 22.8%, and 18.6%, respectively. The cohesion of the soil body decreases linearly as the water content increases, and the cohesion of all three soils decreases by more than 80% after increasing the water content from 10% to 30%.
(3) The stability of the canal slope was analyzed qualitatively using the Taylor diagram method. The safety factor decreases with the increase of the capacity and slope height and decreases significantly with the increase of water content. The slope safety factor of the green soil material decreases to less than 1 when the water content is more than 22% and the volumetric weight is 19 kN/m^3.

The effects of wetting and drying cycles on the mechanical properties of the interface can be further investigated. The results can provide a reference for structural design and construction in expansive soil areas.

ACKNOWLEDGMENTS

This work was supported by the State Grid Gansu Electric Power Company Science and Technology Project (No. SGGSJS00XMYBJS2100062)

REFERENCES

Bin, X. U., Yin, Z. Z., & Liu, S. L. (2011). Experimental study of factors influencing expansive soil strength. Rock & Soil Mechanics, 32(1), 44–50.

Cheng-Gang, B. A. O. (2004). The behavior of unsaturated soil and stability of expansive soil slope. Chinese Journal of Geotechnical Engineering.

Dan, X. U., Chaosheng, T. A. N. G., Ting, L. E. N. G., Yunsheng, L. I., Yan, Z. H. A. N. G., Kan, W. A. N. G., & Bin, S. H. I. (2018). Shear strength of unsaturated expansive soil during wetting drying cycles. Earth Science Frontiers, 25(1), 286.

Gong, B., Cheng, Z., Guo X., et al.: (2011). Research progress on the stability of expansive soil slope in the middle route of the south-to-north water diversion project. Journal of Yangtze River Scientific Research Institute, 28(10): 134–140.

Guo, H.: (2007). Research on deformation prediction and prevention countermeasure of expansive soil landslide. Chang'an University (Xi'an).

Guoqiang, Z., Bin, S., Shuda, Z., Jianliang, W., Yunhua, Y., & Planning, et al. (2014). Landslide genesis of expansive soil slope and slope stability analysis method. Yangtze River.

He, J., Liu, L., Re, Y., et al.: (2013). Research on shear strength characteristic of expansion rock. Journal of Water Resources and Water Engineering, 24(06):92–94.

Hua, L., & Zongze, Y. (2007). Variation of unsaturated soil strength with water contents. Chinese Journal of Rock Mechanics and Engineering.

Jia, L. Z., Hu, C. L., & Yang, X. (2014). Design of supports for deep foundation pits considering attenuation properties of the strength of expansive soil shear. Chinese Journal of Geotechnical Engineering.

Jian, Z., Hong-Zhong, X. U., & Wen-Jie, H. U. (2013). Impact of wetting-drying cycle effects on the stability of expansive soil slopes. Chinese Journal of Geotechnical Engineering, 35(zk2), 152–156.

Jiang, S. Z., & Kuan, M. Z. (2004). Coupled analyses of seepage and deformation of expansive soil slopes during rainfall. Hydro-Science and Engineering (3), 7–11.

Kassiff, G., Alpan, I.: (2011). A Slope Failure in Swelling Clay. Canadian Geotechnical Journal, 10(3): 531–536.

Kong, L. W., & Tan, L. R. (2000). Study on shear strength and swelling-shrinkage characteristic of compacted expansive soil. 1st Asian Conference on Unsaturated Soils.

Lei, X. U., Yang, L. U., Zong, J. M., & Song, Y. J. (2015b). Direct shear test on expansive clay in Nanyang under freeze-thaw cycles. South-to-North Water Transfers and Water Science & Technology.

Li, J. J., Kong, L. W., & Mu, K. (2017b). In-situ borehole shear test on expansive soil and its strength characteristics. Yantu Lixue/Rock and Soil Mechanics, 38(2), 453–461.

Liu, T.: (1997). The expansive soil in engineering. China Architecture & Building Press, Beijing.

Lu, Z. J., Zhang, H. M., Chen, J. H., & F Man. (1992). Shear strength and swelling pressure of unsaturated soil. Chinese Journal of Geotechnical Engineering.

Ma, H., He, X., Zhang, S., et al.: (2020). Analysis of the physical issues of the expanded geotechnical channel in the middle line of south-to-north water diversion. Yellow River, 02: 1–4.

Ma, S., Huang, M., Fan, Q.: (2009). Strength indexes of saturated soil and its application.Chinese Journal of Rock Mechanics and Engineering, 28(3): 635–640.

Miao, L., Zhong, X., & Yin, Z. (1999). The relationship between strength and water content of expansive soil. ROCK AND SOIL MECHANICS.

Michalowski, R. L. (2002). Stability charts for uniform slopes. Journal of Geotechnical and Geoenvironmental Engineering, 128(4), 351–355.

Rao, S. M., Reddy, B. V. V., & Muttharam, M. (2001). The impact of cyclic wetting and drying on the swelling behaviour of stabilized expansive soils. Engineering Geology, 60(1–4), 223–233.

SL 237—1999, (1999). Specification of a soil test. China Water & Power Press, Beijing, China.

Tang, H., Tang, X., Wang, X., et al.: (2018a). Shear strength of expansive soils under wet-dry cycles with loading. Rock and Soil Mechanics, 39(7): 2311–2317, (2018).

Taylor, D. W. (1937). Stability of earth slopes.

Wang, X.: (2011a). Research on Strength Characteristics and Constitutive Relationship of Strongly Expansive Soft-rock in Yanji Basin. China University of Geoscience (Beijing).

Wu, J. H., & Yang, S. (2017a). Experimental study of matric suction measurement and its impact on shear strength under drying-wetting cycles for expansive soils. Rock & Soil Mechanics, 38(3), 678–684.

Zhang, L., Xiong, W. U., Xie, Y., & Chenliang, W. U. (2015a). Influence of water content and overlying pressure on shear strength of remolded expansive soil. The Chinese Journal of Geological Hazard and Control.

Zheng-yin, C. A. I., Hao, C. H. E. N., & HUANG, Y. H. (2019). The failure mechanism of canal slopes of expansive soils considering the action of wetting-drying cycles. Chinese Journal of Geotechnical Engineering, 41(11), 1977–1982.

Advances in Mineral Resources, Geotechnology and Geological Exploration – Rashid & Zhang (eds)

Study on regional landslide hazard supported by multi-source data

Xikui Lu*
Department of Surveying and Mapping Engineering, Henan Geology Mineral College, Zhengzhou, Henan, China

Yongyi Yuan
Guizhou Liupanshui Sanlida Technology Co., Ltd., Liupanshui, China

ABSTRACT: In recent years, landslide disasters have occurred frequently, and the risk analysis of landslide disasters is an important basis for disaster prevention and reduction; hence, it is of great significance to carry out regional landslide risk research. Taking Panzhou City, Guizhou Province as research area, with the support of multi-source data, using the random forest model of regional landslide hazard analysis, the results show that the high-risk area is located in the low elevation area, the areas of the landslide have frequent human activities, which is mainly distributed by soft rock and soft hard alternate with class two of petrofabric rock engineering. It covers an area of 214.2 km^2, accounting for 5.3% of the total area of the study area. The prediction results are basically consistent with the actual situation.

1 INTRODUCTION

In recent years, the frequency of landslides has increased. In 1984, Varnes first proposed the concept of landslide risk and defined the risk calculation formula as the product of vulnerability, landslide risk, and the value of a disaster-bearing body (Varnes 1984). With continuous innovations in geological hazard risk theory, landslide hazard risk assessment has achieved fruitful results (Erener 2012; Hao 2019; Lai 2015; Li 2018; Qiu 2017). Using remote sensing image technology, some missing data in the landslide area database can be supplemented with better remote sensing images. This paper takes Panzhou city of Guizhou Province as the research area and uses high-resolution remote sensing images to extract the element information of landslide-bearing bodies. With the support of multi-source data, regional landslide risk analysis is realized by combining random forest models, providing a reference for regional landslide risk analysis and research.

The research area is located in Panzhou City, its geographical coordinates are as follows: 104°17′E – 104°57′E, 25°19′N – 26°17′N, Panzhou covers an area of 4056 km^2 with a total population of about 1.2 million. The terrain is high in the northwest and low in the southeast, uplifting in the central and southern regions, with dense rivers and an altitude of 1500 m–2000 m. The study area is located in the subtropical monsoon climate zone and has a mild climate throughout the year without obvious cold and hot summer. The rainfall is continuous and concentrated between June and September, and the rainy season mostly comprises heavy rain, with an average annual rainfall of about 1400 mm. According to the data, there are 463 potential landslide points in the area, and most of them are in an unstable state, at risk of harm.

2 ANALYSIS OF LANDSLIDE SUSCEPTIBILITY

Based on the city's geological data, basic geographic data, and remote sensing image data, meteorological data, such as basic data, combined with previous research results, extracted elevation,

*Corresponding Author: 78627083@qq.com

DOI 10.1201/9781003319412-38

engineering geological petrofabric, the distance from drainage, rainfall, land use, in addition to road distance, with the distance from population centers and NDVI eight factors as evaluation indexes, including various evaluation factors, such as shown in Table 1, a 30 m × 30 m grid unit is taken as the evaluation unit for the purpose of this research. A random forest model is used to predict the spatial probability of landslides.

Table 1. Landslide susceptibility analysis factors.

The evaluation index	Parameter values		
	Scope	Average	Standard deviation
Elevation/m	764∼2806	1804	223
Engineering geology rock group	Hard rock, hard and soft rock, soft rock, loose accumulation body		
Distance from water system/m	0∼5574	829	839
Rainfall/mm	856∼1064	964	50
The land use	Construction land, cultivated land, forest land, and water body		
According to road distance/m	0∼8272	1275	1142
According to the residential distance/m	0∼4726	962	576
NDVI	−0.52∼0.68	0.1	0.11

Random forest is a model constructed by Breiman after combining his classification tree algorithm proposed in 1984. Using the sample evaluation index attribute values of landslide points as samples, ArcGIS software was used to conduct random sampling on landslide points. The number of landslide hazard points that are 100 m away from the landslide hazard point is taken as a negative sample, and the same sampling of various types of evaluation index attribute values is used to combine the sample data. Out of these a total of 926 samples of plus and minus data are taken, and 70% of the data are randomly selected as training samples and 30% as validation samples. The random forest model was built using Python language. After several tests, the model was found to have good generalization ability under the conditions that the number of iterations is 150, the maximum depth is 5, the minimum sample number of internal node division is 60, and the minimum sample number of a leaf node is 10.

After the random forest model was established, the evaluation index values of 4,034,261 grid cells in the study area were imported into the random forest model to obtain the probability of landslide in each grid. The landslide susceptibility obtained by ArcGIS is shown in Figure 1. As

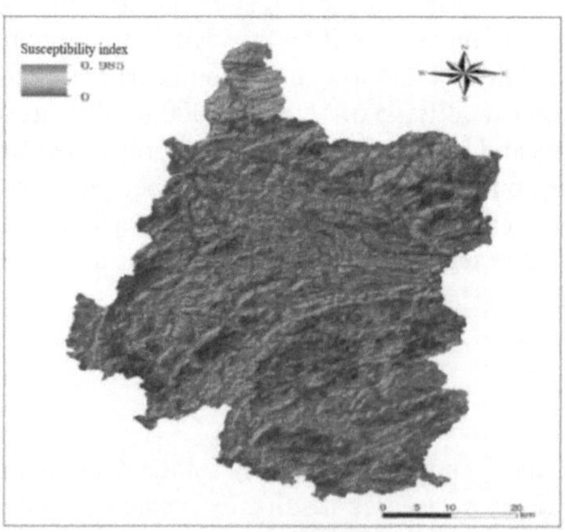

Figure 1. Landslide susceptibility.

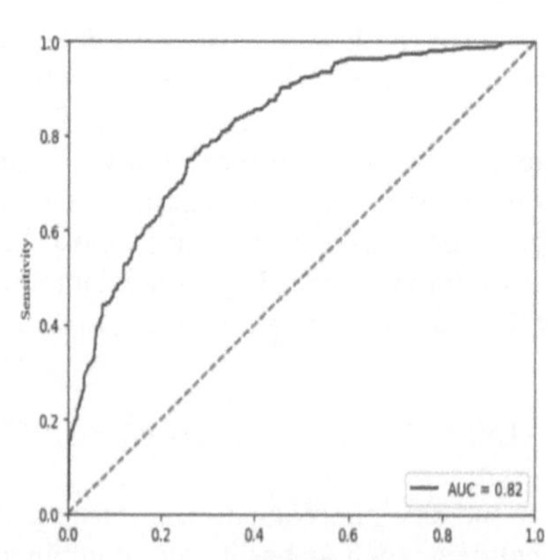

Figure 2. ROC curve of validation landslides.

shown in Figure 1, high-susceptibility areas are mainly concentrated in areas with less vegetation, concentrated human engineering activities, and dense rivers. The low-risk areas are mainly concentrated in areas with high vegetation coverage and sparse rivers.

To verify the test sample by model prediction ability, the confusion matrix method was used for the evaluation of classification model prediction problem of a matrix. The most intuitive and simple way is adopted to show the algorithm performance, wherein it contains category classification based on correct or incorrect classification of the number of samples. The verification sample confusion matrix is shown in Table 2.

Table 2. Confusion matrix of validation data.

	The real value		
Predictive Value	0	1	Error rate
0	102	37	0.263
1	33	106	0.234

As can be seen from Table 2, the accuracy of the model is 0.752 and 0.766, respectively. To further verify the prediction accuracy of the model, the Operating Characteristic (ROC) Curve and Area Under Curve (AUC) were adopted. The ROC curve is shown in Figure 2. AUC from Figure 2 is 0.82. It shows that the random forest model has strong prediction ability and high prediction accuracy, that is, the prediction results of the model can provide more accurate data that support the follow-up.

3 TIME-FREQUENCY AND STRENGTH ANALYSIS

3.1 *Time frequency*

Through the probability of a landslide on a specific slope in a certain period of time, the frequency of landslide occurrence under corresponding environmental conditions in the future can be deduced. (PENG L 2013). The occurrence probability curve of historical landslides in the degree of vulnerability in the study area is shown in Figure 3. The average frequency curve of landslides in the study area is obtained by polynomial fitting of historical landslide probability, and the probability of future landslides in the study area is calculated based on the average probability curve.

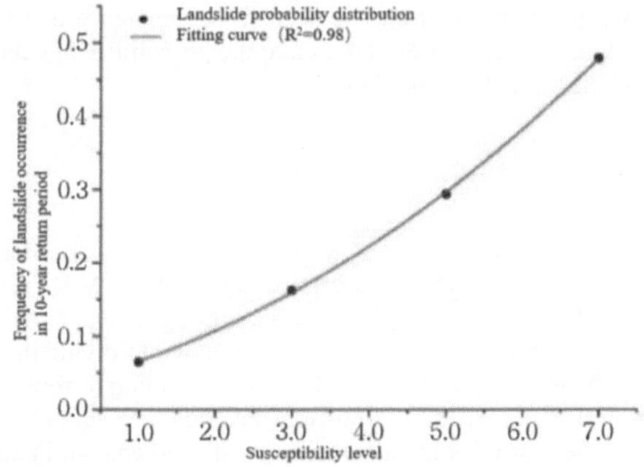

Figure 3. Historical landslide frequency in each susceptibility class and fitted frequency curve.

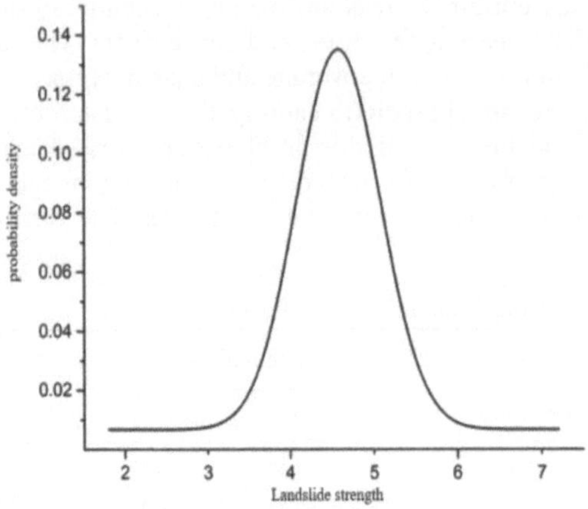

Figure 4. The curve of landslide magnitude probability density.

3.2 *Strength analysis*

In this paper, landslide volume parameters are used to obtain landslide strength, and the formula is as follows:

$$m_L = logV_L \tag{1}$$

In the formula, V_L is the volume of a single landslide, the unit is M^3, and m_L is the intensity of a single landslide; the intensity of each landslide in the study area was calculated by Formula (1), and then the calculated landslide intensity was statistically analyzed. The fitting function was obtained by fitting the frequency distribution of landslide intensity with the Gauss function. On the basis of the fitting function, the landslide intensity frequency was taken as a factor, and the probability function of landslide intensity was obtained as follows:

$$p(m_L) = \left[3.104 + \frac{98.680}{1.098 \times \sqrt{\frac{\pi}{2}}} \times \exp\left(-2 \times \frac{(m_L - 4.579)^2}{1.098^2}\right)\right] / N \tag{2}$$

Where mL represents landslide strength and N represents the total number of landslides in the study area. The probability density curve of landslide strength in the study area is calculated by Formula (2), as shown in Figure 4. It can be seen from the figure that the strength range of landslides in the area is concentrated in grades 4–5, and the probability of occurrence of landslides with strength in this range is 0.52 after calculation.

4 LANDSLIDE HAZARD ANALYSIS

The landslide risk calculation formula H = P_S×P_T×P_M is used, where P_S is the spatial probability (susceptibility), P_T is the time-frequency of landslide, P_M is the strength of landslide, and H is the landslide risk. The results of landslide risk were calculated, and the value ranged from 0 to 0.2389. The dynamic threshold setting method was adopted to divide the landslide risk into 0 to 0.003, 0.003 to 0.03, 0.03 to 0.1, and 0.1 to 0.2389 corresponding to four categories of low-risk area, medium risk area, high-risk area, and ultra-high-risk area, respectively.

As shown in Figure 5, the risk of landslide changes with the change in susceptibility, and the result is similar to the result of susceptibility, indicating that the risk of landslide occurrence is affected by the result of susceptibility. (1) The high-risk area of landslide is 13% of the total area of the study area; the elevation in this area is low; human activity is intense. (2) High-risk areas

Figure 5. Landslide hazard zones.

of landslide account for 24.4% of the total area of the study area; human engineering activities are strong. (3) Landslide hazard area, accounting for 30.6% of the total area of the study area; the surface vegetation coverage is high. (4) The landslide low-risk area accounts for 32.2% of the total area of the study area, and there are few human engineering activities.

5 CONCLUSION

In this paper, eight factors in the study area were taken as evaluation indexes, and the random forest model is used to predict regional landslide susceptibility. Through analysis of landslide time-frequency and landslide intensity, the results show that the intensity of historical landslides in the study area is concentrated in grades 4–5, and the probability of repeated landslides is 0.52. The area with high risk of landslides is located in the area with low elevation and frequent human activities, and is mainly distributed on the rock strata of soft rock and soft and hard alternate rock.

ACKNOWLEDGMENTS

This study was jointly supported by the Science and Technology Foundation of Guizhou Province (No. [2020]4001).

REFERENCES

Akgun A, Kincal C, Pradhan B. Application of remote sensing data and GIS for landslide risk assessment as an environmental threat to Izmir city (west Turkey) [J]. Environ Monit Assess, 184(9): 5453–70 (2012).

Erener A, Duzgun H. A Regional Scale quantitative Risk assessment for Landslides: the case of Kumluca watershed in Bartin, Turkey[J]. Landslides, 10(1): 55–73(2012).

Hao G. Evaluation of landslide susceptibility based on random forest model in Shangnan County [D]. Xi'an University of Science and Technology, (2019).

Lai C, Chen X. Flood risk assessment model based on random forest and its application[J]. Journal of Hydraulic Engineering, 46(1): 58–66(2015).
Li Y. Dynamic risk assessment of landslide disaster in Batu Section of Tibetan-China Interconnection Project [D]. Chengdu University of Technology, (2018).
Peng L, Niu R. Regional landslide disaster risk assessment: a case study of Zigui County in the three gorges reservoir area ofYangtze River [J]. Journal of Jilin University (earth science edition),43(03):891–901(2013).
Peng L. Risk Assessment of landslide disaster in the Three Gorges Reservoir Area [D]. China University of Geosciences, (2013).
Qiu D. Landslide risk analysis based on multi-source Data fusion [D]. China University of Geosciences, (2017).
Varnes D J. Landslide hazard zonation: a review of principles and practice[M]. Paris: UNESCOPRESS, (1984).

Advances in Mineral Resources, Geotechnology and Geological Exploration – Rashid & Zhang (eds)
© 2023 Copyright the Author(s), ISBN 978-1-032-33377-9

Research on dynamic characteristics of bottom hole assembly

Jianqing Ma*
School of Mechanical Engineering, Xi'an Shiyou University, Xi'an, Shanxi, China

ABSTRACT: In this paper, the bottom hole assembly is simplified to a continuous beam model, and the matrix transfer method is used to solve the vibration problem of the continuous beam under axial compression. Study on the influence of vibration frequency under axial pressure in the two common combinations is analyzed, and amplitude modes of each order are given.

1 INTRODUCTION

Drill string mechanics is an important theoretical foundation of modern drilling technology, and it is the mechanical foundation to understand the working mechanism of drill string downhole. One of its main research objects is the bottom hole assembly. The force and deformation analysis of the bottom hole assembly is the basis of the wellbore trajectory control technology. The research on the vibration characteristics of the axially loaded beam model has a long history. Bokaian (Bokaian 1988, 1990) gave the vibration frequency and mode expression of a beam under different boundary conditions under axial tension and compression and analyzed the influence of the axial force effect and boundary conditions on the inherent characteristics of the beam. The drill tool assembly structure is simplified into a continuous beam structure. When there are many spans, the dynamic response analysis will be quite complicated. Bai Jiazhi et al. (Bai & Lin 1985) regarded the drill tool assembly as a famous fulcrum subject to vertical and horizontal bending of continuous beams and columns. The beam was separated into several simply supported beams and columns, and then a set of "three bending moment equations" was established to get the internal bending moment at each centralizer, the lateral force on the drill bit, and the restraining reaction force added to each centralizer. Li Junqiang et al. (Li & Fang 1998) discussed the influence of axial pressure on the frequency equation and mode function of lateral vibration. This method simplified the calculation process but did not reduce the calculation workload. Wang Zanzhi et al. (Wang et al. 2010) established a wave analysis method for the free vibration of a continuous beam. Push-the-bit rotary steering is nowadays commercially applied on a large scale. Guan Zhichuan et al. (Guan et al. 2019; Wang et al. 2018) optimized the flexible sub for better build-up performance of RSBHA in order to master the effects of the flex sub. Bi Yantao et al. (Bi et al. 2020) investigated the effect of flexible pup joints on the steering capacity of BHA used in the push-the-bit rotary steering drilling process through a finite element analysis and summarized their action laws. Li Jun et al. (Li et al. 2018) established a variable cross-section BHA (bottom hole assembly) mechanical model and well deviation trend angle analysis model for push-the-bit rotary steering tool by means of the BHA beam-column method. Based on this, they analyzed the influential laws of BHA structure parameters, weight on bit (WOB), bit anisotropy, and rock drillability on the build-up capacity of the tool.

In this paper, the bottom hole assembly is simplified to a statically indeterminate continuous beam structure, and each span is regarded as a simply supported beam structure. Boundary conditions and continuity conditions are satisfied, and the natural frequency and mode shape function of the transverse vibration of the drill assembly is obtained. The influence of weight-on-bit on vibration is analyzed, and the influence of stabilizer placement on vibration frequency and amplitude are discussed.

*Corresponding Author: majianqing2000@163.com

DOI 10.1201/9781003319412-39

2 BASIC EQUATION

The structure of the bottom hole assembly is Φ215.9mm PDC drill+controllable bend joint rotary steering drilling tool +Φ213 mm stabilizer +Φ172mm drill collar. The left side represents the drill bit. L1 indicates the distance between the drill bit and the stabilizer, and Li indicates the distance between two stabilizers. Different bottom hole combinations can be obtained by changing the spacing between stabilizers and the spacing between stabilizers and the drill bit. Modulus of elasticity of drill collar E=210GPa, density $\rho = 7850$ kg/m^3. Consider two drill tool assemblies, one is 1# Drill tool assembly: L1=0.8 m, L2=9 m, L3=12 m, another is 2# Drill tool assembly: L1=0.8 m, L2=2 m, L3=8.5 m, L4=9 m.

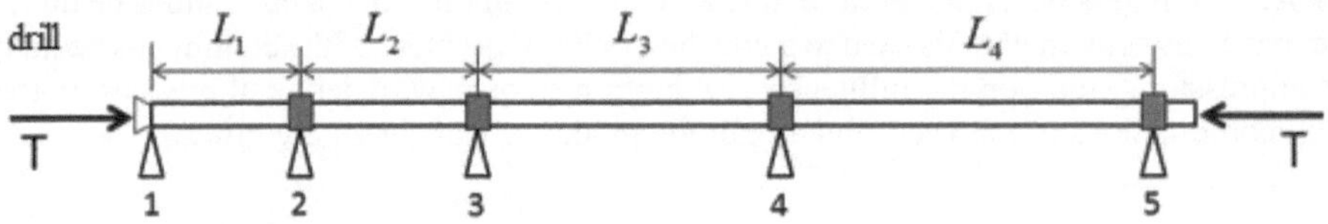

Figure 1. Bottom hole assembly structure.

The flexural vibration equation of any span in the beam is as follows

$$\frac{\partial^2}{\partial x^2}\left[EI\frac{\partial^2 u_i}{\partial x^2}\right] + T\frac{\partial^2 u_i}{\partial x^2} + \rho A\frac{\partial^2 u_i}{\partial t^2} = 0 \tag{1}$$

Namely

$$\frac{\partial^4 u_i}{\partial x^4} + \frac{T}{EI}\frac{\partial^2 u_i}{\partial x^2} + \frac{\rho A}{EI}\cdot\frac{\partial^2 u_i}{\partial t^2} = 0 \tag{2}$$

Let $a = \sqrt{\frac{EI}{\rho A}}$, $N = \sqrt{\frac{T}{EI}}$

Assuming that there are solutions in the form of separate variables, let $u(x,t) = y(x)q(t)$

So $\dfrac{d^2 q(t)}{dt^2} + \omega^2 q(t) = 0$ and $\beta^4 = \dfrac{\omega^2}{a^2}$

The characteristic equation is obtained

$$\frac{d^4 y_i(x)}{dx^4} + N^2\frac{d^2 y_i(x)}{dx^2} - \beta^4 y_i(x) = 0 \tag{3}$$

The general solution of the equation is: $y_i(x) = C_{1i}\sin\lambda_{1i}x + C_{2i}\cos\lambda_{1i}x + C_{3i}sh\lambda_{2i}x + C_{4i}ch\lambda_{2i}x$

$$q(t) = C_5\sin\omega t + C_6\cos\omega t$$

Where

$$\lambda_1 = \sqrt{\frac{N^2}{2} + \sqrt{\frac{N^4}{4} + \beta^4}}, \qquad \lambda_2 = \sqrt{-\frac{N^2}{2} + \sqrt{\frac{N^4}{4} + \beta^4}}$$

The transfer matrix relationship of the i-th span is:

$$\begin{Bmatrix} y \\ \theta \\ M \\ Q \end{Bmatrix}_i^R = \frac{1}{\lambda_1^2 + \lambda_2^2}\cdot[A]_i\cdot\begin{Bmatrix} y \\ \theta \\ M \\ Q \end{Bmatrix}_i^L$$

$A(1,1) = \lambda_2^2\cos\lambda_1 l_i + \lambda_1^2 \mathrm{ch}\lambda_2 l_i$;

$A(2,1) = \lambda_1\lambda_2(-\lambda_2\sin\lambda_1 l_i + \lambda_1 \mathrm{sh}\lambda_2 l_i)$;

$A(3,1) = EI\lambda_1^2\lambda_2^2(-\cos\lambda_1 l_i + \mathrm{ch}\lambda_2 l_i)$;

$A(4,1) = EI\lambda_1^2\lambda_2^2(\lambda_1\sin\lambda_1 l_i + \lambda_2 \mathrm{sh}\lambda_2 l_i)$;

$A(1,2)=\frac{\lambda_2^2}{\lambda_1}\sin\lambda_1 l_i+\frac{\lambda_1^2}{\lambda_2}sh\lambda_2 l_i$; $A(1,3)=\frac{1}{EI}(-\cos\lambda_1 l_i+ch\lambda_2 l_i)$;

$A(2,2)=\lambda_2^2\cos\lambda_1 l_i+\lambda_1^2 ch\lambda_2 l_i$; $A(2,3)=\frac{1}{EI}(\lambda_1\sin\lambda_1 l_i+\lambda_2 sh\lambda_2 l_i)$;

$A(3,2)=EI\lambda_1\lambda_2(-\lambda_2\sin\lambda_1 l_i+\lambda_1 sh\lambda_2 l_i)$; $A(3,3)=\lambda_1^2\cos\lambda_1 l_i+\lambda_2^2 ch\lambda_2 l_i$;

$A(4,2)=EI\lambda_1^2\lambda_2^2(-\cos\lambda_1 l_i+ch\lambda_2 l_i)$; $A(4,3)=-\lambda_1^3\sin\lambda_1 l_i+\lambda_2^3 sh\lambda_2 l_i)$;

$$A(1,4)=\frac{1}{EI}(-\frac{1}{\lambda_1}\sin\lambda_1 l_i+\frac{1}{\lambda_2}sh\lambda_2 l_i);$$

$$A(2,4)=\frac{1}{EI}(-\cos\lambda_1 l_i+ch\lambda_2 l_i);$$

$$A(3,4)=\lambda_1\sin\lambda_1 l_i+\lambda_2 sh\lambda_2 l_i;$$

$$A(4,4)=\lambda_1^2\cos\lambda_1 l_i+\lambda_2^2 ch\lambda_2 l_i;$$

For the first span, the left end is simply supported

$$\begin{Bmatrix} y \\ \theta \\ M \\ Q \end{Bmatrix}_1^R = \frac{1}{\lambda_1^2+\lambda_2^2}\cdot[A]_1\begin{Bmatrix} 0 \\ \theta \\ 0 \\ Q \end{Bmatrix}_1^L = \frac{1}{\lambda_1^2+\lambda_2^2}\begin{bmatrix} A_{12} & A_{14} \\ A_{22} & A_{24} \\ A_{32} & A_{24} \\ A_{42} & A_{24} \end{bmatrix}_1 \begin{Bmatrix} \theta \\ Q \end{Bmatrix}_1^L$$

As $y_1^R=0$, we can get $\theta_1^l=-\frac{A_{14}}{A_{12}}Q_1^l$,

For support 2, the shear forces at both ends meet $Q_2^L=Q_1^R+R_2$.

$$\begin{Bmatrix} y \\ \theta \\ M \\ Q \end{Bmatrix}_2^L = \begin{bmatrix} 0 & 0 \\ A_{24}-A_{22}\cdot\frac{A_{14}}{A_{12}} & 0 \\ A_{34}-A_{32}\cdot\frac{A_{14}}{A_{12}} & 0 \\ A_{44}-A_{42}\cdot\frac{A_{14}}{A_{12}} & 1 \end{bmatrix}_1 \begin{Bmatrix} Q_1^L \\ R_2 \end{Bmatrix} = [F]_1\begin{Bmatrix} Q_1^L \\ R_2 \end{Bmatrix}$$

For the second span,

$$\begin{Bmatrix} y \\ \theta \\ M \\ Q \end{Bmatrix}_2^R = [A]_2\begin{bmatrix} 0 & 0 \\ A_{24}-A_{22}\cdot\frac{A_{14}}{A_{12}} & 0 \\ A_{34}-A_{32}\cdot\frac{A_{14}}{A_{12}} & 0 \\ A_{44}-A_{42}\cdot\frac{A_{14}}{A_{12}} & 1 \end{bmatrix}_1 \begin{Bmatrix} Q_1^L \\ R_2 \end{Bmatrix} = [A]_2[F]_1\begin{Bmatrix} Q_1^L \\ R_2 \end{Bmatrix}$$

By similar derivation, the third span can be obtained

$$\begin{Bmatrix} y \\ \theta \\ M \\ Q \end{Bmatrix}_3^R = [A]_3[F]_2\begin{Bmatrix} Q_1^L \\ R_3 \end{Bmatrix} \text{ and } [P]_{4\times 2}=[A]_3[F]_2.$$

The natural frequency can be satisfied with the equation $\begin{vmatrix} P_{11} & P_{12} \\ P_{31} & P_{32} \end{vmatrix}=0$, so the natural frequency of the lateral vibration can be established.

3 RESULT ANALYSIS

According to the boundary conditions and continuity conditions, the natural frequencies and mode shape functions of the lateral vibration of 1# and 2# drilling tool assemblies are obtained. Tables 1 and 2 respectively show the natural frequencies of the first five orders of lateral vibration of the two drilling tool assemblies under different axial pressures. Figure 1 shows the change curves of the first two natural frequencies of the 2# drilling tool assembly with the axial pressure. It can be seen that the natural frequency of each order decreases with the increase of the axial pressure, and the axial pressure when it approaches zero is the critical load when the drilling tool is buckling.

Table 1. Natural frequency of 1# drill tool assembly with different axial pressure.

Natural frequency (rad/s)	1	2	3	4	5
N=0	42.628	61.653	138.7	171.72	290.05
N=0.1	41.779	61.347	138.03	171.17	289.9
N=0.2	39.111	59.428	135.06	168.61	286.81
N=0.3	34.174	56.056	129.99	164.11	281.48
N=0.4	25.628	50.931	122.48	157.83	274.02

Table 2. Natural frequency of 2# drill tool assembly with different axial pressure.

Natural frequency (rad/s)	1	2	3	4	5
N=0	27.532	49.849	86.538	141.67	179.03
N=0.1	26.749	49.137	85.661	140.73	178.11
N=0.2	24.249	46.936	82.977	137.86	175.34
N=0.3	19.289	42.993	78.298	132.93	170.62
N=0.4	7.7773	36.716	71.251	125.68	163.79

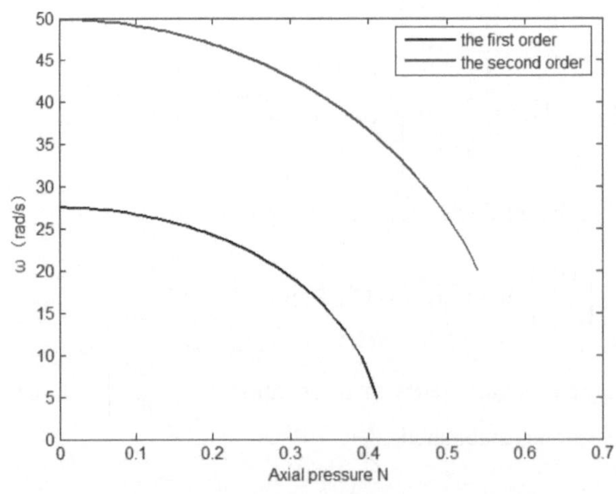

Figure 2. The curve of the first two order natural frequencies with the axial pressure.

Figures 3 and 4 show the first four-order modes of the two drill tool assemblies when the axial pressure is fixed, respectively. For the drill tool assembly, the drill collar is longer, and the greater the higher-order amplitude. For 1# drill tool, the amplitude is related to the length of the drill collar.

For 2# drill tool, the odd-order amplitude is larger, and the amplitude is also related to the position of the drill collar.

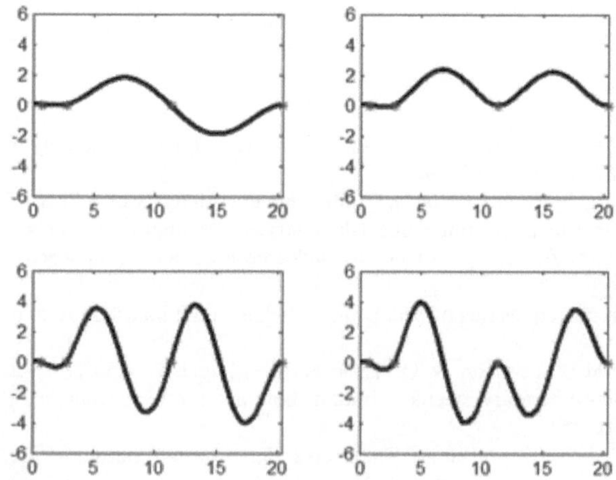

Figure 3. The first four modes of 1# drill tool assembly when N=0.1.

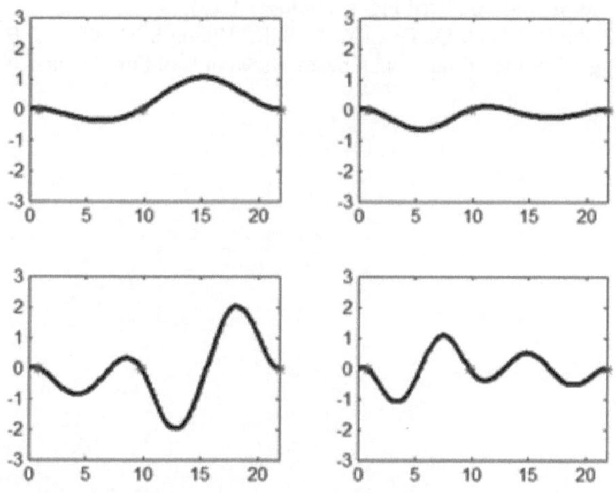

Figure 4. The first four modes of 2# drill tool assembly when N=0.1.

4 CONCLUSION

It is particularly important to control the large-scale vibration of the bottom hole assembly during the drilling process of the drill string. Changing the length of the drill collar and the number of stabilizers will affect the vibration frequency and amplitude. The influence of axial pressure on vibration frequency in two common combinations is analyzed, and the amplitude modes of each order are given. So, the following conclusions are obtained:

(1) The vibration frequency of each order of the drilling tool assembly decreases with the increase of axial pressure to obtain the critical load of instability.
(2) When the drill collar distribution is singular, the vibration amplitude is relatively small.

The vibration problem of the drilling tool assembly is very worth studying in the drilling process, and it is also very important to rationally optimize the position and number of stabilizers in the drilling tool assembly.

REFERENCES

Bai J. Z., Lin Xiao M. (1985). Two-dimensional analysis of Bottom Hole Assembly by beam-column theory. ACTA PETROLEI SINICA, 6(3): 75–64.

Bi Y. T., Liu G. H., Dong G. H., Li J.(2020). Finite Element Analysis of Effect of Flexible Pupjoints on Push-the-Bit Rotary Steering Deflecting Force. Chemistry and Technology of Fuels and Oils,56(6):218–225.

Bokaian A. (1988). Natural frequencies of beams under compressive axial loads. Journal of Sound and Vibration.126(1),49–65.

Bokaian A. (1990). Natural frequencies of beams under tensile axial loads. Journal of Sound and Vibration, 142(3): 481–498.

Guan Z. C., Wang H., Shi Y. C., Chen W. Q., Zhao G. S., Wang J. Y., Cao G. Q.(2019). Dynamic behavior analysis of push-the-bit rotary steerable bottom hole assembly. Journal of Mechanical Science and Technology,33(4):1501–1511.

Li J. Q., Fang T. (1998). Analysis of Natural Transverse Vibration of Continuous Beam Subjected to Axial Force. Mechanical Science and Technology, 17(4):541–547.

Li J., Li D. C., Zhang H., Wang H. (2018). The influencing factors of the inclination ability of push-the-bit rotary guiding tool. Oil Drilling and Production Technology,41(1):1–5.

Wang H., Guan Z. C., Shi Y. C., Bai J., Liang D. Y., Wang J. Y. (2018). Effects of flex sub on build-up performance of push-the-bit RSBHA. Drilling technology, 41(6):19–22.

Wang Z. Z., Hu R. C., Zhang P., Wu H. Q., Deng Y., Ye Y. B., Huang S. B., Chen J. B. (2010). Wave motion solution to free bending vibration of multi-span beam. Sichuan Building Science,36(2):35–37.

Advances in Mineral Resources, Geotechnology and Geological Exploration – Rashid & Zhang (eds)

Single hole dilution method combined with numerical simulation to judge the leakage position of diaphragm wall

Wenkai Wang, Ziyang Cheng, Pengfei Liu & Yong Li
CCCC Tunnel Engineering Company Limited, Beijing, China

Kunyong Zhang*
Key Laboratory of Ministry of Education for Geomechanics and Embankment Engineering, Nanjing, China
Geotechnical Engineering Institute, Hohai University, Nanjing, China

ABSTRACT: In order to ensure the safe construction of the foundation pit of the underground station, this paper takes Suzhou Rail Transit Line 8 as an example and uses the single hole dilution method based on the dilution principle and finite element software to judge the possible leakage position of the underground diaphragm wall and prevent excavation safety accident. The results show that it is accurate and feasible to find out the location of the leakage point of the foundation pit enclosure by measuring the flow rate by the concentration change of the tracer saturated NaCl solution.

1 INTRODUCTION

The underground diaphragm wall leakage problems often occur at the joint position of the groove section and even the wall body (Gao 2017; Liang & Zhang 2018; Li et al. 2020; Zhi et al. 2020), resulting in accidents such as surrounding ground collapse, wall instability, quicksand, and underground pipeline damage. Before excavation of the foundation pit, it is of great guiding significance for the smooth progress of foundation pit construction to detect whether there is any leakage point in the ground connecting wall. Domestic scholars have carried out in-depth research on this. Dong Haizhou (Dong & Chen 2004) used the temperature tracing method to detect the leakage position of the ground connecting wall. Sun Bing (Sun 2012) used the micro-voltage method to detect the leakage of the foundation pit, and expected to realize the non-destructive detection of the leakage of the foundation pit. Wu Tiankai (Wu et al. 2014) tested the leakage position of the underground diaphragm wall based on the principle of a dewatering well, but the result was not very accurate, and only rough judgment could be made.

This paper intends to verify the feasibility of judging whether there are obvious defects in the ground connection wall through the monitoring data of the flow velocity of the dewatering well existing in the foundation pit project and the two-dimensional numerical model.

2 PROJECT OVERVIEW

Hudian Road Station is an underground two-story island station, and its standard section is in the form of a ground connecting wall with a length of 30.3m and a thickness of 0.8m.

*Corresponding Author: 1480810185@qq.com

DOI 10.1201/9781003319412-40

2.1 *Engineering geological conditions*

The surrounding environmental conditions of the project site are relatively complex. The site soil layer includes ①/1Miscellaneous fill layer, ③/1clay, ③/2silty clay, ④/2silt, ⑤/1silty clay, ⑥/1clay, ⑥/2silty clay, ⑥/tClay silt, ⑦/1silty clay, ⑦/2Clay silt mixed with silty clay.

2.2 *Hydrogeological conditions*

The aquifers closely related to the construction of this site are the submerged aquifer, the shallow layer ③/3, the slightly confined aquifer in the ④/2 layer, and the confining aquifer in the middle layer ⑥/t and the deep layer ⑦/2. Pressurized aquifers. During the detailed survey, the stable burial depth of diving measured in the drilling hole is generally 1.3–2.4 m, and its absolute elevation is generally between 1.73–0.54 m. The stable buried depth of the micro-confined water level is 1.01 m, and the absolute elevation is 1.04 m. The buried depth of the confined water level of the ⑦/2 layer is 4.80 m, and the absolute elevation of the water level is correspondingly −2.10 m.

2.3 *Brief description of leakage problem*

Hudian Road Station WQ1-36 floor-to-ceiling wall design measures 144 m^3, the actual pouring is 122 m^3, and the debt is 22 m^3. In order to detect the leakage position of the ground connecting wall of WQ1-36, firstly, drilling holes were arranged on both sides of the ground connecting wall for pumping and tracing detection, and then the numerical model was used to simulate and verify. Figure 1 shows the drilling layout.

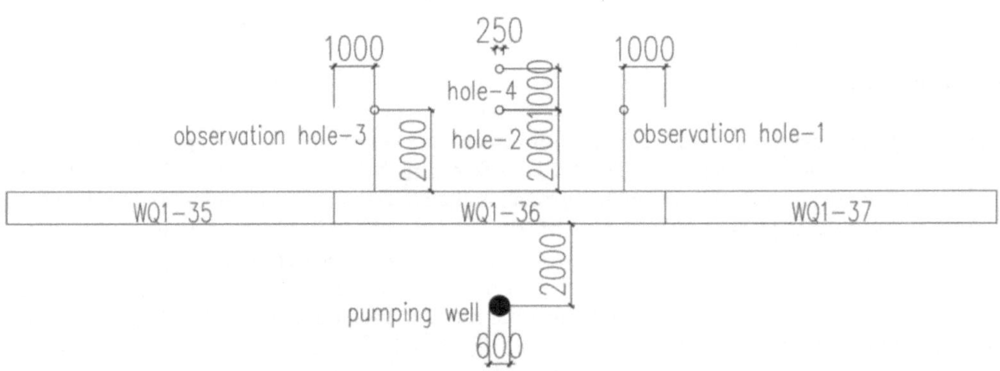

Figure 1. Drilling layout (unit: mm).

The diameter of the pumping well is 600 mm, and it is located 2 m from the ground connecting wall inside the foundation pit. The diameter of the four observation holes is 250 mm, and they are arranged symmetrically on the outside of the foundation pit.

3 DETERMINATION OF FLOW RATE BY SINGLE-HOLE DILUTION METHOD

3.1 *Basic principle of horizontal flow rate measurement*

The tracer is set to be stable, and the tracer is uniformly mixed in the dilution water column from the beginning and the continuation of the test. The axis of the water filter pipe is orthogonal to the streamline of the water flow in the entire thickness of the aquifer. The horizontal flow through the water filter pipe is continuous, uniform, and stable. Without the interference of vertical water flow, the tracer in this section of the diluted water column is diluted by the amount of groundwater in the horizontal flow filter water pipe. The groundwater flow rate can be calculated by the following formula:

$$V_f = \frac{\pi r_1}{2\alpha t} \ln \frac{n_0}{n} \tag{1}$$

where: V_f is the permeation velocity, that is, r_1 the Darcy permeation velocity in the aquifer; α is the correction factor for the flow field distortion caused by the presence of the filter pipe in the aquifer; t is the time difference between two measurements; n_0 is the conductance of the tracer at time t=0, and n is the conductance of the tracer at time t.

4 NUMERICAL SIMULATIONS OF GROUND-TO-WALL LEAKAGE

4.1 *Computational information*

The section where the line connecting the observation hole 2, the observation hole 4, and the pumping well is located is taken as the calculation section. According to the relevant data of the foundation pit enclosure structure design, the calculation parameters of different strata are determined. In order to simulate the possible leakage of the ground connection wall, in the finite element simulation, the retaining wall at the non-leakage place is set as an impermeable material, the retaining wall at the leaking place is set as a highly permeable material, and the height of the permeable material is 0.5 m. The calculation parameters of different formations are shown in Table 1.

Table 1. Soil layer parameters.

Soil layer	Severe ϒ (kN/m^3)	Saturated water content per unit volume	Permeability coefficient (cm/s)
①/1 Filling	18	0.3	5.00E-05
③/1 Clay	19.6	0.424	1.00E-06
③/2 Silty clay	19.2	0.436	2.50E-06
④/2 silt	19.4	0.418	4.00E-03
⑤/1 Silty clay	18.6	0.475	8.00E-06
⑥/1 Clay	19.9	0.405	2.00E-06
⑥/2 Silty clay	19.3	0.436	5.00E-06
⑥/t Clay silt	18.8	0.457	1.00E-04
⑦/1 Silty clay	18.7	0.469	8.00E-06
⑦/2 Clay silt with silty clay	18.7	0.464	1.00E-03
Ground to wall	/	/	1.00E-20
Ground wall leakage	/	/	10

The following calculation conditions include Condition 1 (Integrity of diaphragm wall), Condition 2 (Leakage of diaphragm wall on 2–2.5 m), Condition 3 (Leakage of diaphragm wall on 4–4.5 m), Condition 4 (Leakage of diaphragm wall on 6–6.5 m), Condition 5 (Leakage of diaphragm wall on 8–8.5 m), and Condition 6 (Leakage of diaphragm wall on 10–10.5 m).

4.2 *Model establishment*

The ground connecting wall of the section to be analyzed has a buried depth of 30.3 m and a thickness of 0.8 m. The foundation pit extends 30 m outside the foundation pit and 20 m inside the foundation pit, with the ground connecting wall as the standard. The model size is 50.8×60 (L×H). On the far left is the observation hole 4, followed by the observation hole 2, and on the far right is the pumping well. The grid size is 0.5 m. The pumping water level of the pumping hole is stable at a buried depth of 14 m. The calculation grid of the two-dimensional finite element model is shown in Figure 2.

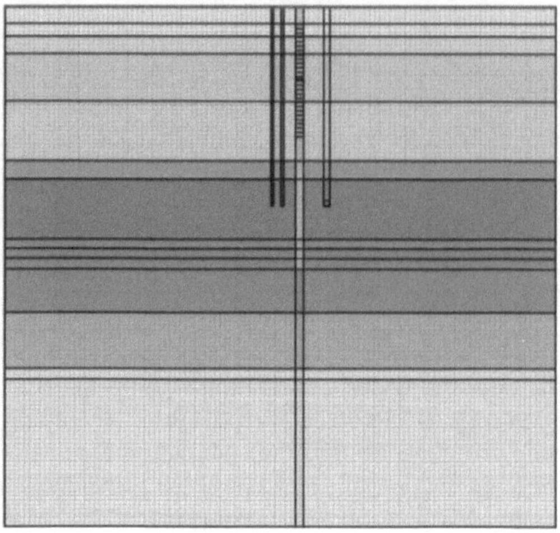

Figure 2. Meshing diagram.

4.3 *Analysis of results*

The calculation results of working Conditions 1 to 6 are analyzed. Figure 3 shows the five working conditions and the flow rate law of observation hole 2 after actual pumping.

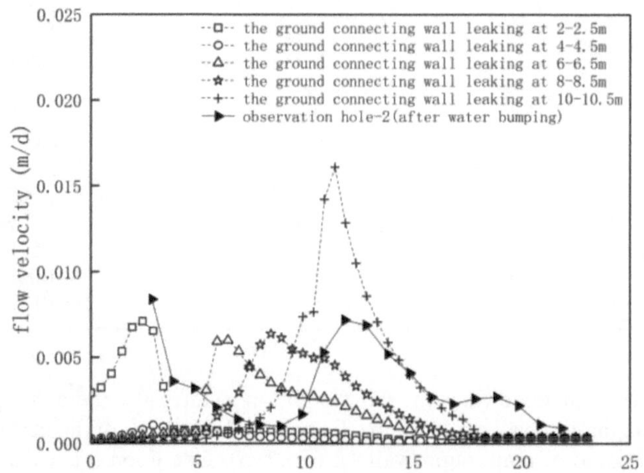

Figure 3. Schematic diagram of the flow velocity of observation well 2 under various working conditions.

The peak velocity of the observation hole 2 increases with the increase of the depth of the leakage position of the ground connecting wall, and the peak depth position is roughly within the depth range of the leakage position. The closer the leakage location is to the pumping water level, the greater the peak flow rate of the observation hole 2. When there is leakage at 4–4.5 m of the ground connecting wall, the maximum flow velocity is 2.9 m, which is shallower than the leakage position, and the corresponding flow velocity is lower, because the leakage occurs in the clay layer. The leakage locations of 6–6.5 m, 8–8.5 m, and 10–10.5 m of the ground connecting wall are in the same stratum, and the maximum flow velocity position is not stable in the "specific location" of the leakage interval. It shows that with the increase of depth, there may be a certain "hysteresis effect" at the position corresponding to the maximum flow velocity of the observation well.

By comparing the numerical results and the monitoring results, it is found that the actual monitoring data peak flow velocity is most consistent with the leakage results of the ground connecting wall at 8-8.5m in the numerical simulation.

5 CONCLUSIONS

Through the monitoring data and numerical simulation results of the single-hole dilution method in the field, a general method for judging the leakage location of the foundation pit wall is obtained. The conclusions are as follows:

(1) The approximate range of leakage is obtained according to the flow velocity characteristics in the water measuring holes outside the foundation pit. In different observation holes, the closer the observation hole is to the leakage position, the shorter the seepage path and the more obvious the change of flow velocity.
(2) Numerical analysis is carried out on the section where the observation hole with the most obvious change in flow velocity is located. For the same observation hole, the closer to the seepage position, the flow velocity reaches its peak value. However, with the increase of the depth of the seepage point and the different permeability of the soil layer, there may be a certain "hysteresis effect" at the peak position of the flow velocity of the observation well.

REFERENCES

Dong Haizhou, Chen Jiansheng. (2004) Detecting foundation pit leakage by temperature tracing method[J]. Chinese Journal of Rock Mechanics and Engineering, (12): 2085–2090.

Gao Qi. (2017) Construction technology for water leakage prevention in deep foundation pit[J]. Railway Construction Technology, (08): 97–100+110.

Li Bo, Yu Zhiqiang, Li Mei, Hu Yong. (2020) Centrifugal model test study on the effect of leakage on the foundation pit of the connecting wall structure of the first-level terrace of the Yangtze River in Wuhan[J]. Journal of Changjiang Academy of Sciences, 37(03): 78–81+89.

Liang Peng, Zhang Lingzhi. (2018) Construction technology for leakage treatment of underground diaphragm wall in cut-and-cover foundation pit[J]. China High-tech, (10): 88–90.

Sun Bing. (2012) Research on simulation test of leakage detection of ground connecting wall by electrical method[D]. Tianjin University.

Wu Tiankai, Liu Ju, Qi Ligang. (2014) Leakage detection method of underground diaphragm wall based on the principle of dewatering well[J]. Tianjin Construction Technology, 24(05): 12–15.

Zhihuiyuan, Li Feng, Gao Jinming. (2020) Risk analysis of the construction of water-rich sand layers in subway stations [J]. Engineering Technology Research, 5(21): 169–170.

Advances in Mineral Resources, Geotechnology and Geological Exploration – Rashid & Zhang (eds)
© 2023 Copyright the Author(s), ISBN 978-1-032-33377-9

Performance evaluation of barite chelate blockage removal agent in Niger Oilfield

Dexiang Duan & Feng Qian
China National Petroleum International Exploration and Development Company Limited, Beijing, China

Shiling Zhang
CNPC Engineering and Technology R&D Company Limited, Beijing, China

Xiuyu Wang*
China University of Petroleum-Beijing, Beijing, China

Ning Jing
China National Petroleum International Exploration and Development Company Limited, Beijing, China

Lingyu Mu
CNPC Engineering and Technology R&D Company Limited, Beijing, China

ABSTRACT: During the drilling of oil and gas wells in the Niger Oilfield, barite is often used as a weighting agent for drilling fluid mud. However, under the combined action of temperature and pressure, precipitation is likely to occur to block the wellbore, and conventional blockage removal agents cannot effectively solve the problem. In this study, aiming at the problem of wellbore blockage removal in the Niger Oilfield, the composition of scale samples was analyzed by the Soxhlet extraction method and the chelating agent LY-CGJ-1 was selected from 6 blockage removal agents. LY-CGJ-1 (10%) was determined as the optimal mass fraction through the experiment of the $BaSO_4$ dissolution rate. At the same time, the reaction threshold temperature of LY-CGJ-1 to $BaSO_4$ was determined to be 50°C, and the time for the second injection of the blockage removal agent was determined to be between 48 to 72 hours. The interaction mechanism between LY-CGJ-1 and Niger Oilfield wellbore scale was analyzed by FT-IR. The LY-CGJ-1 blockage removal agent selected in this study is helpful in improving the efficiency of blockage removing work by more than 50% and enhancing the oil recovery of the Niger Oilfield.

1 INTRODUCTION

During the drilling of oil and gas wells, barite is often used as a weighting agent to increase the specific gravity of drilling fluid mud, thereby preventing the drilling fluid mud from spouting and ensuring the safety of underground operations (Al Hamad 2020; Mahmoud 2018; Zhang 2017). When barite is added to drilling fluid mud, its particle size is required to be no less than 325 meshes to prevent its precipitation in the wellbore (Bageri 2017; Risthaus 2001). At the same time, the specific gravity of barite should be greater than 4.2, and the mass fraction of $BaSO_4$ should not be less than 95%. In the process of oil production, due to the change in temperature and pressure, the thermodynamics of the aqueous solution is unstable, and under the action of chemical incompatibility, it is easy to cause scaling to block the pipeline and formation when the time is too long (Geri 2017). Compared with the carbonate scale, the barium sulfate scale caused by barite is

*Corresponding Author: wangxiuyu@cup.edu.cn

 DOI 10.1201/9781003319412-41

extremely difficult to remove, and there is no effective method (Mahmoud 2018; M. U et al. 2000). Therefore, due to its stable chemical properties, low complexation constant, and low solubility product, the prevention, and removal of barium sulfate scale has always been an international problem.

2 METHODS AND MATERIALS

2.1 *Component analysis of wellbore scale samples*

In this study, the Soxhlet extraction method was used to analyze the composition of wellbore scale samples. The specific experimental steps are as follows: (1) Accurately weigh 2~5g of the dehydrated wellbore scale sample, record it as m1, and put it into the filter paper cylinder; (2) Put the filter paper cylinder into the Soxhlet extractor, and add xylene as a solvent in the round-bottomed flask; (3) Connect the experimental device and heat the solvent to boiling; (4) Continue to extract until the liquid filtered in the filter paper cylinder becomes colorless; (5) When the siphoning of the condensate is completed, the heating is stopped, the filter paper tube is taken out, and dried to a constant weight in a 100°C oven, and the recording quality is m2 and calculate the solid phase content of the wellbore scale sample by equation (1), and record it as w_{solid}; (6) Put the solid phase components into the energy spectrum analyzer to analyze the composition of the components.

$$w_{\text{solid}} = \frac{m_2}{m_1} \times 100\% \tag{1}$$

m_1—mass of wellbore scale sample, g; m_2—mass of solid phase after extraction, g; w_{solid}—solid phase content of scale sample, %.

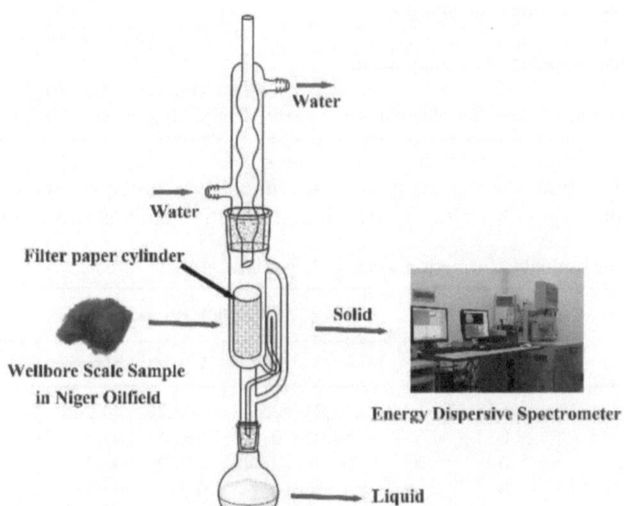

Figure 1. Analysis of wellbore scale samples by Soxhlet extraction.

2.2 *Determination of dissolution rate*

Simulated formation water was used to dissolve 10% of the blockage removal agent; 0.5 g $BaSO_4$ was added to 100 g of the blockage removal agent solution and recorded as m_3. Then, it was placed in a constant temperature drying oven at different temperatures for 24–72h (Dunn K, 1999). The heated residue was filtered and dried at 100°C, the residual $BaSO_4$ mass (m_4) was weighed, and the dissolution rate F was calculated according to equation (2).

$$F = \frac{m_3 - m_4}{m_3} \times 100\% \tag{2}$$

m_1—$BaSO_4$ initial mass, g; m_2—$BaSO_4$ mass after dissolution, g; F—$BaSO_4$ dissolution rate, %.

2.3 *FT-IR infrared spectroscopy*

The $BaSO_4$ particles before and after dissolving were mixed with 200 mg of dried KBr, and then fully ground and placed in a grinding tool. The samples were pressed into transparent flakes using an oil press, and the obtained samples were placed in an infrared spectrometer for testing.

2.4 *Experimental materials*

In this study, the scale samples used were from the production process of the Niger Oilfield, and $BaSO_4$ was used as a simulated scale sample for evaluation. The blockage removal agents used were from various types of products preferred in the Chinese market. The details of the experimental materials are shown in Table 1.

Table 1. Experimental materials.

S/N	Code	Purity	Type	Manufacturer
1	$BaSO_4$	CP	-	Shanghai Macklin Biochemical Co., Ltd
2	YSW-1	Industrial grade	Sulfamate	Anton Oilfield Services Group
3	GTS-1	Industrial grade	Solid hydrochloric acid	Shandong Jinke Chemical Co., Ltd.
4	GTS-2	Industrial grade	Solid hydrochloric acid	Jinan Baike Chemical Co., Ltd.
5	HY-CGJ-1	Industrial grade	Chelating agent	Beijing Huayou Oil and Gas Technology Development Co., Ltd.
6	HY-CGJ-2	Industrial grade	Chelating agent	Beijing Huayou Oil and Gas Technology Development Co., Ltd.
7	LY-CGJ-1	Industrial grade	Complex chelating agent	CNPC Engineering Technology R&D Co., Ltd.

The simulated formation water used in the experiment was prepared according to the actual formation water in the Niger Oilfield. The detailed preparation table is shown in Table 2.

Table 2. Preparation of simulated formation water in Niger Oilfield.

Water type		Content (mg/L)								Total
		Na^+	K^+	Ca^{2+}	Mg^{2+}	Cl^-	SO_4^{2-}	CO_3^{2-}	HCO_3^-	(g/L)
Niger Oilfield Formation Water		660.12	10.24	54.01	25.48	202.28	242.46	0	1414.64	2.61
Simulated formation water	$CaCl_2$	0	0	54.01	0	95.86	0	0	0	0.15
	$MgCl_2$	0	0	0	25.48	74.42	0	0	0	0.10
	KCl	0	10.24	0	0	9.3	0	0	0	0.02
	Na_2SO_4	116.18	0	0	0	0	242.46	0	0	0.36
	$NaHCO_3$	533.39	0	0	0	0	0	0	1414.64	1.95
	NaCl	10.55	0	0	0	22.7	0	0	0	0.03

2.5 *Experimental instruments*

In this study, a variety of experimental instruments such as electronic analytical balance and FT-IR infrared spectrometer were used. The specific details are shown in Table 3.

Other experimental instruments are the Soxhlet extractor, test tubes, medicine spoons, syringes, beakers, droppers, glass rods, water stop clips, catheters, crucibles, etc.

Table 3. Instruments used in experiments.

S/N	Experimental instruments	Types	Manufacturer
1	Electronic analytical balance	FA2004C	Shanghai Yueping Scientific Instrument Co., Ltd.
2	FT-IR infrared spectrometer	MAGNA 560	Thermo Nicolet Corporation
3	Digital electric mixer	JJ-1	Jiangsu Ronghua Instrument Manufacturing Co., Ltd.
4	Electric blast drying oven	DHG	Shanghai Yiheng Scientific Instrument Co., Ltd.
5	Rotary vane vacuum pump	2XZ-2	Linhai Tanshi Vacuum Equipment Co., Ltd.

3 TEST RESULTS AND DISCUSSIONS

3.1 *Component analysis of wellbore scale samples in Niger oilfield*

The Soxhlet extraction method was used to separate the wellbore scale samples of the Niger oilfield, in which the solid phase particles accounted for 78%, and the organic phase accounted for 22%. The elemental composition of the solid phase particle fraction of the scale sample was analyzed by an energy dispersive spectrometer, and the results are shown in Table 4.

Table 4. Elemental composition of solid phase particles in wellbore scale samples in Niger Oilfield.

Element	Ba	S	Ca	Fe	Si	Mg	Zn	Mn	Al	Na	Others
Content (%)	52	25.07	7.35	4.87	3.63	2.39	1.36	1.01	0.77	0.41	1.15

It can be seen from Table 4 that barium has the highest proportion in the wellbore scale-like solid phase particles in the Niger Oilfield, accounting for 52.00%, followed by sulfur at 25.07%. Combined with the development experience of oil and gas wells worldwide and the composition of drilling fluids in Niger Oilfield, the analysis shows that barium and sulfur are mainly derived from barite in drilling fluids. Subsequent evaluation of blockage removal agents in Niger Oilfield needs to focus on solving the $BaSO_4$ problem. Therefore, in this study, $BaSO_4$ was selected as the simulated scale sample to evaluate the performance of the blockage removal agent.

3.2 *Evaluation of dissolving $BaSO_4$ with different blockage removal agents*

Different blockage removal agents were selected to evaluate the dissolution effect of $BaSO_4$ at room temperature. The experimental results are shown in Table 5.

Table 5. Dissolution effect of different types of blockage removal agents on $BaSO_4$.

	Quality (g)			
Type	Before reaction	After reaction	Time (h)	Dissolution rate (%)
YSW-1	0.5	0.495	24	1
GTS-1	0.5	0.498	24	0.4
GTS-2	0.5	0.495	24	1
HY-CGJ-1	0.5	0.497	24	0.6
HY-CGJ-2	0.5	0.496	24	0.8
LY-CGJ-1	0.5	0.493	24	1.4

It can be seen from Table 5 that the dissolution efficiency of various blockage removal agents at room temperature is less than 1.5%, which cannot meet the blockage removal requirements for $BaSO_4$. This is due to the chemical properties of $BaSO_4$ and its stability and extremely low solubility. General oilfield chemical blockage removal agents cannot be effectively dissolved at room temperature. But among these six blockage removal agents, LY-CGJ-1 has higher relative solubility and can reach about 1.4%. Therefore, LY-CGJ-1 was selected for subsequent experiments to study the effects of its mass fraction, reaction temperature, and reaction time on the dissolution rate.

3.3 *Effect of mass fraction of LY-CGJ-1 on the dissolution rate of $BaSO_4$*

At 110°C, the dissolution rate of $BaSO_4$ in LY-CGJ-1 blockage removal agent with different mass fractions was tested, and the experimental results are shown in Figure 2.

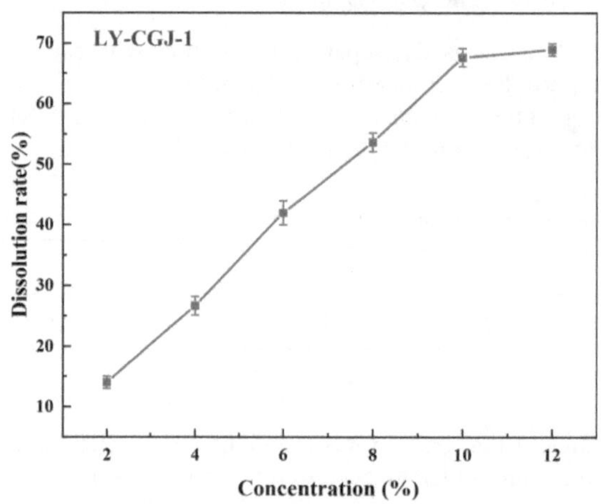

Figure 2. Dissolution rate of $BaSO_4$ by LY-GTS-1 with different mass fractions.

The experimental results in Figure 2 show that the higher the mass fraction of LY-CGJ-1, the higher the dissolution rate of $BaSO_4$, and the highest is about 69%. However, when the mass fraction exceeds 10%, the increasing range of $BaSO_4$ dissolution rate is obviously weakened, and an inflection point appears. The reason for this phenomenon is that the increased mass fraction of LY-CGJ-1 can contact with more $BaSO_4$, thereby accelerating the dissolution reaction, but when the mass fraction increases to a certain extent, the blockage removal agent and $BaSO_4$ have reached sufficient contact. Further increasing its mass fraction can slightly improve the dissolution rate, but it is not cost-effective in terms of economic benefits. Therefore, a 10% mass fraction is the best mass fraction of LY-CGJ-1 for $BaSO_4$.

3.4 *The effect of temperature on the dissolution rate of $BaSO_4$*

In order to analyze the applicability of the LY-CGJ-1 blockage removal agent to the Niger Oilfield, the dissolution rate of $BaSO_4$ at different temperatures was investigated. The experimental results are shown in Figure 3.

As shown in Figure 3, as the reaction temperature increases, the dissolution rate increases continuously. This is because the dissolving reaction of the blockage removal agent to $BaSO_4$ is endothermic. According to the Arrhenius formula, the higher the temperature, the faster the reaction rate, so that the dissolution rate of the blockage removal agent to $BaSO_4$ increases. It can also be found from the experimental results that when the reaction time is 48 h, the dissolution rate increases significantly at 60-110°C relative to the reaction time of 24 h. This is because

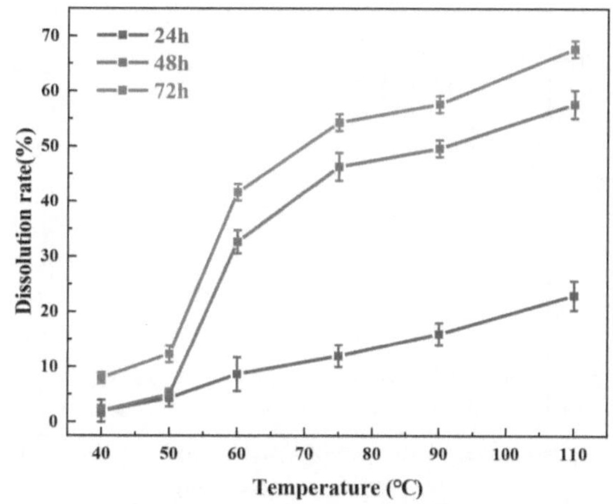

Figure 3. Effect of temperature on the dissolution rate of LY-CGJ-1 blockage removal agent.

prolonging the reaction time can further increase the contact probability of the blockage removal agent with $BaSO_4$, thereby increasing the dissolution rate. However, it can also be found that when the temperature is lower than 60°C, the dissolution rate does not increase significantly, which may be due to the fact that there is a threshold for the reaction temperature of the LY-CGJ-1 blockage removal agent and $BaSO_4$. When it is lower than the threshold, the reaction rate is low, and when the reaction temperature exceeds the threshold, the reaction rate increases significantly, thereby increasing the dissolution rate of $BaSO_4$. When the reaction time was prolonged to 72 h, it could be found that the dissolution rate of LY-CGJ-1 was improved compared with the reaction time of 48 h, but the increase was relatively low. This is due to the continuous reduction of the effective components of the blockage removal agent during the dissolution process, which reduces the amount of $BaSO_4$ dissolved. This phenomenon can also be used to guide the field application. For example, after the first injection of blockage removal agent for 48 hours, the second injection of blockage removal agent is carried out according to the blockage removal situation to improve efficiency.

3.5 *FT-IR analysis of Niger wellbore scale samples before and after the dissolution*

Figure 4 shows the FT-IR spectra of Niger oilfield wellbore scale samples before and after being dissolved by LY-CGJ-1. It can be seen from the FT-IR spectrum before dissolution that there are obvious absorption peaks at 3504 cm^{-1}, 3022 cm^{-1}, 2982 cm^{-1}, 2910 cm^{-1}, 1654 cm^{-1}, 1178 cm^{-1}, 1083 cm^{-1}, 979 cm^{-1}, 606 cm^{-1} and 419 cm^{-1}. The broad absorption peak at 3504cm^{-1} is the -OH stretching vibration of wellbore scale-like adsorbed water. The absorption peaks at 3022 cm^{-1}, 2982 cm^{-1}, and 2910 cm^{-1} are caused by C-H absorption, indicating certain carbohydrates in the wellbore scale samples. 1648 cm^{-1} is the -OH bending vibration of water. The absorption peaks at 1178 cm^{-1}, 1083 cm^{-1}, 979 cm^{-1}, and 609 cm^{-1} are consistent with the characteristic peaks of SO_4^{2-} in barium sulfate, and the absorption peaks are relatively strong, indicating that the scale components are mainly sulfates (Putnis 2008). This is consistent with the results of Soxhlet extraction.

In the infrared spectrum after dissolving the scale sample, compared with the infrared spectrum before dissolving, there is a strong absorption peak at 1622 cm^{-1}, where the absorption peak is the carbonyl C=O stretching vibration in the blocking remover. The new absorption peak at 1438 cm^{-1} may be the N-H bending vibration after the combination of the amino group and the metal ion in the blocking agent. In addition, the absorption peaks at 1178 cm^{-1} and 979 cm^{-1} disappeared, indicating that the SO_4^{2-} group was reduced, and part of the sulfate was destroyed by the blocking

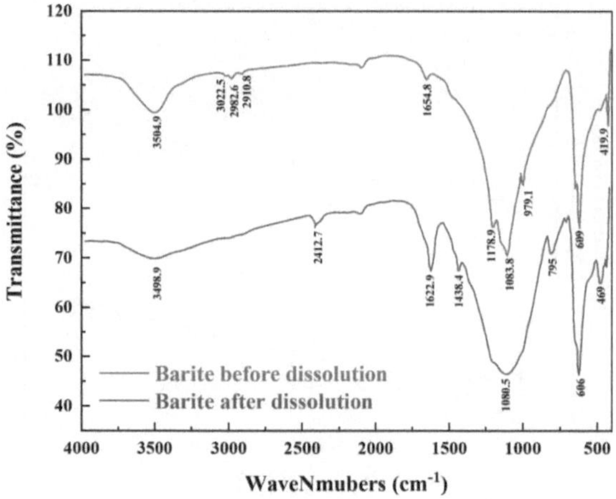

Figure 4. Effect of temperature on the dissolution rate of LY-CGJ-1 blockage removal agent.

agent to form a new compound. This indicates that both C=O and N-H are involved in the formation of new chelates.

4 CONCLUSIONS

According to this study, the efficiency of blockage removal work in Niger Oilfield can be improved by more than 50%, significantly shortening the construction period to help oilfields improve oil recovery. The main conclusions of this study are as follows:

(1) The Niger oilfield wellbore scale samples were separated by the Soxhlet extraction method, and the elemental composition of the solid particles of the scale samples was analyzed by an energy spectrometer. The results showed that the solid particles accounted for 78%, of which barium and sulfur accounted for the highest proportion. The main component is $BaSO_4$.
(2) All blockage removal agents selected in this study are ineffective at room temperature, among which LY-CGJ-1 is relatively effective, with an efficiency of 1.4% for dissolving $BaSO_4$.
(3) At 110°C, the higher the mass fraction of LY-CGJ-1, the higher the dissolution efficiency. When the mass fraction is 12%, the highest dissolution rate is 69%, but this is similar to the result when the mass fraction is 10%. Considering the economy, 10% LY-CGJ-1 is the best.
(4) When the temperature exceeds 50°C, the dissolution efficiency of LY-CGJ-1 increases continuously, indicating that high temperature can promote its reaction rate to $BaSO_4$. When the reaction time exceeds 48 h, the dissolution rate increase rate decreases, indicating that in the actual application process, the second injection of the blockage removal agent should be within 48 to72 hours after the first injection.
(5) The FT-IR spectrum analysis shows that the wellbore scale samples in the Niger Oilfield are mainly $BaSO_4$, which is consistent with the energy spectrum analysis results. The chelating reaction between LY-CGJ-1 and the scale samples formed new chelates, which promoted the dissolution of $BaSO_4$.

ACKNOWLEDGMENTS

This work was supported by the Key Technologies R&D Program from CNPC Niger Petroleum S.A., "The Integrated Study on Improving the Quality and Efficiency of Completion and Stimulation in Agadem Block."

REFERENCES

Al Hamad M, Al Sobhi S A, Onawole A T, et al. (2020) Density-Functional theory investigation of barite scale inhibition using phosphonate and Carboxyl-Based inhibitors[J]. ACS Omega, 5(51): 33323–33328.

Bageri B S, Mahmoud M A, Shawabkeh R A, et al. (2017) Toward a complete removal of barite (Barium Sulfate $BaSO_4$) scale using chelating agents and catalysts[J]. Arabian Journal for Science and Engineering, 42(4): 1667–1674.

Dunn K, Yen T F. (1999) Dissolution of barium sulfate scale deposits by chelating agents[J]. Environmental Science & Technology, 33(16): 2821–2824.

Geri B B, Mahmoud M, Shawabkeh R, et al. (2017) Evaluation of barium sulfate (Barite) solubility using different chelating agents at a high temperature[J]. Journal of Petroleum Science and Technology, 7(1): 42–56.

Mahmoud M A, Elkatatny S. (2018) Removal of Barite-Scale and Barite-Weighted water- or Oil-Based-Drilling-Fluid residue in a single stage[J]. SPE Drilling & Completion, 34(1): 16–26.

Mahmoud M, Ba Geri B, Abdelgawad K, et al. (2018) Evaluation of the reaction kinetics of diethylenetriaminepentaacetic acid chelating agent and a converter with barium sulfate (Barite) using a rotating disk apparatus[J]. Energy & Fuels, 32(9): 9813–9821.

M. U, A. S, T. Y, et al. (2000) Hydrothermal synthesis of needle-like barium sulfate using barium (II)-EDTA chelate precursor and sulfate ions[J]. Journal of Materials Science Letters, 19(15): 1373–1374.

Putnis C V, Kowacz M, Putnis A. (2008) The mechanism and kinetics of DTPA-promoted dissolution of barite[J]. Applied Geochemistry, 23(9): 2778–2788.

Risthaus P, Bosbach D, Becker U, et al. (2001) Barite scale formation and dissolution at high ionic strength studied with atomic force microscopy[J]. Colloids and Surfaces. A, Physicochemical and Engineering Aspects, 191(3): 201–214.

Zhang F, Dai Z, Yan C, et al. (2017) Barite-Scaling risk and inhibition at high temperature[J]. SPE Journal, 22(1): 69–79.

Advances in Mineral Resources, Geotechnology and Geological Exploration – Rashid & Zhang (eds)
© 2023 Copyright the Author(s), ISBN 978-1-032-33377-9

Influence of stress evolution of coal pillar under unilateral water pressure

Tongxu Wang*
College of Energy and Mining Engineering, Shandong University of Science and Technology, Qingdao, Shandong, China
State Key Laboratory of Mining Disaster Prevention and Control Co-founded by Shandong Province and the Ministry of Science and Technology, Shandong University of Science and Technology, Qingdao, Shandong, China

Yumeng Wei
College of Energy and Mining Engineering, Shandong University of Science and Technology, Qingdao, Shandong, China

ABSTRACT: The stress field and the seepage field interact and interact with each other until they reach a dynamic and stable state. This dynamic relationship constitutes a fluid-solid coupling mathematical model. For the actual engineering rock mass, the porosity and permeability coefficient K change dynamically with different stress and strain states of the rock mass. Therefore, in the numerical calculation and analysis, the dynamic change of the permeability coefficient should be considered so that the calculation results can be more in line with the actual engineering. According to the characteristics of the water softening of coal and rock, the fluid-solid coupling calculation related to water content is further carried out. The calculation results show that with the increase of water content, the degree of weakening of the mechanical parameters of coal and rock mass increases, the infiltration range increases, the plastic zone extends to the deep part of the coal pillar, and the range of the plastic zone increases.

1 INTRODUCTION

At present, most of the upper coal seam is about to be mined or has been mined, and the mining of the lower coal seam begins. The overburden structure and caving characteristics of the lower coal seam are different from that of the single coal seam mining, and the production practice faces many technical problems. In recent years, relevant scholars have studied the technology and theory of coal seam group mining and obtained many beneficial results. Frith et al. studied the secondary mining of fully mechanized caving working face, and concluded that the secondary mining of stope overburden of lower coal seam formed a "caving zone superposition" structure, and the peak value of advanced abutment pressure and its influence range of working face decreased (Frith et al. 2019). Mark et al. studied the force transmission mechanism in and under coal rock mass of coal pillars with different widths and believed that with the increase of coal pillar width, the maximum principal stress in coal pillar increased first and then decreased, and furthermore obtained the formula for calculating the stress diffusion angle of coal pillar (Mark 2019). Du et al. studied the stress propagation rule under the left coal pillar of the upper fully mechanized caving mining and analyzed the disturbance range of the coal pillar with different widths to the lower coal seam under the background of the mining of the underlying coal seam group with a depth of 270 m (Du 2020). Through physical simulation and numerical analysis, Liu et al. concluded that the change of floor strain in near coal seams with mining was related to the compaction range of overlying strata, original

*Corresponding Author: 749841061@qq.com

 DOI 10.1201/9781003319412-42

pressure load base, and space boundary (Liu 2020). Liu et al. studied the relationship between coal pillar dislocation and interval rock thickness in the mining section of shallow and close-distance coal seam group, and proposed that the "stress field, displacement field, and fracture field" of coal seam group mining had three evolution and coupling characteristics (Liu 2021). Yang et al. studied that the overburden collapse of the upper coal seam is difficult to form structure, and the increase of separation height is small in the background of the mining of the close seam group (Yang 2020).

The above researchers have carried out a large number of research on overburden collapse characteristics, stress transfer law of coal pillar and crack propagation law, etc. However, there is scant research on the synergistic effect of "coal pillar roof structure" and the stability of coal pillars under the influence of secondary mining in the lower part of the shallow coal seam group. The practice shows that the mining of the upper and lower seams of the shallow coal seam group has a serious influence on each other, the overburden structure is special, and the stress environment is complex. The stress concentration degree of the coal pillar in the lower coal seam is high, and the compression condition and scope of the roadway are uncertain, which is easy to induce a strong mine pressure disaster.

However, the traditional fluid-structure interaction only considers the direct coupling between the fluid and the medium, while in the complete fluid-structure interaction, the change of porosity and permeability coefficient is caused by the change of effective stress. The reduction of porosity will cause the cross-sectional area of the medium, and the decrease in the water content further increases the stiffness of the material, which leads to anisotropy in the permeability coefficient of the porous medium. Therefore, considering the dynamic change of permeability coefficient in practical engineering has important practical significance for the study of coal safety and stability.

2 METHODS AND MATERIALS

For the residual coal pillar, after the influence of multi-face mining, the coal pillar mainly has an elastic zone and plastic yield zone. Previous studies have shown that there are essential differences in permeability between elastic coal and plastic coal. Therefore, it is necessary to distinguish the permeability calculation model of yield block and elastic block in the seepage process. In this paper, the permeability calculation method of the residual coal pillar is adopted in reference. The specific permeability calculation formula is as follows (Zhang 2019):

$$k_E = 7.9549e^{-4.8879\left(1-e^{-0.3888\sigma_1}\right)} \tag{1}$$

$$k_S = 142.2316e^{-2.6349\left(1-e^{-0.3415\sigma_1}\right)} \tag{2}$$

Where, k_E and k_S are permeability in the elastic zone and plastic zone (mD), and $\sigma 1$ is the effective stress on the block (MPa).

Previous studies have shown that the strength of the coal body is significantly weakened under the action of long-term water immersion (Bai 2017).

The coal pillar reaches a saturated state under the action of pore water pressure infiltration (when the coal pillar pore water pressure is greater than 0), and the saturated water content η of the coal pillar is:

$$\eta = \frac{\rho_w \varphi}{\rho_w \varphi + \rho_m} \tag{3}$$

Where, φ is the porosity of the coal sample, ρ_w and ρ_m are the densities of water-containing and dry coal samples (kg/m^3), respectively.

The relationship between porosity and permeability of coal sample is as follows (Wang, 2019)

$$\frac{k}{k_0} = \left(\frac{\varphi}{\varphi_0}\right)^3 \tag{4}$$

Where, φ_0 is the porosity of the block before the deformation; k_0 and k are the permeability of the block before and after deformation (mD).

Taking the narrow coal pillar in the working face of Xiaotun Coal Mine as the research object, the numerical simulation analysis of the water seepage process on one side of the goaf is carried out by using FLAC3D software. Considering the calculation accuracy and calculation speed of the simulation, the size of the model is determined to be 100 m × 47 m (L × H), and the thickness of the overlying rock from the upper part of the model to the ground is 300 m. The additional load (7.5 MPa) is used to simulate, and the mechanical parameters of each rock layer are shown in Tables 1 and 2.

Table 1. Model composition and corresponding parameters.

Rock layer	Thickness /m	Bulk modulus / /GPa	Shear modulus /GPa	Cohesion /MPa	Tensile strength /MPa	Friction angle /°	Density /kg·m^{-3}
Fine sandstone	8	5	4.5	8	3.6	48	2650
Coal	3	3	2.4	4.5	1.8	38	1750
Sandy mudstone	3	4	2.6	5	1.9	45	2100
Grit	5	2.9	1.5	3.5	1.2	43	2000
Siltstone	2	3.5	2.5	4	0.9	35	1900
Grit	13	3	2	3.5	1.2	37	2100
Coal gangue	6	1.4	0.2	1.3	0.7	25	2500

Table 2. Parameters of coal.

Bulk modulus /GPa	Shear modulus /GPa	Cohesion /MPa	Tensile strength /MPa	Friction angle /°	Density /kg·m^{-3}	Permeability coefficient /cm·s^{-1}	Water pressure /MPa
4	2.2	4.5	1.8	38	1790	1e-9	0.03

3 RESULTS AND DISCUSSION

The simulation cases in this paper are mainly divided into three situations:

Case 1: Do not consider the impact of water;

Case 2: Consider the permeability coefficient as a fixed value;

Case 3: Consider the change of permeability coefficient with volumetric strain and the softening effect of water immersion.

The vertical and horizontal displacement cloud diagrams of the coal pillar under different conditions are shown in Figure 1. The cloud map of the seepage field is shown in Figure 2. It can be seen from the figure that when considering the dynamic change of the permeability coefficient, the permeability range is larger than that considering the permeability coefficient as a fixed value. The closer to the goaf, the larger the permeability coefficient, and the final stable state permeability coefficient is 1.27×10^{-8}, an order of magnitude higher than the initial permeability coefficient. The volumetric strain is consistent with the permeability range, indicating that water seepage in the gob on one side of the coal pillar leads to the derivation of fractures and further increases the permeability coefficient (Zhang 2019).

From the displacement of the comparison and analysis, considering the vertical displacement on the dynamic change of permeability coefficient, the deformation quantity was large. If the coefficient was as a fixed value and did not consider the horizontal displacement of water was not considered when deformation and difference were all small. Three kinds of the working condition of horizontal displacement were smaller because the balance after the initial excavation, horizontal displacement, and stress changed. The difference in vertical displacement mainly came from the increase of volume strain caused by the change of permeability coefficient, which should be paid attention to in engineering.

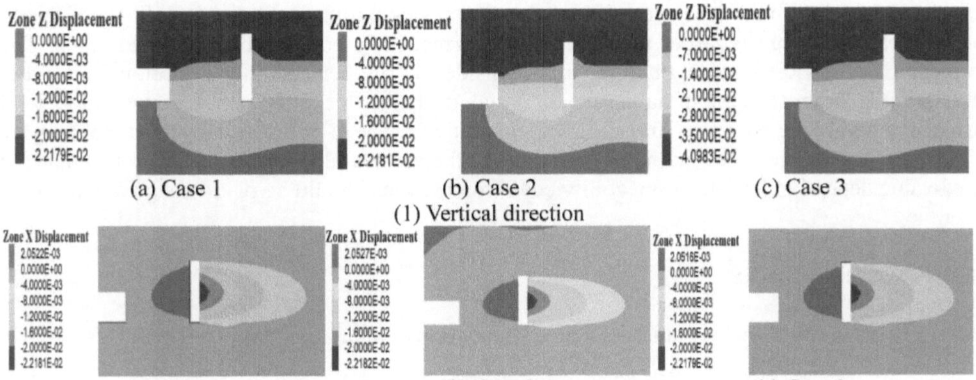

(a) Case 1 (b) Case 2 (c) Case 3

(1) Vertical direction

(a) Case 1 (b) Case 2 (c) Case 3

(2) Horizontal direction

Figure 1. Displacement cloud images in different directions.

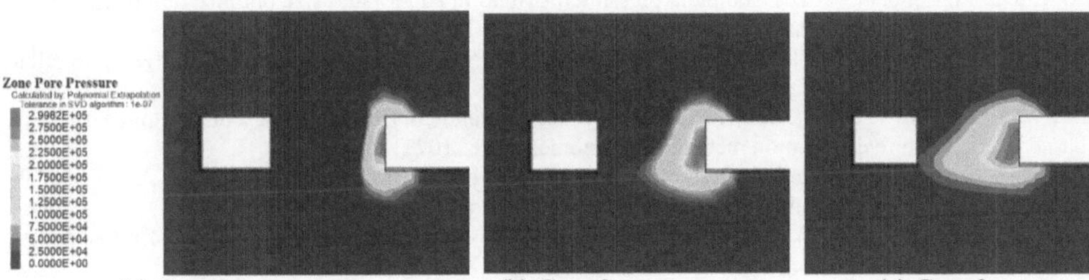

(a) Case 1 (b) Case 2 (c) Case 3

Figure 2. Cloud image of seepage field in different cases.

By comparing the plasti c zone distribution under the three conditions, the plastic distribution range and yield degree of coal pillar were obtained using the permeability dynamic evolution model. The results showed that the change of permeability coefficient aggravates the damage deformation of coal rock to a certain extent, leading to the failure of coal rock. As can be seen from Figure 3, the plastic damage area and permeability increase area were basically in accordance. And in some areas, the coal rock had not yet occurred. However, because the strain change can also lead to increase permeability, it was not enough to calculate the width of a waterproof coal pillar simply depending on whether the coal rock was damaged.

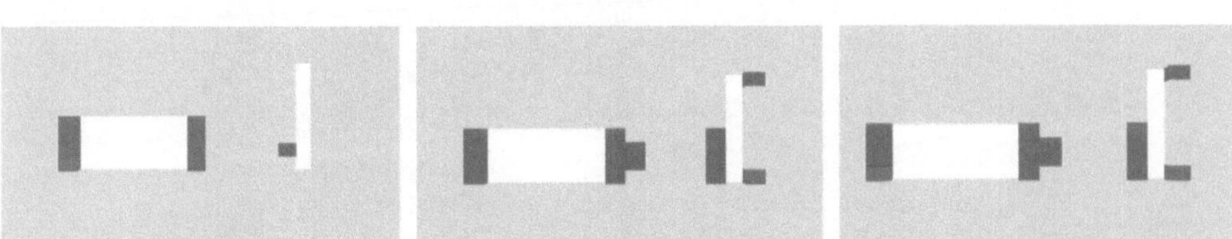

(a) Case 1 (b) Case 2 (c) Case 3

Figure 3. Cloud image of the plastic zone in different cases.

4 CONCLUSION

This paper compares the traditional permeability coefficient as a fixed value and considers the influence of the dynamic evolution of the permeability coefficient on coal pillar stability. The research results show that: when considering the dynamic evolution of the permeability coefficient, the permeability range is larger than taking the permeability coefficient as a fixed value. The permeability coefficient value obtained after considering the permeability evolution is an order of magnitude larger than the initial permeability coefficient value, and the horizontal and vertical displacements are increased by 30% and 20%, respectively. The distribution range and yield degree of the coal rock plastic zone obtained by considering the permeability evolution are greater than the fixed value depending on the permeability coefficient, which should be paid enough attention to.

ACKNOWLEDGMENTS

This study is supported by the Natural Science Foundation of Shandong Province (ZR2016EEM36).

REFERENCES

Bai Q S, Tu S H, Wang F T, et al. (2017) Field and numerical investigations of gate road system failure induced by hard roofs in a long wall top coal caving face. Int. J. Coal Geol., 173: 176–199.

Du B, Liu C, Yang J, et al. (2020) Abutment pressure distribution pattern and size optimization of coal pillar under repeated mining: a case study. Arab. J. Geosci., 13: 1–14.

Frith R, Reed G. (2019) Limitations and potential design risks when applying empirically derived coal pillar strength equations to real-life mine stability problems. Int. J. Min. Sci. Techno., 29: 21–29.

Liu S, Wan Z, Zhang Y, et al. (2020) Research on evaluation and control technology of coal pillar stability based on the fracture digitization method. Measurement, 158: 107713.

Liu A, Liu S. (2021) A fully-coupled water-vapor flow and rock deformation/damage model for shale and coal: Its application for mine stability evaluation. Int. J. Rock Mech. Min., 146: 104880.

Mark C, Agioutantis Z. (2019) Analysis of coal pillar stability (ACPS): A new generation of pillar design software. Int. J. Min. Sci. Techno., 29: 84–88.

Wang F T, Liang N G, Li G, et al. (2019) Damage and failure evolution mechanism for coal pillar dams affected by water immersion in underground reservoirs. Geofluids, 1: 1–12.

Yang K, Gou P. (2020) Research on a reasonable width of coal pillars in high strength mining roadway in Wantugou mine. Geotech. Geol. Eng., 5: 1–9.

Zhang C, Liu J B, Zhao Y X, et al. (2019) A fluid-solid coupling method for the simulation of gas transport in porous coal and rock media. Energy Sci. Eng., 7: 1913–1924.

Zhang C, Tu S H, Zhao Y X, et al. (2019) Discrete element numerical simulation of tri-axial fluid-solid coupling based on seepage experiment. Int. J. Min. Sci. Techno., 4: 27–37.

Remote sensing modeling measurement technology and numerical analysis

Advances in Mineral Resources, Geotechnology and Geological Exploration – Rashid & Zhang (eds)
© 2023 Copyright the Author(s), ISBN 978-1-032-33377-9

Research on the prediction method of coalbed methane production based on GA-BP

Ming Huang* & Tiejun Li
School of Science, Southwest Petroleum University, Chengdu, China

Shuguang Li
China United Coalbed Methane National Engineering Research Center Co., Ltd., Beijing, China

Zixi Guo
State Key Laboratory of Oil and Gas Reservoir Geology and Development Engineering, Southwest Petroleum University, Chengdu, China

ABSTRACT: As clean energy, coalbed methane is a hot spot and focus of research and development at home and abroad. For the prediction of coalbed methane production, the existing methods have a poor prediction effect. In this paper, random forest (RF) is used to sort the importance of the characteristics affecting CBM production, eight main controlling factors are selected, and a CBM production prediction model is established based on the Genetic Algorithm-Optimized BP Neural Network (GA-BP). The GA-BP model predicted the yield with an accuracy of 87.2%, which provides a good method reference for the CBM production forecast.

1 INTRODUCTION

China's coalbed methane resources are rich in reserves. The development of the coalbed methane industry can not only obtain clean energy, but also generate considerable economic benefits while reducing air pollution, which is of great value. However, my country's coalbed methane exploration technology, development technology, and understanding level are still in the early stage of development. Therefore, it is of certain significance to study the prediction method of coalbed methane production (Zhu et al. 2020; Zhang et al. 2020).

At present, scholars at home and abroad have carried out a lot of research on CBM production prediction. Deng Yinger et al. (Liu et al. 2020) proposed a method to predict the change law of CBM production by using the grey model, which is of great significance for the rational development of CBM resources. Zhang Hairu et al. (Zhuang et al. 2013) proposed a method to predict coalbed methane production using the Weibull model, which has a relatively high fitting accuracy. However, the above methods have some shortcomings. First, the grey model does not consider the relationship between the factors, which will lead to large errors, and it will be out of reality as a long-term forecast. Second, the Weibull model does not consider the influence of construction parameters on CBM production.

Therefore, this paper proposes a method for predicting coalbed methane production using GA-BP neural network. RF was used to rank the factors affecting coalbed methane production by feature importance. Eight construction displacement factors with the top eight feature importance, such as swept volume, sand addition, total liquid volume, GR, DEN, specific water-cycle, pre-liquid volume, and the sand ratio were selected, with construction sand ratio as the input and gas production as the output. The prediction accuracy of this method reaches 87.2%, which is better than BP neural network prediction and RBF neural network prediction.

*Corresponding Author: 872868405@qq.com

DOI 10.1201/9781003319412-43

2 MODEL ESTABLISHMENT

2.1 *Random Forest*

Random Forest is an ensemble classifier composed of multiple decision trees. When we classify the input samples, we need to input the samples into each decision tree for classification. Each tree votes on samples, and the class with the most votes is the output of the random forest. The random forest is built in a random way, so it is random. It is mainly reflected in two aspects; one is the random selection of data sets, while the other is the random selection of the features to be selected. These two aspects enable decision trees in random forests to randomly select samples and feature attributes to improve the diversity of the system, thereby improving the performance of classification (Wang et al. 2020; Zhu et al. 2020). The algorithm flow chart of the random forest is shown in Figure 1.

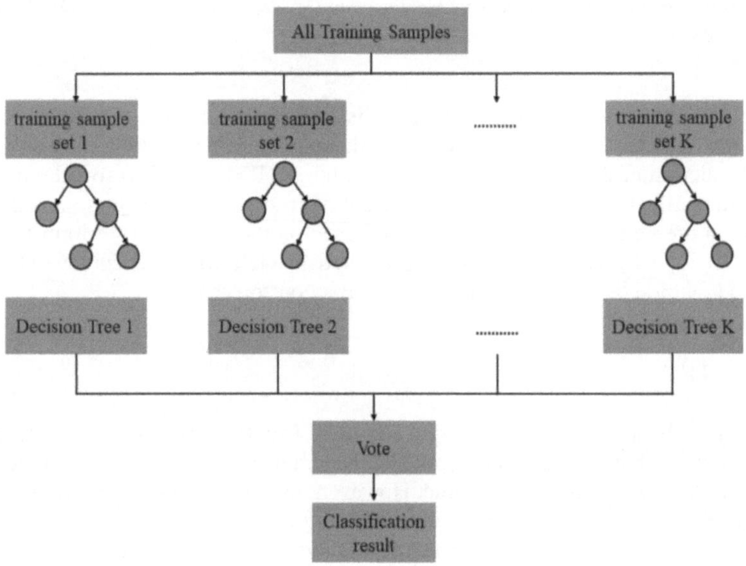

Figure 1. Flow chart of random forest algorithm.

The random forest can rank all features in importance. When we calculate the importance of feature z, we need to select the corresponding out-of-bag data for each decision tree and calculate its error, and then add random noise to the feature z of all samples of the out-of-bag data, and calculate its error again. The sum of the two errors and then removing the total decision tree can get the importance value of the feature z. After adding random noise, the second out-of-bag data error increases, indicating that this feature is more important. By sorting all features to the ascending power, the features we want can be selected (Peng et al. 2020).

2.2 *GA-BP neural network*

The genetic algorithm was proposed by J. Holland in the United States. The algorithm is a random search method evolved from the natural biological genetic and evolutionary mechanism. The algorithm is to first encode and assign individuals and calculate the fitness value, and then select the best individual according to the fitness function. Later, a series of operations can be performed such as selection, crossover, and mutation for all the selected individuals. To this end, the individuals with good fitness are saved, individuals with poor fitness are abandoned, and the new individuals inherit the good genes of the previous generation. This operation is repeated until the requirements are met or the number of iterations is reached (Tian & Ju 2016).

The GA-BP neural network process is shown in Figure 2. The difference between it and the BP neural network is that the initial weights and thresholds of the GA-BP neural network are optimized by genetic operations such as selection, crossover, and mutation of the genetic algorithm, so as to obtain the optimal weights and thresholds.

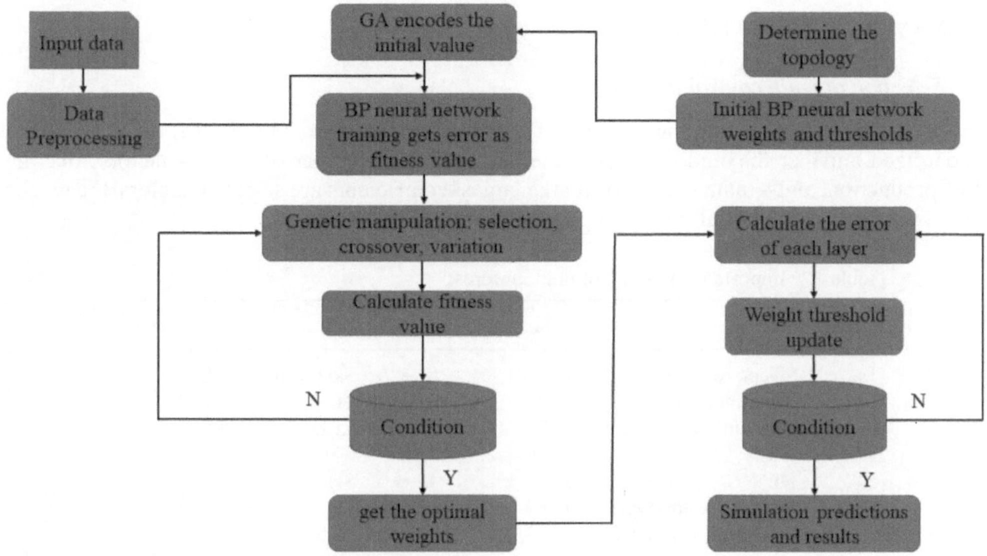

Figure 2. GA-BP neural network algorithm flow

The most important thing for the GA-BP neural network model to predict coalbed methane production is to optimize the network weights and thresholds. The optimization design steps of the network weights and thresholds are as follows;

Step 1: Determine the BP neural network topology.

Step 2: Initialize the population.

Step 3: Determine the fitness function. Generally, the error between the actual output value and the network output value uses $E\ (x_i)$ as the standard, and the fitness is $f\,(x_i)$.

$$f\ (x_i) = \frac{1}{E\ (x_i)} \tag{1}$$

Step 4: Select an action. Select the best individual from the population, multiply offspring (crossover, mutation), continuously obtain new individuals, and continue to select until the selected population gets the best individual. The selection of individuals is mainly based on fitness. Commonly used selection methods include the roulette selection method, fitness ratio method, random traversal sampling method, and so on. In the roulette wheel selection method, the fitness of the individual x_i is $f\ (x_i)$, and the probability of all individuals being selected is $P\ (x_i)$.

$$P\ (x_i) = \frac{f\ (x_i)}{\sum_{i=1}^{N} (x_i)} \tag{2}$$

Step 5: Crossover operation. Randomly select the crossover position and use the real number crossover method to crossover, and judge whether it needs to be recrossed by checking the feasibility of chromosomes.

Step 6: Mutation operation. Randomly select a chromosome for mutation, and use the mutation probability to determine whether the cycle is mutated. If it can be mutated, then randomly select the position of chromosome mutation to mutate according to the mutation formula. Finally, determine the viability of chromosomes.

Step 7: Repeat Steps 4 to 6 until a certain body satisfies the end condition.

3 EXAMPLE APPLICATION

3.1 *Selection of main control factors*

Aiming at the 14 factors summarized from the CBM field data in the LF area, the data is first standardized, and then the random forest is used to sort the importance of these 14 factors affecting CBM production, and 8 main controls with the largest coefficients are selected. factors, the results of random forest sorting of all factors are as follows:

Table 1. Importance ranking of random forest.

Rank	Feature	Importance	Rank	Feature	Importance
1	Swept volume	0.257	8	Sand ratio	0.032
2	Sand addition	0.188	9	AC	0.029
3	Total liquid Volume	0.13	10	LLD	0.025
4	GR	0.116	11	CNL	0.024
5	DEN	0.081	12	SH	0.022
6	Specific water-cycle	0.034	13	SP	0.022
7	Pre-liquid volume	0.033	14	LLS	0.021

It can be seen from Table 1 that the most important factors related to the CBM production obtained by random forest are eight construction factors, including swept volume, sand addition, total liquid volume, GR, DEN, specific water-cycle, pre-liquid volume, and sand ratio. The remaining 6 factors have little effect on gas production. Selecting too many factors will increase the prediction error of the model, and too few factors will also affect the prediction accuracy of the model. Therefore, the top 8 factors are selected as the main control factors, and based on these 8 main control factors as input, the production gas volume is used as the output, and the GA-BP neural network model is constructed.

3.2 *Application and inspection*

Using Matlab software to train and simulate the model, the evolutionary algebra in the GA-BP neural network is set to 50 generations, the population size is set to 10, the crossover probability is set to 0.2, the mutation probability is set to 0.1, and the hidden layer nodes are set to 20. The maximum number of iterations is set to 2 000 steps, the error target is set to 0.001, and the Logsigmod activation function is selected from the input layer to the hidden layer and from the hidden layer to the output layer.

3.3 *Comparative analysis of different methods*

In order to verify the superiority of the coalbed methane production prediction method established in this paper and the rationality of the selection of the main control factors, the main control factors are used as the input. The RBF neural network, BP neural network, and GA-BP neural network are used for comparison, and prediction results of different methods are obtained (Figure 3).

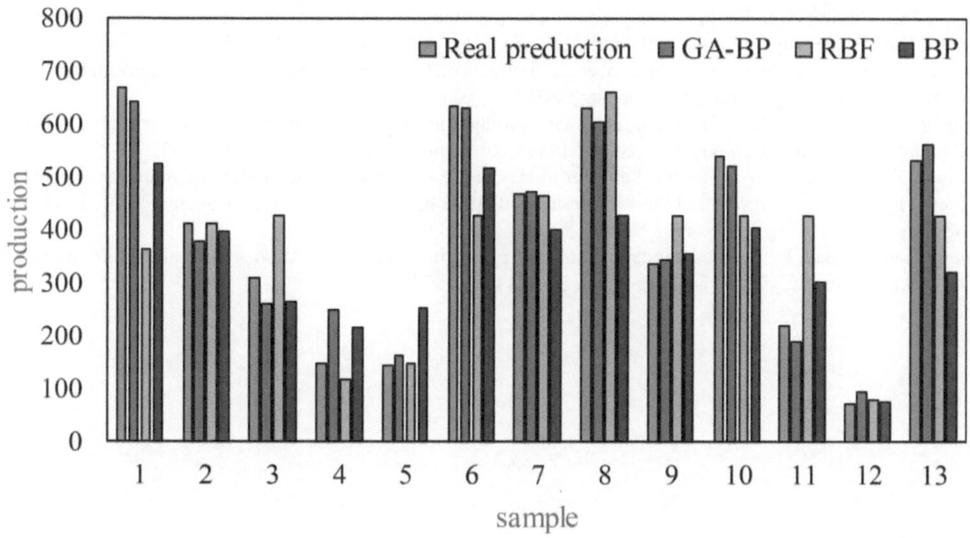

Figure 3. Comparison of real output and predicted output of each model.

Through comparative analysis, GA-BP neural network predicts coalbed methane production better than RBF neural network and BP neural network. This shows that the GA-BP neural network prediction model can better predict production through the relationship between variables, and can be better applied to the prediction of coalbed methane production in actual wells, which provides an effective method for CBM production prediction.

4 CONCLUSIONS

(1) Random forest was used to rank the factors affecting the CBM production in the LF area, and 8 main control factors were obtained through feature selection, namely: swept volume, sand addition, total liquid volume, GR, DEN, specific water-cycle, pre-liquid volume, and sand ratio.
(2) Based on the field data in the LF area, a GA-BP neural network prediction model for coalbed methane production was established. The prediction accuracy of the model was 87.2%, and the effect was better than the traditional BP neural network and RBF network.

ACKNOWLEDGMENTS

This research was funded by the National Science and Technology Major Project of China (No. 2016ZX05065, 2016ZX05042).

REFERENCES

Dameng Liu, Zhengshuai Liu, Yidong Cai. Research progress on the formation mechanism and geological conditions of coalbed methane[J]. Coal Science and Technology, 2020, 48(10): 1–16.

Dejun Wang, Qigang Jiang, Yuanhua LI, et al. Land use classification of farming area based on Sentinel-2A/B time series data and random forest algorithm[J]. Remote Sensing for Land & Resources, 2020, 32(04): 236–243.

Hairu Zhuang, Hao Li. Research on peak coal-bed methane production fitting and production dynamic prediction method[J]. Lithologic Reservoirs, 2013, 25(04): 116–118

Liuya Peng, Huiting Xie, Weidong Feng. Sandy soil liquefaction prediction method based on random forest algorithm[J]. Geophysical and Geochemical Exploration, 2020, 44(06): 1429-1434.

Qingzhong Zhu, Yanhui Yang, Yinqing Zuo, et al. Thoughts on the scientific development of high-rank coalbed methane resources[J]. Natural Gas Industry, 2020, 40(01): 55–60.

Qingzhong Zhu, Qiujia Hu, Haiwei Du, et al. Gas production model of coalbed methane vertical well based on random forest algorithm[J]. Journal of China Coal Society, 2020, 45(08): 2846–2855.

Xianmin Zhang, Haoyu Wu, Qihong Feng, et al. Dynamic response characteristics of multi-layer coalbed methane wells[J]. Journal of China University of Petroleum (Edition of Natural Science), 2020, 44(06): 88–96.

Yapeng Tian, Binshan Ju. Forecast model of shale gas production decline based on genetic algorithm improved BP neural network[J] China Science paper, 2016,11(15): 1710–1715.

Advances in Mineral Resources, Geotechnology and Geological Exploration – Rashid & Zhang (eds)
 ISBN 978-1-032-33377-9

Analysis of biodiversity in Liaoning Province based on GIS

Zhenzhen Zhao
Faculty of Science, Jiamusi University, Jiamusi, China

Zicheng Zhao*
School of Mechanical and Electrical Engineering, Henan University of Technology, Zhengzhou, China

Yawen Yang
School of Foreign Languages, Jiamusi University, Jiamusi, China

ABSTRACT: In this paper, we conducted a biodiversity assessment and analyzed the current situation of ecosystem diversity, species diversity, and genetic resource diversity. Based on the current biodiversity status, resource endowment, socio-economic development level, and development potential of the priority areas, the functional zoning of the priority areas was carried out, and the priority areas were divided into three categories of areas; the conservation priorities of the priority areas were proposed based on the current status of the priority areas, the conservation vacancies existing in the priority areas were proposed, and the layout optimization of the priority areas was carried out.

1 INTRODUCTION

The China Biodiversity Conservation Strategy and Action Plan (CBCSAP) has identified 35 priority areas for biodiversity conservation, taking into account various factors such as the representativeness of the ecosystem type, the degree of endemism, and the special ecological functions, etc (Zhang et al. 2022). Among the 35 priority areas, there are 32 terrestrial biodiversity conservation areas and 3 marine biodiversity conservation priority areas (Yuan et al. 2022). The number and area of nature reserves and nature protection zones have been increased on the basis of the number of vacant theoretical zones and the high proportion of farmland ecosystems in the priority areas for biodiversity conservation in Liaoning Province and will be gradually upgraded according to their importance. The priority area for biodiversity conservation in Liaodong has a high percentage of forest cover and a relatively small percentage of farmland ecosystems (Wang & Xiao 2021), and there are fewer theoretical blocks of conservation gaps in the priority area for biodiversity conservation (Zhao et al. 2021). It is important to strengthen the scientific research and management of biodiversity conservation priority areas in Liaoning Province.

2 BIODIVERSITY CONSERVATION IN CHINA

At present, some important and comprehensive regional delineation work has been carried out in a large area in China. The China Biodiversity Conservation Strategy and Action Plan (CBCSAP) takes various factors into consideration, and this paper constructs a model for identifying priority areas for biodiversity conservation in Liaoning Province and divides the established ecosystem

*Corresponding Author: q1335300187@163.com

DOI 10.1201/9781003319412-44

protection, human impact, and biodiversity conservation areas into three guideline layers (Yang & Zhou 2021). Considering the index systems and weights about priority areas for biodiversity conservation in other provinces, the index systems and weights are designed in conjunction with the focus and characteristics of biodiversity conservation in Liaoning Province (Table 1).

Table 1. Index system and weights of the identification model of priority areas for biodiversity conservation in Liaoning Province.

Composition level	Index system
Ecosystem protection (33.3%)	Native Ecosystems (30%) County Ecosystem Diversity (30%) Key ecological function areas (40%) Human disturbance index (50%)
Human impact (33.3%)	Change in human disturbance over 5 years (50%) Biodiversity conservation zoning (33.3%)
Biodiversity conservation zoning (33.3%)	Biodiversity conservation classification (60%) Nature reserves at or above the provincial level (40%)

3 DATA SOURCE AND PROCESSING

3.1 *Spatial distribution of ecosystem types*

The land use data were obtained from the 30 m precision distribution map of land use in Liaoning Province in 2015; the data on biodiversity zoning were obtained from the national database of ecosystem service functions; the data on nature reserves above the provincial level were obtained from the data managed by the Liaoning Department of Ecology and Environment.

3.2 *Calculation method*

The spatial distribution data (30 m precision) of terrestrial ecosystem types in China in 2015 for Liaoning and the county boundaries of Liaoning Province were used to calculate the ecosystem diversity index of the county in Liaoning Province according to Equation (1), which reflects the complexity of regional ecosystem composition.

$$ESDI = \sum_{i=0}^{m} P_i l_n P_i \tag{1}$$

Where: ESDI is the county ecosystem diversity index; m is the number of ecosystems of each type; Pi is the area share of ecosystem i. and ESDI is the same as SHDI. The larger the value, the more complex the ecosystem type in the region.

The spatial distribution data of land use types in China in 2015 (30 m precision) were taken for the Liaoning part. Land use within each grid was analyzed and human disturbance within each cell grid was calculated using Equation (2).

$$IHD = \sum i = InSi/S \tag{2}$$

Where: IHD is the human disturbance index; Si is the area of the i-th human disturbance type; i is the human disturbance classification code; n is the number of human disturbance index types, and S1, S2, S3, ..., Sn are the area of cultivated land, reservoir area, urban land area (Chen 2021), rural settlement area, transportation land area, industrial land area, industrial and mining site area, planted forest area; S is the total area of the region. The standardized treatment is used for the human disturbance index, and the values are normalized to 0~1, and the calculation formula is

$$Si = \frac{(HDI_i - HDI_{min})}{(HDI_{max} - HDI_{min})} \tag{3}$$

Where: Si is the standardized value of the human disturbance index in the i-th raster, and the value range is 0~1; HDIi is the actual value of the human disturbance index in the i-th raster; HDImax is the maximum value of the human disturbance index in the study area; HDImin is the minimum value of human disturbance index in the study area.

3.3 *Identification of biodiversity conservation priority areas in Liaoning Province*

Ecosystem protection indicators and guideline layers.

The calculation results of ecosystem protection indexes are shown in Figure 1 (left). The distribution of native ecosystems shows that forest ecosystems, grassland ecosystems, and water and wetland ecosystems are widely distributed in the eastern and western parts of Liaoning, and the distribution area in the east is significantly larger than that in the west and central parts of the country (Chen, 2021), while there is almost no large distribution of native ecosystems in the central part of the country except for the Liaohe River Figure 1 (left).

Calculation of the county ecosystem diversity index revealed (Figure 1 (right)) that the ecosystem diversity is higher in western Liaoning and lowest in central Liaoning, which is more urbanized and has more ecological functions influenced by cities. The key ecological function areas in Liaoning province are mainly water and soil conservation, water conservation, and biodiversity maintenance types (Figure 2 (left)).

The value of the ecosystem conservation guideline layer in Liaoning Province was calculated and displayed in a hierarchical manner (Figure 2 (right)). Native ecosystem, ecosystem diversity, and key ecological function zones in Liaoning Province are all positive indicators, and the combined calculation of the three indicators shows that the eastern and western regions of Liaoning are the key areas for ecosystem protection.

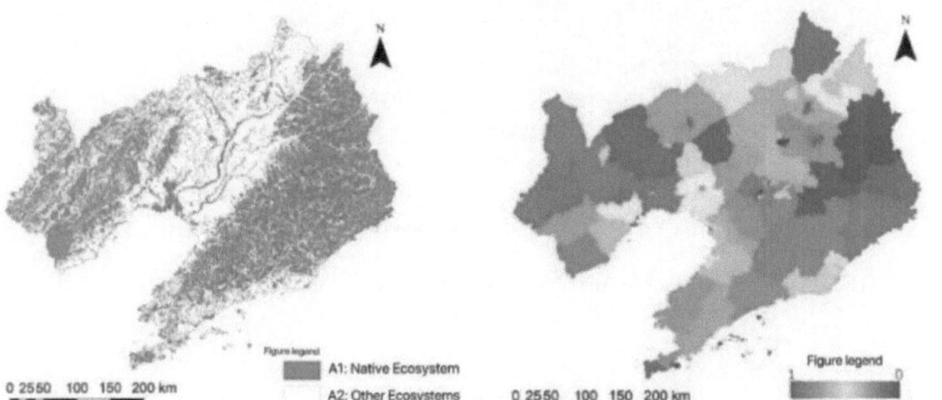

Figure 1. Primary ecosystem indicator results (left) and county ecosystem diversity (right).

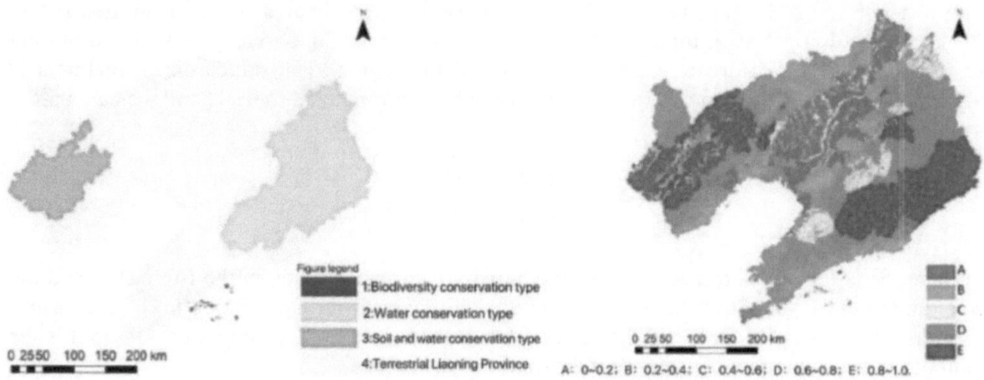

Figure 2. Key ecological function areas (left) and ecosystem conservation guidelines layer values for Liaoning Province (right).

The extent of change in human disturbance over five years was calculated using 2010 and 2015 land use data in Liaoning Province. The green part of the figure indicates that the ecology of some areas has been restored to a certain extent due to measures such as returning farmland to forest, ecological nurturing, and artificial afforestation (Yu & Hu 2022), while the disturbance of some lands has increased and conservation work and ecological restoration are urgently needed. In the western part of the country, the human disturbance index has changed less and, in some areas, there is no change in the level of human disturbance, while in the eastern part of Liaodong there is both positive and negative change.

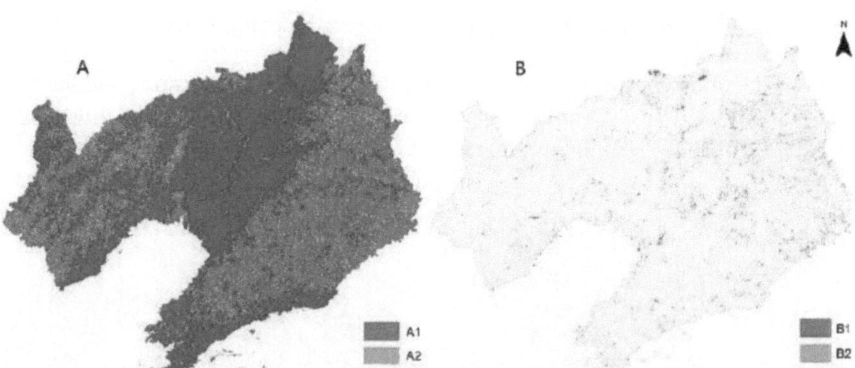

Figure 3. Human impact indicator results for Liaoning Province.

A: Human disturbance index; A1: 0~0.2; A2: 0.2~0.4; A3: 0.4~0.6; A4: 0.6~0.8; A5: 0.8~1.0.
B: 5-year human disturbance change level; B1: -1~0.5; B2: -0.5~-0.1; B3: -0.1~0.1; B4: 0.1~0.5; B5: 0.5~1.0.

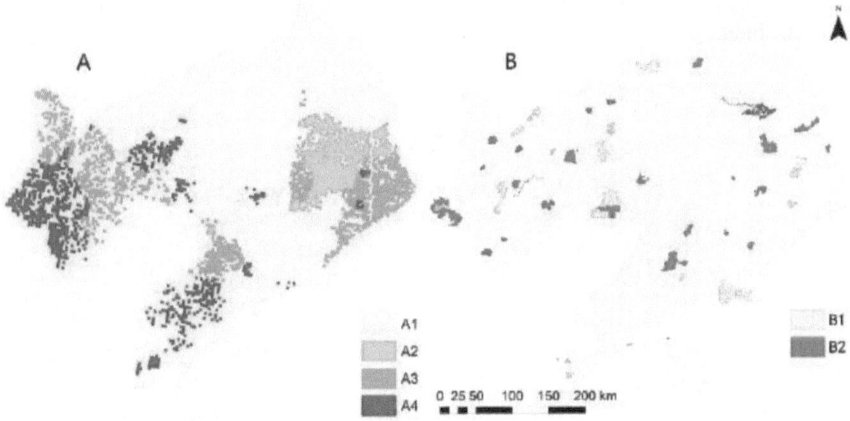

Figure 4. Key ecological function areas.

A: Classification of biodiversity conservation; A1: generally important; A2: moderately important; A3: important; A4: very important.

B: Distribution of nature reserves at or above a provincial level; B1: national nature reserves; B2: provincial nature reserves.

The biodiversity conservation grading indicators (Figure 4-A) were determined based on the national ecosystem service function data released by the China Ecosystem Assessment and Ecological Security Database and were divided into four levels, generally important, moderately important, important, and very important, for the calculation of the guideline layer values. As can be seen from Figure 4-B, the number of national and provincial nature reserves in western Liaoning is high.

4 FUNCTIONAL ZONING OF PRIORITY AREAS

Taking into account the current state of biodiversity and resource endowment, as well as the level of socio-economic development and its development potential, the functions of the priority areas are zoned, and three categories of priority areas are defined: Category I includes nature reserves of all levels and types, world cultural and natural heritage sites, scenic spots, forest and geological wetland parks, special marine reserves, genetic and germplasm reserves, and other areas designated by the State (Lv 2022). Category II: Areas within the priority areas that have not yet been designated for state protection and are characterized by rich biodiversity, fragile ecosystems, or important ecological functions are designated as Category II areas, including key species habitats and key ecosystem distribution areas, etc. Category III: Areas within the priority areas such as built-up urban areas and development zones of all kinds, as well as areas with strong human interference in agriculture, animal husbandry, and fishery production are designated as Category III areas. The areas within the priority areas are designated as Class III areas.

Delineation of Class III areas: Based on the 2015 ecosystem distribution data, the spatial distribution data of Class III zones, such as urban built-up areas, various development zones, and areas with strong human interference in agriculture (Yang & Xu 2022), animal husbandry and fishery production, were extracted from the priority areas to determine the initial distribution range of Class III zones. The GIS software is used to aggregate the patches and deduct the small independent patches (less than 1km2) from the Class III zones to ensure the integrity and continuity of the functional zoning results. The zoning of Class III areas was divided into two cases, one of which did not classify farmland as Class III areas (Figures 5), and the other classified farmland as Class III areas (Figures 6). The initial extracted Class III layers were spatially overlaid with Class I layers, and if there was any spatial overlap between Class III and Class I zones, the overlapping areas were classified as Class I zones. After verification of all the overlapping areas, the boundary of the Class III area was defined.

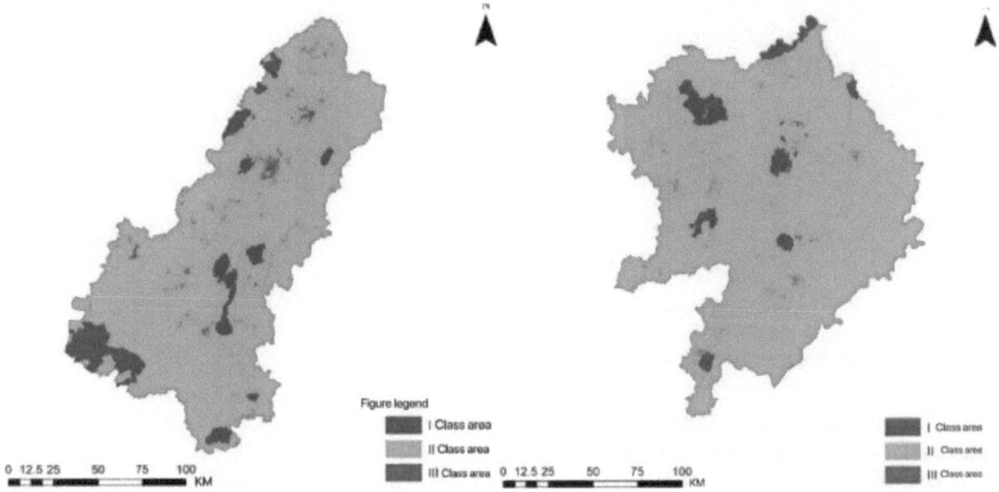

Figure 5. Functional zoning map of the priority area for biodiversity conservation in Liaoxi and Liaodong (Class III area excluding farmland).

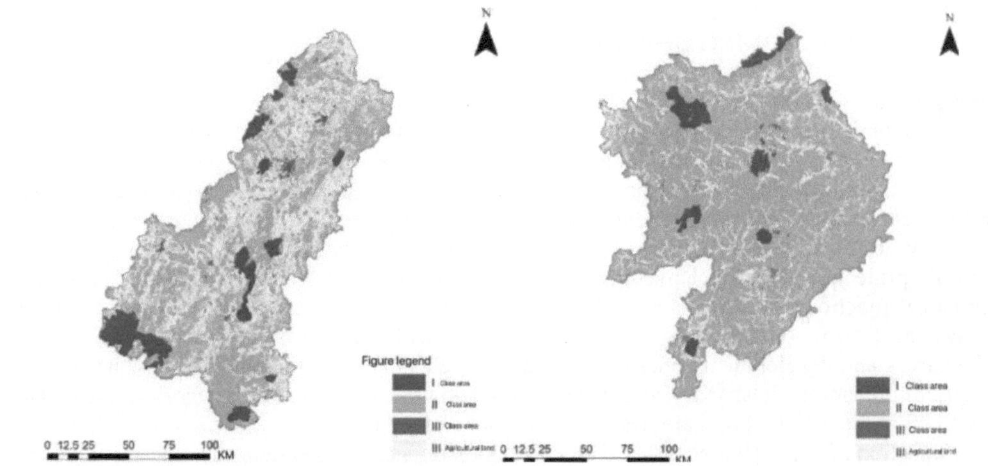

Figure 6. Functional zoning map of the priority area for biodiversity conservation in Liaoxi and Liaodong (Class III area including farmland).

Areas within the priority areas, other than those in categories I and III, are identified as the distribution of category II areas, which are areas of conservation gaps within the priority areas that are not included in the national and local protected area systems and urban development plans and can be considered as conservation gaps within the priority areas.

5 CONCLUSION

The priority area of biodiversity conservation in Liaodong involves the ecological function zone of Huanren-Kuandian Hun River water conservation and biodiversity conservation ecological function zone of ecological function zoning of Liaoning Province, the ecological function zone of biodiversity conservation and water conservation at the source of Qingyuan-Xinbin Hun River, the ecological function zone of water conservation, and hydrological storage at Guanyinge Reservoir of Prince Edward River in Benxi (County). Fushun Dahuofang Reservoir Soil Conservation and Nutrient Conservation, Hydrological Storage Ecological Function Area is the richest and most concentrated area of biodiversity in Liaoning Province, and mainly undertakes the function of water connotation and biodiversity protection. Old Baldingzi, Baishi Lazi, and other areas in the area are the most concentrated distribution areas of important and endangered species. Baishi Lazi National Nature Reserve and Laodudingzi National Nature Reserve are the most important protection units in this priority area, which have the ecological functions of biodiversity protection and water containment, especially in ecosystem protection. This research has injected new ideas into the integration of biodiversity identification with GIS and will continue to be studied in depth.

REFERENCES

Chen Jl. (2021). Analysis of the application of cartography and GIS technology in smart city mapping [J]. Small and medium-sized enterprise management and science and technology (lower journal). (10):191–193.

Chen Yb. (2021). Application of geographic information system in land resource management[J]. Rural Science and Technology. 12(26): 102–104. DOI:10.19345/j.cnki.1674-7909. 2021. 26.040.

Lv J. (2022). Synergistic international rules for addressing climate change and biodiversity conservation: evolution, challenges and China's choice [J/OL]. Journal of Beijing University of Technology (Social Science Edition). 1–13. DOI:10.15918/j.jbitss1009-3370.2022.0194.

Wang, Bz, Xiao Zy. (2021). Channel attention multi-branching networks for fine-grained image recognition[J]. Advances in Lasers and Optoelectronics. 58(22):172–180.

Yang Qw, Zhou K. (2021). Pressure plate state identification based on an improved bilinear fine-grained model[J]. Advances in Laser and Optoelectronics. 58(20):146–155.

Yang Yt, Xu H. (2022). Feasibility study on incorporating biodiversity into environmental impact assessment for planning in China[J]. Environmental Impact Assessment. 44(01):33–37+41. DOI:10.14068/j.ceia.2022.01.006.

Yu Wx, Hu Zh. (2022). The improvement strategy of biodiversity rule of law in the context of ecological civilization [J/OL]. Journal of Beijing University of Technology (Social Science Edition). 1–11. DOI:10.15918/j.jbitss1009-3370.2022.0294.

Yuan Ps, Song J, Xu Hl. (2022). Fish image recognition based on residual network and small sample learning[J/OL]. Journal of Agricultural Machinery.1–11. http://kns.cnki.net/kcms/detail/11.1964.S.20211231.1122.006.html

Zhang Bl, Wang Hy, Wang Bc. (2022). Exploring the application of geographic information system in forestry management [J]. Journal of Intelligent Agriculture. 2(01):116–118.

Zhao Jr, Li L, Su Zy. (2021). Application of geographic information system in the construction of geological disaster information system[J]. Engineering Technology Research. 6(21):251–252. DOI:10.19537/j.cnki.2096-2789.2021.21.117.

Advances in Mineral Resources, Geotechnology and Geological Exploration – Rashid & Zhang (eds)

GIS-based analysis of the spatial pattern of grassland vegetation cover

Zhenzhen Zhao
Faculty of Science, Jiamusi University, Jiamusi, China

Zicheng Zhao*
School of Mechanical and Electrical Engineering, Henan University of Technology, Zhengzhou, China

Yawen Yang
School of Foreign Languages, Jiamusi University, Jiamusi, China

ABSTRACT: The landscape index is a quantitative reflection of different landscape structural components and spatial pattern characteristics through highly concentrated landscape pattern information. The landscape pattern index can be divided into three levels: Patch metrics, Class metrics, and Landscape metrics, among which the patch level index is the basis for calculating other landscape indices, reflecting the landscape pattern by calculating the area, perimeter, number, shape and other characteristics of each patch in the landscape. There are differences in the landscape indices of different families of plants, the range of plant species involved in our study is small, and future studies can add different plant species as the dominant grassland communities. Also in future studies, the ecological implications of small-scale patterns should be explored under different vegetation cover, climate, soil, topography, and vegetation types.

1 INTRODUCTION

Different types of vegetation have different shape characteristics, and different landscape indices can be obtained by taking vegetation photos in the field combined with indoor image analysis software to analyze their correlation with the macroscopic characteristics of vegetation (Li 2022), detect different types of vegetation characteristics, and provide a reference for the optimal allocation and rational use of grassland resources in the future. In this paper, we analyze the spatial distribution characteristics of non-vegetation patches and vegetation patches of different artificially improved grasslands in Guizhou Province at the image element scale, and by analyzing the relationship between different landscape indices and vegetation characteristics (Zhang 2021), we conclude that the respective characteristics can be detected in the future by taking photos in the field combined with using image analysis software indoors.

2 OVERVIEW OF THE STUDY AREA

Guizhou Province is located on the slope of the Yunnan-Guizhou Plateau extending from the remnants of the Qinghai-Tibet Plateau in China, with high topography in the west and low topography in the east, large undulations, and large differences in elevation. The highest elevation of the seven sampling sites in the study area is 2652 m in Leekiping, Hezhang County, in the northwest, and the lowest is 963 m in Xujiaba Grassland, Sinan County, in the northeast (Table 1) (Huang 2021). In the FAO 1990 soil classification system, the soil types of Xujiaba Grassland in Sinan County, Maiping Grassland in Huaxi District, and Yaosuhe Ranch in Dafang County are mainly Haplic Alisol and Haplic Luvisols, and the soil types of Yunding Grassland and Wumeng Prairie are mainly Dystric Regosols.

*Corresponding Author: q1335300187@163.com

DOI 10.1201/9781003319412-45

All seven sampling sites are artificially improved grasslands, mainly used for grazing and tourism, each with an area of more than 600 mu, and the vegetation coverage of the grasslands is in the medium to high coverage range (Gao 2021), with the lowest coverage in Yunding Grassland (48.07%±11.27%) and the highest in Scorched Garden Grassland (96.50%±2.75%, Table 1). Most of the meadow community establishment plants were artificially improved grass species and a few were wild vegetation, mainly including Trifolium repens, Lolium perenne, Paspalum thunbergii, Dactylis glomerata, Arthraxon hispidus, Bothriochloa ischaemum, Plantago depressa, Festuca rubra, Eragrostis pilosa), Potentilla chinensis, Miscanthus sinensi), etc.

Table 1. Sampling plots.

Sampling plots	Precipitation (mm)	Temperature (°C)	Altitude (m)	Dominant species	Plant cover (%)
Yusugi River Ranch	1085.40	12.06	1736.55	Trifolium repens Plantago depressa	69.09±9.41
Xujiaba Grassland	118.50	17.49	963.40	Trifolium repens Lolium perenne	88.07+4.01
Maiping Grassland	1104.70	14.98	1260.50	Paspalum thunbergii Dactylis glomerata	83.23+7.47
Scorched Garden Grassland	859.40	10.79	2414.10	Festuca rubra Eragrostis pilosa	96.50+2.75
Leek ping	832.90	13.51	2651.70	Miscanthus sinensis Eragrostis pilosa	78.30+11.56
Genting Grassland	1104.70	14.98	1587.30	Arthraxon hispidus Bothriochloa ischaemum	48.07+11.27
Wumeng Grassland	1383.40	15.3	2505.10	Eragrostis pilosa Potentilla chinensis	76.04+16.81

3 SAMPLING AND DATA ANALYSIS TEXT AND INDENTING

3.1 *Field measurements of grassland vegetation cover*

Type primary headings in capital letters roman (Heading 1 tag) and secondary and tertiary headings in lower case italics (Headings 2 and 3 tags). Headings are set flush against the left margin (Chen 2021). The tag will give two blank lines (24 pt) above and one (12 pt) beneath the primary headings, 1½ blank lines (18 pt) above and a ½ blank line (6 pt) beneath the secondary headings, and one blank line (12 pt) above the tertiary headings. Headings are not indented and neither are the first lines of text following the heading indented. If a primary heading is directly followed by a secondary heading, only a ½ blank line should be set between the two headings.

In the Word programme this has to be done manually as follows: Place the cursor on the primary heading, select Paragraph in the Format menu (Cao 2022), and change the setting for spacing after, from 12 pt to 0 pt. In the same way, the setting in the secondary heading for spacing before should be changed from 18 pt to 6 pt.

3.2 *Grassland vegetation cover landscape pattern index calculation*

The landscape pattern index can be divided into three levels: Patch metrics, Class metrics, and Landscape metrics, of which the Patch metrics are the basis for calculating other landscape indices.

It reflects the landscape pattern by calculating the area, perimeter, number, shape, and other characteristics of each patch in the landscape (Sun 2021). Fragstats software is used to calculate landscape spatial pattern indices from patch, classification, and landscape levels. The selected landscape indices include Patch Area, AREA, Number of Patches, NP, Shape Index, SHAPE, Fractal Dimension Index, FRAC, Largest Patch Index, LPI, Contagion, CONTAG. Effective Mesh Size, MESH, Aggregation Index, AI, Landscape Division Index, DIVISION, Shannon's Diversity Index, SHDI, Simpson Diversity Index, SIDI, Shannon's Evenness Index, SHEI, and Simpson's Evenness Index, SIEI. Area (AREA) is the basic unit of the landscape and is the basis for calculating other indices (Equation1) (Meng 2021). When AREA is close to 0, it indicates that the patches of this type are closer to a single image element (Yu 2021), and when Area is close to the whole sample landscape area, it indicates that the landscape is mainly composed of one kind of patches.

$$AREA = \sum_{j=1}^{n} a_{ij} \frac{1}{10000} \tag{1}$$

In the formula, n denotes the number of pixels in the patch, i denotes the number of patch types, j denotes the number of patches, and denotes the area of the j-th patch in the i-th patch type.

The number of patches reflects the fragmentation degree of the whole landscape by counting the number of patches in a certain shape or area range.

Shape Index, SHAPE is the ratio of the transformed patch edge length to the area, and the complexity of a patch is measured by calculating the difference between its shape and a circle or a square of the same area; the smaller the SHAPE value, the simpler the patch shape, and the more complex the opposite is. Equation 2 takes the smallest value when the patch shape is a circle, and Equation 3 takes the smallest value when the patch shape is a square.

$$SHAPE = \frac{P}{2\sqrt{\pi A}} \tag{2}$$

$$SHAPE = \frac{0.25P}{\sqrt{A}} \tag{3}$$

The formula P is the perimeter of the patch, A is the area of the patch.

Fractal Dimension Index, FRAC, is the non-integer dimension of irregular geometry (Equation 3). The fractal dimension can be used to measure the complexity of individual plaque shapes. The more regular the patch, the closer the FRAC is to 1, and the more complex the patch, the closer the FRAC is to 2. The fractal dimension index reflects the influence of the external environment on the spatial pattern of the grassland to some extent, and the larger the fractal dimension value, the less disturbing it is, and vice versa.

3.3 *Field survey of grassland community character is Remote Sensing Data*

Remote sensing images of the Landsat (Landsat 8 OLI) series without clouds or with low clouds in the study area were downloaded from two websites, the Geospatial Data Cloud of the Chinese Academy of Sciences Computer Network Information Center (http://www.gscloud.cn) and the USGS Earth Resources Observation and Science Center (http://glovis.usgs.gov/) (Liu 2020). The time span is 5 years (2015-2019) with strip numbers 126, 127, 128, 129, and row numbers 41 and 42. Images were selected to synchronize with field sampling time as much as possible, and insufficient images were supplemented with remote sensing images from the period of maximum biomass (July-September) in the last five years (2015-2019). The images were corrected and pre-processed in Arc Map software, and the vegetation indices of the images were calculated and extracted by the layout of the text.

Table 2. Specifications of the Landsat 8 Images.

Sampling plots	Years	Date (Month/Day)	Path/Row	Sensor
Yusugi River Ranch	2016	7/6	128/41	OLI
Xujiaba Grassland	2019	7/21	126/41	OLI
Maiping Grassland	2019	7/29	127/42	OLI
Scorched Garden Grassland	2019	8/6	129/41	OLI
Leek ping	2015	9/1	128/42	OLI
Genting Grassland	2019	9/21	127/42	OLI
Wumeng Grassland	2015	9/27	128/42	OLI

Vegetation indices are values that are indicative of the vegetation growth status calculated by combining the spectral data from remote sensing sensors (Zhang 2022). The vegetation indices used for remote sensing monitoring of vegetation cover broadly include Normalized Difference Vegetation Index, NDVI, Enhanced Vegetation Index, EVI, Difference Vegetation Index, DVI, Ratio Vegetation Index, RVI, Modified Vegetation Index, MVI, GreenNDVI, Visible Atmospherically Resistant Index, VARI, Modified Soil Adjusted Vegetation Index, MSAVI, Atmospherically Resistant Vegetation Index, ARVI, Soil Adjusted Vegetation Index, SAVI, and Green Vegetation Index, GVI, etc.

NDVI is the most widely used vegetation index because it can eliminate part of the solar altitude angle and atmospheric noise, it is less disturbed by topography and community shadows, and has a strong correlation with vegetation cover, which can quantitatively reflect vegetation growth and distribution. The higher the value, the higher the coverage.

$$NAVI = (NIR - RED)/(NIR + RED) \tag{4}$$

RVI is the ratio of the near-infrared light band to the visible red-light band (Equation 7) commonly used to estimate vegetation biomass.

$$RIV = NIR/Red \tag{5}$$

Green NDVI is the ratio of the difference between the near-infrared light band and the blue light band to the sum of the two (Equation 8)

$$GreenNDVI = (NDVI - Green)/(NDVI + Green) \tag{6}$$

The VARI is the ratio of the difference between the blue and red bands to the sum of the blue, red and green bands (Equation 10). This index is sensitive to vegetation in all coverage areas and is able to reduce the sensitivity to the atmosphere.

$$VARI = (Green - Red)/(Green + Red + Blue) \tag{7}$$

The corresponding bands in Equation 4 to Equation 7 are shown in Table 3.

Table 3. Bands combination of Landsat 8 Images.

Band	Name	Wave length (μm)	Resolution (m)
Band 1	Coastal	0.43–0.45	30
Band 2	Green	0.45–0.51	30
Band 3	Blue	0.53–0.59	30
Band 4	Red	0.64–0.67	30
Band 5	NIR	0.85–0.88	30
Band 6	SWIR1	1.57–1.65	30

(continued)

Table 3. Continued.

Band	Name	Wave length (μm)	Resolution (m)
Band 7	SWIR2	2.11–2.29	30
Band 8	Pan	0.50–0.68	15
Band 9	Cirrus	1.36–1.38	30

3.4 *Data analysis*

Data were collated using Microsoft Excel (version 2016) for statistical purposes, IBM SPSS Statistics (version 20.0, IBM, Armonk, NY, USA) for paired t-tests for significant differences between variables ($P < 0.05$), one-way ANOVA for significant differences between variables ($P < 0.05$), and regression analysis for correlations between variables ($P < 0.05$). -way ANOVA) to detect significant differences between variables ($P < 0.05$), and regression analysis to count significant correlations between variables of interest ($P < 0.05$) using SigmaPlot (version 10.0), Adobe Photoshop CS6 Extended (version 20.0, Adobe Systems, San Jose, CA, USA) and Arc Map10.2 software were used for mapping Headings.

Type primary headings in capital letters roman (Heading 1 tag) and secondary and tertiary headings in lower case italics (Headings 2 and 3 tags). Headings are set flush against the left margin. The tag will give two blank lines (24 pt) above and one (12 pt) beneath the primary headings, 1½ blank lines (18 pt) above and a ½ blank line (6 pt) beneath the secondary headings, and one blank line (12 pt) above the tertiary headings. Headings are not indented and neither are the first lines of text following the heading indented. If a primary heading is directly followed by a secondary heading, only a ½ blank line should be set between the two headings.

4 ANALYSIS OF RESULTS

The total number of non-vegetated and vegetated patches was 17299.27±12744.69 (mean±standard error) and 7782.59±8190.41, respectively. the total number of patches was related to the vegetation cover of the sample, and the total number of patches tended to increase and then decrease with the increase of vegetation cover, and the maximum number of patches was reached at about 50% vegetation cover (52.40%) (Yuan 2022). The number of patches reached a maximum of 32671 at about 50% vegetation cover (52.40%). The non-vegetation and vegetation were divided into seven ranges according to patch size as follows.

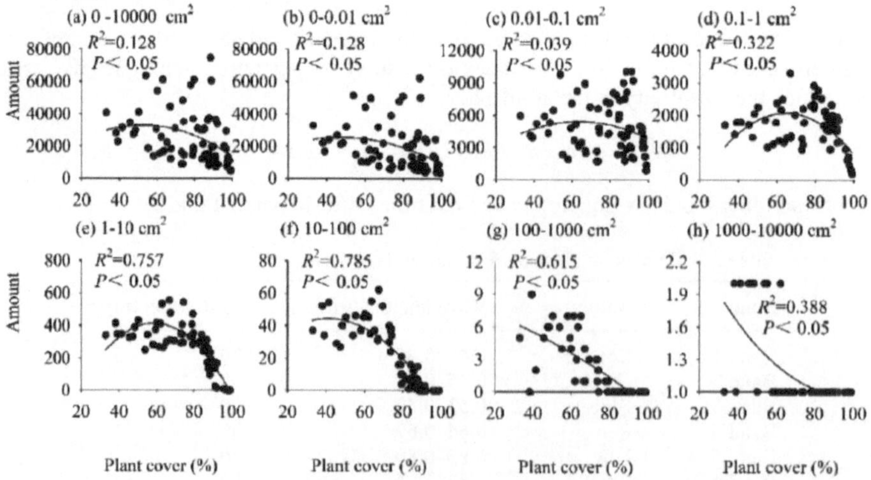

Figure 1. The relationship between patch amount and plant cover.

The number of plaques was also related to plaque size. As plaque size increased, the number of plaques decreased rapidly as an exponential function (y = 217.522e-1.147x, R2 = 0.999), while the plaque area tended to increase. The proportions of small (area < 0.1 cm^2), medium (area 0.1 cm^2-100 cm^2) and large (area >100 cm^2) plaques were 90.83% ± 4.56%, 9.16% ± 4.56% and 0.01% ± 0.01%, respectively, while the proportions of plaque area were 2.14% ± 1.04%, 15.51% ± 9.39% and 82.35%, respectively. 9.39% and 82.35%±9.64% respectively.

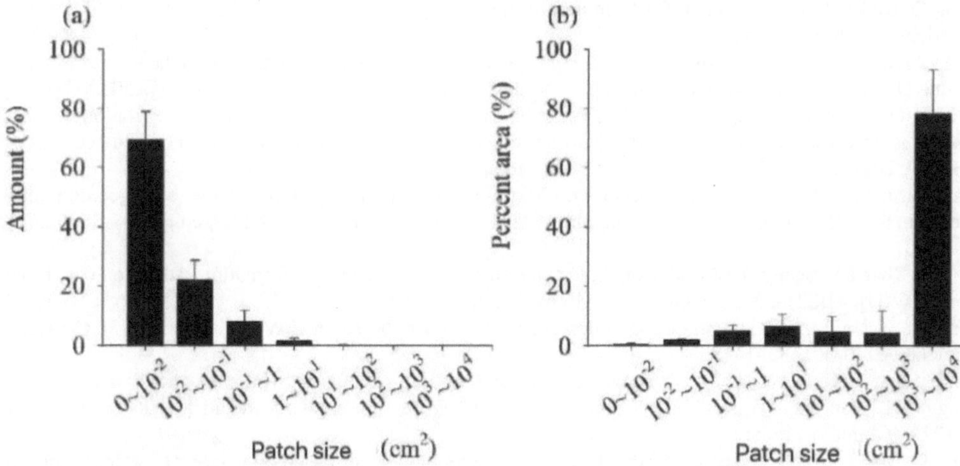

Figure 2. Patch amount and area occupied distribution for non-plant and plant patches.

The number of plaques was also related to the plaque size. As the plaque size increased, the number of plaques decreased rapidly as an exponential function (y = 217.522e-1.147x, R2 = 0.999), while the plaque area tended to increase. The proportions of small (area < 0.1 cm^2), medium (area 0.1 cm^2-100 cm^2) and large (area >100 cm^2) plaques were 90.83%±4.56%, 9.16%±4.56% and 0.01%±0.01%, respectively, while the proportions of plaque area were 2.14%±1.04%, 15.51%±9.39% and 82.35%±9.64%, respectively.

5 CONCLUSION

With the increase of vegetation cover, the number of non-vegetation patches increases and then decreases in a quadratic polynomial, but the change rates of different size patches are different, and large patches reach the peak faster than small patches. With the increase of vegetation cover, the area of non-vegetation patches decreases, and the larger patches will be decomposed into several smaller patches, and the smaller patches will be decomposed into several smaller patches, and so on repeatedly until they finally disappear. The scattered vegetation patches are connected as the vegetation coverage increases, and the large patches are obviously easier to connect than the small patches. Small-scale vegetation distribution patterns can be well associated with plant characteristics. Leguminosae, Gramineae, and other families differed significantly in macroscopic characteristics such as above-ground biomass, below-ground biomass, soil capacity, and plant height, while the shape index of Leguminosae was significantly lower than that of Gramineae and other families. However, no significant correlations existed between species number, biomass, plant height, and any of the above-mentioned indicators at the patch level, taxonomic level, and landscape level. In future studies, different vegetation characteristics can be analyzed by taking vegetation photos and image processing software classes.

REFERENCES

Cao Yx, Mao Donglei, Xue Jie, Su Sl, Cai Fy. (2022). Vegetation cover dynamics and its driving factors in the oasis-desert transition zone: the case of Cele, Xinjiang [J/OL]. Arid Zone Research. 1–13.

Chen F. (2021). Dynamic change analysis of vegetation cover in Jining City based on Landsat TM/OIL[J]. Henan Science and Technology. 40(22): 129–133.

Gao C, Wang Lw. (2021). Remote sensing monitoring and analysis of dynamic change of vegetation cover in Linyi City from 2015-2021[J]. Science and Technology Innovation. (35): 29–31.

Huang Q, Du Yy. (2021). Locust infestation monitoring in Egypt based on Sentinel-2 satellite data[J]. China Agricultural Information. 33(04): 13–20.

Li Zh. (2022). Research on the inversion model of vegetation cover based on Landsat-8OLI data–an example of Yanta District, Xi'an[J]. Agriculture and Technology. 42(04): 83–87. DOI:10.19754/j.nyyjs.20220228121.

Liu X, Yang G, Lin Q, Zhang Ly, Chen Hy, Wang N, Liu F, Liu C. (2020). Prediction of vegetation cover of Tribulus terrestris based on continuous wavelet transform[J]. China Agricultural Science and Technology Herald, 2021, 23(10): 107–116. DOI:10.13304/j.nykjdb. 0268.

Meng J, Liu TQ, Zhou H, Zhang Xq, Gao Xj. (2021). Monitoring the dynamic change of vegetation cover in Yulin City[J]. Beijing Survey and Mapping. 35(11):1400–1404. DOI:10.19580 /j.cnki.1007-3000.2021.11.008.

Sun MH, Zhang X, Shang G. (2021). Coffee. Extraction of vegetation cover information in Jinzhou City based on NDVI[J]. 41(23): 50–56. DOI:10.19754/j.nyyjs.20211215015.

Yu H, Wu Yf, Niu Lt. (2021). Estimation of desert grassland cover based on UAV visible images[J]. Grassland Science. 38(08): 1432–1438.

Yuan P, Song J, Xu Hl. (2022). Fish image recognition based on residual networks and small sample learning [J/OL]. Journal of Agricultural Machinery. 1–11. http://kns.cnki.net/kcms/detail/11.1964.S.20211231.1122.006.html

Zhang C, Ma C. (2022). Research on the inversion of vegetation cover based on time-series remote sensing data[J]. Electronic Technology. 51(02): 31–33.

Zhang W, Wan Hl, Chen C, Zhang Lj. (2021). A study on the spatial and temporal changes of vegetation cover based on normalized vegetation index - taking the Cangzhou section of the Grand Canal Cultural Belt as an example[J]. Journal of Hebei Conservancy and Electric Power Institute, 2021, 31(04): 72–78. DOI:10.16046/j.cnki.issn2096-5680. 04.012.

Advances in Mineral Resources, Geotechnology and Geological Exploration – Rashid & Zhang (eds)
© 2023 Copyright the Author(s), ISBN 978-1-032-33377-9

Early warning method of water inrush in mining area based on a curve similarity analysis model

Xingguo Qiu, Yuge Si & Zhen Liu
College of Computer Science and Engineering, Xi'an University of Science and Technology, Xi'an, Shaanxi, China

ABSTRACT: The hydrogeological conditions of coal resource exploitation are complex. Therefore, the research on effective water control technology is of great significance for the mining of coal mines. This paper proposes a method of water inrush early warning based on the curve similarity analysis model (CSAM). Firstly, the data variation curve based on the advancing distance is plotted according to the monitoring data of two adjacent mine working faces. Secondly, based on the derivative dynamic time warping (DDTW) algorithm integrating Gaussian function and Procrustes analysis combined with discrete fréchet distance, the similarity of distance and trend of the curve is measured respectively. The experiment shows that there is an obvious difference in the similarity of the water inflow curve between normal mining and water inrush in the working face. The method based on curve similarity measurement proposed in this paper provides a new idea for early warning of water inrush in the coal mining areas.

1 INTRODUCTION

Mine water disaster accident refers to the disaster accident caused by an uncontrolled influx of surface water and groundwater into the mine working face through cracks, faults, collapse areas, and other channels due to inadequate water control measures in the process of mine construction and production (Dong et al. 2020). Frequent mine water inrush accidents not only cause casualties but also seriously affect the normal production of coal mines.

At present, the domestic coal mine water inrush monitoring and early warning mainly include the traditional water inrush prediction methods, such as the water inrush coefficient method and water resistance coefficient method (Shi et al. 2014), the method of predicting coal mine water disaster by studying the mechanism of coal mine water inrush, using geophysical information, and the method of machine learning (Chen et al. 2021). However, the above methods have their shortcomings and limitations. The water inrush coefficient formula contains only one-factor affecting water inrush from the coal floor, so the prediction accuracy of multi-factor influence is insufficient. However, the exploration methods of drilling and in-kind exploration, which rely on hydrogeological personnel, not only have low timeliness but also occupy a large number of human resources. Using machine learning method to predict water inrush is usually to model the water inrush data of the mining area in previous years or the water inrush data of other mining areas, which has the problems of poor universality and low accuracy of model prediction results when the sample size of water inrush data is small. For mining areas with complex geological conditions, the development of some small fault structures and hidden collapse columns is difficult to grasp in the exploration stage, so the existing methods are not accurate in predicting local water inrush. Moreover, the traditional coal mine water inrush early warning method based on multi-source hydrological related data cannot integrate the monitoring of continuous hydrological data and water inrush trend, and cannot fully meet the safety needs of the mine.

DOI 10.1201/9781003319412-46

Based on the above problems, this paper proposes a curve similarity to the mine water inrush early warning method based on distance and trend weighted analysis. Different from the traditional water inrush prediction for the whole mine, this method carries out the water inrush early warning of distance and trend weighted analysis for the working face in a mining area, solves the problem of low reliability of single factor water inrush prediction, and avoids the insufficient accuracy caused by using machine learning method in the case of a small number of data samples.

2 RELATED THEORIES AND ALGORITHMS

2.1 *The classic dynamic time warping algorithm*

Dynamic time warping (DTW) is a technology to measure the similarity of two groups of unequal length sequences (Senin 2008). For two sequences A and B, the DTW distance is a path to minimize the cumulative distance of the two curves, as shown in Figure 1. This path is defined as a warping path (WP) and represented by W, the first element of W is defined as $W_k = (i, j)_k$, which defines the mapping of sequences A and B. There are:

$$W_k = \left(w_1, w_2, \ldots, w_k, \ldots, w_K\right), \max(m, n) \leq K < m + n - 1 \tag{1}$$

Specifically, DTW first needs to find a regular function to minimize the cumulative distance between the two curves, and the resulting minimum value is the distance between the two curves.

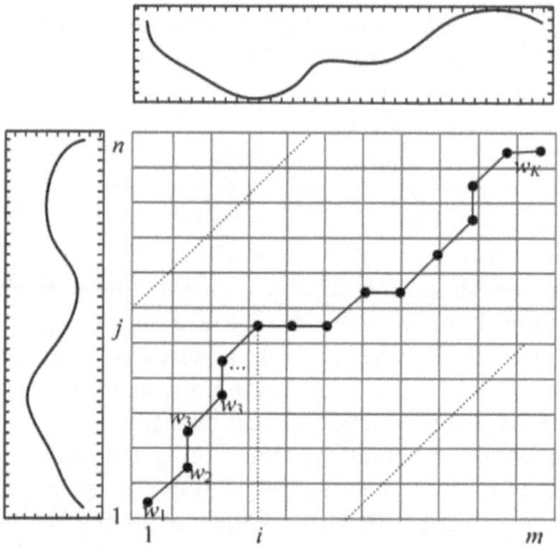

Figure 1. DTW distance calculation.

Resetting a cumulative distance $W_s = (i, j)$ to represent the cumulative distance of step k; dist (i, j) is the distance difference between the current grid point (i, j), that is, the Euclidean distance (similarity) of the points A_i and B_j. The iterative process for calculating the cumulative distance is as follows:

$$\begin{cases} W_s(i, 1) = \sum_{k=1}^{n} dist(k, 1), i \in [1, m] \\ W_s(1, j) = \sum_{k=1}^{m} dist(1, k), j \in [1, m] n \\ W_s(i, l) = dist(i, j) + \min \begin{cases} W_s(i-1, j-1) \\ W_s(i-1, j), \\ W_s(i, j-1) \end{cases} \quad i \in [2, m], j \in [2, n] \end{cases} \tag{2}$$

The final DTW distance of two sequences A and B is as follows:

$$D_{DTW} = dist\,(A, B) = DTW\,(A, B) = W_s\,(m, n) \tag{3}$$

Although DTW has been successfully applied in many areas, however, there are two problems: one is that DTW may cause a single point in one-time series to be mapped to a large sub-part of the other time series by warping the x-axis to align the curves; second, in the process of curve matching, DTW may not be able to find obvious contrast between the two groups because of the difference of one feature (such as inflexion and wave crest). The Derivative dynamic time warping(DDTW) algorithm (Keogh&Pazzani 2001) was proposed to solve the problems in DTW.

2.2 *Derivative dynamic time warping algorithm*

The derivative dynamic time warping algorithm obtains the information about the shape by considering the first derivative of the sequence.

Construct an n×m matrix, where the (i,j) element of the matrix contains the distance $d(q_i,c_j)$ between two points q_i and c_j. DDTW replaces the distance formula of the DTW algorithm by estimating the square of the derivative difference between q_i and c_j. The estimated derivative formula is as follows:

$$D_x[q] = \frac{(q_i - q_{i-1}) + ((q_i - q_{i-1})/2)}{2}, 1 < i < m \tag{4}$$

This estimated derivative formula does not include the first and last points, but uses the second point and the penultimate point instead, as follows:

$$\begin{cases} D_q[i] = \frac{(q_i - q_{i-1}) + ((q_i - q_{i-1})/2)}{2}, 1 < i < m \\ D_q[0] = D_q[1] \\ D_q[m] = D_q[m-1] \end{cases} \tag{5}$$

The derivative dynamic time warping algorithm improves the problem of singularity caused by the traditional dynamic time warping algorithm and can get a more accurate distance measure between two curves.

2.3 *Procrustes analysis*

2.3.1 *Translation*

Procrustes analysis is a rigid shape analysis that uses isomorphic scaling, translation, and rotation to find the best match between two or more iconic shapes.

By translating the object so that the mean value (i.e., its centroid) of all points of the object is located at the origin, the translation component on the object can be removed. Take k points $((x_1, y_1), (x_2, y_2), \ldots, (x_k, y_k))$ in the two-dimensional space, and the average value of these points is $(\bar{x}, \bar{y})$. When $\bar{x} = \frac{x_1 + x_2 + \ldots + x_k}{k}$ and $\bar{y} = \frac{y_1 + y_2 + \ldots + y_k}{k}$. Now translate these points to make their mean value shift to the origin: $(x, y) \rightarrow (x - \bar{x}, y - \bar{y})$ to get a new point $(x_1 - \bar{x}, y_1 - \bar{y}), \ldots,$ $(x_k - \bar{x}, y_k - \bar{y})$.

2.3.2 *Scaling*

The scaling component can be removed by scaling the object so that the root mean square distance (RMSD) from the point to the translated origin is 1. This RMSD is a statistical measure of the scale or size of an object:

$$s = \sqrt{\frac{(x_1 - \bar{x})^2 + (y_1 - \bar{y})^2 + \ldots + (x_k - \bar{x})^2 + (y_k - \bar{y})^2}{k}} \tag{6}$$

Therefore, new points can be obtained: $\left(\frac{x_1-\bar{x}}{s}, \frac{y_1-\bar{y}}{s}\right), \ldots, \left(\frac{x_k-\bar{x}}{s}, \frac{y_k-\bar{y}}{s}\right)$.

2.4 *Discrete fréchet distance*

Fréchet distance (Bringmann & Mulzer 2016) is a measure of similarity between curves. It takes into account the position and order of points on the curve.

The curve is defined as a continuous mapping: $f:[a,b]\rightarrow V$, when $a,b\in R$ and $a\leq b$, where $(V,d)$is a metric space.

Given two curves $f:[a,b]\rightarrow V$ and $g:[a',b']\rightarrow V$, their fréchet distance is defined as follows (Alt&Godau, 1995):

$$\delta_F(f,g)=\inf_{\alpha,\beta,t\in[0,1]}\max d\left(f\left(\alpha\left(t\right)\right),g\left(\beta\left(t\right)\right)\right) \tag{7}$$

Where $\alpha\,(resp.\beta)$ is an arbitrary continuous non-decreasing function from [0,1] to $[a,b]\,(resp.[a',b'])$.

The discrete fréchet distance is defined as follows (Buchin et al., 2008):

Firstly, the continuous curve is discretized. Let curve P be composed of p trajectory points and curve Q be composed of q trajectory points.

If $\sigma\ (P)$ and $\sigma\ (Q)$ are used to represent the sequential set of two-track points respectively, there are $\sigma\ (P)=(\mu_1,\mu_2,\ldots,\mu_p)$ and $\sigma\ (Q)=(\nu_1,\nu_2,\ldots,\nu_p)$. At the same time, the following sequence point pair L can be obtained:

$$(\mu_{a1},\nu_{b1}),(\mu_{a2},\nu_{b2}),\ldots,(\mu_{am},\nu_{bm}) \tag{8}$$

The length $\|L\|$ between sequence pairs between P and Q is defined as the maximum Euclidean distance of each sequence pair, i.e

$$\|L\|=\max_{i=1,2,\ldots,m} d\left(\mu_{a1},\nu_{b1}\right) \tag{9}$$

So, the discrete fréchet distance of P and Q is defined as

$$\delta_{\mathrm{dF}}(P,Q)=\min\left\{\|L\|\,|\,\mathrm{L\ is\ a\ coupling\ between\ P\ and\ Q}\right\} \tag{10}$$

3 CURVE SIMILARITY ANALYSIS MODEL (CSAM)

Due to the complexity of geological conditions and the irregularity of mining activities, it is difficult to fully explain the mechanism of the occurrence and development of all water inrush accidents. However, with the help of a large amount of perceptual information, we can find the internal law of the data and realize the prediction and early warning of mine water inrush. In general, the geological structure and hydrological conditions of the same mining area are roughly the same, so the variation laws of mine hydraulic discharge, water stage, and other indicators during mining are basically the same. In this paper, based on the mine hydrological parameter curve monitored by the sensor, the similarity value is obtained by comparing it with the hydrological parameter curve of the adjacent working face for early warning.

The calculation process of similarity is as follows:

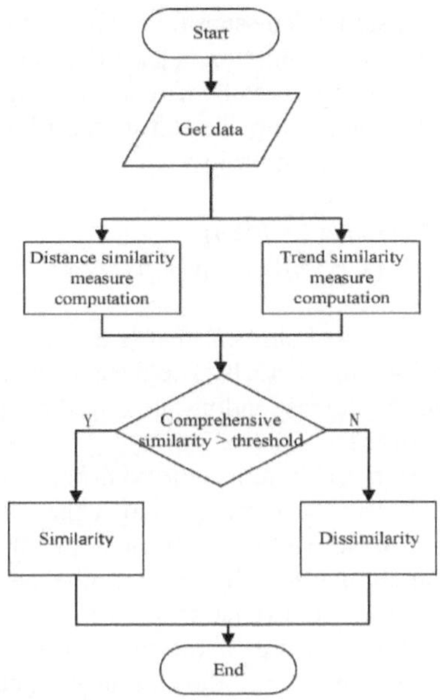

Figure 2. Flow chart of curve similarity determination.

This paper analyzes and calculates the similarity between the two dimensions of trend and distance. For trend similarity, the method of combining Procrustes analysis and discrete fréchet distance is adopted. Firstly, Procrustes analysis is used to normalize the curves, and then the discrete fréchet distance between the curves is calculated.

Discrete fréchet distance is a measure of similarity between curves, which considers the position and order of points along the curve. The fréchet distance between two curves is the shortest path length of two curves from the beginning to the end. Firstly, the Procrustes analysis is used to translate and scale the two curves that need to be compared, and the normalization operation is carried out to eliminate the influence of dimension and reduce the absolute value difference of data. Unify them into an approximate range, and then focus on their changes and trends to judge the similarity of trends better. The steps to calculate the trend similarity are as follows:

(1) Two new curves P1 and P2 are obtained by Procrustes normalization of the two curves respectively;

(2) Calculate the average length AvgCL of the two normalized curves:

$$\begin{cases} AvgCL = \sqrt{CL\,(P1) \times CL\,(P2)} \\ CL\,(P) = \sum_{i=0}^{n-2} \sqrt{(x_i - x_{i+1})^2 + (y_i - y_{i+1})^2} \end{cases} \tag{11}$$

The final similarity measurement is:

$$TS = \max\left\{1 - \frac{frechet\,(P1, P2)}{\frac{AvgCL}{\sqrt{2}}}, 0\right\} \tag{12}$$

For distance similarity, the similarity distance of curve matching can be used as the basis of curve similarity measurement. Different from trend similarity, it is an absolute distance measurement.

It needs a threshold to determine whether the two curves are similar, and it cannot quantitatively measure the similarity between curves. To convert an absolute measure of distance into a relative measure. In this paper, a one-dimensional Gaussian function is introduced to convert the distance and threshold, and the deformed Gaussian function is used to calculate the similarity.

Let the similarity distance of the two curves be d, and the judgment threshold be δ, d is the distance between two curves calculated using the ddtw algorithm. At this point, the gaussian function is used to convert the distance to similarity:

$$\begin{cases} DS = N\left(d, \delta^2, \eta\right) \\ N\left(x, \sigma^2, \eta\right) = \exp\left(-\eta x^2/2\sigma^2\right) \end{cases} \tag{13}$$

In the formula, the determination threshold is δ, that is the standard deviation of the Gaussian function $N\left(x, \sigma^2, \eta\right)$. The similarity range is [0,1]. The distance similarity is related to the similar distance between two curves and the corresponding judging threshold. Where δ is constant, the smaller d is, the higher the similarity is.

$N\left(x, \sigma^2, \eta\right)$ is a deformed gaussian function. η is the control parameter. According to the nature of Gaussian function, when d and δ are the same, the larger the value of η, the smaller the value of $N\left(x, \sigma^2, \eta\right)$. As long as δ and η work together, they can adjust the variance of the center of the Gaussian function. Choosing different η is actually to determine the average similarity of the same similar distance at different scales. Here, the value of η is 1.4.

In this way, the absolute distance measure of the two curves is transformed into a relative similarity measure through the transformation of the Gaussian function. To sum up, the similarity determination process is as follows:

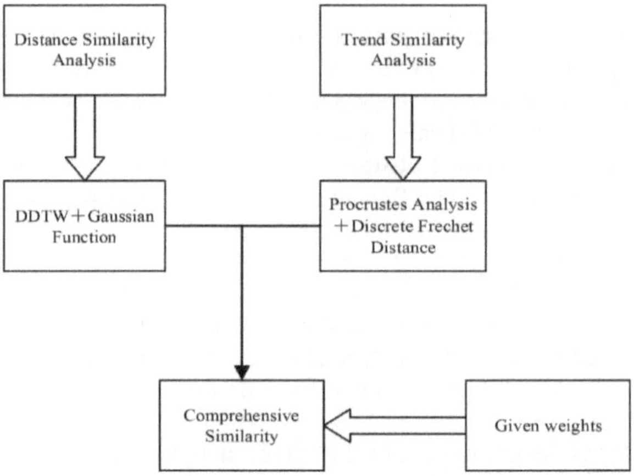

Figure 3. The comprehensive similarity determination process.

When the overall similarity judgment threshold is given, we can judge whether the curve is similar. For example, when the curve similarity is greater than 0.60, the two curves are considered to be similar. If the calculated similarity is greater than 0.60, the two curves can be considered to be similar. When the threshold of global similarity judgment is given, the two curves calculated by this method can be considered similar as long as they are more than the threshold of similarity.

4 EXPERIMENTAL ANALYSIS

D coal mine is located in Shandong Province. The current production level is –660m. It is divided into seven mining areas, namely mining areas 1, 3, 4, 5, 6, 7, and 14. In the process of mine construction and production, the mine discharge is more than 300m^3/h for four times, and the maximum discharge is 570m^3/h, which has a great impact on mining.

From 1989 to 2009, the average annual discharge of coal mines was 127.85 ~ 270.62m^3/h, with an average of 185.01m^3/h. From 2010 to 2015, the average mine water discharge was 185.00m^3/h. From 2016 to 2018, the average mine discharge was 181.59m^3/h, and the maximum monthly discharge is 226.55m^3/h. The mine discharge from 1989 to 2018 is shown in Figure 4.

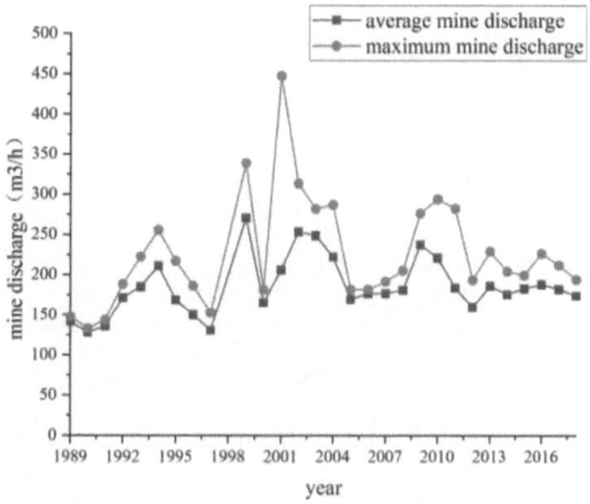

Figure 4. Maximum and average discharge of D coal mine over the years.

In this paper, the experimental data is the measured water discharge data monitored in the mining process of working face in the D coal mine, and the similarity measurement is carried out with the advancing distance of 50m as the query window.

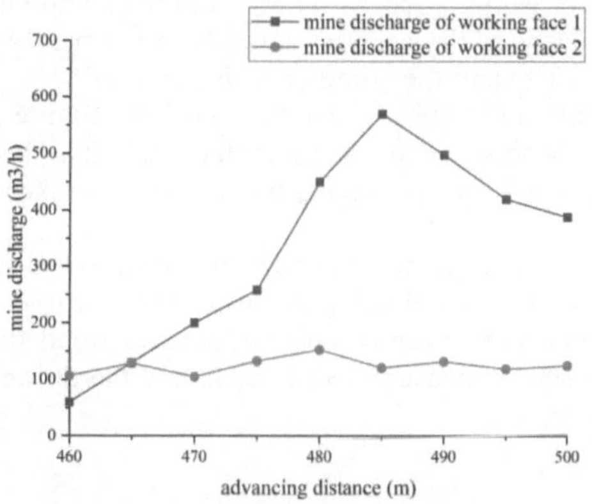

Figure 5. Contrast curve of water discharge between 4308 and its adjacent working face.

On July 4, 1999, during the mining process of 4308 working faces in the D coal mine, there was water pouring on the roof, and the water inflow was 6.0m^3/h. The first water inrush peak is on July 8, and the water inflow is 450.0m^3/h; The water inflow of the second water inrush peak is 570.0m^3/h. On July 29, the mining task of the working face was completed, and the mine water discharge decreased to 23.0m^3/h. In Figure 5, the calculation shows that the trend similarity of the water discharge curve of the two working faces is 72.6% and the distance measurement is 715.8 when the water inrush occurs.

On July 30, 2001, a water inrush occurred under the influence of mining in 14301 working face. The initial water inflow was 25m^3/h, and the water discharge increased with the advance of the working face. By September 25, the maximum water discharge was 540m^3/h. After that, the water inflow began to decline and remained at 110m^3/h until the mining of the working face was completed on October 18. When water inrush occurs, the trend similarity of the mine water inflow curve of the two working faces is 65.8%, and the distance measurement is 593.7.

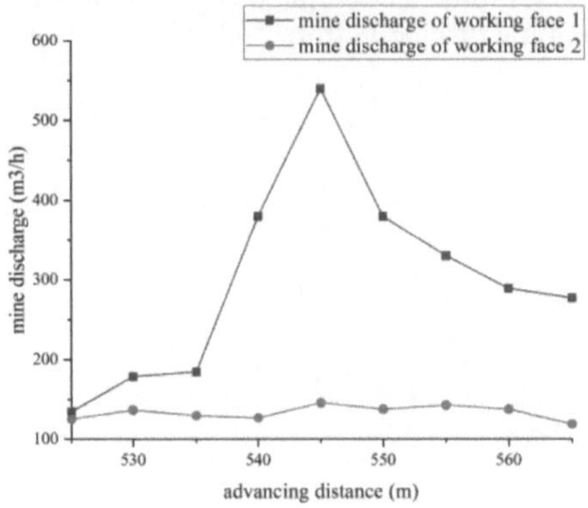

Figure 6. Contrast curve of water output between 14301 and its adjacent working face.

By analyzing the similarity between the above working face and its adjacent working face when water inrush, the distance measurement threshold is set to 655, and the weights of distance similarity and trend similarity are 0.5 respectively. That is, the calculation formula of comprehensive similarity (ComS) is:

$$ComS = 0.5TS + 0.5DS \tag{14}$$

According to the formula, the distance similarity (DS) of the comparison curve between the 4308 working face and its adjacent working face is 43.3%, and the comprehensive similarity (ComS) is 58%. The distance similarity of the contrast curve between the 14301 working face and its adjacent working face is 56.3%, and the comprehensive similarity is 61.5%. The average value of their comprehensive similarity is 60%. It can be seen from Figure 2 that the comprehensive similarity is negatively correlated with the water inrush probability. Therefore, in order to give an early warning of water inrush in the actual mining process, 65% is taken as the threshold for water inrush judgment.

The 6306 working face of the D coal mine has been mined since February 2020, and its adjacent working face is 6305 working face. Comparing the water inflow curves of the two working faces under the same advancing distance, it can be obtained that the trend similarity of the two curves in Figure 7 is 81.0%, the distance measurement is 56.2, and the distance similarity is 99%. The comprehensive similarity is 90%.

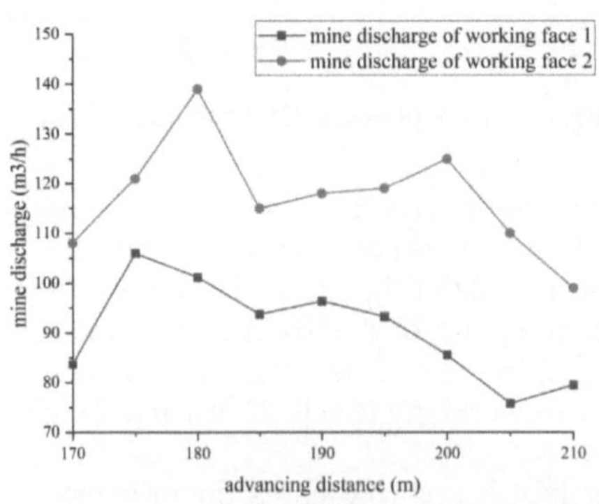

Figure 7. Comparison curve 1 of water discharge of 6306 and 6305 working face.

In Figure 8, the trend similarity of the two curves is 78.1%, the distance measurement is 38.5, and the distance similarity is 99%. So, the comprehensive similarity is 88.5%.

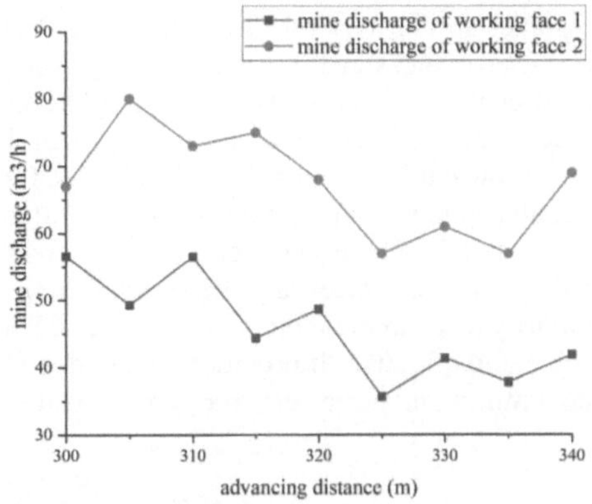

Figure 8. Comparison curve 2 of water discharge of 6306 and 6305 working face.

The comprehensive similarity of the above two sets of curves is less than the threshold value, which is the normal case of non-water inrush. This method identified the water inrush accident correctly in the mine in October 1999.

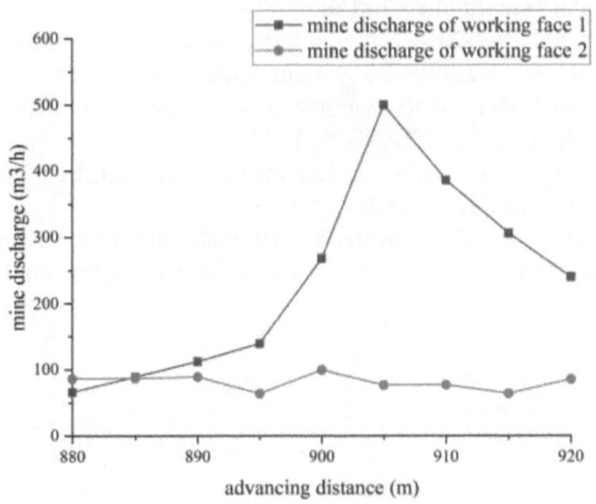

Figure 9. Contrast curve of water discharge between 14309 and adjacent working face.

According to the actual information of the mine, a water inrush accident occurred in the 14309 working face in 1999. The mine discharge of the working face reaches $500m^3/h$ and decreases to $7m^3/h$ after stabilization. The result shows that the method that this paper proposed can successfully judge the occurrence of water inrush. In Figure 9, the trend similarity of the two curves is 59.2%. The distance measurement and distance similarities are 678 and 47.2% respectively. The comprehensive similarity is 53.2%, which is less than the given threshold of 65%, so there is a very high risk of water inrush. It can be seen that the similarity obtained by this method is used to analyze the water inrush of the mine, which is consistent with the actual situation. Based on the above experimental analysis, a set of mine water inrush early warning systems based on the curve similarity analysis model can be established in the actual mining process.

5 CONCLUSION

A curve similarity analysis model (CSAM) based on both distance and trend is proposed in this paper. In this paper, the ddtw algorithm combined with the Gaussian function is used to convert the distance measurement into distance similarity. At the same time, the Procrustes analysis combined with discrete fréchet distance is used to measure the trend similarity. And this model is applied to the prediction of water inrush in the coal mining areas. By comparing the similarity of mine discharge curves monitored by two adjacent mining faces, water inrush may occur if the similarity is less than the threshold. The mine water inflow data curves of 4308, 14301, 14309, 6306, and 6305 working faces of the D coal mine are analyzed in the experiment. The similarities of mine discharge curves are 57.8%, 60.9%, 53.2%, 90 % and 88.5% respectively. This method can accurately distinguish the water inrush in the first three working faces and the normal mining in the last two working faces by setting the curve similarity measurement threshold value as 65%. The experimental results show that the method is feasible and effective. It provides a new idea for the prediction and early warning of water inrush in coal mines and has a certain reference value.

REFERENCES

Alt H and Godau M 1995 Computing the Fréchet distance between two polygonal curves International Journal of Computational Geometry & Applications 5 75–91.

Bringmann K and Mulzer W 2016 Approximability of the discrete Fréchet distance Journal of Computational Geometry 7 46–76-46–76

Buchin K, Buchin M and Wenk C 2008 Computing the Fréchet distance between simple polygons Computational Geometry 41 2–20.

Chen J, Wang C and Wang X 2021 Coal mine floor water inrush prediction based on CNN neural network The Chinese Journal of Geological Hazard and Control 32 50–7.

Dong S, Zheng L, Tang S and Shi P 2020 A Scientometric Analysis of Trends in Coal Mine Water Inrush Prevention and Control for the Period 2000–2019 Mine Water and the Environment 39 3–12.

Keogh E J and Pazzani M J 2001 Derivative dynamic time warping. In: Proceedings of the 2001 SIAM international conference on data mining: SIAM) pp 1–11.

Senin P 2008 Dynamic time warping algorithm review Information and the Computer Science Department University of Hawaii at Manoa Honolulu, USA 855 40

Shi L, Qiu M, Wei W, Xu D and Han J 2014 Water inrush evaluation of coal seam floor by integrating the water inrush coefficient and the information of water abundance International Journal of Mining Science and Technology.

Advances in Mineral Resources, Geotechnology and Geological Exploration – Rashid & Zhang (eds)
© 2023 Copyright the Author(s), ISBN 978-1-032-33377-9

Experimental study on single pile model under combined load in loess area

Shixu Zhang
School of Civil Engineering, Xi'an University of Architecture and Technology, Xi'an, Shaanxi, China

Weiping Cao
Key Lab. of Geotechnical & Underground Space Engineering, Xi'an, Shaanxi, China

ABSTRACT: In order to reveal the influence law of single pile behavior under the combined action of downforce load and horizontal load in loess, the straight pile model test is carried out indoors, the pile body strain of straight pile is obtained, the soil resistance and corresponding horizontal displacement are calculated, and the theoretical p-y curve of the single pile under combined load is established and verified. The results show that the maximum bending moment appears at the 3D depth below the mud surface. When the horizontal load is small, the soil around the pile is in the elastic stage. When the horizontal load increases, the soil around the pile begins to enter the plastic stage; the bending moment, soil reaction, and horizontal displacement all appear in the upper half of the pile body and tend to be zero at the bottom of the pile; the hyperbolic function distribution is more consistent with the measured p-y curve of the model test.

1 INTRODUCTION

With the implementation of the western development strategy, various long-span bridges and high-rise buildings have sprung up, and the engineering problems in the loess area are becoming increasingly prominent. In these deep foundation pit projects, pile foundation is widely used as a form of deep foundation. In addition to bearing a vertical load, a pile foundation also bears a large horizontal load, such as a wind load or water load (Gerolymos et al. 2008; Jones et al. 2022). In the design of pile foundation, not only the bearing capacity of pile foundation should be determined, but also the distribution law of internal force of pile body should be studied. The p-y curve method considers the nonlinear effect of soil and can better describe the pile-soil interaction, so it has been widely used.

At present, domestic and foreign scholars have studied the p-y curve of the single pile through model tests and field tests. Mcclelland and Focht (1958) first established a method to calculate the modulus of pile-soil by the indoor model test of pipe pile in 1958 and deduced the relationship between the stress-strain curve and secant modulus in the p-y curve. Based on the elastic foundation beam, Zhang et al. (2021) analyzed the pile element stress of horizontal bearing pile under vertical load and established the theoretical calculation model of pile side bearing capacity under vertical load. Chortis et al. (2020) carried out centrifugal model tests on the single pile model under different scouring conditions, obtained the p-y curve according to the measured data of the model pile, and corrected the p-y curve. Lu and Zhang (2020) used the plastic constitutive model to simulate the pile bearing the combined load and established the p-y curve of the pile under the combined action of vertical and horizontal load in the sand. Wang et al. (2021) carried out two groups of field tests of test piles in clay and established a modified p-y curve in clay based on the influence of pile diameter and formation depth on piles. Cao et al. (2019) established a hyperbolic p-y curve of the horizontally loaded inclined pile in sandy soil. It can be seen from the above that at present, most studies on the single pile are single load, and there are few studies on single pile under the combined action of compressive load and horizontal load, and the foundation soil where the test

DOI 10.1201/9781003319412-47

pile is located is mostly sandy soil and cohesive soil, and there is little research on the p-y curve of loess foundation.

In this paper, the model test of a single pile under combined load in loess foundation is carried out, the variation laws of single pile bending moment, pile side soil resistance, and pile displacement along the pile body are analyzed, and the measured p-y curve in loess foundation is established, which has a certain reference value for the actual project of the single pile under combined load in loess foundation.

2 MATERIALS AND METHODS

The test device used in this paper consists of three parts: model slot, loading system, and measurement system (see Figure 1). The size of the model groove is 1.47 m × 0.8 m × 2.9 m (length × wide × high), the outside of the model groove is composed of tempered glass and angle steel. The loading system is mainly composed of loading weight, fixed pulley, hydraulic jack, steel slide rail, dowel bar, and reaction beam. The strain, load, and displacement of the pile body are measured by an electronic stress-strain gauge, tension-compression sensor, and dial gauge respectively.

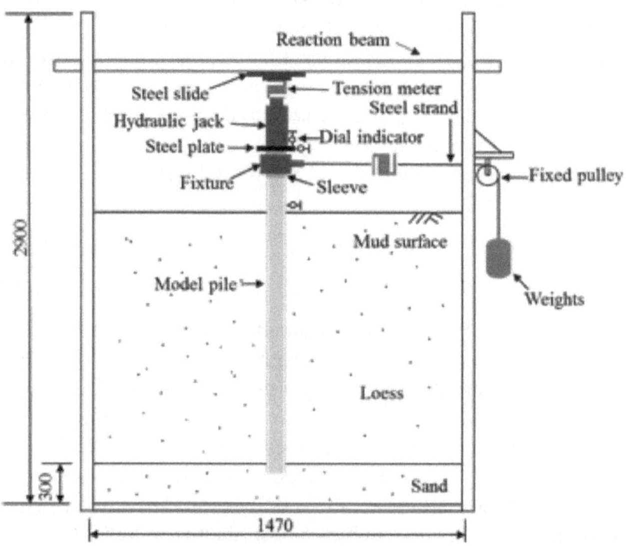

Figure 1. Layout diagram of the model test.

The prototype test pile corresponding to the model test is a concrete cylindrical pile with a diameter of 800mm and a pile length of 22m. The model test pile has an outer diameter of 40mm, a wall thickness of 2mm, and a pile length of 1.1m. The geometric similarity ratio was determined to be 1:20. A total of 14 corresponding variants are arranged on the inner wall from top to bottom (see Figure 2).

The Loess taken from the model test comes from Lantian county, Shaanxi province. The overall color is yellowish-brown and mixed with fine sand and gravel. The density of soil mass is 1.68 g/m^3, the moisture content is 0, and the specific gravity is 2.69.

Fix the model pile to the predetermined position, first fill the model groove with a 300mm thick sand layer as the bearing layer of the test model pile, and then fill the loess layer by layer until the design mud surface position. The filling thickness of each layer is about 200mm. After each layer is filled, level the surface, and vibrate the soil with a trowel machine. Through the pre-test, the ultimate bearing capacity of the vertical pile foundation is 500N, and the slow maintenance load method is used to apply the load step by step. It is needed to apply the download first, and then apply the horizontal load until the download reaches the ultimate bearing capacity until the pile is damaged.

In the model test carried out in this paper, the slow maintenance load method is used to apply the load step by step. First, we need to apply the download to the model pile, and then apply the horizontal load until the download reaches the ultimate bearing capacity until the pile body is damaged. Through the pre-test, the ultimate bearing capacity of the vertical pile foundation is 500N. Therefore, first, it is attempted to load the download to 500N step by step, and then apply the horizontal load step by step.

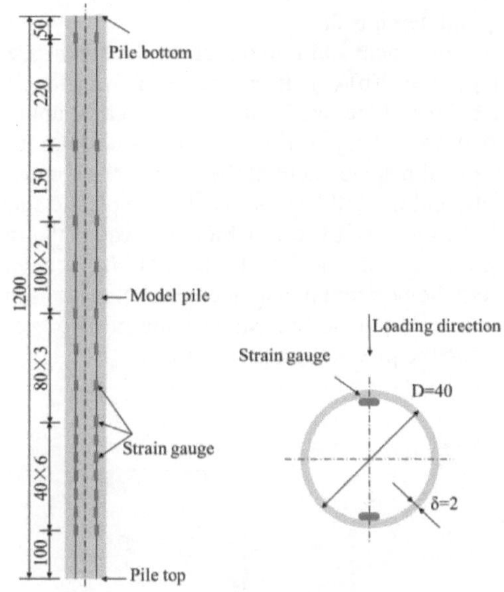

Figure 2. Longitudinal and transverse sections of the model pile.

3 RESULTS

3.1 *Experimental data processing*

In this paper, the model test can directly obtain a set of strains at the same position of the pile section, a total of 14 sets of strain values. Combined with the basic theory of elastic beam, the relationship between the bending moment and strain of the pile is obtained as formula (1).

$$M(z) = \frac{(\varepsilon_t - \varepsilon_c) \cdot EI}{D} \tag{1}$$

where $M(z)$ = bending moment at the depth of z; ε_t and ε_c = tensile and compressive strains on both sides of the pile; D = pile diameter; EI = bending stiffness of the pile.

At present, many domestic and foreign scholars (Moayedi et al. 2020; Lee et al. 2019; Liu et al. 2021) usually fit the measured bending moment of the pile with five to seven polynomial fitting, with the purpose to obtain the bending moment at each depth of the pile, so as to derive the continuous distribution of soil resistance and pile deformation function of the model pile. According to the principle of the least square method, the bending moment scatter points at different sections of the measured pile are fitted. In this paper, the sixth-order polynomial is selected for fitting.

In this paper, the lateral soil resistance of pile is deduced by twice differential of pile bending moment, and the horizontal displacement of pile is obtained by twice integral.

$$p(z) = \frac{d^2M(z)}{dz^2} \tag{2}$$

$$y(z) = \frac{1}{EI} \iint M(z)dz^2 \tag{3}$$

where $p(z)$ = soil reaction force; and $y(z)$ = horizontal displacement.

Two integral constants are determined by boundary conditions. The first is that the pile body is long and the deformation of the pile bottom is small. It is assumed that the soil resistance and horizontal displacement of the pile bottom are 0; The second is that the horizontal displacement of the pile body at the neutral point depth is 0.

The bending moment, soil resistance, and horizontal displacement are calculated according to Formulas (1) (2) (3) (see Figure 3(a)(b)(c)). It can be seen from the figure that the maximum bending moment of the pile body and the maximum soil resistance appear at the 3D depth below the mud surface. When the load is small, the soil around the pile is in the elastic stage. When the load gradually increases, the horizontal displacement of the mud surface increases rapidly, the bending moment also increases rapidly, and the soil begins to enter the plastic stage. The soil resistance of the pile body increases with the increase of load, which is because the increase of soil resistance at the pile side can resist the increase of load. The horizontal displacement of the pile body at the mud surface is the largest, and the horizontal displacement decreases gradually with the increase of depth. The pile bending moment, pile-soil resistance, and pile horizontal displacement appear in the upper half of the pile, and the pile bottom tends to zero.

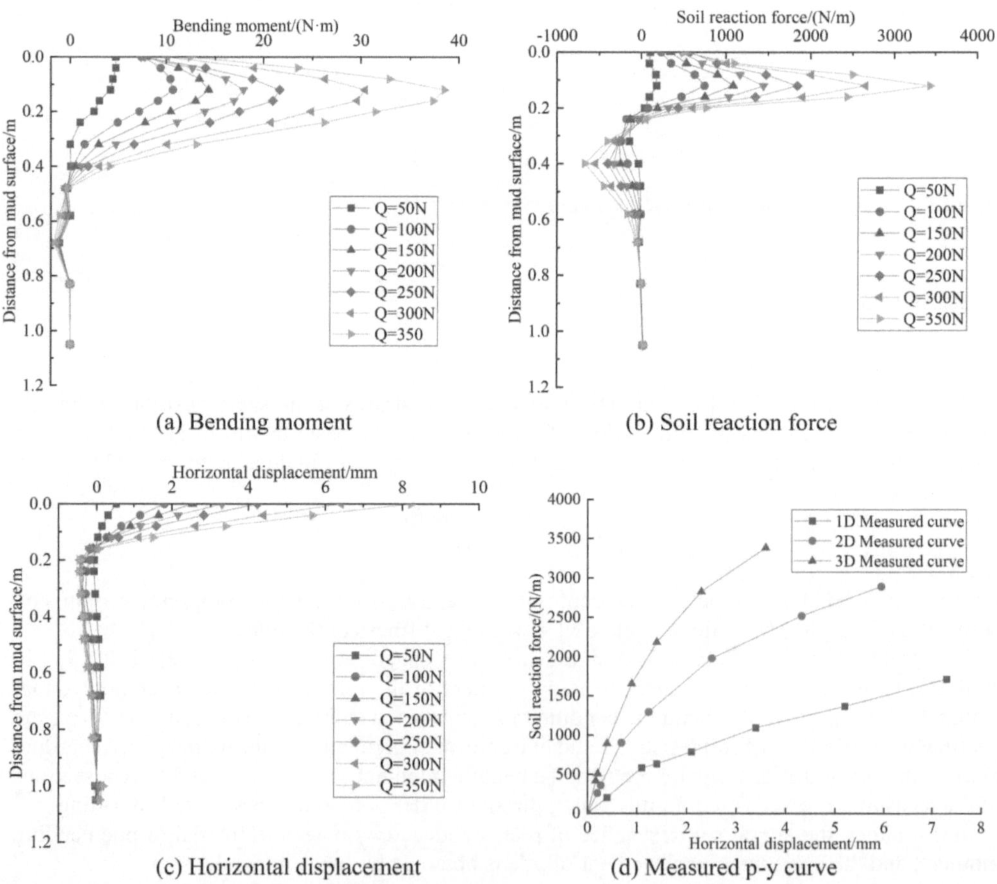

(a) Bending moment (b) Soil reaction force

(c) Horizontal displacement (d) Measured p-y curve

Figure 3. Analysis diagram of measured data.

3.2 *Establishment of the measured p-y curve*

Taking horizontal displacement as abscissa and pile-soil resistance as ordinate, the p-y curves of straight piles at depths of 4mm, 8mm, 12mm, and 16, namely 1D, 2D, and 3D, were drawn (see Figure 3(d)). As can be seen from the figure, the measured p-y curve of pile shaft soil reaction at each depth increases with the increase of horizontal displacement, and the p-y curve becomes more and more gentle with the increase of horizontal displacement. Finally, the soil resistance tends to an extreme value, and the corresponding pile shaft soil resistance is the ultimate soil resistance. It can also be seen from the figure that the slope of the p-y curve gradually increases with the increase of pile depth, that is, the p-y curve gradually becomes steeper.

3.3 *Establishment of the theoretical p-y curve*

By observing the measured p-y curve, it can be seen that the curve has hyperbolic function characteristics. Therefore, the measured P-Y curve of the model test straight pile is fitted by the least square method (see Figure 4). By fitting the measured p-y curve, the mathematical expressions of the depth 1D, 2D, and 3D between the straight pile and the mud surface are respectively obtained, as shown in Table 1.

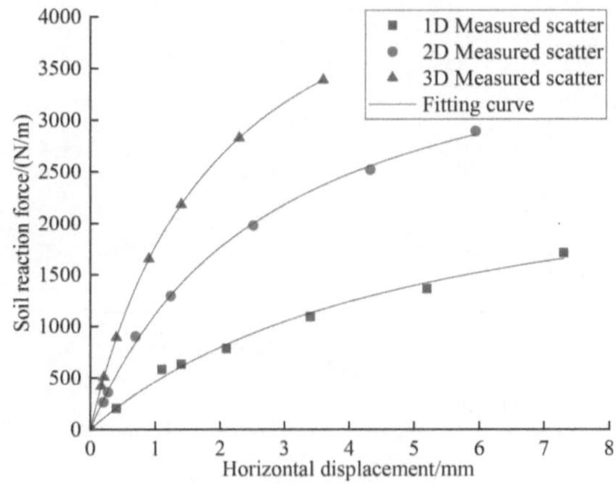

Figure 4. Measured fitting p-y curve.

Table 1. Mathematical expressions of p-y curves at different depths.

Depth	Mathematical expression
1D	p=2788×y/(5.0+y)
2D	p=4174×y/(2.7+y)
3D	p=5206×y/(1.9+y)

It can be seen from Table 1 that the mathematical expression has a general form.

$$p = \frac{ay}{b + y} \tag{4}$$

where a and b = determined parameters.

It can be seen from the mathematical expression that the fitting curve has the distribution characteristics of the hyperbolic function. According to the values of P_u and Y_{50} at the depths of

1D, 2D, and 3D from the mud surface, combined with the theoretical formula of the hyperbolic p-y curve, the values of undetermined parameters a and B at each depth can be obtained, and then the average values of a and B at three depths can be taken to obtain the expression of the hyperbolic p-y curve.

$$p=\begin{cases} \frac{0.09P_u y}{0.18Y_{50}+y}, y/Y_{50} \leq 8 \\ P_u, y/Y_{50} > 8 \end{cases} \tag{5}$$

where P_u = the ultimate soil reaction; and Y_{50} = the horizontal displacement of the pile when the soil reaction is half of the ultimate soil reaction.

The calculated p-y curve is compared with the measured p-y curve (see Figure 5). Within a certain error range, the theoretical p-y curve is very close to the measured p-y curve, so it is reasonable to use the hyperbolic p-y curve to fit the measured data.

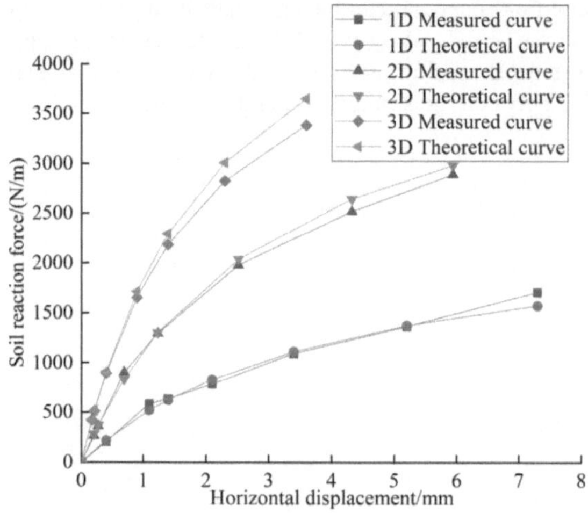

Figure 5. P-y curve comparison diagram.

4 CONCLUSIONS

The single pile model test under combined load is carried out in loess foundation, the distribution law of bending moment, soil resistance, and horizontal displacement along depth is analyzed, and the p-y curve of the single pile under combined load is established. The following conclusions are obtained:

(1) The maximum bending moment of the pile appears at about 3D depth under the mud surface. When the horizontal load is small, the soil around the pile is in the elastic stage. When the horizontal load increases, the bending moment increases rapidly, and the soil around the pile begins to enter the plastic stage.
(2) The pile bending moment, soil reaction and horizontal displacement all appear in the upper half of the pile and tend to be zero at the bottom of the pile.
(3) In this paper, the hyperbolic function is used to fit the measured p-y curve, and it is verified that the theoretical p-y curve obtained according to the model test is reasonable.

REFERENCES

Cao, W. P., Xia, B., Ge, X. 2019. Formation and application of hyperbolic p-y curves for horizontally loaded single batter piles. *Journal of Zhejiang University* 53(10):1946–1954.

Chortisa, G., Askarinejad, A., Prendergast, L.J. et al. 2020. Influence of scour depth and type on p–y curves for monopiles in the sand under monotonic lateral loading in a geotechnical centrifuge. *Ocean Engineering*197(2):106838.

Gerolymos, N., Giannakou, A., Anastasopoulos, L., et al. 2008. Evidence of Beneficial Role of Inclined Piles: Observations and Summary of Numerical Analyses. *Bulletin of Earthquake Engineering* 6(4):705–722.

Jones, K., Sun, M., Lin, C. 2022. Numerical analysis of group effects of a large pile group under lateral loading. Computers and Geotechnics144,104660.

Lee, M., Bae, K., Lee, I., et al. 2019. Cyclic p-y Curves of Monopiles in Dense Dry Sand Using Centrifuge Model Tests. Applied Sciences 1641(9):1–17.

Liu, X. Y., C, G. J., Liu, L. L. 2021. Improved p-y curve models for large diameter and super-long cast-in-place piles using piezocone penetration test data. *Computers and Geotechnics*130(2):103911.

Lu, W., Zhang, G. 2020. New p-y curve model considering vertical loading for piles of an offshore wind turbine in the sand. Ocean Engineering203:107228.

Mcclelland, B. & Focht, J. A. 1958. Soil modulus for laterally loaded piles. *Transactions of the American Society of Civil Engineers*123(1),1049–1063.

Moayedi, H., Nazir, R., Mesut Gör, et al. 2020. A new real-time monitoring technique in the calculation of the p-y curve of single thin steel piles considering the influence of driven energy and using strain gauge sensors. Measurement 153(1):107365.

Wang, W., Yang, J. Y., Liu, J. P. 2021. Study on p-y models of large-diameter pile foundation based on in-situ tests of offshore wind power. *Chinese Journal of Geotechnical Engineering* 43(6): 1131–1138.

Zhang, X. L., Xue, J. Y., Xu, C. S., et al. 2021. An analysis method for lateral capacity of pile foundation under existing vertical loads. *Soil Dynamics and Earthquake Engineering* 142(2):106547.

Advances in Mineral Resources, Geotechnology and Geological Exploration – Rashid & Zhang (eds)
© 2023 Copyright the Author(s), ISBN 978-1-032-33377-9

Numerical simulation of karst collapse based on hydrodynamic – Mechanical coupling method

Dongqin Chen*
Changjiang Polytechnic, Hubei, China

ABSTRACT: Karst collapse is one of the geological disasters of the more serious dangers, prediction of Karst collapse is a widely recognized conundrum. This study is based on Terzaghi consolidation theory and GMS software to establish a hydrodynamic – mechanics coupling model. The hydrogeological conditions in Fenghua village, Qingling township, Wuhan, were used as the background for establishing the hydrodynamic – mechanical coupling model. Simulation analyzed the cave collapse at the soil-rock junction in Karst collapse area in Wuhan city Qingling town Fenghuo village. The variation value of earth pressure, displacement value of soil, and soil damage are closely related to the groundwater flow field. During periods of high water, the pressure value of the soil is relatively small, but the displacement of the soil is relatively large. When the groundwater level drops, the soil produces the largest damage area.

1 INTRODUCTION

Analyzing the research results of predecessors, the mechanism of Karst collapse is mainly composed of erosion, burst, vacuum cavitation, and disintegration, but most of the collapses are caused by the fluctuation of the groundwater table. Therefore, the groundwater flow model of the Karst collapse area is established in this research, and, according to the consolidation theory of Terzaghi, the calculation method of hydrodynamic-mechanical coupling is realized (Sun et al. 2017). GMS software and self-programming are used to simulate and analyze the collapse mechanism caused by the change in the groundwater table, extraction of karst water, and rainfall infiltration. The prediction of Karst ground collapse of typical Karst cavity area in Fenghuo village, Qingling township, Wuhan was carried out. The rationality of the simulated data is verified by the monitored data, and the above simulation methods will be promoted and applied in similar projects.

2 GEOLOGICAL CONDITIONS OF COVERED KARST IN THE RESEARCH AREA

The research area belongs to the Fenghuo Village, Qingling Township, Hongshan District, Wuhan, The Quaternary Holocene Stratigraphy overground Karst collapse area in Fenghuo Village is formed by river alluvium, which gradually thickened from top to bottom and has a binary structure. The vertical strata are divided into four layers, the total thickness of the caprock is 22.8–30m. The thickness of the first layer of artificial filling soil is 1–2m, the second layer of clay is 1.5–3m; the third layer of sandy soil is 22.8–25m; the fourth layer is local Neogene(N) strata, which includes green and gray-yellow clay rock, gravel-bearing clay rock, semi-consolidated, pilotaxitic and amicable flake structure, the general thickness is 0.0–1.0m. Below the fourth layer is the carbonate rock strata of Karst development and the thickness is over 100m.

By analyzing the vertical distribution elevation of the Karst caves, the borehole rate is 6.4% in the depth range of 30–37m, and the borehole rate in the most developed Karst section in the

*Corresponding Author: 3chendq@163.com

 DOI 10.1201/9781003319412-48

depth range of 37–42m is 29.56%. The borehole rates in the less developed Karst section in the depth range of 42–47m and 42–52m are 18.72% and 10.34% respectively. The borehole rate is 8.37% in the depth range of 52–57m. In the depth range of 57–67m, Karst and Karst's caves are relatively few.

3 MATHEMATICAL MODEL

The numerical simulation method of Karst collapse in Qingling Township, Wuhan is the hydrodynamic-mechanical coupling model based on Terzaghi consolidation theory and GMS software. With the calculation of the groundwater flow field by GMS software, the hydrodynamic-mechanics coupling simulation is formed.

3.1 *Mathematical model of groundwater flow*

According to Darcy's law and water balance principle, the three-dimensional flow mathematical model which neglects the influence of density change on groundwater is:

$$\begin{cases} \frac{\partial}{\partial x}\left(K_{xx}\frac{\partial h}{\partial x}\right)+\frac{\partial}{\partial y}\left(K_{yy}\frac{\partial h}{\partial y}\right)+\frac{\partial}{\partial z}\left(K_{zz}\frac{\partial h}{\partial z}\right)+w=\mu_s\frac{\partial h}{\partial t} \\ h(x,y,z,t)|_{t=0}=h_0(x,y,z) \\ h(x,y,z,t)|_{(x,y,z)\in B_1}=h_t(x,y,z,t) \\ -K_n\frac{\partial h}{\partial n}|_{(x,y,z)\in B_2}=q(x,y,z,t) \end{cases} \tag{1}$$

Where: $h(x,y,z,t)$ stands for the groundwater table [L] at a certain time point t; K_{xx}, K_{yy} *and* K_{zz} respectively represent the permeability coefficient [L/T] along the main directions x,y和z; w stands for the amount of water injected or withdrawn within per unit time and per unit volume [1/T]; μ_s stands for unit water storage coefficient [1/L]; t stands for time [T]; $h_0(x,y,z)$ stands for the initial water level of the flow field [L]; B_1 stands for the first kind of boundary conditions; B_2 stands for the second kind of boundary conditions; $q(x,y,z,t)$ stands for the known flow rate (L / T) per unit area in the second kind of boundary.

3.2 *Mechanics model*

It is assumed that the vertical total stress of strata caused by the change of earth pressure remains unchanged, the lateral deformation of the soil is neglected and only vertical deformation occurs. The effective stress changes and vertical strain of the soil are:

The effective stress changes are:

$$\Delta\sigma_x=\Delta\sigma_y=-\frac{v}{1-v}\alpha_P\Delta P \tag{2}$$

$$\Delta\sigma_z=-\alpha_p\Delta P \tag{3}$$

The mean effective stress change is [15]:

$$\Delta\sigma_M=\frac{1}{3}\frac{(1+v)}{(1-v)}\alpha_P\Delta P \tag{4}$$

The vertical strain is:

$$\varepsilon_z=\alpha_P\Delta P\frac{(1+v)}{3K(1-v)} \tag{5}$$

Where: ΔP stands for the pressure increment, KPa; $\Delta\sigma_x$, $\Delta\sigma_y$ and $\Delta\sigma_z$ respectively stand for effective stress changes along three directions of the x, y, and z, KPa; $\Delta\sigma_M$ stands for the mean effective stress, $\sigma_M=(\sigma_x+\sigma_y+\sigma_z)/3$; v stands for Poisson's ratio; K stands for the volume modulus, GPa; α_P stands for Biot coefficients; ε_z For vertical strain, mm.

3.3 *The method of mechanical analysis*

As the water table rises and falls, the soil has undergone multiple compressions and rebounds. The following factors should be considered for the soil pressure increment ΔP in the equation: As the groundwater table decreases, the soil is compacted and weighted. the osmotic pressure of groundwater flow. When the groundwater level rises or falls, vacuum erosion or blasting. When the groundwater level rises due to rainfall, the hydrostatic load should be considered. The hydrodynamic pressure generated by rainwater infiltration on unit soil particles.

3.4 *Coupling method*

The GMS software is used to simulate the flow field value of the groundwater in different periods and calculate the stress in different periods. The difference in stress in different periods is the increment of the pressure, ΔP; the effective stress difference between periods are $\Delta\sigma_x$ and $\Delta\sigma_z$; and the stress and strain of the soil at different periods are calculated. The effective stress expression for saturated soils is:

$$\begin{cases} \sigma_x = \Delta\sigma_x + u \\ \sigma_z = \Delta\sigma_z + u \end{cases} \tag{6}$$

Where, σ_x and σ_z stand for the effective stress in the x and z directions, KPa; u stands for pore water pressure, KPa.

3.5 *Criteria for soil damage (Comanici, Barsanescu, 2018)*

Mohr-Coulomb failure criterion is used to determine whether stress failure occurs:

$$\begin{array}{ll} \sigma_1 > \sigma_3 tg^2\left(45^0 + \frac{\varphi}{2}\right) + 2ctg\left(45^0 + \frac{\varphi}{2}\right) & \text{destroyed} \\ \sigma_1 < \sigma_3 tg^2\left(45^0 + \frac{\varphi}{2}\right) + 2ctg\left(45^0 + \frac{\varphi}{2}\right) & \text{safe} \end{array} \tag{7}$$

Where: c stands for soil cohesion, kpa; φ stands for the internal soil friction angle and degree. σ_1, σ_3 respectively stand for the largest and smallest principal stress, kpa.

The maximum stress and minimum principal stress for each element volume are:

$$\sigma_1 = \frac{\sigma_z + \sigma_x}{2} + \sqrt{\left(\frac{\sigma_z - \sigma_x}{2}\right)^2 + \tau^2} \qquad \sigma_3 = \frac{\sigma_z + \sigma_x}{2} - \sqrt{\left(\frac{\sigma_z - \sigma_x}{2}\right)^2 + \tau^2} \tag{8}$$

τ stands for soil shear strength, kpa.

4 NUMERICAL MODEL CONSTRUCTION

4.1 *Construction of a conceptual model of geological structure in the research area (Panno Samuel et al. 2018)*

The numerical simulation of a collapse site in the collapse area in Fenghuo village, Qingling Township, Wuhan is carried out. With a total thickness of 50m, the model is divided into three layers. The thickness of the first clay layer is 5m; the second layer of sandy soil is 25m; the third layer of bedrock clay and limestone is 0-20m (Figure 1). In the process of numerical simulation, according to the section of Fenghuo village, a soil hole was set up, the diameter of this soil hole is 11.5m, the arch height is 7m, and the distance between the roof and ground is 13m. The length of this model is 100m and the thickness is 50m. The coordinate system shall be established by taking the direction perpendicular to the Yangtze River as the X-axis, and the depth of the rock and soil strata as the Z-axis.

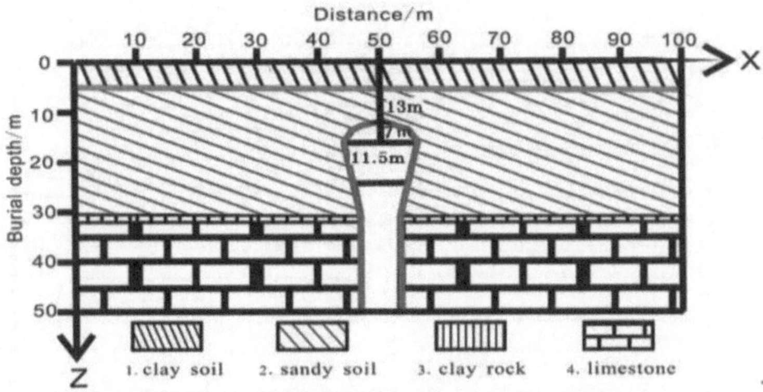

Figure 1. The conceptual model of the geological structure.

4.2 *Model boundary conditions and grid generation*

The east and west of the lateral boundary of the simulated area are defined as the boundary of fixed water level. The bottom boundary is selected at the depth of 20m of the limestone, it as the boundary of water-resistant strata. The free water surface of the phreatic aquifer of the upper boundary works as the upper boundary of the system.

According to the geological generalization model of Fenghuo village, the calculation model meshes, the upper part of the grid plane is divided into 50×50, the single grid area is $1\times2m^2$. The vertical height is determined by the top and bottom elevation of the aquifer, the total simulation depth is 50m. the caprock groundwater flow model is generalized as a heterogeneous isotropic and unsteady flow model. The bedrock strata is defined as a heterogeneous anisotropic, unsteady flow model.

4.3 *Model parameter values*

4.3.1 *Selection of fundamental values of parameters*

(1) Permeability coefficient, elastic water storage coefficient, water yield.

Permeability coefficient, elastic water storage coefficient, and water yield are key to establishing a water flow model correctly. The current model is based on the equivalent thickness method, by combining drilling, geological section, observation data, and the research results of many predecessors in the research area, the initial parameter values can be determined and further adjustment in the model identification stage can be made.

(2) Initial Selection of mechanical parameters

Mechanical parameters are obtained by testing the sample rock and soil through drilling, which contains volume modulus, natural unit weight, saturated unit weight, Poisson's ratio, cohesion, friction angle, and so on.

4.3.2 *Model identification*

The total flow field in January 2016 in this district was selected as the initial flow field of the model identification; the flow field, soil pressure, and deformation monitoring data in July 2020 were used as the basis for the fitting test.

(1) flow field fitting

According to the initial flow field in January 2016 (Figure 2), calculate the initial strain and failure distribution area of the soil.

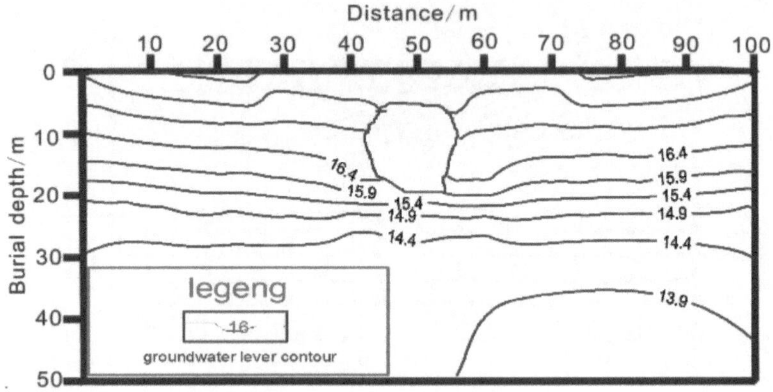

Figure 2. groundwater flow field contours in January 2016.

When the groundwater table drops, the negative pressure is not considered in the karst cave, simulated calculation of soil strain and soil damage area distribution is shown in Figures (3–4) in January 2016.

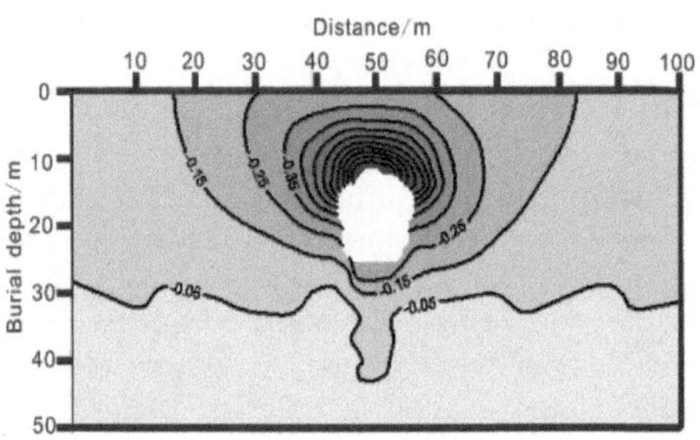

Figure 3. Vertical strain in January 2016.

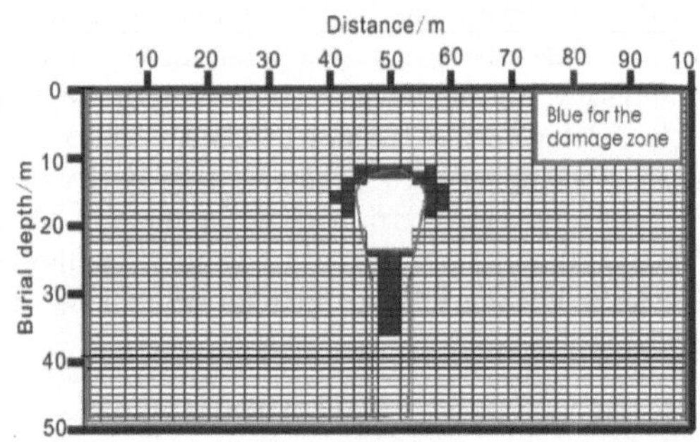

Figure 4. Destroyed zone in January 2016.

When the groundwater table drops, consider the negative pressure in the karst cave, The simulation calculation of soil strain and soil damage area distribution is shown in the Figures (5–6) in January 2016.

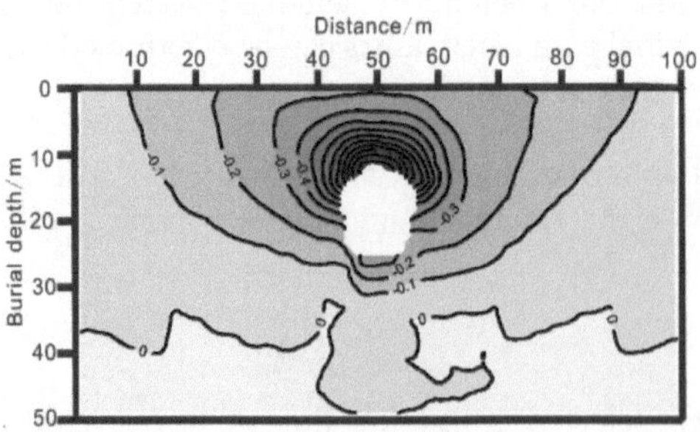

Figure 5. Vertical strain in January 2016.

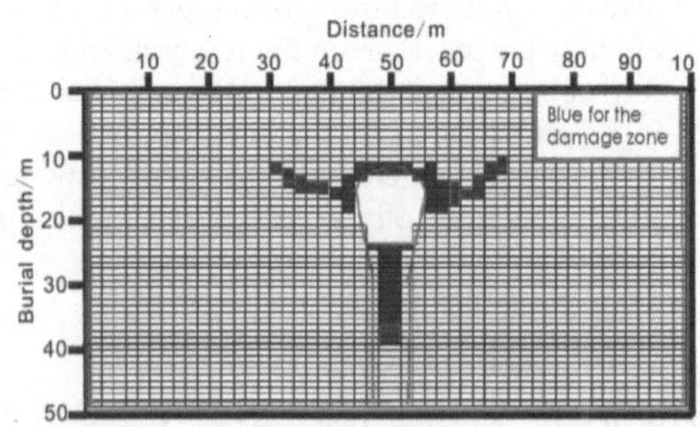

Figure 6. Destroyed zone in January 2016.

January is the dry season in Wuhan, July is the wet season, the water level rise in the Yangtze River, and rainfall is concentrated in July. The measured values and calculated values of the flow field in July 2020 are compared (Figure 7).

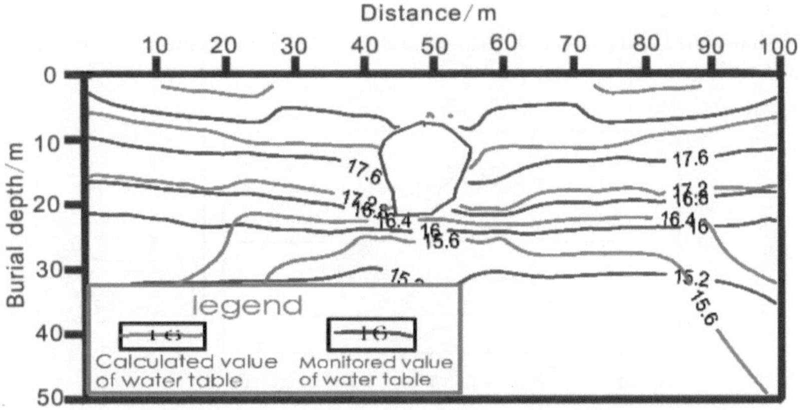

Figure 7. Comparison of measured and calculated values for groundwater flow field contours in July 2020.

The soil strain is calculated and the damage distribution zone of the soil is determined according to the groundwater flow field value in July 2020.

No consideration is given to rises in the Karst cave. The vertical strain and destroyed zone are shown in the following Figures (8–9) in July 2020.

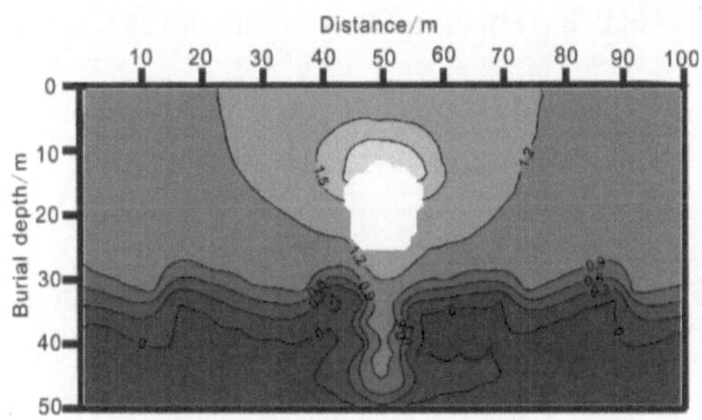

Figure 8. Vertical strain in July 2020.

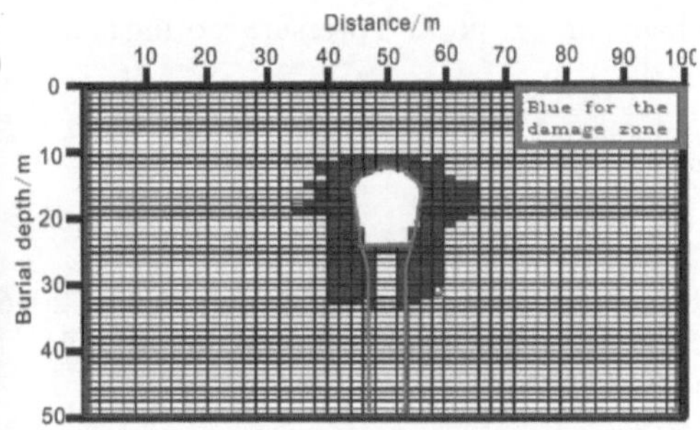

Figure 9. Destroyed zone in July 2020.

Consideration is given to the water, gas positive pressure top support, and impact blasting effect when the groundwater table rises in the Karst cave. An atmospheric pressure value (101.325kpa) is added when the simulation calculation. The vertical strain and destroyed zone are shown in the following Figures (10–11) in July 2020.

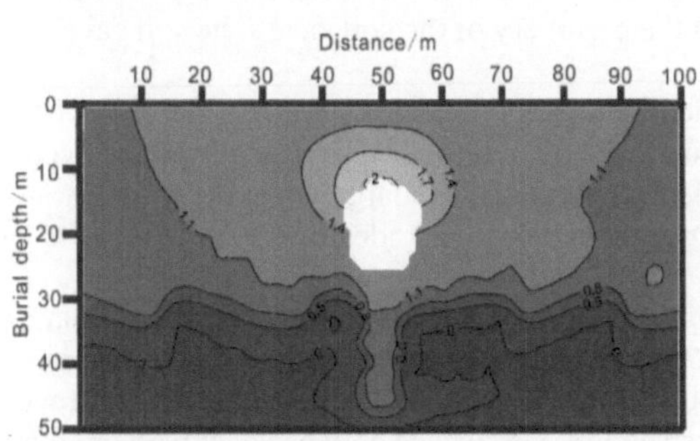

Figure 10. Vertical strain in July 2020.

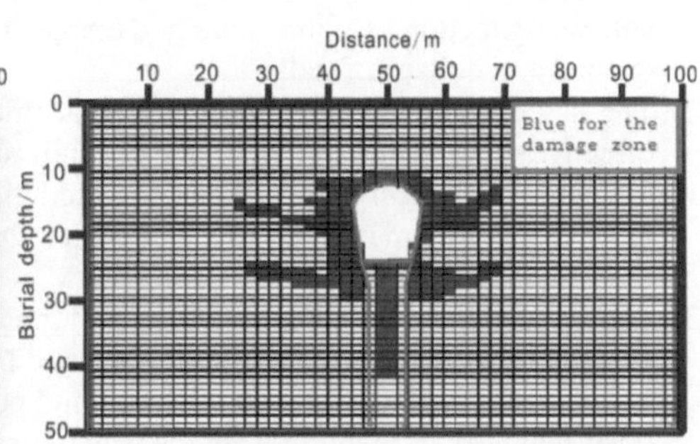

Figure 11. Destroyed zone in July 2020.

(2) Stress and deformation fitting

Figure 12 shows the fitting effect of earth pressure and soil deformation of monitoring points in 2020 (Xiao et al. 2018).

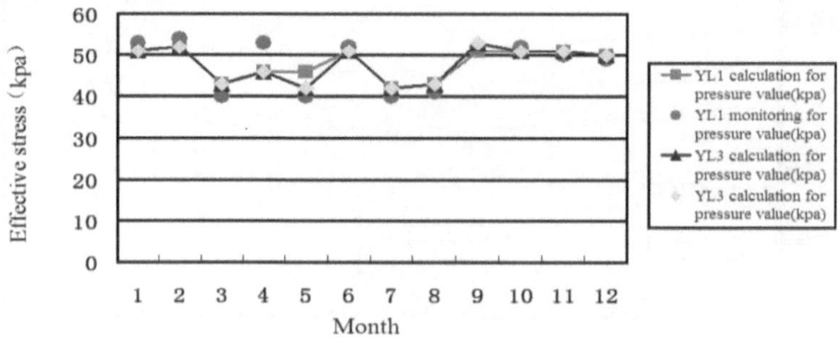

Figure 12. Earth pressure fitting of monitoring points.

Figure 12 selects the simulation data considering vacuum suction erosion and top support impact blasting effect to fit the soil pressure and soil deformation values. with the monitored values. that can analyze from the results of the fitting. When the groundwater table rises, there are gas positive pressure top support and impact blasting effect in the karst cave; When the groundwater table drops, there is a vacuum suction erosion effect in the karst cave. The important collapse factors for karst collapse are gas positive pressure roofing and impact blasting effects as well as vacuum absorption effects.

5 CONCLUSION

The simulated soil deformation area in July 2016 had a close relationship with the change in the groundwater flow field. It is feasible to simulate the soil deformation and Karst collapse by using the flow field value.

(1) The time law of soil pressure change is that the effective stress value of soil is relatively low in July and August, which is the result of the rise of the groundwater table and the occurrence of floating force. During the year of simulation, the soil pressure compaction rebounded with the change in the groundwater table. After several rounds, the plastic deformation occurred in the soil, and finally, the Karst collapsed. The law of the spatial change of soil pressure is that on both sides of the soil cave and karst pipeline at the soil-rock junction occurred stress concentration. When the groundwater table changes, the stress state of the soil around the soil cave changes too. The most significant change is the two sides of the karst pipeline at the soil-rock junction, which results in damage at the periphery of the soil cave. The soil cave will expand and collapse finally.

(2) The law of soil displacement, the displacement of the soil changes with the increase of the effective stress of the soil. When the groundwater level rises, the maximum displacement occurs at the top of the soil cavern. The generated displacement value is large when the groundwater table drops. The maximum displacement value appears on both sides of the karst pipeline at the soil-rock junction.

(3) Laws of soil mass failure and activity and transport of groundwater will facilitate the formation of soil caves by eroding the strata. The sand layer on the top of the soil cave will form funnel-like damage, with the change of stress, and both sides of the soil cave are gradually destroyed (Wei et al. 2017). Viewing from the way of destruction, the surface collapse occurs in the form of discs, which is consistent with the surface collapse in Fenghuo Village.

ACKNOWLEDGMENTS

This work was financially supported by the 2021 Construction science and Technology plan project and the building energy conservation demonstration project of Hubei Province. [2021]. 2075.

REFERENCES

Comanici, A. M.; Barsanescu. P. D. Modification of Mohr's criterion to consider the effect of the intermediate principal stress. Int J Plasticity. (2018), 108, 40–54.

Panino Samuel, V.; Luman Donald, E. Characterization of cover-collapse sinkhole morphology on a groundwater basin-wide scale using lidar elevation data: A new conceptual model for sinkhole evolution. Geomorphology. (2018), 318, 1–17.

Sun, Z. X.; Zhang, X.; Xu Y.; Cai, M.Y. Numerical simulation of the heat extraction in EGS with thermal-hydraulic-mechanical coupling method based on discrete fractures model. Energy. (2017), 120, 20–33.

Wei, Y. Y.; Sun, S. L. Study on formation and expansion condition of hidden soil cavity under the condition of groundwater exploitation in karst areas. Environ Earth Sci. 2017, 76:282.

Xiao, X. X., Xu. M., Ding, Q. Z., Kang, X. B., Xia, Q., Du, F. Experimental study investigating deformation behavior in land overlying a karst cave caused by groundwater level changes. Environ Earth Sci. 2018, 77,64.

Advances in Mineral Resources, Geotechnology and Geological Exploration – Rashid & Zhang (eds)

Study on the optimal design of deep roadway support based on three-dimensional in-situ stress measurement

Kun Zhang, Sen Zhang*, Jianxi Ren, Man Wang & Gui Yi
School of Architecture and Civil Engineering, Xi'an University of Science and Technology, Xi'an, China

ABSTRACT: The geological conditions of deep-buried roadway are complex and the horizontal stress is large, resulting in severe deformation, difficult support, and frequent repair of the working face along the trench. Therefore, it is of great significance to carry out in-situ stress measurements in deep mining areas and analyze the distribution characteristics of the in-situ stress field for roadway optimization design and maintenance. Taking the 2407 working face of Yuhua Coal Mine as an example, the core stress relief method is used to measure the three-dimensional in-situ stress of the south wing roadway, obtain the parameters such as the magnitude and direction of the in-situ stress in the horizontal section, and provide in-situ stress parameters for the support design. The support optimization scheme is put forward, and the field industrial test is carried out to verify the rationality of the support scheme. The results show that the main stress direction of the Yuhua Coal Mine is east-west, and the tectonic movement has a great influence on the in-situ stress, which belongs to the control area of tectonic movement; the included angle between the maximum principal stress direction and the axial direction of 2407 working face is large, which has a significant impact on roadway excavation; the field industrial test shows that the application effect of the support optimization scheme is good, the deformation of roadway surrounding rock is effectively controlled, and the safe production of the working face is ensured.

1 INTRODUCTION

In-situ stress is the initial stress existing in the earth's crust and the main cause of serious deformation and damage to roadways. Accurate measurement of in-situ stress provides necessary basic data for excavation design of underground engineering, stability analysis of surrounding rock, and scientific selection of support methods (Brady & Brown 2004; Cai et al. 2000; 2010). Especially with the continuous expansion of the mining scale and the continuous development to the deep, the influence of in-situ stress in construction is becoming more and more obvious. The design and construction without considering the influence of in-situ stress will lead to the occurrence of mine dynamic disasters such as stope collapse and damage, underground roadway and rock burst, and even serious production and personal safety accidents (Qi et al. 2016). Therefore, scholars at home and abroad have carried out research on the law of in-situ stress distribution. Literature (Cai et al. 2013; Kang et al. 2016; Liu & Liu 2012; Li 2017; Wang & Chang 2018; Wang et al. 2014; Xiao et al. 2013; Zhang et al. 2008) analyzed the distribution law of coal mine in-situ stress according to the in-situ stress measurement results, applied the studied in-situ stress distribution law to the stability evaluation of coal and rock mass, roadway and stope support design, rockburst disaster prevention and control, and achieved good support effect. Tao Wenbin et al. (Tao et al. 2020) took the transportation roadway of 1621 working face of Pansan Coal Mine in Huainan as an example, measured the in-situ stress by using the stress relief method, installed force measuring bolts at key positions of the roadway, monitored and recorded the axial force of bolts in real-time

*Corresponding Author: zhanghbf97@163.com

 DOI 10.1201/9781003319412-49

and compared the support schemes before and after optimization in combination with numerical simulation. Wang Jiong et al. (Wang et al. 2019) conducted in-situ measurement of in-situ stress in two main roadways in the mining area by using the hollow core inclusion stress relief method to study the asymmetric large deformation failure mechanism of main roadways and carry out targeted the treatment. Liu Chang, et al. (Liu et al. 2018) measured the in-situ stress by using the stress relief method, and applied three-dimensional numerical simulation to design different schemes to study the stress and displacement characteristics of roadway surrounding rock and the distribution of plastic zone, respectively calculated and analyzed the stability of roadway with different strike and different section, and determined the optimal design scheme to maintain the minimum deformation of roadway through analysis and comparison.

Due to the shallow research on the deep in-situ stress in the Jiaoping mining area of Tongchuan, with the trend of turning to deep mining in the Jiaoping mining area, the transportation chute and return air chute of 2407 working face have asymmetric deformation. In view of the serious deformation and difficult support of the roadway, the roadway repair is frequent and the safety accidents of anchor rod loss of anchor are significantly increased. In order to analyze the failure mechanism of the surrounding rock along the channel of 2407 working face, it is urgent to measure the in-situ stress in its area, so as to obtain the parameters such as the magnitude and direction of in-situ stress, so as to provide a basis for the optimal design of support and on-site industrial test.

2 GEOLOGICAL STRUCTURE AND ENGINEERING BACKGROUND OF MINE FIELD

Yuhua Coal Mine is located in the Jiaoping mining area in the east section of Huanglong coalfield. The stratum in the area is flat, the dip angle of the rock stratum is $3° \sim 5°$, and the structure is simple. It is basically Southwest dip syncline. The Cretaceous coal measure strata are less exposed, and most of the area is covered by tertiary and quaternary laterite. It is basically controlled by the Longwang Yuhua syncline structure with axial NE~SW. At least 6 groups of sub-NW wide and gentle small anticlines and small synclines are developed in the north wing of the syncline, all of which are arranged in the echelon; In the south wing of the syncline, there is a secondary NE trending wide gentle small syncline structure.

The designed minable strike length of the working face is 1850 m, the inclined length of the working face is 240 m, the thickness of the coal seam is 3~9 m, the average thickness is 5.9 m, the buried depth of the working face is 533.9~628.1 m, and the average buried depth is 581 m. During the excavation period, 0.5-1 m bottom coal excavation is reserved, with an average minable thickness of 5 m. The fully mechanized top coal caving mining process is selected, with a mining height of 3.0 m and a top coal caving thickness of 2 m. See Table 1 for the operation of the top and bottom plates.

Table 1. Histogram of along channel drilling for transportation of 2407 working face.

Name	Rock name	Thickness (m)	Lithology and physical and mechanical properties
Main roof	Fine and medium sandstone	12.9	It is mainly composed of quartz and feldspar, containing pyrite nodules and coal debris, argillaceous calcareous cementation, dense and hard, with inclined and wavy bedding. The hardness is grade 5~7, belonging to a medium stable roof.
Immediate roof	Fine sandstone	5.0	Silty fine sandstone contains a small number of pyrite nodules, plant fossils, and coal debris. The hardness is grade 4~5, belonging to a medium solid to the extremely unstable roof. The bottom of the immediate roof is partially a false roof of 0.6m thick carbonaceous mudstone.

(continued)

Table 1. Continued.

Name	Rock name	Thickness (m)	Lithology and physical and mechanical properties
Coal seam	4^{-2} Coal	5.9	It is black, massive, and terminal, the coal seam has strong compression and weathering resistance, and the luster of glass or asphalt. Shell shaped or stepped fracture, with joint fissures, banded or linear structure, and layered structure.
Bottom	Carbonaceous mudstone	1.1	Carbonaceous mudstone is easy to expand in case of water, and the lower part of carbonaceous mudstone is root-soil rock, which is agglomerated and contains oolitic particles and plant fossils. The coal seam floor belongs to an unstable to the extremely unstable floor.

Original support parameters of 2407 working face: (1) roof and side bolts are adopted Φ20 × 2200 mm left-handed deformed steel bolt with equal strength without longitudinal reinforcement, and the row spacing is 800×800 mm, 7 roof bolts in each row and 8 side bolts in each row. Each bolt adopts one K2335 and two Z2335 resin anchoring agents. The anchor tray is a Q 235 steel plate with the specification of 150×150×8 mm; (2) The material of the anchor cable is Φ15.24 × 7100 mm steel strand with row spacing of 1800×2400 mm, "2-3-2" arrangement, with 2 and 3 staggered arrangement of anchor cables. The anchor cable supporting plate is 16# channel steel with a length of 350 mm, and each anchor cable adopts two K2335 and one Z2335 resin anchoring agent; (3) Selection of reinforcement mesh Φ 6 mm steel bars are welded, and the grid is 100×100 mm, net width 800×1500 mm, connected by 16 # iron wire in space, with an overlapping length of 100 mm.

The 2407 working face of the Yuhua mine has a certain dip angle (5–10 degrees), the mining completion time of the adjacent 2408 working face is short, the ground pressure appears violently, and the smooth and smooth deformation is serious. The maximum subsidence of the roof reaches 0.5 m, and the roof is lifted many times; the convergence of the two sides is large, and the slice side is serious; the bottom heave disaster is serious, the rise of the trough bottom plate is affected by 0.5 m, and the safety production of 2407 working face is affected. Therefore, it is urgent to analyze the serious deformation of the 2407 working faces and study its repair and support parameters.

3 MEASURING STRATA STRESS

3.1 *In situ stress test method*

The main mining 4^{-2} coal is located at the bottom of the coal-bearing formation Yan'an formation and enters the deep mining with a buried depth of more than 600 m. The deep rock mass has experienced shear compression of multiple tectonic movements, and there are high ground stresses. After the roadway is excavated, the prominent contradiction between the high ground stress borne by the surrounding rock and its own low strength leads to the rapid deterioration of the surrounding rock from the outside to the inside, the rapid initiation, and expansion of cracks, fractures, fragmentation, swelling, deformation, and instability. The lining support of the deep roadway in the Tongchuan Mining Area is often seriously damaged and the bottom heave is greatly deformed, so it has to be repaired many times, which seriously restricts the safe and efficient production of the mine. The existing roadway deformation is asymmetric, with large deformation and serious convergence of two sides. In order to analyze the failure law of roadways and provide the basis for the design of roadway support parameters, it is urgent to explore the distribution law of in-situ stress in the mining areas.

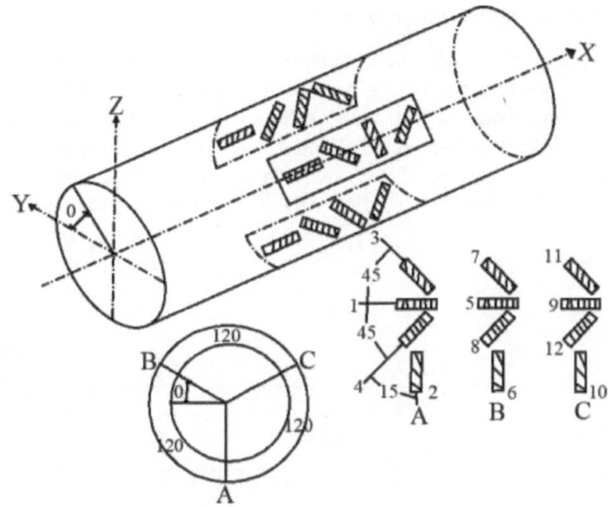

Figure 1. Distribution of resistance strain flower.

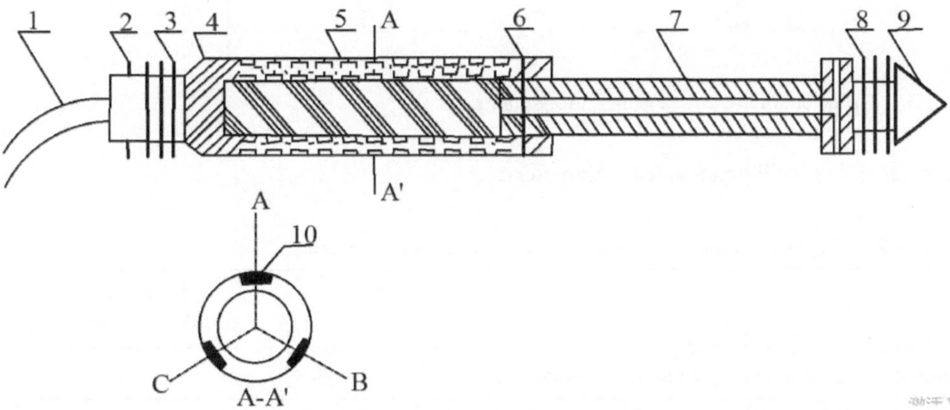

(1. Cable; 2. Directional pin; 3、 8. Sealing ring; 4. Epoxy resin cylinder; 5. Adhesive; 6. Fixing pin; 7. Plunger; 9. Guide rod; 10. Rosette)

Figure 2. Structural diagram of KX-2002 air-core inclusion full stress gauge.

Research shows that stress is the main factor controlling roadway deformation and failure. According to different measurement principles, in-situ stress measurement methods can be divided into stress recovery method, stress removal method, strain recovery method, strain relief method, hydraulic fracturing method, acoustic emission method, X-ray method, gravity method, etc (Ma et al. 2020; QIU et al. 2004; WANG et al. 2014). Coring stress relief method, hydraulic fracturing method, and rheological stress recovery method are commonly used measurement methods. Due to the advantages of three-dimensional stress measurement and relatively high measurement accuracy, the coring stress relief method is the most widely used and mature in-situ stress measurement method at home and abroad. In order to study the distribution law of in-situ stress in the Tongchuan Mining Area, the in-situ stress test was carried out by the core stress relief method on site, as shown in Figures 1 and 2.

3.2 *Layout of in-situ stress measuring points*

Three measuring points are selected by the core stress relief method, and one borehole is drilled at each measuring point. Among them, 1~2 # measuring point is arranged in the south wing track roadway, 1 # is 260 m away from the entrance of the south wing track roadway, is located in the south of the roadway, the distance between 2 # and 1 # is 100 m, 3 # measuring point is arranged in the south wing return air roadway, 320 m away from the roadway opening, and is located in the north of the roadway. The specific location of the in-situ stress measuring point is shown in Figure 3:

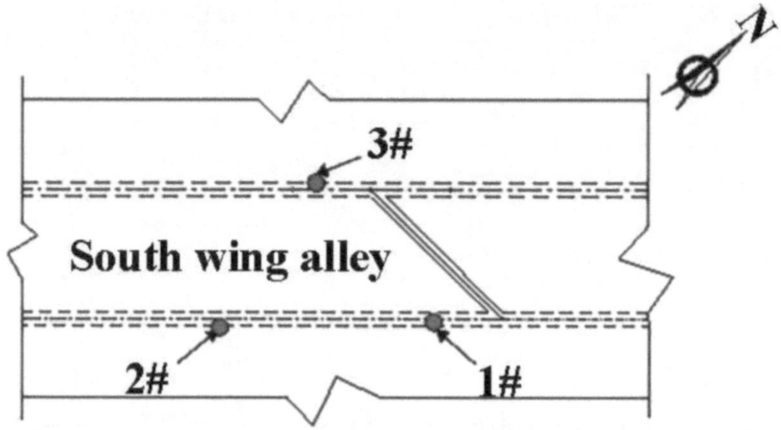

Figure 3. The layout of in-situ stress measuring points.

See Table 2 for drilling technical parameters.

Table 2. Technical parameters of drilling with core stress relief method.

Measuring point	Buried depth/m	Hole depth/m	Azimuth/(°)	Dip angle/(°)	Sensor mounting angle/(°)	Coordinates (Beijing coordinate system)	
						X	Y
1#	695.25	12.1	134.5	7	90	3912329	36580647
2#	696.84	12.1	135.2	5	90	3912241	36580568
3#	698.24	12.1	312	5	270	3912411	36580559

3.3 *Result analysis of core stress relief method*

The installation enclosure is placed in the prefabricated borehole. After 20 hours, the epoxy resin is completely solidified and the enclosure is completely cemented with the surrounding rock mass. use the core set of Φ 130 mm thin-walled drill bit is relieved. After the surrounding pressure is relieved, the core will recover elastically, and the inclusion strain gauge will change accordingly. The reading will be taken after the footage is relieved for 3 cm each time until the footage exceeds the working length of the inclusion, and the stress relief curve between the instrument reading and the relief depth will be obtained. In the process of field release, the data of strain and release distance are recorded by means of strain gauge and field reading respectively. The strain relief distance curve can be drawn from the original data, and the strain relief distance curve of 1~3 # measuring points is shown in Figure 4.

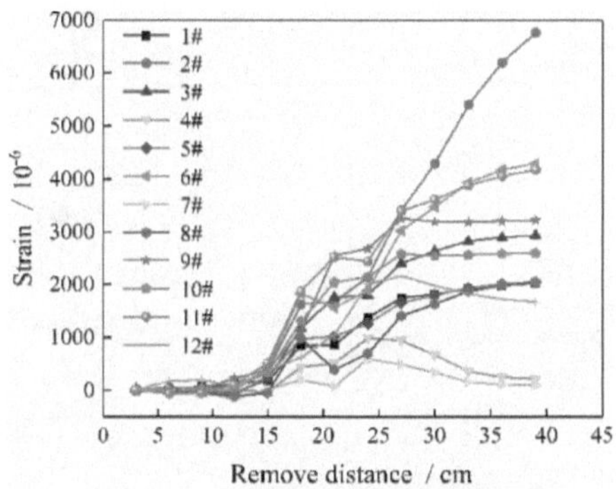

(a) 1 # Measuring point

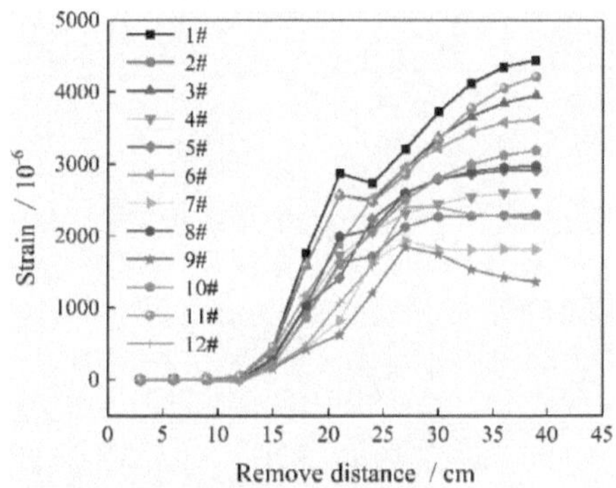

(b) 2 # Measuring point

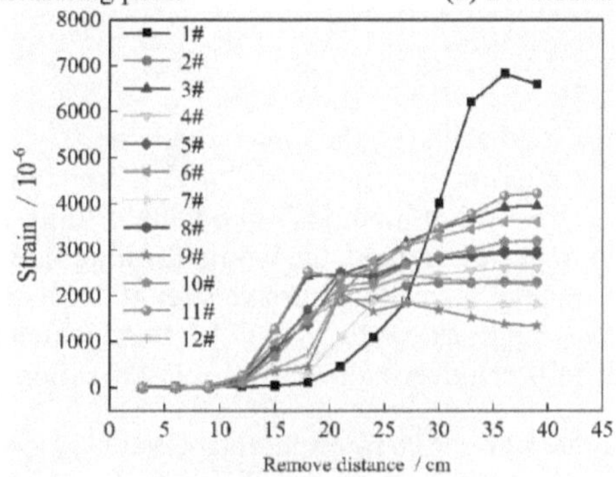

(c) 3 # Measuring point

Figure 4. Channel strain relief distance curve of each measuring point.

3.4 *Determination of elastic modulus and Poisson's ratio of rock*

The rock samples obtained from drilling and coring are processed into standard specimens in the laboratory, and at least three groups of standard specimens are processed at each measuring point to reduce the error. The specimens with small discreteness are selected by measuring the longitudinal wave velocity of rock, and the uniaxial compression test is carried out by MTS electro-hydraulic triaxial servo test system to measure the uniaxial compressive strength, elastic modulus, and Poisson's ratio of rock. The specific experimental process is shown in Figure 5.

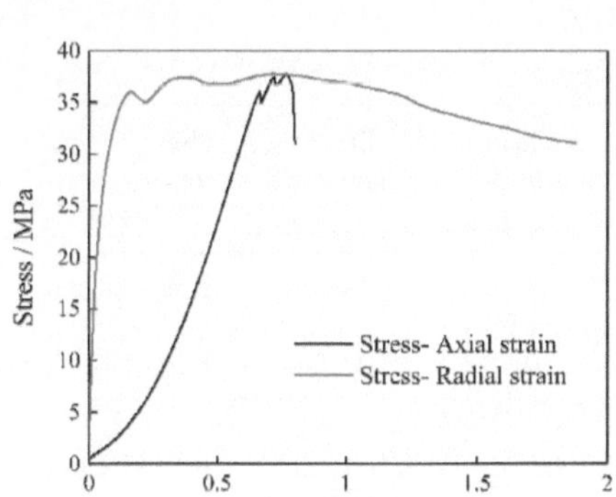

(a) 1 # Measuring point

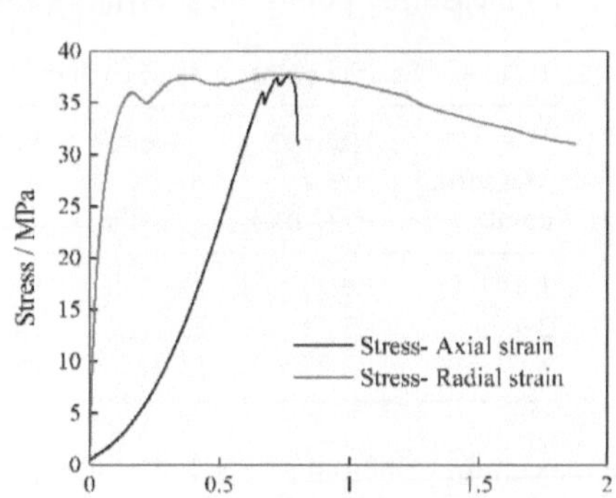

(b) 2 # Measuring point

Figure 5. Stress-strain curve of the core at each measuring point.

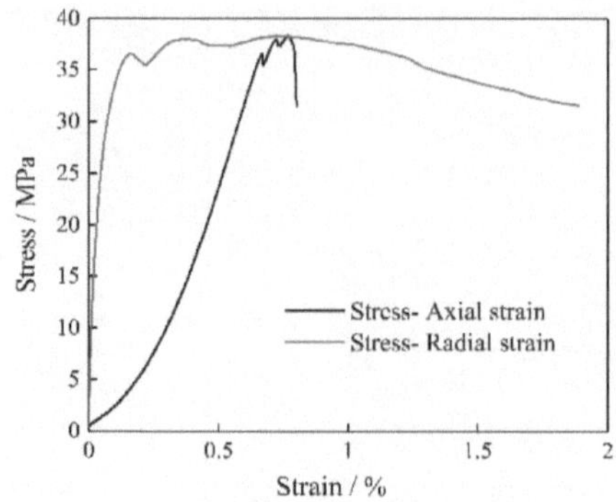

(c) 3 # Measuring point

Figure 5. Continued.

Through formula calculation and analysis, the uniaxial compressive strength, elastic modulus, and Poisson's ratio of rocks at measuring points 1 #, 2 #, 3 # are shown in Table 3. The on-site drilling condition of the core stress relief method is good, the drilling integrity of the measuring points in the south wing track roadway is good, the whole forming rate is high, and the rock core is relatively complete; the average uniaxial compressive strength of rock samples measured at the measuring points of the south wing track roadway is 37.82 MPa, which belongs to relatively hard rock according to the standard for engineering rock mass classification (GBT 50218–2014).

Table 3. Elastic modulus and Poisson's ratio of rocks at 1 #, 2 #, 3 # measuring points.

Measuring point	Uniaxial compressive strength σ_c/MPa	Modulus of elasticity E/GPa	Poisson's ratio
1#	37.56	7.58	0.24
2#	37.54	7.91	0.27
3#	38.37	7.74	0.23

3.5 *Analysis of in-situ stress measurement results*

Combined with rock mechanics parameters, drilling parameters, and sensor parameters, by calculating the principal stress of each measuring point, the stress components and principal stresses of 1 #, 2 #, 3 # measuring points are shown in Table 4.

Table 4. Stress components of each measuring point.

Measuring point	Normal stress σ_x (MPa)	Normal stress σ_y (MPa)	Normal stress σ_z (MPa)	Shear stress τ_{xy} (MPa)	Shear stress τ_{yz} (MPa)	Shear stress τ_{xz} (MPa)
1 #	29.27	12.11	17.15	2.67	0.70	2.59
2 #	27.31	13.14	14.30	–1.04	–2.62	1.72
3 #	22.29	15.72	15.77	0.54	–1.78	1.59

The maximum principal stress value of each measuring point is between 22.66 MPa and 30.22 MPa, and the inclination angle is $-1.6° \sim -12.8°$, close to the horizontal direction. There is little difference between the intermediate principal stress value and the vertical stress value, and the range is within 2 MPa. The in-situ stress field presents a stress magnitude relationship of $\sigma_1 > \sigma_v > \sigma_3$.

Table 5. The principal stress measurement results of each measuring point.

Measuring point	Principal stress	Measured value (MPa)	Azimuth (°)	Dip angle (°)
1 #	σ_1	30.22	90.2	–11.52
	σ_2	16.63	4.5	77.98
	σ_3	11.69	168.4	3.39
	σ_v	17.15	—	—
2 #	σ_1	27.38	93.7	–1.6
	σ_2	16.39	3.7	–51.6
	σ_3	11.08	183.7	–38.4
	σ_v	14.30	—	—
3 #	σ_1	22.66	88.0	–12.8
	σ_2	17.42	13.4	–41.5
	σ_3	13.70	165.5	–45.7
	σ_v	15.77	—	—

For the convenience of analysis, the vertical stress, the maximum horizontal principal stress, and the minimum horizontal principal stress are calculated from the six stress components under the geodetic coordinate system. The calculated stress components and their quantitative relationship of 1 #, 2 #, and 3 # measuring points are shown in Table 6.

Table 6. Comparison of results of horizontal principal stress and vertical stress at each measuring point.

Measuring point	Maximum horizontal principal stress σ_H/ MPa	Minimum horizontal principal stress σ_h/ MPa	Vertical stress σ_v/ MPa	Azimuth/ (°)	σ_H/σ_v	σ_H/σ_h
1 #	29.27	12.11	17.15	90.2	1.71	2.42
2 #	27.31	13.14	14.30	93.7	1.91	2.08
3 #	22.29	15.72	15.77	88.0	1.41	1.42

Through the in-situ stress test of three measuring points in the south wing roadway, the vertical stress value of each measuring point is 14.30~17.15 MPa. The self-weight stress is basically equal to or greater than the weight of overburden per unit area (2.4 t/m^3 for γ_h and γ). The ratio of the maximum horizontal principal stress σ_h to the minimum horizontal principal stress σ_h is 1.42~2.42, and its ratio changes greatly. The higher the maximum principal stress is, the smaller the corresponding minimum principal stress is, indicating that the influence of the stress field in the minefield on roadway excavation has obvious directionality. The maximum horizontal principal stress of the Yuhua Coal Mine is 22.29 ~29.27 MPa. The azimuth is between 88.0° ~ 93.7°, with an average of 90.6°. The whole is in the E-W direction. The maximum horizontal principal stress is significantly greater than the vertical stress. The maximum horizontal principal stress σ_h is 1.41~1.91 times the vertical stress. Therefore, the horizontal stress is dominant at the three measuring points. The tectonic movement has a great impact on the in-situ stress and belongs to the tectonic movement control area.

Figure 6 shows the in-situ stress measurement results. The included angle between the azimuth of the maximum principal stress σ_1 and the axial direction of the roadway is 35° ~ 72°, and the average included angle is 53.5°, which seriously affects the roadway excavation construction, so that the horizontal stress is concentrated in the roadway excavation direction, and the horizontal stress concentration on the left side is the most significant. Especially in the drilling area of 2407 transportation roadway gas control roadway, the included angle between the roadway and the maximum principal stress is as high as 72°, and the horizontal stress concentration is particularly significant. Through the actual measurement of in-situ stress and roadway deformation in the Yuhua mine, it is found that there is a high proportion correlation between roadway deformation and in-situ stress, roadway axis, and maximum principal stress direction. Therefore, higher requirements are put forward for bolt support design of high-stress soft rock roadway.

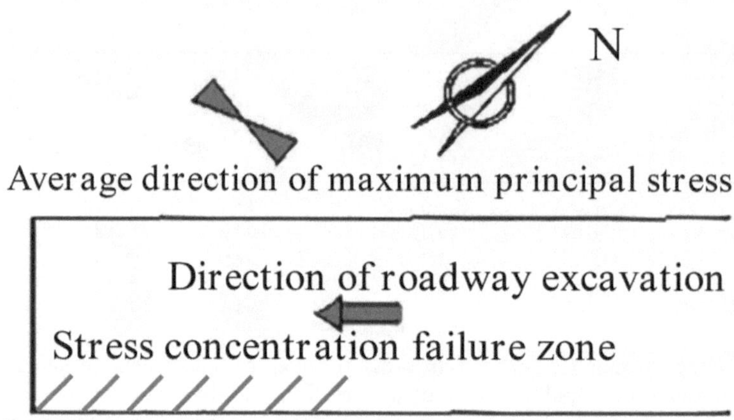

Figure 6. Influence of maximum horizontal principal stress on roadway heading orientation.

4 NUMERICAL SIMULATION ANALYSIS AFTER SUPPORT OPTIMIZATION

4.1 *Anchor rod (cable) design optimization*

Tectonic stress type stress field has a significant impact on the stability of underground roadways. Due to the large horizontal stress, the junction of roadway roof and floor and sidewall is easy to produce shear failure, and the floor is affected by the concentration of tensile stress to produce floor heave. This is also the reason for the asymmetric deformation of surrounding rock and severe deformation of the roof in the 2407 working face. In order to prevent uncontrollable severe deformation of roadway surrounding rock, countermeasures must be taken to minimize the adverse impact of structural stress on the roadway. Based on this situation, the support optimization scheme is proposed under the original support of 2407 working faces:

(1) Φ 20×2600 mm twist head steel anchor is used as a sidewall anchor bolt, with 8 rows (two sides and 4 roots) in each row, a spacing of 800 mm, and a row spacing of 800 mm. The top anchor rod Φ 20×2600 mm left-handed deformed steel bolt without longitudinal reinforcement is selected, with 5 in each row, a spacing of 800 mm, and a row spacing of 800 mm.
(2) Specification of bolt support material: the bolt spacing and row spacing are 800×800 mm. Side upper use Φ 20×2600 mm twist head steel anchor, 150×150×8 mm steel supporting plate, 2 K 3530 anchoring agents, 700 mm anchoring length, and 50 kN design anchoring force. The roof uses Φ 20×2600 mm left-hand threaded steel bolt without longitudinal reinforcement, and a 150×150×8 mm arch prestressed steel supporting plate, with an anchorage length of 700 mm and a design anchorage force of 100 kN.
(3) Anchor cable specification: use of cable Φ 21.6×8000 mm prestressed steel strand, anchorage length of 1.4 m, design anchoring force of 200 kN, anchor cable support arrangement mode is three in each row, and the row spacing is 1600×1600 mm; the anchor cable joist is WD 180–3.2 W steel belt, and the two rows of anchor cables are linked along the roadway direction, with a length of 1800 mm.

4.2 *Numerical simulation verification*

To verify the above support theory and provide a theoretical basis for roadway support parameter design, FLAC3D software is used for simulation. In the simulation, the production geological conditions of transportation roadway in 2407 working face of Yuhua Coal Mine are taken as the background, and the model size is 80 × 20 × 83 m, the roadway section is three core arch, and the section size is 5 × 3.6 m, the height of the arch crown is 1.2 m, and the rock and soil mass adopt the solid element model. After the model is established, a total of 165000 units and 176841 nodes are generated. The specific numerical model is shown in Figure 7.

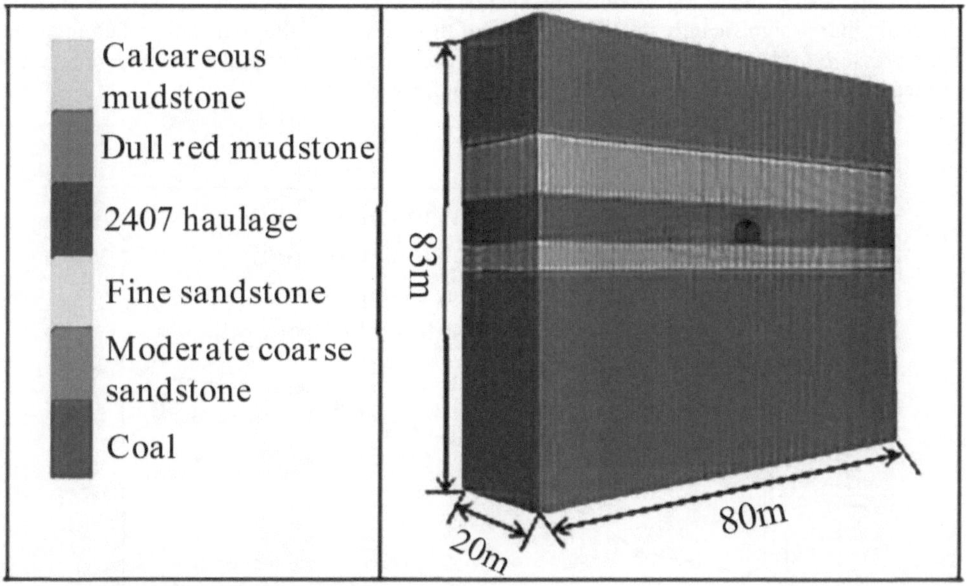

Figure 7. Establishment of the calculation model.

The Mohr-Coulomb model is used in the simulation. The initial stress of 12 MPa is applied at the top, the horizontal stress of 27 MPa is applied in the horizontal direction, the preload of 20 kN is applied to the original anchor cable, and the preload of the optimized support scheme is 50 kN. Displacement constraints are applied to the horizontal direction and the vertical direction of the bottom. Coal and rock mass parameters are shown in Table 7.

Table 7. Physical and mechanical parameters of overburden and coal seam.

Serial number	Stratum name	Thickness (m)	Density (kg/m^3)	Bulk modulus (GPa)	Shear modulus (GPa)	Internal friction angle (°)	Cohesion (MPa)	Tensile strength (MPa)
1	Medium coarse sandstone	60	2580	8.24	5.87	38	61	2.5
2	Fine siltstone	10	2390	8.14	4.87	41	34	1.2
3	4^{-2} coal	8	1370	0.565	0.367	17	4.8	0.43
4	Carbonaceous mudstone	5	2080	9.43	6.56	28	32	0.81
5	Dark red mudstone	40	2050	8.43	6.16	23	25	1.1

4.3 *Field industrial test*

Figure 8 shows the variation curve of the measured force value of the anchor rod (cable) with the distance from the mining face. With the advance of the mining face, the tension value of the anchor rod (cable) increases continuously, especially within 30~50 m from the mining face. It shows that within this range, the roof is compressed, the anchor cable tension increases, and the increase of anchor cable tension slows down within 30 m from the mining face. The main reason is that after the shed is erected, the roof pressure decreases and the roof load borne by the anchor cable decreases, but the anchor cable tension still increases gradually. The maximum anchor cable tension is 166 kN, the design anchoring force of the anchor cable is 200 kN, and the maximum value is 83% of the design anchoring force of the anchor cable. The bolt tension value of the sidewall of the coal pillar is significantly greater than that of the sidewall of the coal wall, which indicates that the pressure of the sidewall of the coal pillar is greater and also explains the phenomenon of asymmetric deformation of the surrounding rock along the channel.

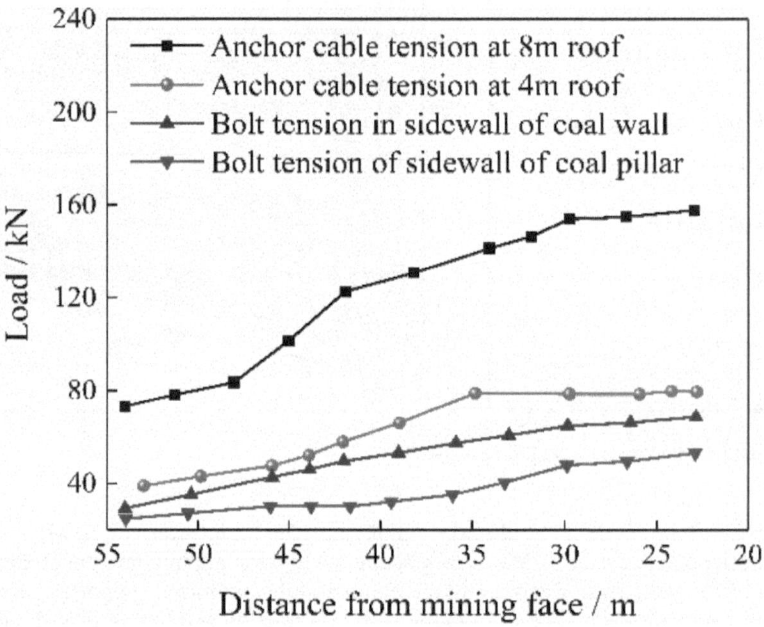

Figure 8. variation curve of measured force value of anchor bolt (cable) surface with the distance of mining face.

4.4 *Comparison of treatment effects*

According to the comparative analysis in Figure 9, after the implementation of the new support optimization scheme, the roof cutting phenomenon on the left of the roof is restrained to a certain extent, the flatness of the roadway roof is improved, the roadway is well-formed, and the phenomenon of a mesh pocket is rare; on the contrary, in the original support scheme, the roof subsidence is serious, the roof cutting occurs from time to time, the roadway formation is poor, and the bolt and anchor cable are broken. Therefore, the effect of support optimization design in the test section along the trough of 2407 fully mechanized top coal caving face is good, and the pressure effect is remarkable, which ensures the safe mining of the working face.

(a) Original support scheme

(b) New support scheme

Figure 9. Field comparison of new and old support schemes.

5 CONCLUSION

Based on the measured results of in-situ stress in the - 698 m horizontal section of Yuhua Coal Mine, the optimization research of along channel support in 2407 working face is carried out, and the following conclusions are obtained:

1) Through the core stress solution method, the in-situ stress of 4^{-2} coal seam is mainly horizontal stress, which is significantly greater than the vertical stress. The direction of the maximum horizontal principal stress is east-west, with a large included angle with 2407 along the channel, with an average of 53.5°, resulting in serious asymmetric deformation of the surrounding rock along the channel, which needs frequent repair and support, indicating that the layout of the channel is not reasonable;
2) According to the measured tension value of anchor bolt and anchor cable, the tension of anchor bolt and anchor cable is within the range of its design value, which shows that the optimized support design meets the support requirements;
3) Through the actual measurement of in-situ stress and roadway deformation in the Yuhua mine, it is found that the roadway deformation has a high correlation with the magnitude of in-situ stress, the axis of the roadway, and the direction of maximum principal stress, so the requirements for bolt support design of high-stress soft rock roadway are higher;
4) The field industrial test shows that the roof is basically within the controllable range, and the optimized support scheme effectively governs the deformation of roadway surrounding rock and ensures the normal and safe mining of the working face.

REFERENCES

Brady B H G. & Brown E T. (2004). Rock Mechanics for Underground Mining (3rd ed) M. Springer-Verlag New York Inc.

Cai M F, et al. (2000). Results of in situ stress measurement and their application to mining design at five metal mines. J. International Journal of Rock Mechanics and Mining Sciences. 37 (3), 509–515.

Cai M F, et al. (2010). In-situ stress measurement at the deep position of linglong gold mine and distribution law of in-situ stress field in the mine area. J. Chinese Journal of Rock Mechanics and Engineering. 29 (2), 227–233.

Cai M F, et al. (2013). In situ stress measurement and its application in the 10th Mine of Pingdingshan Coal Group. J. Journal of University of Science and Technology Beijing. 35 (11), 1399–1406.

Kang H P, et al. (2016). Characteristics of underground in-situ stress distribution in shallow coal mines and its applications. J. Journal of China Coal Society. 41 (6), 1332–1340.

Li B. (2017). Distribution characteristics of the crustal stress in the northeast of Guangxi Basin. J. Chinese Journal of Rock Mechanics and Engineering. 36 (S1), 3475–3484.

Liu C, et al. (2018). Study on Drift Stability Optimization Based on 3D Geo-stress Measurement. J. Chinese Journal of Underground Space and Engineering. 14 (5), 1372–1380.
Liu Q S & Liu K D. (2012). Characteristics of in-situ stress field for deep levels in Huainan coal mine. J. Rock and Soil Mechanics. 33 (7), 2089–2096.
Ma CH D, et al. (2020). Comparative study of stress relief method and acoustic emission method in In-situ stress measurement in the deep area of xincheng gold mine. J. Gold Science and Technology. 28 (3), 401–410.
Qi H G & Zhou G. (2016). Study and Application of Hollow Inclusion Stress Measuring Technique. J. Safety in Coal Mines. 47 (12), 141–144.
Qiu X D, et al. (2004). The technology of stress relief by overcoring used on dangerous rock slopes. J. Journal of Chongqing University (Natural Science Edition). (11), 97–99.
Tao W B, et al. (2020). In-situ stress characteristics and optimal support design of the deep roadway. J. Journal of South China University of Technology (Natural Science Edition). 48 (4), 28–37.
Wang J, et al. (2019). In-situ stress distribution characteristics and their relationship with geological structure in Xingcun mine. J. Journal of Mining & Safety Engineering. 36 (6), 1240–1246.
Wang Y C, et al. (2014). Study of distribution regularities and regional division of in-site stresses for PingDingShan mining area. J. Chinese Journal of Rock Mechanics and Engineering. (S1), 2620–2627.
Wang Z & Chang X Y. (2018). Study on distribution laws of ground stress at mining depth of -750 m level in No.2 Xinji Mine. J. Coal Science and Technology. 46 (10), 143–148.
Xiao T Q, et al. (2013). Relationship between ground stress distribution and roadway stability in the deep tectonic region. J. Journal of Mining & Safety Engineering. 30 (5), 659–664.
Yang R S, et al. (2014). Research of influencing factors of coal mine ground stress test precision by stress relieving method. J. China Mining Magazine. (8), 136–139.
Zhang B H, et al. (2008). Study of 3D in-situ stress measurement and stability of roadways in depth. J. Rock and Soil Mechanics. 29 (9), 2547–2555.

Advances in Mineral Resources, Geotechnology and Geological Exploration – Rashid & Zhang (eds)

Study on the secondary consolidation coefficient of soft soil based on various influencing factors

Hong-Jun Liu*, Cheng-Ling He & Hong Lu
School of Civil Engineering and Architecture, Wuyi University, Jiangmen, China

ABSTRACT: Through the indoor one-dimensional consolidation and compression test of soft soil samples with different burial depths in the western Pearl River Delta region, this study evaluated the variation characteristics of the secondary consolidation coefficient with the consolidation pressure, load ratio, soft soil depth, the direction of the compression, consolidation time, compression index, and other factors like changes in the characteristics and correlation. The test results show that the secondary consolidation coefficient of soft soil is mainly affected by the load ratio. When the loading ratio remains constant or increases, the secondary consolidation coefficient increases with the increase in consolidation pressure. When the load ratio decreases, the secondary consolidation coefficient with the increase in consolidation pressure is gradually decreased. Moreover, the variation process is less affected by sampling depth and compression direction. The ratio of the secondary consolidation coefficient to the compressibility index is in a very limited range, indicating their linear relationship. The secondary consolidation coefficient has a time effect and decreases with the increase in the t_2 value. Finally, the research findings provide a basis for the design and research of soft soil foundations in this area.

1 INTRODUCTION

With the implementation of the Guangdong-Hong Kong-Macao Greater Bay Area Planning Outline, the construction scale, grade, and usage requirements of the Greater Bay Area will be increasingly higher. The location of the Greater Bay Area is the river entering the sea. In this case, the geological conditions will be widely distributed in marine soft soil characterized by its low strength, high compressibility, and more fluid state (Bc. O'Kelly 2006; Luo 2017; Pusch 2010; Zhou 2014). Under the action of the load, these soft foundations will produce excessive residual deformation, which will be marked by continuous uneven settlement after work. Additionally, the settlement stability time is long, occasionally lasting for several years or even decades. When the settlement is severe, it will result in foundation instability and safety hazards (Tai 2002; Zhou 2016). Concerning road engineering, the large settlement of thc soft soil foundation and the jumping of the bridgehead sometimes occur.

The final settlement of saturated soft soil is analyzed based on the mechanism and consists of three parts, namely the immediate settlement, primary consolidation settlement, and secondary consolidation settlement (Hu 2011; Wan 2017; Xiang 2005). In the post-construction settlement of soft soil, the secondary consolidation settlement is not negligible (Yin 2003); (Liao 2002; Wang 2018). As a result, it is particularly important to predict the secondary consolidation of soft soil and determine its secondary consolidation coefficient. Based on this, the author executed the research on the secondary consolidation coefficient of soft soil in this region. The methodology involves carrying out a one-dimensional consolidation compression test indoors using undisturbed soil samples. Then, the discussion explores the effects of sampling depth, consolidation pressure, compression

*Corresponding Author: liuhongjun_0821@163.com

DOI 10.1201/9781003319412-50

direction, compression index, and other factors on the secondary consolidation coefficient of soft soil. The experimental results can provide a reasonable and reliable reference for the calculation of the consolidation settlement of this project.

2 EXPERIMENTAL RESEARCH PROTOCOL

The test soil samples were taken from K0+400, Houde Road, Xinhui District, Jiangmen City, Guangdong Province. The area belongs to the alluvial plain of the Xijiang Delta. Additionally, the Quaternary Holocene intercontinental sedimentary soft soil is widely distributed in the area (Chen 2011; Lai 2006), which is a poor foundation soil. The soft soil layer of the road section has a buried depth of 1.80-6.20 m, a layer thickness of 8.00-17.95 m, and an average thickness of 12.18 m. The soil sample is grayish-black, saturated, and fluidized. Based on The Test Methods of Soils for Highway Engineering (JTG3430-2020), the fundamental physical properties of the soil samples were tested, and the relevant physical property indicators were measured (Table 1).

Table 1. Basic physical properties of soil samples.

Depth of soil/m		ω / %	e	G_s	ρ_0 / $g \cdot cm^{-3}$	S_r / %	a_{1-2} / MPa^{-1}	E_{s1-2}/ MPa	I_l
5.20–5.40	Vertical	91.29	2.40	2.72	1.53	100	2.86	1.19	2.22
	Landscape	80.99	2.28	2.72	1.50	100	2.13	1.54	1.88
8.20–8.40	Vertical	76.54	2.04	2.72	1.58	100	2.36	1.29	1.74
	Landscape	76.37	2.14	2.72	1.53	100	2.25	1.40	1.73
11.40–11.60	Vertical	76.36	2.14	2.72	1.53	100	2.46	1.27	1.73
	Landscape	76.66	2.20	2.72	1.50	100	2.22	1.44	1.74

Table 1 details that the soft soil in this area has a high natural moisture content, which is greater than the liquid limit of the soil. Also, the buried depth is relatively large and in a fluid-plastic state. The void is as high as 2.0 or larger, and the compression coefficient is 2.0 Mpa^{-1} or more, demonstrating that it is a high compressibility soil.

In order to study the secondary consolidation coefficient of soft soil in the western part of the Pearl River Delta, it should be noted that it is affected by factors like sampling depth, drainage consolidation direction, loading process, and loading time, respectively, at the depths of 5.20-5.40 m, 8.20-8.40 m, and 11.40-11.60 m. The collected soil samples of each depth are respectively subjected to vertical and lateral compression, that is, compression in and perpendicular to the direction of their own weight. Each soil sample is graded and loaded, and the load ratings of the consolidation pressure are 12.5 kPa, 25 kPa, 50 kPa, 100 kPa, 200 kPa, 300 kPa, 400 kPa, 600 kPa, and 800 kPa. Taking into account the time effect of secondary consolidation, the load compression time for each stage is up to 48 hours.

3 ANALYSIS OF TEST RESULTS

The results demonstrate that the $e \sim \lg t$ curve curvature is small, and the primary and secondary consolidation boundary points become evidently insufficient. There are two main reasons. First, with a small load, the soft soil has a large void, is not dense enough, and is easily compressed. As the consolidation pressure increases, the soil becomes increasingly dense and harder. Also, the compressibility will gradually decrease, and the primary and secondary consolidation boundary points become less and less obvious. The second reason is the influence of the loading ratio, which refers to the difference between the front and rear loads and the previous load. By definition, when the consolidation pressure is less than or equal to 200 kPa, the loading ratio is $\Delta p/p=1$. When the consolidation pressure is greater than 200 kPa, the loading ratios of the latter four stages are $\Delta p/p=0.5, 0.33, 0.5$, and 0.33, respectively. Given that these values are less than 1, it is one of the reasons why the post-four-stage load compression increment is reduced, and the primary and secondary consolidation divisions are not obvious.

3.1 *Effects of various factors on the secondary consolidation coefficient*

To study the influence of sampling depth, compression direction, loading ratio, and other factors on the secondary consolidation coefficient of soft soil, its curve with consolidation pressure is plotted under each sampling depth, as shown in Figure 1. Figure 1(a) and 1(b) illustrate the compression results in the vertical and horizontal directions, respectively.

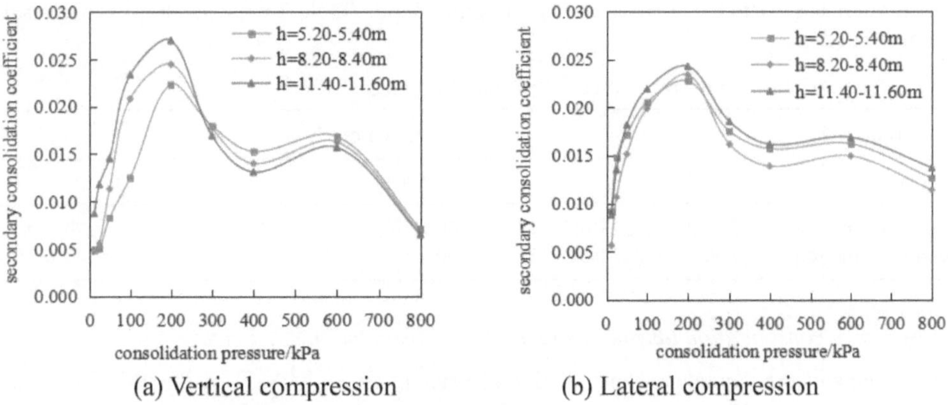

Figure 1. Curve of soft soil secondary consolidation coefficient with consolidation pressure and compression depth.

Based on the curve spread characteristics on the two graphs, changing the sampling depth and the compression direction have no evident influence on the shape and the changing trend of the curve, and the whole curve is the same. With regards to the influence of the sampling depth on the secondary consolidation coefficient, the vertical compression is more apparent than the lateral compression, especially when the consolidation pressure is less than 200 kPa. According to Figure 1(a) on the secondary consolidation coefficient obtained by vertical compression of soft soil, when the pressure is small, the buried depth is correspondingly small. When the consolidation pressure reaches 300 kPa, the inflection point occurs, and the test result is the opposite. Figure 1(b) presents that the secondary consolidation coefficient of the buried depth of the soft soil remains large, the buried depth is small, and the state is maintained as the consolidation pressure increases, which is relatively stable.

The curve shape of the soft soil secondary consolidation coefficient with the consolidation pressure is primarily affected by the loading ratio. When the consolidation pressure is less than or equal to 200 kPa, the loading ratio is $\Delta p/p=1$, which consistently maintains a constant value and does not decrease. The secondary consolidation coefficient gradually increases as the consolidation pressure increases. First, the consolidation pressure is increased from 200 kPa to 300 kPa, the loading ratio is $\Delta p/p=0.5$ (smaller than 1), and the secondary consolidation coefficient is reduced. Subsequently, the consolidation pressure increased from 300 kPa to 400 kPa, the loading ratio is $\Delta p/p=0.33$ (less than 0.5), and the secondary consolidation coefficient decreased. Then, the consolidation pressure is increased from 400 kPa to 600 kPa, the loading ratio is $\Delta p/p=0.5$ (larger than 0.33), and the secondary consolidation coefficient is increased. Finally, the consolidation pressure increased from 600 kPa to 800 kPa, the loading ratio is $\Delta p/p=0.33$ (less than 0.5), and the secondary consolidation coefficient decreased. According to these outcomes, the loading ratio can be kept constant or gradually increased, and the secondary consolidation coefficient of the soft soil increases with the increase in the consolidation pressure. When the loading ratio decreases, the opposite result is obtained.

The influence of the secondary consolidation coefficient of soft soil on the consolidation pressure, sampling depth, and compression direction is studied. Table 2 shows the variation range of the soft soil secondary consolidation coefficient under the entire consolidation load level. It also notes that the variation range of the vertical compression of the secondary consolidation coefficient is significantly larger than that of the transverse compression. This indicates that the secondary consolidation of the vertical compression of the soft soil in this region is affected by the consolidation pressure to a greater extent. The maximum value of the soft soil secondary consolidation coefficient of the vertical and horizontal compression increases with the increase in sampling depth. In this context, the variation range of the secondary consolidation coefficient of the vertical compression gradually increases with the increase in the sampling depth. Table 2 presents the specific results and statistical data.

Table 2. Range of variation of soft soil secondary consolidation coefficient.

Compression	Sampling depth/m		
direction	5.2~5.4	8.2~8.4	11.4~11.6
Vertical compression	0.0049~0.0223(0.0174)	0.0048~0.0245(0.0197)	0.0065~0.0270(0.0205)
Lateral compression	0.0092~0.0227(0.0135)	0.0056~0.0234(0.0178)	0.0088~0.0242(0.0154)

3.2 *Effects of consolidation time on the secondary consolidation coefficient*

In the formula of the secondary consolidation coefficient $C_a = \frac{e_1-e_2}{\lg t_2-\lg t_c}$, t_2 is the time for the secondary consolidation. The previous research findings revealed that the secondary consolidation deformation has a linear relationship with the logarithm of time (Hu 2017; Lei 2002), but this is only basic, not absolute. As a result, it is necessary to further explore whether the t_2 value affects the secondary consolidation coefficient and the degree of influence of soft soil. Figure 2 shows the relevant research results. Furthermore, it can be concluded that the secondary consolidation coefficient of the soft soil under vertical and lateral compression gradually decreases with the increase in t_2. The maximum variation range is at the consolidation pressure of 300 kPa, and the variation range is 0.0179~0.0304, and 0.0175~0.0340. The minimum variation range is at the consolidation pressure of 12.5 kPa, and the range of variation is 0.0049~0.0069, 0.0092-0.0176.

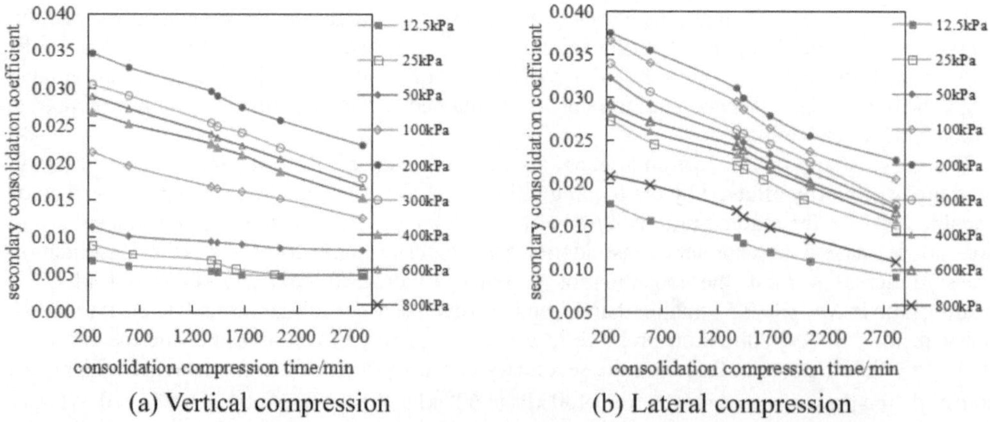

(a) Vertical compression (b) Lateral compression

Figure 2. Curve of soft soil secondary consolidation coefficient with time from 5.2–5.4 m depth.

3.3 *Relationship between the secondary consolidation coefficient and the compression index*

After summarizing the experimental results of the existing literature, it is found that the ratio of the secondary consolidation coefficient *Ca* and the compression index *Cc* of the same undisturbed soil is fundamentally constant, and the ratio is maintained between 0.025 and 0.10.

In this test, the ratio of the soft soil secondary consolidation coefficient *Ca* to the compression index *Cc* ranges between 0.018 and 0.027. The small range of this ratio indicates that the secondary consolidation coefficient *Ca* has a linear relationship with the compression index *Cc*. In this context, the empirical formula is

$$Ca = k \cdot Cc \quad (k = 0.018 \sim 0.027) \tag{1}$$

Figure 3 exhibits that the equation has a high degree of fitting, and the correlation coefficient can reach 0.82 or more.

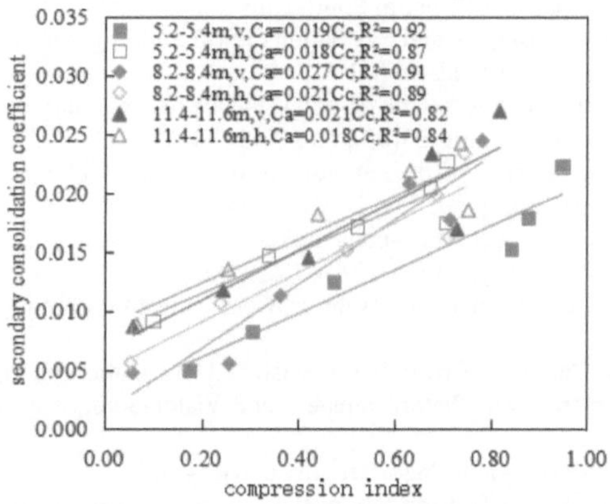

Figure 3. Relationship between the secondary consolidation coefficient and the compression index.

4 CONCLUSION

Laboratory test results indicate that the secondary consolidation coefficient of soft soil is mainly affected by the load ratio. With the increase in consolidation time, the secondary consolidation coefficient gradually decreases, and the secondary consolidation coefficient has a linear relationship with the compression index.

(1) The variation of soft soil secondary consolidation coefficient with consolidation pressure is less affected by the sampling depth and compression direction. It is predominantly affected by the loading ratio. When the loading ratio is kept constant or increased, the secondary consolidation coefficient remains at a fixed value. The junction pressure increases, the loading ratio is reduced, and the secondary consolidation coefficient is gradually decreasing as the consolidation pressure increases.
(2) Using the mapping method, the time effect of soft soil secondary consolidation coefficient is studied. It is concluded that the vertical and lateral compression soil samples decrease under the consolidation pressure of each stage, and the secondary consolidation coefficient decreases with the increase in consolidation time.
(3) The ratio of the soft soil secondary consolidation coefficient to compression index ranges between 0.018 and 0.027. The small range denotes that the secondary consolidation coefficient is basically linear with the compression index, and the empirical formula is $Ca = k \cdot Cc \quad (k = 0.018 \sim 0.027)$.

(4) Based on the research results of this paper, a multi-dimensional mathematical model can be established between the secondary consolidation coefficient and other parameters.

REFERENCES

BC. O'kelly: Compression and consolidation anisotropy of some soft soils [J]. Geotechnieal and Geological Engineering. 24: 1715–1728 (2006).

Chen, X.P: Consolidation effect of soft soil in interactive marine and terrestrial deposit[J]. Chinese Journal of Geotechnical Engineering. 33(04): 520–528 (2011).

Chief Editor, Highway Science Research Institute, Ministry of Communications: The Methods of Soils for Highway Engineering JTG 3430–2020[M]. Beijing: People's Communications Publishing House (2021).

Hu, Y.Y: Settlement of soft ground induced by cycle load considering secondary settlement[J]. Journal of Zhejiang University (Engineering Science). 45(01):106–111 (2011).

Hu, Y.Y, Yang, P: Secondary settlement estimation in surcharge preload subject to time effect of secondary consolidation coefficient[J].Journal of Central South University. 24 (02): 341–352 (2017).

Luo, J.H, Mi, D.C, Liu, X.L, et al: Research and Evaluation of Soft Soil Engineering Characteristics[J]. Journal of China & Foreign Highway. 37(06): 42–46 (2017).

Liao, HJ, Su, LJ, Shirako Hiroaki, et al: Research on the influence of secondary consolidation on compression time curve[J]. Rock and Soil Mechanics. 23(05): 536–540 (2002).

Lai, Y.M: Analysis of the characteristics of the soft soils of the sea-land interactive facies in the Pearl River Delta[J]. West-china Exploration Engineering. 18(07): 169–172 (2006).

Lei, H.Y, Xiao, SH.F: Study on secondary-consolidation deformation characteristics of soft in Tianjin[J].Journal of Engineering Geology. 10(04): 385–389 (2002).

R, Pusch, Zhang, L, R, Adey, et al: Rheology of an artificial smectitic clay [J]. Applied Clay Science.47: 120–126 (2010).

Tai, Q, Han, GC: Analysis of Causes to Errors in Forecasting Deformation of Expressway Soft Ground and Countermeasures[J]. South-to-North Water Transfers and Water Science & Technology. 91(03): 40–42 (2002).

Wan, Y.Y: Research on Soft Soil Settlement Characteristics and Engineering Application in the Pearl River Delta Region[J]. Journal of Highway and Transportation Research and Development.13(06): 136–137 (2017).

Wang, CH.M, Zhang, S.Y, Li, SH: Secondary Consolidation Characteristics of Yitong Soft Soil [J]. Journal of Jilin University: Earth Science Edition. 48(03): 799–804 (2018).

Xiang, X.CH, Wang, R, Zhu, CH. Q: Analysis of soft foundation instant settlement in seaside areas[J]. Rock and Soil Mechanics. 26(07): 1095–1098 (2005).

Yin, Z.Z, Zhang, H.B, Zhu, J.G, et al: Secondary consolidation of soft soils[J]. Chinese Journal of Geotechnical Engineering. 25(05): 521–526 (2003).

Zhou, H: Analysis of Distribution Characteristics and Cause of the Geological and Hydrological Conditions of PRD's Soft Soil [J]. Guangdong Architecture Civil Engineering. 21(07): 36–38 (2014).

Zhou, X.W, Chen, L.W, Cheng, ZH. L, et al: Stability controlling standard for embankment construction on soft soil ground[J]. Rock and Soil Mechanics. 37(09): 2631–2635 (2016).

Advances in Mineral Resources, Geotechnology and Geological Exploration – Rashid & Zhang (eds)
© 2023 Copyright the Author(s), ISBN 978-1-032-33377-9

Large-scale piping detection method based on thermal imaging and computer technology

Jianzhong Qiu
Department of Mechanical Engineering, Qilu University of Technology (Shandong Academy of Sciences), Jinan, China
Jiangxi Wuda Yangfan Technology Co., Ltd, Nanchang, China
Sichuan Machinery Research and Design Institute Co., Ltd. Chengdu, China

Guoyong Wan*
China Railway Water Resources and Hydropower Planning and Design Group Co., Ltd, Nanchang, China
Jiangxi Wuda Yangfan Technology Co., Ltd, Nanchang, China

Caiwei Liu, Yanjin Liu & Jun Wu*
Department of Mechanical Engineering, Qilu University of Technology (Shandong Academy of Sciences), Jinan, China
Sichuan Machinery Research and Design Institute Co., Ltd. Chengdu, China

Pengcheng Huang
Department of Mechanical Engineering, Qilu University of Technology (Shandong Academy of Sciences), Jinan, China

Zhijian Zhong
China Railway Water Resources and Hydropower Planning and Design Group Co., Ltd, Nanchang, China
Jiangxi Wuda Yangfan Technology Co., Ltd, Nanchang, China

Yuying Chen
Department of Mechanical Engineering, Qilu University of Technology (Shandong Academy of Sciences), Jinan, China

ABSTRACT: Piping is the external manifestation of defects in traditional Chinese medicines in dams. Preventive detection, investigation, and repair are required to avoid water conservancy safety accidents. In view of the reality of the long cycle, low efficiency, and limited detection methods, which are mainly used in the inspection of piping, there are also problems such as missed inspections and the inability to comprehensively analyze the potential risks of piping. After analyzing a variety of piping detection methods, this paper proposes a thermal infrared temperature difference piping detection method based on deep learning and computer vision and uses YOLOv3 to achieve rapid detection of piping infrared images. By analyzing the detection results, an evaluation model for the danger of piping is established, which provides a new technical solution for the detection of piping.

1 INTRODUCTION

Water conservancy engineering facilities are a major event related to my country's national economy and people's livelihood. Most of the dams on the rivers and lakes in my country are built on alluvial plains, and the impermeable layer on the surface of the dam is relatively thin, while the lower layer

*Corresponding Authors: wgy596@163.com, wujun@qlu.edu.cn

DOI 10.1201/9781003319412-51

of the dam is composed of thicker fine silt or gravel, which has good water permeability and is prone to seepage leak. The phenomenon of piping usually occurs on the dams based on multi-texture structures. The upper part of the dam structure is a clay or clay layer that is not permeable to water (Li et al. 2019), and the lower part is mainly composed of fine sand, sand and gravel, and other permeable substances, Figure 1(a) is a schematic diagram of the dam structure. When the flood season comes, the water level in the dam suddenly rises, and the outer soil layer of the dam will be subjected to huge water pressure. When the water pressure is so great that it breaks through the topsoil layer, the fine sand structure in the lower part will seep out and flow away with the water flow breaking through the surface, resulting in the formation of hole channels under the dam. If this kind of piping problem cannot be dealt with in time, the channel will gradually become larger with the loss of the lower part of the fine sand, and even multiple channels will merge, which will eventually lead to the collapse of the dam. This paper analyzes the generation and emergency measures of the piping phenomenon and proposes a piping detection method based on computer vision and image processing.

At this stage, the detection of piping is basically based on the analysis of the structure of the dam, to obtain the location of piping. For example, Su HZ (Su & Zhou 2022) and others made a detailed introduction to the working principle, technical equipment, and development process of the leakage risk detection method for earth-rock dams and compared the application scenarios and operation requirements of various technologies. Fan SF (Fan 2012) and others used the underground magnetic fluid detection method to detect the leakage of characteristic information of dams carried by the underground natural electromagnetic field and invert the stratigraphic structure according to the detection results, to achieve the purpose of detection. Liu ZR (Liu 2017) and others studied the piping and its formation process and put forward a series of control measures according to the principle of emergency protection when the piping occurred and summarized the occurrence conditions of the piping in detail. Wang Ping (Wang 2017) et al. combined engineering examples and used surface wave exploration technology to discuss the practical application of surface wave exploration in detecting the disturbance range and influencing the depth of piping in deep foundation pits and proved the applicability of this technology to solve such problems sex and accuracy. Jiang L (Jiang 2020) and others conducted in-depth research on the dynamic change law of three-dimensional electric field characteristics during the evolution of earth-rock dam piping through physical model tests, and then dynamically tracked the evolution trend of piping. At this stage, people have more research experience on the formation mechanism and prevention methods of piping. However, the research on the rapid positioning of piping hazards after the formation is still in its infancy, and the use of safety factor evaluation methods to evaluate the level of hazards is also in a blank stage (Qiu et al. 1966).

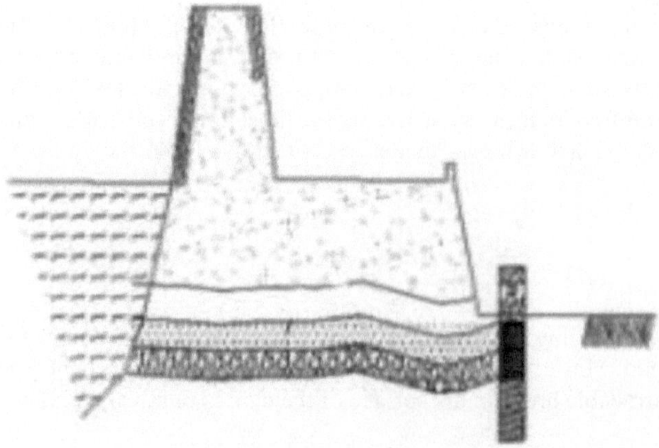

Figure 1. Schematic diagram of the dam.

2 PIPING DETECTION METHOD

The generation of piping in flood season has the characteristics of concealment, randomness in time and space, etc. The time from the occurrence of leakage danger to the serious damage to dams is very short. Timely discovery and accurate positioning of piping in flood season is the top priority to ensure the safety of dams. Nowadays, technologies including electricity and electromagnetism, vibration waves, water flow, heat, and light have been formed to detect piping.

2.1 *Existing piping detection methods*

The existing detection methods for piping mainly include electrical and electromagnetic detection methods, vibration wave detection methods, water flow detection methods, and thermal detection methods, but these detection methods cannot meet the needs of large-scale piping detection.

2.2 *The piping detection method adopted in this paper*

The purpose of this paper is to investigate the dangers of piping in a wide range of areas, especially after the occurrence of floods. The current investigation methods for the dangers of piping are mainly based on the reporting of personnel after discovery, supplemented by the gradual exploration of bare feet in key areas. This mode is time-consuming and labor-intensive, and the troubleshooting efficiency is low. According to the research, the formed piping has temperature characteristics, and there is a temperature difference of 2–4°C between the temperature of the center point of the piping and the temperature of the surrounding water. Therefore, the temperature difference analysis method can effectively detect the danger of piping. Figure 1(a) and (b) is the actual image of the piping and the simulated image of the temperature difference of the piping. Figure 1(c) is the processed image of the piping.

Because there is no public infrared piping data set, this paper uses its own data set for piping collection. This paper uses LabelImg to manually label the above collected and enhanced piping infrared images and uses the Pytorch-framework for labeling training. The image is trained for the piping recognition model.

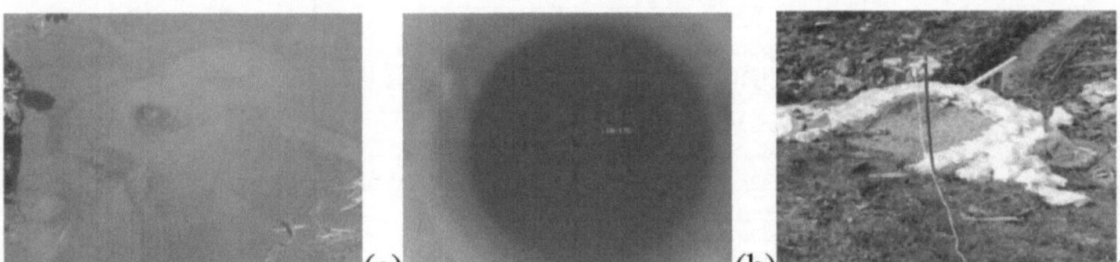

(a) (b) (c)

Figure 2. Piping live images.

3 RELATED WORKS

In this paper, with the help of deep learning and computer vision technology, the YOLO convolutional neural network is used as the core of the algorithm, and the UAV is equipped with a long-wave infrared camera and an ordinary optical camera to realize real-time detection of suspicious areas. The long-wave infrared camera realizes the temperature field detection, and the ordinary optical camera realizes the scene recording function. The latitude and longitude position information are recorded by the UAV, and the infrared data information, real scene information, and latitude and longitude position information are transmitted to the analysis server through the image information transmission function of the UAV. The analysis server analyzes the video frame in the infrared image. Whether there is a piping danger, combine the latitude and longitude information to further determine the location of the danger, and jointly determine the danger coefficient of the place based on multiple data such as real scene information.

3.1 *Visual detection algorithm based on deep learning*

Machine vision originated from computer vision. Its early research can be traced back to the research on pattern recognition of two-dimensional images in the 1950s (Guo 2011). The purpose is to recognize and classify objects in images, and target detection is one of them. Target detection is an image processing technology for identifying and locating single or multiple target objects in an image. Given an image, it can solve the two problems of what object exists in the image and where the object is. In 2014, Girshick et al. applied the convolutional neural network (CNN) in the field of target detection for the first time and proposed the R-CNN model, which became the beginning of the target detection algorithm based on the convolutional neural network. It cannot meet the needs of real-time detection, especially in the detection of piping. Therefore, this paper adopts the one-stage detection algorithm-YOLO. Redmon (Redmon et al. 2016) et al. understand target detection as solving the regression problem. For the image to be detected, the model can directly predict the bounding box and class probability of the target end-to-end with only one network calculation. The YOLO detection algorithm has gone through 4 stages. This paper adopts the most classic YOLOv3.

	type	filters	size	output
	convolutional	32	3*3	256*256
	convolutional	64	3*3/2	128*128
	convolutional	32	1*1	
1*	convolutional	64	3*3	
	residual			128*128
	convolutional	128	3*3/2	64*64
	convolutional	64	1*1	
2*	convolutional	128	3*3	
	residual			64*64
	convolutional	256	3*3/2	32*32
	convolutional	128	1*1	
8*	convolutional	256	3*3	
	residual			32*32
	convolutional	512	3*3/2	16*16
	convolutional	256	1*1	
8*	convolutional	512	3*3	
	residual			16*16
	convolutional	1024	3*3/2	8*8
	convolutional	512	1*1	
4*	convolutional	1024	3*3	
	residual			8*8
	avgpool			global
	connected			1000
	softmax			

Figure 3. YOLOv3 network architecture diagram.

3.2 *Dangerous situation safety evaluation model*

This paper proposes an evaluation model for the danger of piping based on a percentage system. Through the collection of a large amount of information, this paper concludes that the parameters affecting the safety evaluation model of dangerous situations are mainly composed of the following aspects.

Date: The current date mainly affects the safety hazards of dams in the form of general climate types. For example, the flood season in Jiangxi Province, China is from June to September. If the flood occurs in this period, the index will account for a relatively high proportion of the risk safety evaluation model.

Current and future weather: Weather is the main factor affecting the danger of piping. For example, if it is sunny now and it will be sunny in the future, the correlation of this item to the danger of forming piping is low.

Types of dams: The type of dams is an important factor in determining whether to cause the danger of piping. The dangers of piping mostly occur in old earth-rock dams, that is, the impermeable water layer that is in direct contact with the water body is thin, and the material below is easy to penetrate the sand. composed of embankments.

At the same time, this paper gives the safety factor evaluation formula.

$$\varepsilon = \sum_{i,j,\ldots}^{n} a_i x_i + a_j x_j + \ldots \tag{1}$$

In the formula, a is the weight coefficient of different constraints, and the sum of a is 1; x is the constraint value, which is in the range of [0, 100]; the weight coefficient changes with the scene change, and the minimum value of ε is 0, and the maximum value is 100.

4 CONCLUSION

This paper firstly analyses the reasons for the formation of piping and the detection methods of piping in China and proposes a large-scale piping detection method based on machine vision. The surrounding environment of the piping is determined by mapping and matching with the knowledge base of piping danger to determine whether there is a piping danger in the area. Then, according to the test results, a safety factor evaluation model based on the danger of piping is proposed. The next work is mainly to collect more thermal infrared images of the piping scene, enrich the piping feature information database according to the actual temperature data collected by the thermal infrared camera, fully study the danger safety evaluation model, add more limiting factors, and improve the detection accuracy and response speed of the spring model.

ACKNOWLEDGMENT

This paper is mainly supported by the Jiangxi Provincial Department of Water Resources Project (202123YBK07). The Sichuan Provincial Science and Technology Department Project (2020YFS0366), and the Shandong Provincial Natural Science Foundation (ZR202109280010) also provided valuable suggestions for the research. Finally, the authors express special thanks to China Railway Water Resources and Hydropower Planning and Design Group Co., Ltd.

REFERENCES

Fan SF. Research on key technologies and methods of dam leakage monitoring [D]. Hunan University of Science and Technology, (2012).

Guo ZF. Overview of Deep Learning Algorithms for Image Object Detection [J]. Mechanical Engineering and Automation, 220–222, 224 (2011).

Jiang L. Experimental study on the dynamic change law of three-dimensional electric field characteristics during the evolution of earth-rock dam piping [D]. Chongqing Jiao tong University. DOI: 10.27671/d.cnki.gcjtc.2020.000988 (2020).

Liu ZR, Jiang KZ. Research on the piping problems of embankments and dams [J]. China Water Transport (Second Half Month), 197–199 (2017).

Qiu JZ, Wu J, Wan GY, et al. A Target Detection and Evaluation Method for Safety Protection Equipment[J]. Journal of Physics: Conference Series, 1966 (2021).
Redmon J, Divvala S, Girshick R, et al. You only look once: Unified, real-time object detection[C]. Proceedings of the IEEE conference on computer vision and pattern recognition. 779–788 (2016).
Su HZ, Zhou RL. Research progress on leakage detection modes and methods of earth-rock dams [J]. Advances in Water Resources and Hydropower Science and Technology, 1–10+39 (2022).
Wang P. Application of surface wave exploration technology in piping detection [J]. Building Supervision, Inspection, and Cost, 56–59+65 (2017).
Yang QP. Causes and emergency measures of embankment piping of earth-rock dams [J]. Science and Technology Innovation, 106–107 (2019).

Advances in Mineral Resources, Geotechnology and Geological Exploration – Rashid & Zhang (eds)
© 2023 Copyright the Author(s), ISBN 978-1-032-33377-9

Study on geological deformation prediction of shield tunnel based on multi-source information fusion

Yang Wang*
Merchants Chongqing Highway Engineering Testing Center Co., Ltd., Chongqing, China

ABSTRACT: With the rapid development of urban construction in China, tunnel construction will become an important part of urban underground space development in the 21st century. The shield method is widely used in urban tunnel construction because of its obvious advantages such as strong stratum adaptability, high speed, and guaranteed construction quality. Shield construction will inevitably disturb the surrounding soil, causing stratum movement and deformation, which is most directly manifested as surface subsidence or uplift. Therefore, it is necessary to effectively predict the ground deformation caused by shield tunnel construction in order to formulate corresponding control measures to prevent engineering accidents. There are many influencing factors of shield tunnel geological deformation, which are nonlinear and difficult to predict, so it is necessary to choose an effective prediction method. Geological deformation prediction of shield tunnels plays a key role in improving the reliability of shield tunnel construction. Based on the multi-source information fusion method, this paper studies the prediction of stratum deformation caused by tunnel shield construction and constructs the calculation model of geological deformation of shield tunnels.

1 INTRODUCTION

The rapid development of tunnel technology and the continuous construction of various tunnels have not only brought great convenience to highway traffic and economic development but also brought new tests to the safe operation of tunnels (Xing et al 2021). Tunnel construction is not only an effective way to solve the problem of urban traffic congestion, but also to solve the increasingly serious shortage of urban land resources and ensure the sustainable development of the city (Ye et al. 2017). A Shield tunnel is a complex three-dimensional underground structure, which can be regarded as a slender tubular underground structure formed by splicing in turn. Because the longitudinal stiffness of the tunnel is small, the longitudinal stress and deformation are closely related to the performance of the transverse structure. Due to the limitations of technical means and investment funds, and the complexity of geological bodies, the geological data obtained before construction are often very rough and limited, which is difficult to fully meet the requirements of construction, especially for deeply buried tunnels (Wei et al. 2017). The shield tunnel is affected by factors such as Jack thrust, grouting pressure, and brush extrusion pressure at the shield tail during the construction period, and under the influence of adjacent building construction, soil pressure, and trainload during the operation period, the tunnel will have longitudinal deformation, which will affect the transverse structure performance of the tunnel (Rui et al. 2020). Shield construction will inevitably disturb the surrounding soil, causing stratum movement and deformation, which is most directly manifested as surface settlement or uplift (Wang et al. 2018). When the stratum deformation caused by tunnel construction is too large, it will affect the safe use and normal operation of surface buildings.

Due to high speed, large flow, poor light, poor air quality, and high environmental noise, the tunnel is more prone to traffic accidents than ordinary sections. Therefore, it is necessary to effectively predict the stratum deformation caused by shield tunnel construction, so as to formulate

*Corresponding Author: 790693032@qq.om

DOI 10.1201/9781003319412-52

corresponding control measures to prevent engineering accidents and ensure that the impact of shield construction on the surrounding environment is minimized (Lei et al. 2018). There are many kinds, large capacity, and complex levels of information in the tunnel. The collected vehicle, road, and environmental information must be analyzed quickly to improve the comprehensive processing ability of data (Zhu et al. 2018). In the tunnel monitoring system, data acquisition and processing occupy an important position, ranging from the formulation of a control scheme to the opening and closing control of monitoring equipment, which should be based on the results of data acquisition and processing. Based on the multi-source information fusion method, this paper studies the prediction of stratum deformation caused by tunnel shield construction and constructs the calculation model of geological deformation of shield tunnels (Keye & Gammon 2018).

2 INFORMATION FUSION

2.1 *Information fusion system model*

Generally, it is impossible to get a conclusion from a single sensor by using the information fusion method. It can also be said that the fusion data from each sensor, which is higher than that of each sensor, is the "group effect" data of each sensor after mathematical processing. Multi-sensor is the hardware foundation of information fusion, multi-data is the processing object of information fusion, and coordination, optimization, and comprehensive processing are the core of information fusion. Theoretically, it has been proved that using multiple sensors to detect targets together is more accurate than using a single sensor. Multi-sensor information fusion is just like the comprehensive processing of information by the human brain. Its basic principle is to make full use of multi-sensor resources and to combine the redundant or complementary data of multi-sensors in space or time according to a certain criterion by reasonably controlling and using these sensors and observation data, so as to obtain a consistent explanation or description of the measured object (Wang et al. 2020). Based on the principle of discretization, the numerical method decomposes or discretizes a wide range of complex problems into many smaller equivalent elements, and then carries out analysis and calculation. It is based on establishing a certain mechanical constitutive model and geometric model, and fully considering the deformation and displacement characteristics of rock and soil, it calculates the stress and strain of surrounding rock (Tsuno et al. 2020).

For a specific fusion system, the data it receives can be data of a single level of data of several levels. The basic strategy of fusion is to fuse the data of the same level to obtain higher-level data information, and then import it into the corresponding information fusion level. The hierarchical structure model of information fusion is shown in Figure 1.

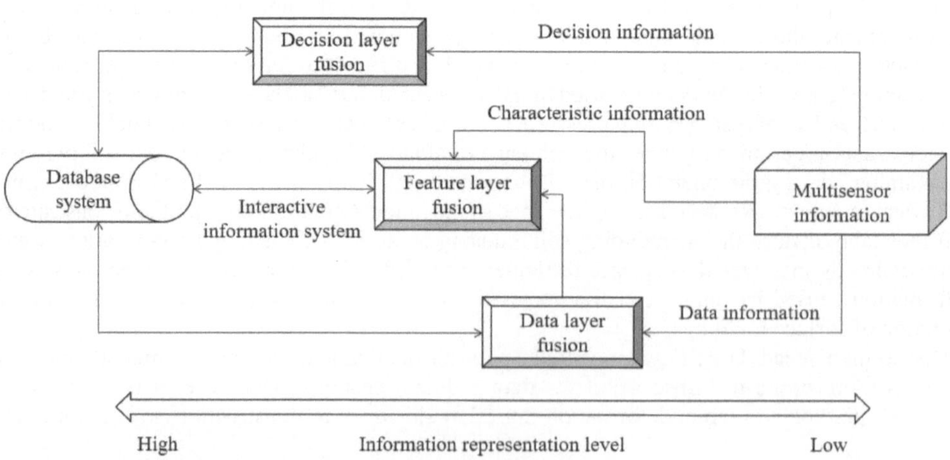

Figure 1. Hierarchical structure model of information fusion.

The purpose of information fusion is to deduce more information through data combination rather than any individual information appearing in input data, so as to obtain the best synergy result. That is to say, the joint operation of multiple sensors can be utilized to improve the effectiveness of the sensor system and eliminate the limitations of a single sensor or a few sensors. That is to say, information fusion is to use the computer technology to automatically analyze and synthesize the observation data of several sensors according to time sequence under certain criteria, so as to complete the data processing process needed for identification, decision-making, and estimation tasks (Zhang et al. 2020). Before the measurement data of each sensor is sent to the fusion center, it is processed locally by its own data processor, and then the processed data is sent to the fusion center to form a global estimate. Information fusion technology can also be used to discover, verify and recover sensor errors, and give higher accuracy detection results. When multiple sensors of the same type work at the same time, because of the different error characteristics of the sensors, their detection data all have certain errors, and the method of information fusion can produce higher precision detection results (Wei et al. 2020).

2.2 *Information fusion classification*

The data collected by the sensor system are various, and it is necessary to classify the data in order to make the data be used in a unified and coordinated way. After the early tunnel geological disaster risk assessment and stability analysis of surrounding rock, unexpected geological problems will still be encountered in the process of tunnel excavation. Redundant data refers to multiple pieces of data provided by multiple independent sensors about the same feature in environmental data, or it can also be data obtained by multiple measurements of a certain sensor within a certain period of time. The fusion of redundant data can reduce the uncertainty caused by measurement noise and improve the accuracy of the whole system. The consistency of sensor data should be checked before the redundant information of the same feature is fused. The spatial distribution patterns of various traffic detectors are different according to the location, importance, and accident rate of road sections (Wei et al. 2020). As for the complementary data in a multi-sensor system, the environmental characteristics provided by each sensor are independent of each other, that is, they perceive different sides of the environment. The fusion of complementary data reduces the ambiguity of environmental understanding caused by the lack of certain environmental characteristics, improves the integrity and correctness of the system description environment, and enhances the ability of the system to make correct decisions. When the local seismic wave encounters the interface where the wave impedance of rock changes greatly, part of the energy of the seismic wave is scattered back. The propagation time of the scattering signal is proportional to the distance of the scattering interface, so it can be used as a direct measurement method of geological body position when the accurate wave velocity of surrounding rock is obtained.

3 GEOLOGICAL DEFORMATION PREDICTION OF SHIELD TUNNEL BASED ON MULTI-INFORMATION FUSION

The support force of the excavation face plays a key role in maintaining the stability of the excavation face. If the support force of the excavation face is too large, it will cause the surface uplift, while if the support force of the excavation face is too small, it will cause the surface subsidence. Therefore, to stabilize the excavation face is to keep the basic balance between the support force of the excavation face and the initial stress of the stratum in front of the cutter head. The purpose of data calibration is to unify the time and space reference points of each detector. Because each detector works independently and asynchronously, it is necessary to calibrate in time and space. For spatial calibration, in order to meet the requirement that the fused loop coil and video detection signal are synchronized, it is necessary to synchronize the loop coil sensor with the virtual coil in video detection.

The tunnel monitoring system is a typical multi-source information system. A large number of sensors are widely deployed in the limited tunnel space. Traffic parameters are detected by detection equipment, so as to realize real-time monitoring of tunnel operation and provide the basis for control. Information fusion is a multi-level and multi-level data processing process, which mainly completes the automatic detection, correlation, correlation, estimation, and combination of data from multiple data sources. The information fusion model is shown in Figure 2.

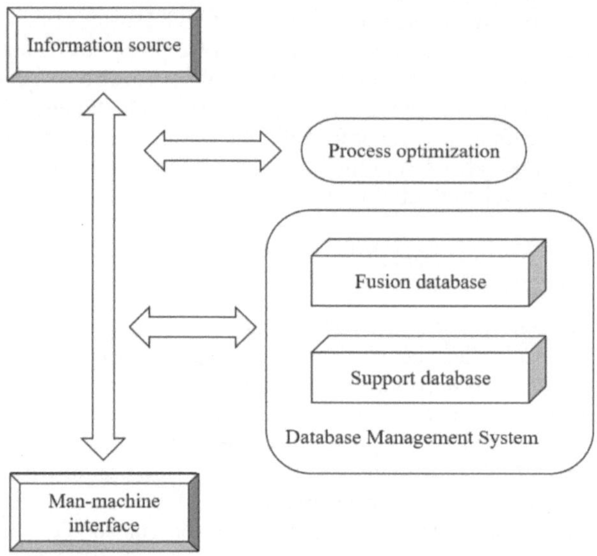

Figure 2. Information fusion model.

Shield construction is a three-dimensional problem, and it is a complex and dynamic construction process. Influenced by many factors, it is difficult to fully and accurately reflect the real state of the shield construction process only by engineering empirical formulas and model test research. At the end of each scan, the detector fuses the new observation results with the original results, estimates the target parameters according to the observation values, and uses these estimates to predict the magnitude of the parameters in the next scan. Due to the effect of earth pressure on the tunnel structure and the influence of tunnel excavation, the pipe re is damaged to a different degree. The damage development areas of the first few rings of the tunnel are mainly concentrated near the vault and arch bottom, and the damaged area of the pipe ring gradually decreases along the axis direction. In the process of shield excavation, the disturbance of surrounding soil will destroy the original stress state of the soil, which will increase the stress of the soil unit, cause the surrounding strata to produce undrained deformation, and then cause the displacement of soil. At the same time, due to the change of stress state, excess pore water pressure zone will be produced around the shield tunnel. When the shield machine leaves this area, the pore water pressure around the tunnel begins to decrease and the pore water flows out, which causes the soil around the shield machine to continuously displace for a considerable period of time after the excavation is completed.

Multi-sensor information fusion technology is to improve the measurement accuracy of each parameter by monitoring multiple parameters and adopting certain data processing methods. The information fusion process is completed in two steps. The first step is low-level processing. For data-level fusion, the output is features, attributes, and states. The rotation of the cutter and the contact of the soil in front will produce friction, which will disturb the surrounding strata. The calculation formula for geological deformation of shield tunnel in this paper is as follows:

$$U(x) = U_{\max} e^{-\frac{x^2}{2k^2}} \tag{1}$$

$$U_{\max} = \frac{L_{loss}}{k\sqrt{2\pi}} = \frac{L_{loss}}{2.5k} \quad (2)$$

$$L_{loss} = \pi R^2 \lambda \quad (3)$$

$$L_{loss} = \frac{h}{\sqrt{2\pi}\tan\left(\frac{\pi}{4} - \frac{\varphi}{2}\right)} \quad (4)$$

Among them, $U(x)$ represents the surface deformation, x represents the horizontal displacement of the shield tunnel, $U_{\max}$ represents the maximum surface deformation, k represents the settlement tank factor, R represents the equivalent radius of the shield tunnel, L_{loss} represents the formation loss per unit length, h represents the depth of the tunnel, and λ represents the volume loss rate.

During the jacking process of the shield machine, the contact between the shield shell and the surrounding soil layer generates friction. When the shield moves forward, the friction force generated is forward, the deformation of the soil around the shield shell is forward or upward, and the surface deformation is realized as uplift or settlement. After passing through the camera, the optical scene information of traffic is converted into a digital video image signal, but when a frame of image is input into the traffic collection information, the video detector system is triggered to perform an image processing operation. Synchronous grouting behind the wall can effectively fill the gap of the shield tail to reduce stratum loss and prevent ground subsidence. Stratum deformation is closely related to grouting amount and grouting pressure in the shield tail gap. Excessive grouting pressure and grouting amount will lead to stratum uplift deformation. On the contrary, insufficient grouting amount or too small grouting pressure will lead to incomplete filling of shield tail gap, which will lead to stratum settlement.

4 CONCLUSIONS

Stratum loss is the main influencing factor of stratum deformation in shield construction. Stratum loss caused by the influence of excavation face and shield tail clearance should be strictly controlled. The risk assessment of tunnel geological disasters, stability analysis of tunnel surrounding rock, and tunnel geological advance prediction are progressive relations. The whole process of research is from macroscopic to concrete, and the risk assessment of tunnel geological disasters is carried out, and the risk zoning of geological disasters is obtained. Because each detector works independently and asynchronously, it is necessary to calibrate in time and space. For spatial calibration, in order to meet the requirement that the fused loop coil and video detection signal are synchronized, it is necessary to synchronize the loop coil sensor with the virtual coil in video detection. During the jacking process of the shield machine, the contact between the shield shell and the surrounding soil layer generates friction. When the shield moves forward, the friction force generated is forward, the deformation of the soil around the shield shell is forward or upward, and the surface deformation is realized as uplift or settlement. In the traffic flow database, it is very necessary to add attributes such as road surface humidity in the tunnel. In the model prediction, the traffic data with the same or similar attributes will be selected to predict, and the prediction effect will be more reasonable and the accuracy will be further improved.

REFERENCES

Keye S, Gammon M R. (2018) Development of Deformed Computer-Aided Design Geometries for the Sixth Drag Prediction Workshop[J]. Journal of Aircraft, 55(4): 1401–1405.

Lei Zhiqiu, Zhang Tonggang, Liu Xiaohua, (2018). Segmentation of typical elements of shield tunnel point cloud [J]. Journal of Surveying and Mapping Science and Technology, 035(004): 395–399.

Rui R, Cheng C, Zhai YX.(2020) Model tests on earth pressure and settlement of shield tunnel crossing adjacent underground retaining structures[J]. Yantu Gongcheng Xuebao/Chinese Journal of Geotechnical Engineering, 42(5): 864–872.

Tsuno K, Kinoshita K, Ushida T.(2020) Investigation of Deformation of Shield Tunnel Based on Large-scale Model Test[J]. Quarterly Report of RTRI,61(1): 16–21.
Wang J, Lin G, Tang X, (2018). Face Stability Analysis of Shield Tunnel in Sandy Ground Using 3D DEM[J]. Xinan Jiaotong Daxue Xuebao/Journal of Southwest Jiaotong University,53(2):312–321.
Wang Jinan, Zhou Jiaxing, Li Fei, Xin Zhensheng.(2020) Research on fluid-structure coupling effect of large-diameter underwater shield tunnel excavation [J]. People's Yangtze River, 51(9):8.
Wei Gang, Yang Bo, Wu Huajun, (2020). Research on longitudinal deformation of existing shield tunnels caused by shield crossing [J]. Chinese Journal of Underground Space and Engineering, 16(6):10.
Wei Gang, Yu Haoqiang, Chen Chunlai, (2017). Research on the influence of double-lane parallel shield tunnels through adjacent frame buildings [J]. Science and Technology Bulletin, 33(2):6.
Xing Huitang, Xu Qianwei, Liu Hao (2021), et al. Research on the construction impact and control of shield tunnels on the upper span of existing tunnels [J]. Chinese Journal of Railway Engineering, 38(9):7.
Ye Zhi, Liu Huabei, Liu Wen. (2017) Analysis of the influence of water gushing on the excavation surface of shield tunnels on the surface settlement and internal force of segments [J]. Tunnel Construction, 37(10):11.
Zhang Kaiwei, Fu Wen, Yin Xiaodong, (2020). Damage identification of shield tunnel structure based on ARMA model [J]. Journal of Civil Engineering and Management, 35(5):6.
Zhu Y, Zhang Z, Huang X,(2018). Prototype Loading Tests on Full-Ring Segmental Lining of Rectangular Shield Tunnel[J]. Journal of Shanghai Jiaotong University (Science), 23(6): 746–757.

Advances in Mineral Resources, Geotechnology and Geological Exploration – Rashid & Zhang (eds)
© 2023 Copyright the Author(s), ISBN 978-1-032-33377-9

Design and application of intelligent monitoring based on MEMS sensors

Xiaodong Wu, Yan Gao* & Yibin Zhou
School of Earth Sciences and Engineering, Sun Yat-sen University, Guangdong, Zhuhai, China
Southern Marine Science and Engineering Guangdong Laboratory (Zhuhai), Guangdong, Zhuhai, China

ABSTRACT: In this study, MEMS (Micro Electro Mechanical Systems) sensors are used to fabricate intelligent sensing equipment to achieve continuous and automated monitoring of the acceleration, angular velocity, and magnetic induction intensity induced by the deformation of the target structure. The accelerometers of AdiADXL345 and BoschBMA280, and the motion tracking devices InvenSenseMPU6050 and InvenSenseMPU9150 are used as the sensing sensors to collect motion-related data. Raspberry Pi 3B is utilized as the processing microcomputer to control the intelligent sensing equipment. The data acquisition, conversion, transmission, storage, and judgment of the target motion state are realized by the programming of the relevant program interface, in which the target acceleration data, gyroscope data, motion posture, location, and other information can be checked in real-time. The digital filtering and deviation correction methods are adopted to improve the accuracy of the sensing data. The experiments using these four intelligent sensing devices are conducted to examine the measuring displacement of the moving target. It is found that the smart sensing equipment can determine whether the monitoring target slides or not with high accuracy; however, it is difficult to directly obtain the absolute displacement value with a tolerable error based on the collected acceleration data.

1 INSTRUCTION

Engineering activities are often faced with various structural deformations, such as landslides, ground subsidence, dam breakages, etc. These disasters are generally due to the deformation of the structure exceeding its allowable range, which seriously endangers the safety of human life and property and induces huge losses. To ensure the safety of the engineering structure, it is necessary to conduct deformation monitoring. Traditional deformation monitoring is mainly divided into conventional ground measurement and special measurement. The conventional ground measurement refers to the utilization of conventional measuring instruments such as total station, theodolite, and level; and the special measurement mainly refers to strain measurement, collimation measurement, and tilt measurement. The conventional ground measurement is relatively mature and can provide the overall deformation state of the deformed body with different accuracy requirements, while the related fieldwork is large and it is difficult to realize automatic monitoring. With the development of technology, some new deformation monitoring technologies have been applied, such as close-range photogrammetry, GPS (global positioning system), distributed fiber optic sensing technology, and terrestrial laser scanning technology. These emerging automated measurement technologies are applied to various engineering deformation monitoring, effectively improving the efficiency of deformation monitoring. However, the large-scale application is limited due to the high cost of the equipment.

*Corresponding Author: gaoyan25@mail.sysu.edu.cn

DOI 10.1201/9781003319412-53

A MEMS sensor is a sensor made by integrating a sensor, actuator, and processing circuit using MEMS technology (Liu et al, 2010). Compared with traditional sensors, it has the advantages of small weight, small size, high sensitivity, low power consumption, good linearity, good repeatability, etc. There are many kinds of MEMS sensors. MEMS motion sensor is the most widely used at present. In this study, based on four different MEMS sensing sensors, the intelligent sensing equipment with high measurement accuracy is set up; the monitored deformation data is transported through the MQTT transmission protocol to a remote server for storage; a visualization program is written to examine the monitoring deformation information and the operating status of the monitoring equipment in real-time; then, multiple sensing equipment can form a wireless sensor network (Liang et al. 2010; He and Shen 2013), which can be used for landslide warning, underground tunnel deformation, etc.

2 CONSTRUCTION OF INTELLIGENT SENSING EQUIPMENT BASED ON MEMS SENSORS

The smart sensing device made in this study includes two modules as shown in Figure 1, namely the sensing module and the control module. The sensing module consists of a MEMS sensor and its protection circuit and other components, which adopt accelerometer AdiADXL345 and BoschBMA280, motion tracking device InvenSenseMPU6050 and InvenSenseMPU9150, respectively. The accelerometer can obtain three-axis acceleration while the motion tracker InvenSenseMPU6050 can obtain three-axis acceleration and three-axis angular velocity and the motion tracker InvenSenseMPU9150 can obtain three-axis acceleration, three-axis angular velocity, and three-axis magnetic induction intensity information. The control module, which is an ARM-based microcomputer made in the Raspberry Pi 3B+ (Raspberry Pi 3B+) in the shape of a credit card size, is responsible for recording, remote transmission, and processing the output signal of the sensing module. Considering that the working environment of deformation monitoring is generally harsh and external factors such as dust and water seepage may affect the normal operation of the instrument, a waterproof box made of corrosive ABS material is used to protect the module to make the intelligent sensing device work stably for a long time. In addition to the battery, the solar module is added to realize the dual power supply of the intelligent sensing equipment.

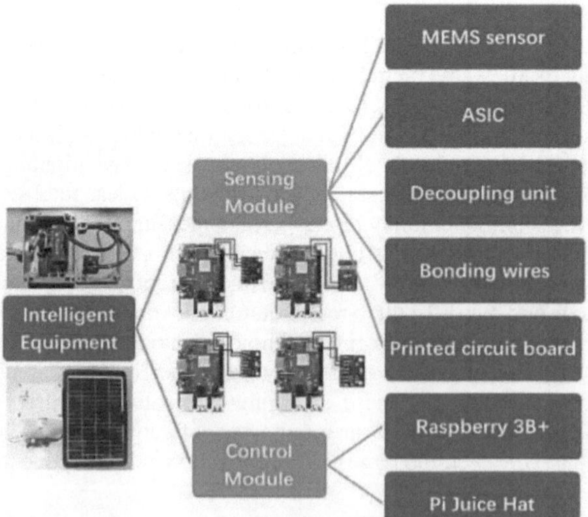

Figure 1. The architectural design of smart sensing design based on MEMS sensor.

3 SOFTWARE FUNCTIONAL ARCHITECTURE

The software architecture of the intelligent sensing equipment mainly realizes three functions: (1) data transmission, which converts the digital signal transmitted from the sensor module to the control module into readable data, and then transmits it to the remote server; (2) real-time monitoring, which can illustrate the collected data in real-time; (3) analysis and judgment the target motion status, which use the Support Vector Machine model to determine whether the motion target is sliding or not. The visualization GUI of the intelligent sensing equipment is compiled to examine the working conditions of the equipment and visually display the real-time data collected by the intelligent sensing equipment, including acceleration, angular velocity, and motion attitude in three orthogonal directions, and also the equipment location information. The overall interface of the visualization program is shown in Figure 2.

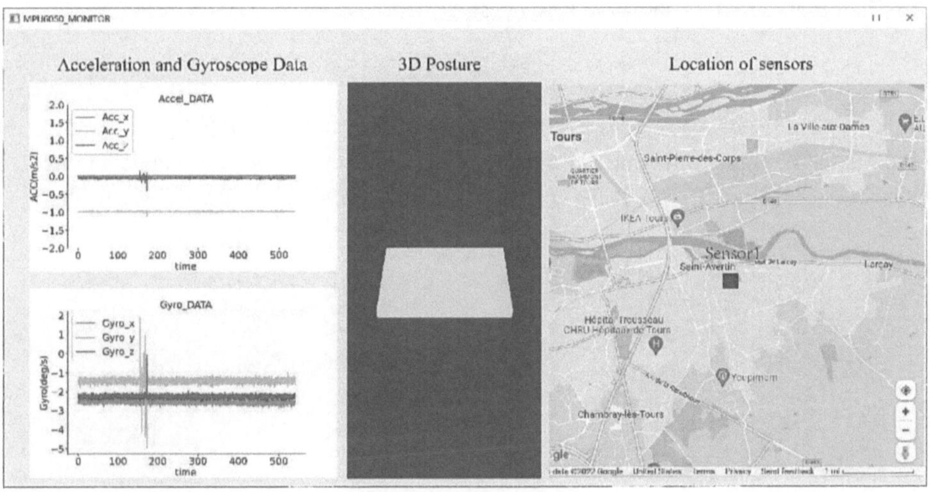

Figure 2. Visualization of the overall interface.

In the field of landslide warning, support vector machine (SVM) models and the related optimization models are widely used (Fu 2008; Li, et al. 2020; Tian & Li 2021; Xia et al. 2018). In this study, the SVM classification algorithm determines whether the monitoring objective is sliding. The basic idea is that a hyperplane in N-dimensional space (N, the number of features) is found according to different characteristic attribute values of obtained data which satisfies the requirement of the maximum geometric distance between the nearest sample points and the plane and also meets the principle of structural risk minimization.

4 ACCURACY DETERMINATION OF INTELLIGENT SENSOR EQUIPMENT

To test the static and dynamic response of the intelligent sensing equipment and the accuracy of the data, the experimental tests are conducted based on the intelligent sensing equipment to acquire the displacement. The intelligent sensing equipment is fixed on the test car which can move on the track, as shown in Figure 3. The track length is 1.3m. Except for the length of the car itself, the moving distance of the test car for each test is 1.25m. The smart sensing devices made of ADXL345, BMA280, MPU6050, and MPU9150 are used for the experiments, respectively. During the test process, the test car is first set up at a fixed starting point, then the intelligent sensing equipment starts to collect data and stands still for a certain period to collect initial data; after this, the test car begins to move to a definite endpoint and also stands still for a certain period until stopping data collection.

The errors of MEMS accelerometers and gyroscopes mainly include fixed errors, random noise, temperature effects, etc. (Yin et al. 2014; Xiong 2017). Hence, the MEMS sensor data need to be corrected before use. Among them, the six-position calibration method (Zhang et al. 2016) is used to correct the zero drift and scale factor. The original data and the corrected data are shown in Figure 4. The correction of random noise is removed by the Kalman filtering algorithm (Cai and Hu 2021; Chen et al. 2018; Duan 2014; Fu et al. 2015; Liu 2015). Then data fusion is performed, and in this study, Mahony complementary filtering is used to improve the accuracy of the pose solution, which has an improvement for the accuracy only by a single filtering algorithm (Xiong 2017; Xu et al. 2019). Finally, the processed acceleration data are used for quadratic integration to obtain the trajectory (Liu et al. 2009; Li et al. 2017; Qiu & Peng 2016; Qiu et al. 2020; Zheng 2015). The data from the MEMS acceleration sensor is discrete and an iterative integration form is used (Tian et al. 2018), as shown in Eqs. (1) and (2),

$$v(t) = v(t-1) + \frac{a(t)+a(t-1)}{2}\Delta t \tag{1}$$

$$s(t) = s(t-1) + v(t-1)\Delta t + \frac{1}{4}[a(t) + a(t-1)]\Delta t^2 \tag{2}$$

where Δt is the sampling time interval, $v(t)$ is the instantaneous velocity at moment t, $a(t)$ is the instantaneous acceleration at moment t, and $s(t)$ is the displacement from 0 to t moments.

(a) (b)

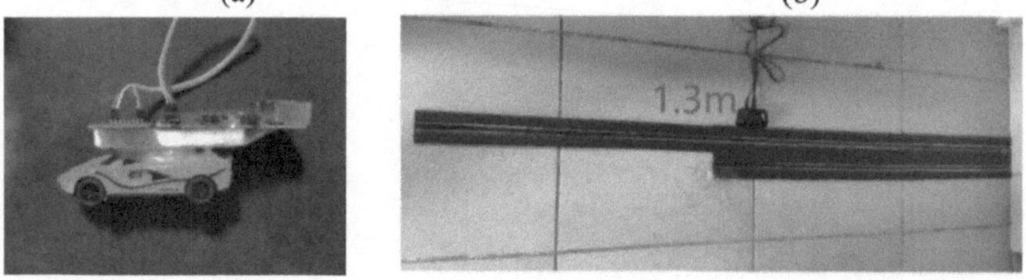

Figure 3. The test car equipped with intelligent sensing equipment and a moving track.

The displacement trajectory obtained after integration is shown in Figure 5, in which the solid line is the trajectory obtained by the intelligent sensing equipment and the dashed line is the theoretical trajectory. The origin of the coordinates is the starting point of the trajectory, and the hollow point is the endpoint of the trajectory obtained by the intelligent sensing equipment.

(a) (b)

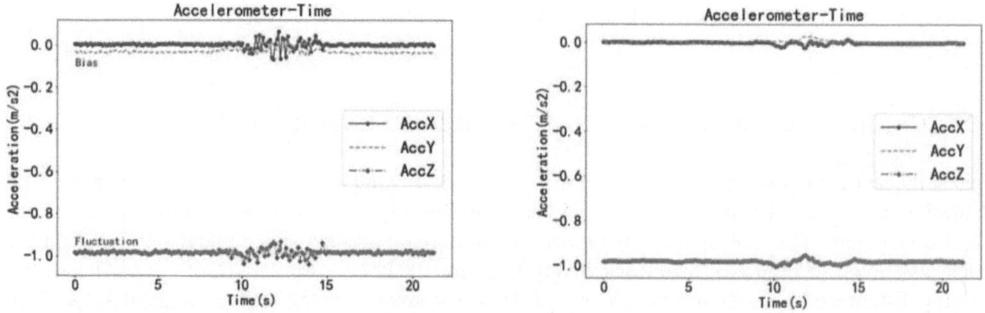

Figure 4. MEMS monitoring data: (a) raw data; (b) data after zero drift correction and Kalman filter processing (take the accelerometer ADXL345 data as an example).

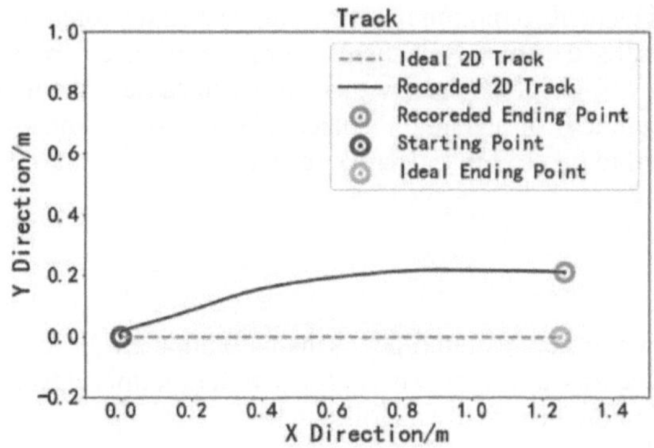

Figure 5. Two-dimensional trajectory diagram.

By repeating the test several times, the trajectory endpoints obtained from each test are recorded, and the test results are illustrated in Figure 6. Each diamond-shaped point represents the end of the trajectory obtained in one experimental test, the dot represents the theoretical end of the trajectory, and the dotted circle represents the 20% error range. As can be seen from Figure 6, the trajectory error range calculated by four different intelligent sensing equipment is large, and the

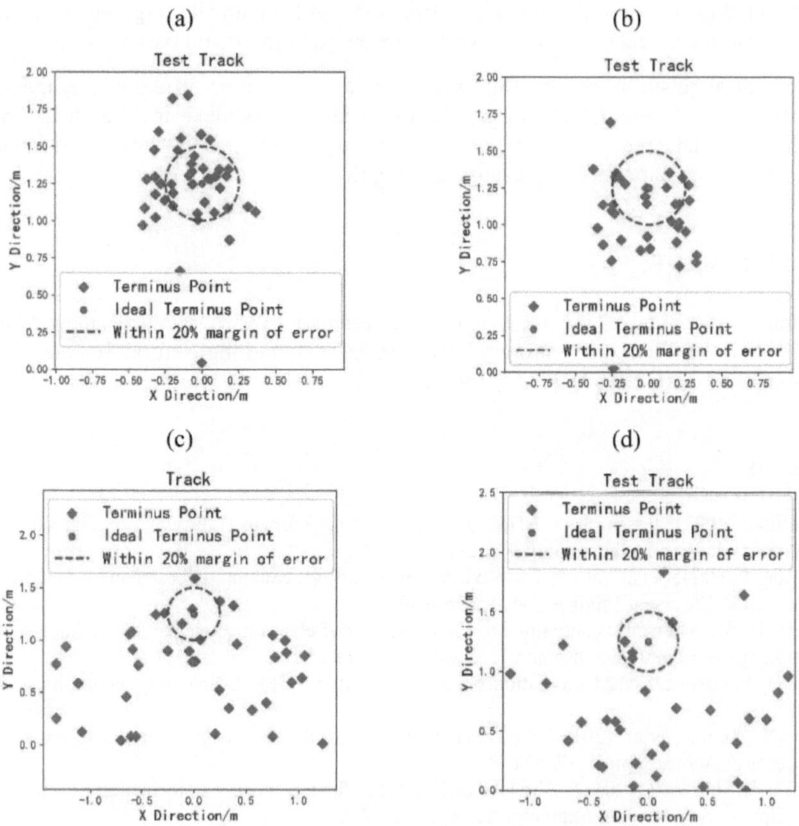

Figure 6. Test results: (a) ADXL345; (b) BMA280; (c) MPU6050; (d) MPU9150.

trajectory error obtained by using a motion tracker is larger than that obtained by an accelerometer, which is presumed to be caused by the insufficient accuracy of the accelerometer built-in MPU6050 and MPU9150. The four MEMS chips used in this study can capture the moving phenomenon but have insufficient test accuracy for absolute displacement measurement. Maybe higher precision MEMS chips are required to be used in trajectory recovery.

5 CONCLUSION

This study designs and manufactures intelligent sensing equipment based on MEMS technology, and realizes a low-cost, accurate and applicable deformation monitoring technology in a wide range of applications. The main conclusions are as follows.

(1) Based on MEMS, an intelligent sensing device comprises a sensing module and a control module, in which four different MEMS sensors, namely accelerometer AdiADXL345, accelerometer BoschBMA280, motion tracking device InvenSenseMPU6050 and motion tracking device InvenSenseMPU9150, are used to make the sensing module, and Raspberry Pi is used as the control module. ABS waterproof box is used for package design to ensure that smart sensing equipment can work normally in harsh environments. The solar power supply module is added to enable the intelligent sensing device to work for a long time.
(2) The related programs are completed which can realize intelligent automatic monitoring including the data conversion, data transmission, real-time remote monitoring, and judgment of whether the monitoring target is sliding, etc.
(3) The collected data is optimized by bias correction and Kalman filtering, which can make the data closer to the actual situation to a certain extent for subsequent data analysis.

Displacement acquisition experiments were carried out on four different intelligent sensing equipment produced. It is found that the smart sensing devices can make an accurate determination of the dynamic and static motion state of the monitored target, but can obtain measured displacement within the error range remains to be discussed in depth.

ACKNOWLEDGMENTS

The research was supported by the Science and Technology Program of Guangzhou, China (201904010415), the Guangdong project (2017ZT07Z066), and the Natural Science Foundation of Guangdong Province (2018A0303130154).

REFERENCES

Cai, Y., & Hu, J. (2021). Research on AHRS system based on Kalman filter and Complementary filter. J. Computer Knowledge and Technology. 17, 230–232.

Chen, G., Zhou, P., Hu, Z., et al. (2018). Four-axis hardware attitude calculation based on MPU6050. J. Journal of Mechanical & Electrical Engineering. 35, 95–100.

Duan, D. (2014). A master thesis submitted to the university of electronic science and technology of China. D. University of Electronic Science and Technology of China.

Fu, W. (2008). Landslide hazard evaluation based on GIS and SVM. J. Scientia Geographica Sinica. 28, 838–841.

Fu, Z., Zhu, H., Sun, J., et al. (2015). Study on filtering algorithm based on inertial sensors MPU6050. J. Piezoelectrics & Acoustooptics. 37, 821–825+829.

He, B., JI, Y., & SHEN, R. (2013). Wireless inclinometer for monitoring deformation of an underground tunnel. J. Optics and Precision Engineering. 21, 1464–1471.

Li, B., Xu, Z., & Huang, C. (2017). Research on human posture detection system based on MEMS inertial Sensor. D. Nanjing University of Science and Technology.

Li, L., Zhang, X., Deng, H., et al. (2020). Assessment of landslide susceptibility based on SVM-LR Model: A case of Shanyang county. J. Science Technology and Engineering. 20, 10618–10625.
Liang, S., Hu, Y., Wang, K., et al. (2010). Design of an early-warning system based on a wireless sensor network for a landslide. J. Chinese Journal of Sensors and Actuators. 23, 1184–1188.
Liu, B., Hu, S., & Song, Q. (2009). A new3D-tracking system based on an accelerometer. J. Journal of Huazhong University of Science and Technology (Natural Science Edition). 37, 40–43.
Liu, C., Yin, X., & Zhang, B. (2015). Analysis and prediction of landslide deformations based on data fusion technology of Kalman filter. J. The Chinese Journal of Geological Hazard and Control. 26, 30–35+42.
Liu, Y., Ju, W., & Liu, Y. (2010). Research progress of gauge check technology of MEMS accelerometers. J. Metrology & Measurement Technology. 30, 5–8.
Qiu, H., She, S., Lin, L., et al. (2020). Posture detection and trajectory tracking system for a two-dimensional moving target. J. Industrial Technology Innovation. 07, 63–68.
Qiu, L., & Peng, S. (2016). Three-dimensional space trajectory extraction method based mems sensor. J. Computer Applications and Software. 33, 292–295.
Tian, H., Sun, Y., & Liu, H. (2018). 3D motion trajectory recovery based on improved Mahony complementary filtering algorithm. J. Transducer and Microsystem Technologies. 37, 118–121.
Tian, W., & Li, Q. (2021). Application of the entropy method and the GWO-SVM coupling model in landslide warning. J. Journal of the China University of Metrology. 32, 253–259.
Xia, H., Yin, K., Liang, et al. (2018). Landslide susceptibility assessment based on SVM-ANN models: A case study for Wushan county in the three gorges reservoir. J. The Chinese Journal of Geological Hazard and Control. 29, 13–19.
Xiong, B. (2017). Research on the modeling and correction technology of random drift error of low-cost MEMS gyroscope. D. Southwest University.
Xu, E., Lu, W., Liu, Y., et al. (2019). Research on attitude calculation and application based on Mahony filter. J. Intelligent Computer and Applications. 9, 80–83.
Yin, H., Zhang, W., & Yuan, L. (2014). An error analysis and calibration method of MEMS accelerometer. J. Chinese Journal of Sensors and Actuators. 27. 866–869.
Zhang, W., Zhang, P., & Zhai, Z. (2016). Research on an accelerometer calibration method based on the six-posture model. J. Transducer and Microsystem Technologies. 35, 37–39+48.
Zheng, S. (2015). The detection of attitude angle and displacement based on MEMS inertial sensors. D. Jinan University.

Advances in Mineral Resources, Geotechnology and Geological Exploration – Rashid & Zhang (eds)

Measuring discharge of the Yangtze river based on coastal acoustic tomography

Z. Zhang, H. Zheng*, Y.F. Tang & Z.W. Chen
Marine Acoustics and Remote Sensing Laboratory, Zhejiang Ocean University, Zhoushan, Zhejiang, China

ABSTRACT: In recent years, coastal acoustic tomography technology has been widely used for measuring the discharge of river channels and offshore harbors in the world. Compared with conventional measurement methods, it has the advantage of continuous measurements without interference from complex environments. In this experiment, three stations were set up on both sides of the riverbank near the Datong Hydrological Station for cumulative flow velocity monitoring of two survey lines. According to the principle of reciprocal propagation of acoustic signals, the flow velocity of the survey line is calculated, the flow velocity of the river is obtained by the method of reconstructing the vector based on projection, and the cross-sectional flow velocities were calculated according to the angular relationship between the river flow direction and the cross-section. The measured flow velocity of the river was around 1.5m/s during the experiment period, and the error of the cross-sectional discharge calculation was kept within 10% when compared with the data published by the Yangtze River Hydrographic Network. The results demonstrate that the new method is more effective than the method of inverting a single survey line and provides better guidance for application to subsequent river flow monitoring.

1 INTRODUCTION

The Datong Hydrographic Station is located in the lower reaches of the Yangtze River Basin in Chizhou City, Anhui Province, the upper boundary of the rivers accessible by the rising tides of the East China Sea. As a hydrographic station with long-term observations in the lowermost reaches of the Yangtze River (Zhang et al. 2006), the monthly average discharge of the Datong hydrographic station is close to the actual sea inflow measured at the Xuliujing hydrographic station, so it can basically represent the inlet flow (Chen et al. 2011). By measuring the river discharge of Datong Hydrological Station, continuous monitoring of the sea inflow of the Yangtze River can be realized, which in turn enables the study of the physic-chemical and biological characteristics of the basin, providing a valuable reference for flow monitoring and estuary management in the estuarine coastal region.

Ocean Acoustic Tomography (OAT) is a large-scale marine environment detection technology, which was first proposed by American scientists Munk and Wunsch (1979). Thanks to the unique propagation characteristics of sound in seawater, the acoustic tomography technology allows the monitoring of complex horizontal and vertical flow field distributions over large areas with only a few acoustic stations. Later Coastal Acoustic Tomography (Zheng et al. 1997) was developed by scholars at Hiroshima University in Japan for small-scale temperature and current measurements in the offshore area, which is more suitable for monitoring shallow marine environments such as harbors and channels than traditional anchoring and shipboard current measurement methods in terms of accuracy and implementability (Liao et al. 2010). The accuracy of measurement is higher

*Corresponding Author: seahzheng@msn.com

DOI 10.1201/9781003319412-54

than that of OAT. In recent years, CAT technology has been widely used in the coastal areas, and many scholars in China have achieved many results in the research of acoustic tomograph positive inversion methods (Liao et al. 2008; Lin et al. 2005; Qiu et al. 2001).

Currently the commonly used methods of measuring river discharge include current meter or Acoustic Doppler Current Profiler walk-through measurements, whose frequent sampling at multiple points results in the inability to obtain real-time flow velocity at the cross-section (Zhang et al. 2021), and bad weather conditions can affect the measurement. Coastal acoustic tomography solves these problems with the advantage of fixed-point measurements, and the accuracy of the data collected in real-time is guaranteed. In addition, this study differs from the method of inversion of single survey line flow velocities to derive the cross-sectional discharge, by using two survey lines of flow velocities as the projection reconstruction vector method to calculate the cross-sectional discharge, based on the comparison of the two results, the latter data error is smaller, and can be used as a new method for future river discharge measurement.

2 MATERIALS AND METHODS

2.1 *Experiment introduction*

Coastal acoustic tomography experiments are based on the principle of reciprocal propagation of acoustic signals and require two or more stations on both sides of the experimental area to be measured, the greater the number of stations, the greater the accuracy of the data results. The transmission and reception of acoustic signals are controlled by the CAT system. The built-in GPS clock unit in the system ensures the synchronization of the time of the transmission and receipt of the signal between the two stations, the speed of sound is taken as 1500 m / s. The velocity value of the survey line between the two stations can be obtained from the reciprocal time difference and the propagation distance of the acoustic signals at the two stations. The method of reconstructing the vector by multiple projections based on the flow velocity of multiple survey lines can obtain the continuous flow velocity of the river and enables cross-sectional discharge monitoring.

The experimental equipment used in this experiment was developed by Hiroshima University in Japan and has been used in coastal experiments many times with high stability and accuracy. The schematic diagram of the experimental system arrangement is shown in Figure 1. The CAT system is arranged near the bank and connected to a 12V battery to maintain operation. The transducer responsible for transceiving is placed in the experimental water via a cable, and the transducer is powered by two 12V batteries connected in series. The transmitting power of the transducer is 50 kHz. To improve the experimental accuracy, an external GPS clock module is used (Kawanisi et al. 2010) and the internal clock is used as a backup.

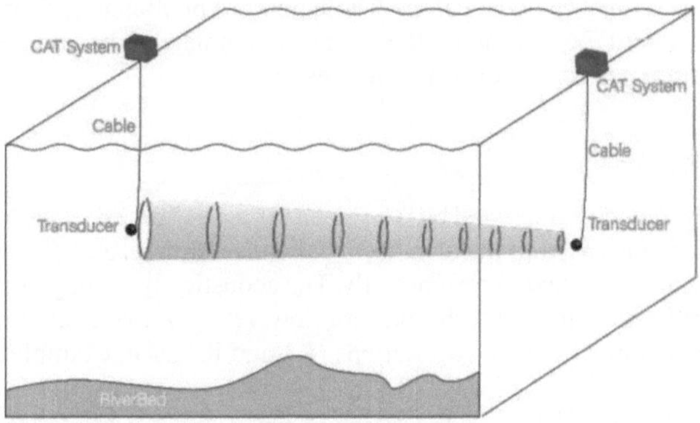

Figure 1. Schematic diagram of CAT system layout.

The area of this experiment was the Yangtze River in Meizhou Town, Chizhou City, Anhui Province, near the Datong Hydrographic Station. The experimental team set up three measurements on both sides of the riverbank, as shown in Figure 2. Station K1 is placed on a small fixed boat on the north side of the riverbank, Station K2 is located at the tourist pier on the south bank of the Yangtze, while K3 is placed on the right side of K2 by the Maritime Authority pier. The distance between K1 and K2 is 2628m, and the distance between K1 and K3 is 2414 m. The location information of each site is shown in Table 1.

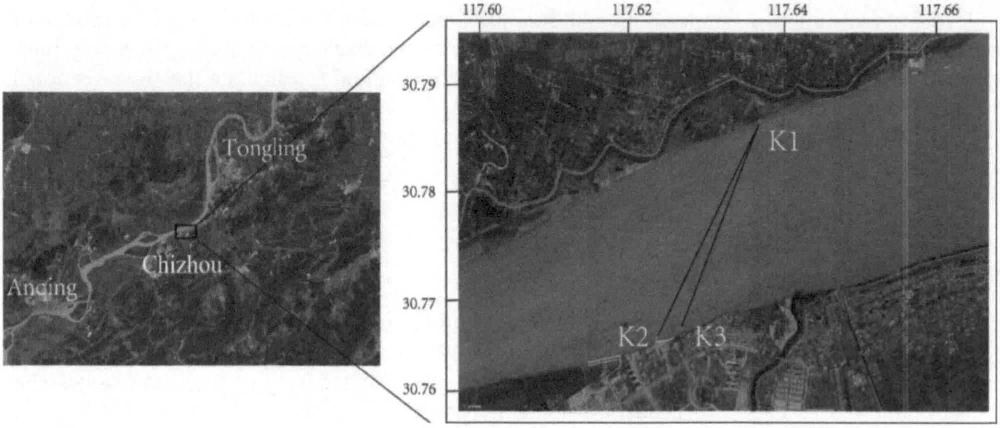

Figure 2. Site location of the experimental area.

Table 1. GPS coordinates of the experimental site.

Stations	Latitude (°′)	Longitude (°′)	Latitude (°)	Longitude (°)
K1	30 47.2022E	117 38.2051 N	30.7867E	117.6368N
K2	30 45.9686E	117 37.3961 N	30.7661E	117.6233N
K3	30 46.0108E	117 37.6019 N	30.7668E	117.6267N

The experiment is carried out on 19 July 2019, with a test phase from 9:57 to 10:30. The experiment officially started at 11:09 after confirming that the test results were correct, and ended at 13:36 for a total of about two and a half hours, with signals being sent and received every three minutes and data being collected more completely. In addition, the official gauging data (the water level and cross-sectional flow data at the whole point of publication) from the Yangtze River hydrographic network about the Datong hydrographic station need to be obtained instantly during the experiment as a reference for the experimental results.

2.2 *Measurement method*

The calculation of flow velocity is derived by relying on the time difference between the signals sent and received at the two stations, and the accuracy of the inversion results depends on whether the arrival time of the signals can be extracted correctly. The acoustic ray propagation time tomography method is a mature and efficient method for current flow velocity inversion (Zhu et al. 2002). The signal reciprocity propagation time for two stations (K1 and K2 as an example) is expressed as:

$$\begin{cases} t_1 = \oint_{\Gamma} \frac{ds}{c+u} \approx \frac{L}{c+u} \\ t_2 = \oint_{\Gamma} \frac{ds}{c-u} \approx \frac{L}{c-u} \end{cases} \tag{1}$$

Where c is the speed of sound; u is the velocity component of the river flow on the survey line; t_1 and t_2 are the times from K2 to K1 and K1 to K2. Then the flow component u can be expressed as:

$$u = \frac{L}{2}\left(\frac{1}{t_1} - \frac{1}{t_2}\right) \tag{2}$$

The horizontal rectangular coordinate system is established with the due east direction as the positive direction of the x-axis, then the river runoff velocity v and the survey line velocity u are expressed as:

$$v = (v_x, v_y) = (v\cos\theta_v, v\sin\theta_v) \tag{3}$$

$$\vec{u} = u \bullet \vec{r_u} = u(\cos\theta_u, \sin\theta_u) \tag{4}$$

where θ_v is the angle between the river flow direction and the x-axis, and θ_u is the angle between the survey line and the x-axis. The following geometric relationship is commonly used for inversion when measuring river flow velocities using coastal acoustic tomography (Bahreinimotlagh et al. 2016):

$$v = \frac{u}{\cos(\theta_v - \theta_u)} \tag{5}$$

The method of vector reconstruction is based on the relationship between the two survey lines and the flow direction of the river:

$$\begin{cases} \vec{v} \bullet \vec{r_{u_1}} = u_1 \\ \vec{v} \bullet \vec{r_{u_2}} = u_2 \end{cases} \tag{6}$$

Converting the above formula to matrix form, the river flow velocity by generalized inverse solution is obtained:

$$\begin{bmatrix} \cos\theta_1 & \sin\theta_1 \\ \cos\theta_2 & \sin\theta_2 \end{bmatrix}\begin{bmatrix} v_x \\ v_y \end{bmatrix} = \begin{bmatrix} u_1 \\ u_2 \end{bmatrix} \tag{7}$$

Where θ_1 and θ_2 are the angles between the two lines and the river flow direction respectively, we can get $\vec{v} = (v_x, v_y)$. And the cross-sectional flow is expressed by the following equation (Kawanisi et al., 2013):

$$Q = A(H)v\sin\theta \tag{8}$$

where $A(H)$ is the cross-sectional area, and θ is the angle between the flow direction and the section of the river.

3 RESULT & DISCUSSION

3.1 *Exact the arrival time of the acoustic signal*

When the station starts transmitting signals after a short delay into the reception state, the signal received by the station is the serial number of the M10 sequence. By autocorrelating the received raw signal with the M10 sequence signal used in the transmission, it is possible to obtain a three-dimensional histogram as shown in Figure 4, where the horizontal axis is the propagation time, the vertical axis is the time of the signal transmission and the vertical axis on the left is the magnitude of SNR. Where the left-hand side of the group (a) indicates the arrival time of the signal received by K1 transmitting K2 and the right-hand side is the arrival time of the signal received by station K2 transmitting K1, and the same for the group (b).

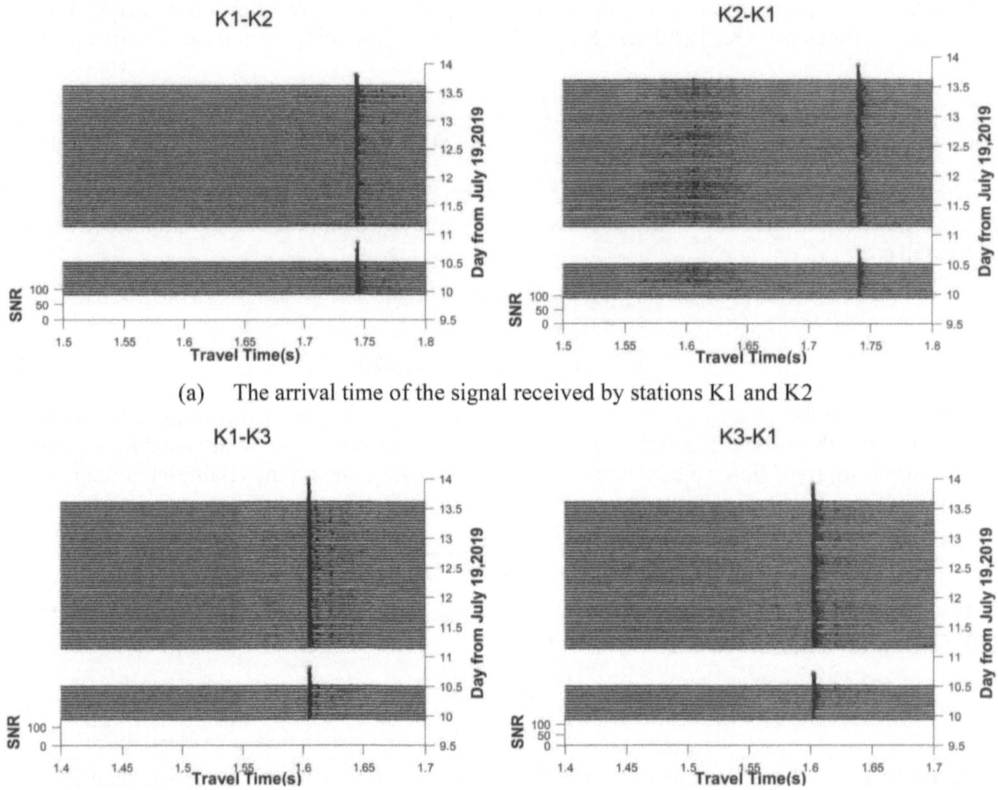

(a) The arrival time of the signal received by stations K1 and K2

(b) The arrival time of the signal received by stations K1 and K3

Figure 3. The arrival time of the acoustic signal received by stations.

The selected arrival times are annotated with red dots in the diagram and were chosen to correspond to the time of arrival of the acoustic signal at SNR > 12. It can be observed that the arrival times of the line signals by stations K1 and K2 are roughly between 1.74s and 1.75s, and the arrival times of the line signals received by stations K1 and K3 are roughly between 1.60s and 1.61s. The time points marked in red on the left and right graphs of each of the two groups (a) and (b) remain basically the same. Slight left-right fluctuations may be due to ambient noise or flow velocity perturbations to the sound propagation, but do not affect the processing of subsequent results.

3.2 *Calculation of flow velocity*

Figure 5 shows the flow velocity of the two survey lines, of which the left side is the flow velocity of the K1 to K2 stations, and the right side is the flow velocity of the K1 to K3 stations. The blue line is the flow velocity of the original three-minute data inversion. Considering that there are many disturbances in the process of acoustic signal transmission, including complex environmental noises such as the traffic of vessels and uncertainties such as abnormal system operation, the data was optimized by 30-minute moving average processing and normal distribution to eliminate outliers, while the red line is the optimized data, it can be found that the overall performance of the two graphs is good, except for some individual time periods with up and down fluctuations. The flow velocity in both lines remained between 1.1 and 1.3m/s, with flow velocity data more stable in lines K1 to K3 than in the other line.

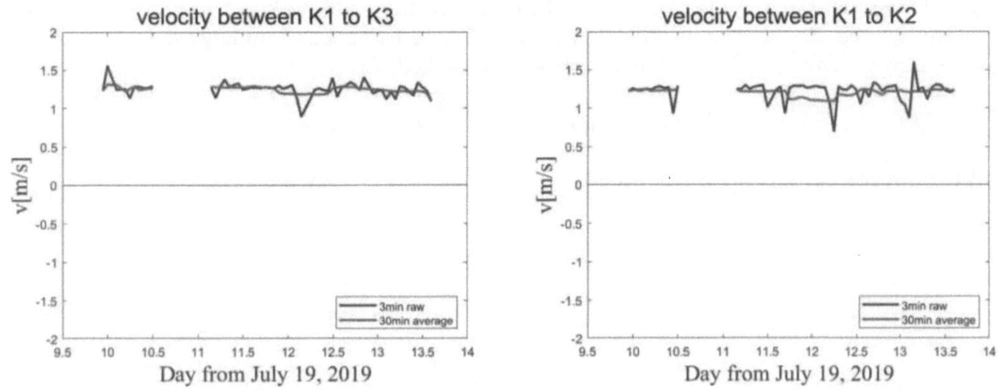

Figure 4. The flow velocity of the survey line.

After reconstructing the vector based on the velocity components of the two survey lines, Figure 6 shows the flow velocities at the Datong hydrographic station during the experiment. The blue line indicates that the river flow velocity is roughly stable at around 1.5m/s. The flow velocity of the survey line K1 to K3 is more biased towards the river flow than the other lateral line, and therefore the value is higher than the latter.

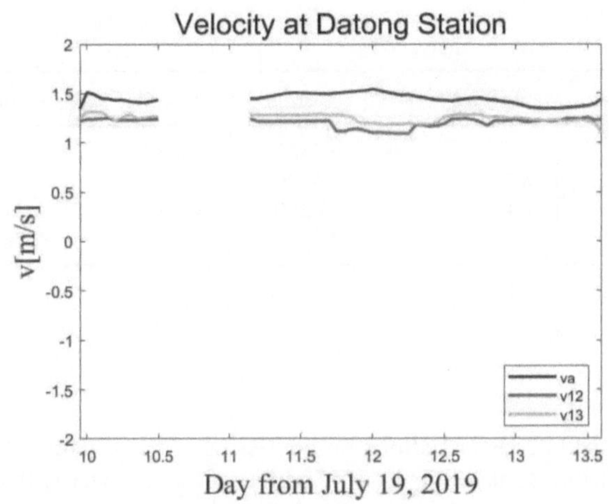

Figure 5. Real-time flow rates at Datong hydrographic station.

3.3 *Calculation of cross-sectional flow*

The calculation of the cross-sectional flow requires the cross-section velocity v and the cross-section area S, while the cross-sectional flow velocity can be obtained by converting the river flow direction to the angle of the cross-section, and the cross-sectional area $S = 43000\ m^2$, based on the water level of 14.4m.

The inversion of the cross-sectional flow at the Datong hydrographic station is shown in Figure 7, where the red and green lines represent the cross-sectional discharge of the two survey lines according to their respective geometric relationships, while the blue line represents the discharge after vector reconstruction. The dots in the figure indicate the results of the whole point discharge released by the Yangtze River Hydrographic Official Website, and it can be observed that the discharge after inversion of the two measurement lines is basically the same, but the deviation is slightly larger compared to the result of vector reconstruction, and the specific data comparison is shown in Table 2. The overall error was kept within 10 percent and no large deviations were observed, and the trend of the data curve was well smoothed.

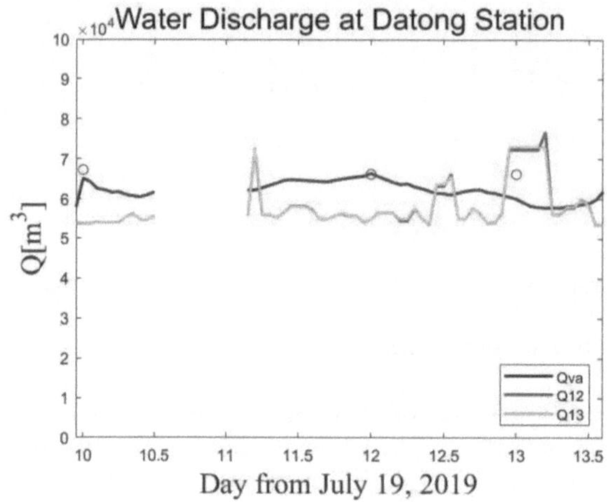

Figure 6. Cross-sectional flow map at Datong Hydrographic Station.

Table 2. Flow error analysis.

Time	Water level (m)	Q(m^3)	Q_{va}(m^3)	error (%)
10	14.46	67300	65095	3.27
11	14.45	66300	——	——
12	14.44	66200	66410	0.31
13	14.44	66200	60052	9.28

4 CONCLUSIONS

By arranging three coastal acoustic stations on both sides of the riverbank near the river cross-section, continuous measurements of the flow velocities of the two survey lines are achieved at an interval of three minutes between each experiment, and then the inversion of the cross-sectional discharge is achieved based on a projection reconstruction vector method. After data analysis and optimization, the method was found to be more effective than single line inversion, and the error could be controlled to within 10 percent to meet the accuracy requirements of future river surveys. Most of the current common river measurement methods in China are walking voyage carrying instrumentation for multiple measurements, which is not only economically costly but also cumbersome in terms of experimental steps. A frequent sampling at multiple points also does not allow accurate real-time flow data to be obtained for the entire cross-section. The method summarized in this paper allows continuous data collection at a fixed location and is easy to experiment with.

The reconstruction vector method used in this experiment involves only two survey lines of data, and it remains to be seen if increasing the number of lines will improve the accuracy of the final data. This experiment is a new attempt at a method for testing flow in coastal acoustic laminar rivers, and further improvements in subsequent experimental refinements and data optimization are needed to accommodate more complex situations.

ACKNOWLEDGMENT

Thanks to Chizhou Maritime Safety Administration for its help in data collection. This work was supported by the National Key Research and Development Program of China (No. 2019YFC1408404).

REFERENCES

Bahreinimotlagh, M. et al. (2016). Application of shallow-water acoustic tomography to measure flow direction and river discharge. Flow Meas. Instrum. 51, 30–39.

Chen, Q. J. et al. (2011). Determination of critical flow for saline water intrusion into water source area of Yangtze River estuary and guarantee measures. Yangtze River. 42 (18), 68–72.

Kawanisi, K. & A. Kaneko (2010). New acoustic system for continuous measurement of river discharge and water temperature. Water Sci. Eng. 3(1), 47–55.

Kawanisi, K. et al. (2013). Continuous monitoring of a dam flush in a shallow river using two crossing ultrasonic transmission lines. Meas. Sci. Technol. 24(5), 055303.

Liao, G. H. et al. (2008). Overview of the applications and observations of ocean acoustic tomography. Progress In Geophysics. 23(6), 1782–1790.

Liao, G. H. et al. (2010). Observation technology and methods of ocean acoustic tomography. Acta Oceanol. Sin. 32(3), 14–22.

Lin, J et al. (2005). Accurate imaging and prediction of Kanmon Strait tidal current structures by the coastal acoustic tomography data. Geophys. Res. Lett. 32(14).

Munk, W. & C. Wunsch (1979). Ocean acoustic tomography: A scheme for large-scale monitoring. Deep-Sea Research Part A. Deep-Sea Res., Part A. 26(2), 123–161.

Qiu, X. F. et al. (2001). Inversion Method for Sound Velocity Profile of Eddy in Deep Ocean. China Ocean Eng. 15(4), 589–596.

Zhang, G. X. et al. (2021). Multi-layered velocity measurement and discharge calculation method of river section based on H-ADCP. Yangtze River. 52(8), 78–83.

Zhang, R. et al. (2006). Analyses with Wavelet and Hilbert-Huang Transform on Monthly Water Discharges at Datong Station, Yangtze River. Journal of Nanjing University (Natural Sciences). 42(4), 423–434.

Zheng, H. et al. (1997). Reciprocal sound transmission experiment for current measurement in the Seto Inland Sea, Japan. J. Oceanogr. 53, 117–128.

Zhu, Y. & L. G. Lu (2002). Basic principle and application of ocean acoustic tomography. Adv. Mar. Sci. 20(4), 70–75.

Advances in Mineral Resources, Geotechnology and Geological Exploration – Rashid & Zhang (eds)
© 2023 Copyright the Author(s), ISBN 978-1-032-33377-9

Intelligent advanced geological prediction model of tunnel based on wavelet neural network

Ming Zhong & Yang Wang*
China Merchants Chongqing Highway Engineering Testing Center Co., Ltd., Chongqing, China

ABSTRACT: Advanced geological forecasting is indispensable for tunnel construction and plays a significant role in tunnel informatization construction, disaster prevention, and safety assurance. With continuous development of high-grade highways and railway tunnels, the need for accuracy and real-time performance of advanced geological forecasting is high. To adapt to the new situations arising during highway tunnel development, improve the level of informatization and intelligence of advanced geological forecasting, and fully learn from past experience and lessons, on the basis of an extensive collection of various advanced forecasting methods and geological disaster forecasting methods, a new method is developed. A set of auxiliary decision-making systems for advanced geological prediction of tunnel geological hazards is put forth. The advanced forecasting methods commonly used in tunnels are discussed in detail, the data required by each method and the content that can be forecasted are analyzed, and the intelligent management of advanced geological forecasting data is realized on this basis.

1 INTRODUCTION

Tunnel engineering is part of typical geological engineering, and the design, construction, construction period, and cost of the tunnel are all subject to the constraints arising in engineering geological conditions (Liu et al. 2018). Particularly in the process of tunnel construction, all kinds of geological disasters induced by excavation are non-selective, complex, special, and sudden, and often become the most important factor restricting tunnel construction (Gou & Xie 2017). With the construction of a large number of dangerous road projects in the future, geological work during dangerous road construction, especially advanced geological forecast, will be widely popularized as one of the essential processes of dangerous road construction (Xue et al. 2017). Before construction, due to limitations in technical means and investment funds, and the complexity of geological bodies, the geological data obtained are often very rough and limited. This makes it difficult to fulfill the requirements of construction, especially for deep-buried tunnels (Qiang et al. 2017).

Karst areas in my country are widely distributed, and the distribution area of soluble rocks accounts for about one-third of the total land area, among which karsts are most developed in parts of Yunnan, Guizhou, Guihe, Sichuan, Hubei, and Hunan in the southwest (Lenz et al. 2017). With increase in constructions in China due to urbanization, large-scale geological disasters such as water inrush, rock bursts, landslides, gas, and mud gushing have been largely reported during underground projects, such as tunnels, hydropower stations, cross-basin water transfer, and deep mining in strong karst areas. Therefore, many scholars use fuzzy mathematics, analytic hierarchy process, and traditional mathematical model of a neural network to assess the safety risk of karst tunnels to ensure safe construction of tunnels (Li et al. 2016). However, due to the complexity, uncertainty, and nonlinearity of safety risk assessment of karst tunnels, traditional methods for

*Corresponding Author: zxbtianyi@vip.com

 DOI 10.1201/9781003319412-55

safety risk assessment are complicated in operation and low in precision. Cudrigh-Maislinger S (Cudrigh-Maislinger 2018) and others combined artificial neural network theory, wavelet analysis, and fuzzy evaluation method, and established a karst tunnel safety risk assessment model based on a fuzzy wavelet neural network. The fuzzy evaluation method is used to quantify the indicators of risk factors, the output of the fuzzy system is used as the input of the neural network, and the fuzzy wavelet neural network is constructed and trained, which can solve defects of the existing evaluation methods, such as high subjective arbitrariness and fuzzy conclusions, and has higher fitting accuracy and faster convergence than BP neural network.

Currently, there are many geological advanced prediction methods used in tunnel construction, such as TSP seismic detection, geological analysis, geological radar, infrared water detection, advanced horizontal drilling, and more. However, most tunnels have many types of strata, complex geological structures, and uneven surrounding rocks, and different geological advance prediction methods have their own scope of application, advantages, and disadvantages (Maslova et al. 2016). The comprehensive advanced prediction technology of tunnel construction has been put forward, but it is not systematic enough. According to different geological conditions, determining different risk levels, formulating different prediction schemes, and combining with the proximity alarm technology of construction geological disasters can solve the problem of comprehensive advanced prediction technology being unsystematic (Song et al. 2016). To sum up, the research on risk assessment and comprehensive advanced geological prediction technology of karst tunnel based on fuzzy wavelet neural network has certain theoretical and practical significance.

2 OVERVIEW OF COMMON ADVANCED GEOLOGICAL PREDICTION METHODS

2.1 *Geological method*

Advanced geological prediction of tunnel construction has become one of the main contents of tunnel engineering geological work (Djerbouai et al. 2016). In the construction project, various technical methods and means are used to timely and effectively predict the surrounding rock conditions in front of the construction operator to be aware, and take construction countermeasures in advance to avoid the occurrence of geological disasters or reduce disaster losses as much as possible. Providing good conditions for safe construction is the main work content of advanced geological prediction of tunnel construction (Wei et al. 2019). The geological conditions at the tunnel site are complex and different, and the engineering effects to be achieved by the advanced geological forecast during the construction period are also different. This has led to different advanced geological forecasting methods and technical means (Panwar 2017). There are many classification methods for advanced geological forecasting. At this stage, the methods used for forecasting are mainly divided into the following categories: geological method, advanced geological drilling method, geophysical detection method, and a combination of two or more forecasting methods. Advanced geological prediction method, etc. (Spitsyn et al. 2016).

The geological method is the most basic and commonly used method in the advanced geological prediction of tunnel construction. Geological method for tunnel-oriented advanced geological prediction is mainly based on the geological survey results of the surface, tunnel, and tunnel face, that is, according to the occurrence of rock structural plane exposed on the surface and its development and fracture status, weathering and metamorphism degree of rock and other characteristics, through geological mapping and structural correlation analysis. Based on the above characteristics, the exposed position of the rock mass structural plane at the tunnel crossing depth is inferred, and the possible geological structure in front of the tunnel is inferred according to the occurrence, development, and fragmentation of the rock mass structural plane exposed by the excavation of the tunnel face, the degree of weathering, and metamorphism of the rock (Weifang S et al. 2017). The process of advanced geological prediction by the geological method is shown in Figure 1. According to the geological survey inside and outside the tunnel and the geological sketch of the construction face, through the statistical analysis of joints and fissures, structural correlation analysis, geological mapping of the face and necessary tests, the lithology, deformation trend, the possibility, and form of instability and failure, the location, scale, and development of the possible fracture zone in front of the face are inferred. The strength characteristics of surrounding rock mass and the distribution of in situ stress are the basic principles and main working methods of geological prediction in tunnel construction.

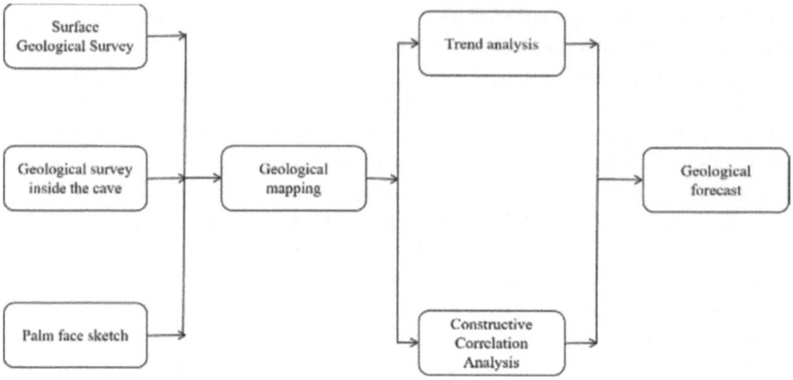

Figure 1. Flow chart of the geological method for tunnel-oriented advanced geological prediction.

The surface survey method is to conduct extensive and meticulous geological surveys at dangerous sites, review geological survey data provided by the survey and design unit, and master the surrounding rock lithology and its occurrence, geological structure, hydrogeology, and the possible existence of the tunnel site. The unfavorable geological conditions of the tunnel can be used to provide directional guidance for the advanced geological forecasting work in the tunnel. This qualitative prediction method is based on the abovementioned geological survey data, combined with the surrounding rock lithology and geological structure in the dangerous site area obtained from the surface survey, and the geological survey provided by the survey and design unit through the geological mapping method. Data are reviewed, improved, and supplemented. The scope and content of the investigation focus on the strata lithology and its occurrence characteristics, the development of joints and fissures, the location, shape, scale, and trend of possibly unfavorable geological bodies within the entrance and exit of the tunnel and on both sides of the midline. Surface surveys can be divided into two categories: key surveys and comprehensive surveys, according to the degree of conformity between the previous geophysical work and the actual geological conditions. The key survey is suitable for tunnels where the geological survey work is in good agreement with the actual situation, such as major dangerous road projects of long trunk lines. The latter is suitable for tunnels where the geological survey work is poorly done, and the design drawings and the geological conditions revealed by the construction are not well or seriously inconsistent fall in the category to be fully investigated.

2.2 *Comprehensive geological advance prediction method*

Due to the complexity of engineering geology and hydrogeology and the limitations of a single advanced prediction method, using a certain advanced geological prediction method is far from achieving good prediction results. The applicability and economy should be considered in the selection of advanced prediction methods.

The comprehensive advanced geological prediction method that combines two or more prediction methods overcomes the limitations of a single method, improves the prediction accuracy, and achieves good engineering results. It has become the main method used in advanced geological prediction. The common combination methods of advanced geological prediction are: (1) the combination of geological method and geophysical method; (2) the combination of advanced geological drilling and geophysical prospecting methods; (3) the combination of geophysical exploration methods, etc.

(1) Combination of geological method and geophysical method

The common combination methods of geological method and geophysical method are: (1) the combination of geological survey inside and outside the tunnel, tunnel face sketch, and geological radar method; (2) geological survey inside and outside the tunnel, a combination of face sketch and method, etc.

Geological methods, such as geological surveys inside and outside the tunnel and tunnel face sketches, require forecasters that yield rich geological work experience. Through detailed geological survey inside and outside the tunnel and tunnel face recording, they help determine faults, rock stratum occurrence, fracture zone, and other conditions not identified in the original survey data, and obtain the fault position and scale in the dangerous tunnel through structural plane graphic methods, such as stereographic projection. However, due to the unstable occurrence of faults and fracture zones, it is difficult for forecasters to determine their representative occurrence, which affects the positioning on the dangerous road, and the method has a number of limitations. The detection results of geophysical methods, such as geological radar method and geological method, generally have multiple solutions. Forecasters should combine their rich experience and geological knowledge to eliminate interference and make accurate prediction.

The combination of these two methods is based on the geological survey to assist the interpretation of geophysical exploration results, avoid the blindness of wave reflection method in judging interface properties, give full play to the advantage of geophysical exploration method in long detection distance, and improve the accuracy of prediction. At present, the comprehensive advanced geological prediction method of dangerous roads based on geological method and geophysical method has become the consensus of tunnel engineering circles and has been widely recognized and popularized.

(2) Combination of advanced geological drilling and geophysical prospecting methods

Advanced geological drilling is time-consuming and labor-intensive, and the combination of advanced geological drilling and geophysical prospecting methods generally has its specific situation. When the geophysical prospecting method cannot accurately determine whether there is an unfavorable geological body in front of the tunnel face, or cannot determine what kind of unfavorable geological body it is, the direct method can be used for advanced geological drilling verification, which can improve the understanding of the unknown surrounding rock in front and avoid construction risks.

(3) The combination of geophysical methods and geophysical methods

The principles of various geophysical prospecting methods are different, and their wave reflection degree to geological bodies is also different, each of which has its own usability. Common combination methods include the combination of ground penetrating radar (GPR) method and infrared water detection method, the combination of infrared water detection method and acoustic reflection method and GPR method. Geological radar and the detection effect and resolution of the weak interlayer, fault fissure zone, and karst cave in front of the face are better, but the detection of the water body is far from infrared water exploration. The infrared water detection method is sensitive to water-bearing structures, but cannot detect other unfavorable geological bodies. The combination of different geophysical prospecting methods can effectively exert their respective advantages and complement each other.

3 NEURAL NETWORK PREDICTION MODEL

3.1 *Fuzzy neural network comprehensive prediction*

A fuzzy system and a neural network system have different characteristics. The neural network has strong adaptive learning ability to the change of environment, but its learning mode is a typical black box type. When learning is complete, the input and output obtained by the neural network cannot be expressed in an easily accepted way. The modeling process of a fuzzy system is easy to be accepted, but the establishment of membership function and fuzzy rules of a fuzzy system is a subjective process. Fuzzy Neural Network (FNN) combines the advantages of neural networks and fuzzy systems to make up for their respective shortcomings. The FNN formed by the combination of the two has the advantage that fuzzy logic is easy to express human knowledge and distributed information storage and learning ability of the neural network. Common fusion forms of fuzzy systems and neural networks include loose combination, parallel combination, series combination, network learning combination, and more. In this paper, the series combination is adopted, that is, the fuzzy system and the neural network are connected in series in the system, that is, the output of the fuzzy system becomes the input of the neural network, which makes the network converge more easily during the learning process. Series FNN is adopted, and the neural network is a three-layer BP network. The first layer of the network is preceded by fuzzy quantitative evaluation, and an appropriate membership function is constructed. After the third layer of the network is output, fuzzy restoration is carried out. The overall structure of the fuzzy neural network is shown in Figure 2. Taking the fuzzy neural network comprehensive prediction of water-rich rock mass and broken rock mass as an example, the network model is illustrated:

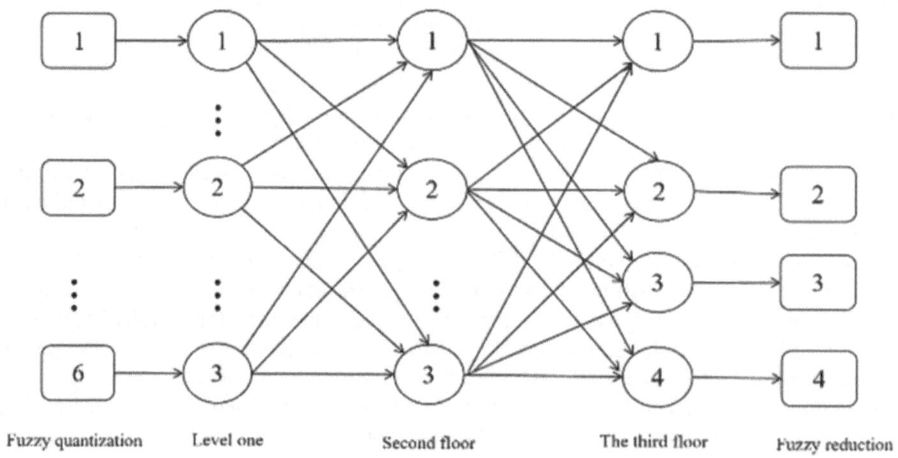

Figure 2. Structure diagram of fuzzy neural network.

(1) Fuzzy neural network model for water-rich rock mass

Input layer: By constructing an appropriate membership function, the membership degree of each prediction index of the water-rich rock mass can be calculated as the input node of the neural network. There are six prediction indexes for water-rich rock mass, corresponding to four kinds of the water-rich rock mass. The total number of neurons in the network input layer is $6 \times 4 = 24$.

Hidden layer: The number of nodes in the hidden layer does not strictly stipulate in theory. Generally, when the number of hidden neurons is small, the learning time of the network each time is short, but the network may not remember all the learning contents due to insufficient learning. The more the number of hidden layer neurons is, the more accurate the network will be and the longer the training time will be. However, it is not easy to select too many hidden layer neurons, otherwise, the network storage capacity will increase, resulting in the decline of the induction ability of the network to unknown inputs and the anti-noise ability of the network. Through empirical calculation, it is appropriate to set the number of neurons in the hidden layer between 10 and 16.

Output layer: The result of the output layer of the network is a fuzzy amount, and it is difficult to adjust the network weights, so the results of the output layer are quantified, and finally fuzzy restoration is performed. The output layer is set with four neurons, and the corresponding output results are a large number (1000), a medium (0100), a small amount (0010), and a dry (0001).

(2) Fuzzy neural network model of broken rock mass

Input layer: By constructing an appropriate membership function, the membership degree of each prediction index of the broken rock mass is calculated and used as the input node of the neural network. There are five prediction indicators of broken rock mass, corresponding to four different result situations (complete, relatively complete, relatively broken, broken), and the total number of neurons in the input layer of the network is $5 \times 4 = 20$.

Hidden layer: The number of layers and neurons in the hidden layer is basically the same as that of the FNN in the water-rich condition of the rock mass.

Output layer: Quantify the results of the output layer and restore them by blur. There are four neurons in the output layer, and the corresponding output results are: complete (1000), more complete (0100), more broken (0010), and broken (0001).

3.2 *BP wavelet neural network prediction model*

The basis of a wavelet neural network is wavelet analysis. When combining wavelet analysis and neural network to construct a wavelet neural network, a set of wavelet function systems are obtained by translating and expanding the wavelet decomposition, which can be used to represent or approximate a certain wavelet function system. Moreover, since the wavelet neural network introduces translation and scaling factors, the wavelet neural network has better function approximation ability, which provides the basis for us to try to predict the time series of monthly average temperature changes. The structure of the wavelet neural network is shown in Figure 3.

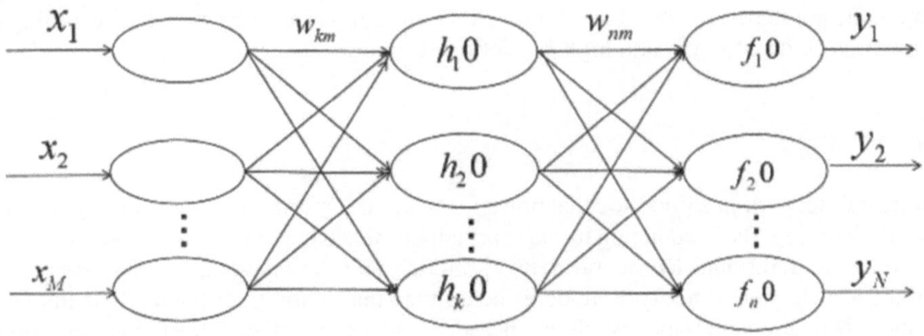

Figure 3. Structure of wavelet neural network.

The establishment of the BP wavelet neural network prediction model mainly includes the following three steps:

Step 1: Initialize the network parameters. Determine the number of input nodes $M = 5$, output nodes $N = 1$, neurons $n = 8$, training error $e_{goal} = 0.001$, momentum factor $\mu = 0, \gamma = 0.2, L = 1.15$, and learning rate $1r_1 = 1r_2 = 0.3$. Assign the wavelet expansion factor a, translation factor b, and network weight $\omega_{i,j}, \omega_{j,k}$ to initial values.

Step 2: Calculate the gradient vector. The gradient vector is calculated as

$$\frac{\partial E}{\partial \omega_{i,j}} = \sum_{p=1}^{P} (d_2^p - y_i^p)\Psi_{a,b}(net_j^p) \tag{1}$$

$$\frac{\partial E}{\partial \omega_{j,k}} = -\sum_{p=1}^{P}\sum_{i=1}^{N} (d_i^p - y_i^p)\, \omega_{i,j}\Psi'_{a,b}(net_j^p)x_k^p / a_j \tag{2}$$

$$\frac{\partial E}{\partial b_j} = \sum_{p=1}^{P}\sum_{i=1}^{N} [(d_i^p - y_i^p)\, \omega_{i,j}\Psi'_{a,b}(net_j^p) / a_j \tag{3}$$

To control the training effect, the network output error after each cycle is expressed by the following formula, and the relative error E is calculated, i.e.,

$$E = \sum_{i=1}^{N} \left(s(t) - \bar{s}(t) \right) / \sum_{i=1}^{N} s(t)^2 \tag{4}$$

Step 3, when the relative error E is greater than e_{goal}, the learning rate is corrected by $\eta = 1 \times \eta_{i-1}$; Otherwise, $\eta = g \times \eta_{i-1}$ is used to modify the learning rate. Then, the network parameters are modified, and the calculation formula is

$$a_j(t+1) = a_j(t) - \eta \frac{\partial E}{\partial a_j} + \eta \Delta a_j(t) \tag{5}$$

$$b_j(t+1) = b_j(t) - \eta \frac{\partial E}{\partial b_j} + \eta \Delta b_j(t) \tag{6}$$

To prevent the program from being unstable due to the random assignment of some network parameters in step (2), when the relative error E is less than e_{goal}, it is needed to assign the value of the network parameter at this time to the network parameter corresponding to step (2) and go to step (3).

Since previous prediction results are added to the prediction samples to predict the later deformation data, the prediction with the prediction data will lead to error superposition and a large deviation of the prediction results. Therefore, prediction samples must be adjusted according to the on-site monitoring data to maintain high prediction accuracy.

4 CONCLUSIONS

Wavelet neural network has good localization of wavelet transform and the advantages of neural network self-learning. By introducing its basic principles, this paper compares its prediction results with those of the traditional BP neural network model, and the results show that the prediction effect of the wavelet neural network model is better than that of the traditional BP neural network model, and it is in good agreement with the measured values, which can meet the requirements of engineering and control. Therefore, the wavelet neural network model can be used for advanced geological prediction of high-speed railway tunnels, providing a reliable theoretical basis for the safe construction and normal operation of tunnels in the future. For constructors, the advanced geological detection technology in tunnel construction is expected to be better applied. Constructors should combine the structural characteristics of tunnel engineering, adopt reasonable detection technology, continuously improve the accuracy of detection results, and promote the better development of tunnel engineering in China.

REFERENCES

Cudrigh-Maislinger S 3D geological modelling - Through the example of Karawanken Tunnel project, northern section: Geologische 3D-Modellierung am Beispiel des Karawankentunnels, Abschnitt Nord. Geomechanik und Tunnelbau, vol. 11, no. 5, pp. 530–536, 2018.

Djerbouai S, Souag-Gamane D Drought Forecasting Using Neural Networks, Wavelet Neural Networks, and Stochastic Models: Case of the Algerois Basin in North Algeria. Water Resources Management, vol. 30, no. 7, pp. 2445–2464, 2016.

Gou D Q, Xie X G Common Geological Prediction Methods and Its Application in Karst Tunnel. Journal of Railway Engineering Society, vol. 34, no. 1, pp. 75–80, 2017.

Lenz G, Kluckner A, Holzer R et al. Prediction of fault zones based on geological and geotechnical observations during tunnel construction. Geomechanik und Tunnelbau, vol. 10, no. 4, pp. 366–379, 2017.

Li H, Zhu P, Ji G et al. Modified image algorithm to simulate seismic channel waves in 3D tunnel model with rugged free surfaces. Geophysical Prospecting, vol. 64, no. 5, pp. 1259–1274, 2016.

Liu B F Zhang, Li S et al. Forward modelling and imaging of ground-penetrating radar in the tunnel ahead geological prospecting. Geophysical Prospecting, vol. 66, no. 4, pp. 784–797, 2018.

Maslova I, Ticlavilca A, Mckee M Adjusting wavelet-based multiresolution analysis boundary conditions for long-term streamflow forecasting. Hydrological Processes, vol. 30, no. 1, pp. 57–74, 2016.

Panwar V Wavelet neural network-based H∞ trajectory tracking for robot manipulators using fast terminal sliding mode control. Robotica, vol. 35, no. 7, pp. 1–16, 2017.

Qiang, Xia, Mo, et al. A dynamic modeling approach to simulate groundwater discharges into a tunnel from typical heterogeneous geological media during continuing excavation. KSCE Journal of Civil Engineering, vol. 22, no. 1, pp. 1–10, 2017.

Song L, Peng W, Goel L A Novel Wavelet-Based Ensemble Method for Short-Term Load Forecasting with Hybrid Neural Networks and Feature Selection. IEEE Transactions on Power Systems, vol. 31, no. 3, pp. 1788–1798, 2016.

Spitsyn V G, Bolotova Y A, Phan N H et al. Using a haar wavelet transform, principal component analysis and neural networks for OCR in the presence of impulse noise. Computer Optics, vol. 40, no. 2, pp. 249–257, 2016.

Wei B, Yuan D, Li H et al. Combination forecast model for concrete dam displacement considering residual correction. Structural Health Monitoring, vol. 18, no. 1, pp. 232–244, 2019.

Weifang S, Bin Y, Nianyin Z et al. An Intelligent Gear Fault Diagnosis Methodology Using a Complex Wavelet Enhanced Convolutional Neural Network. Materials, vol. 10, no. 7, pp. 790, 2017.

Xue Y, Wang D, Li S et al. A risk prediction method for water or mud inrush from water-bearing faults in subsea tunnel based on cusp catastrophe model. Ksce Journal of Civil Engineering, vol. 21, no. 7, pp. 2607–2614, 2017.

Advances in Mineral Resources, Geotechnology and Geological Exploration – Rashid & Zhang (eds)
© 2023 Copyright the Author(s), ISBN 978-1-032-33377-9

Comparison and numerical simulation analysis of treatment methods for airport alluvial lacustrine soft foundation

Bin Yan*, Zhi Heng Shang & Yuan Feng Liu
China Civil Aviation Zhongnan Airport Design Research Institute, Guangzhou, China

ABSTRACT: Long plastic vertical drainage plate-short cement-soil mixing pile (D-M method for short) is a new type of composite foundation treatment method, which has no precedent in airport foundation treatment engineering. To solve the problem of settlement and deformation of alluvial lacustrine soft soil foundation in the mountains of Plateau Airport, this paper selects typical filling sections for settlement and deformation analysis, numerically simulates four working conditions: untreated method, plastic vertical drainage plate (PVD method) treatment, cement-soil mixing pile (CMP method) treatment, and D-M method treatment, analyzes the action mechanism and deformation law, defines the reasonable treatment scheme, and explores the best design index.

1 INTRODUCTION

The site of the Lijiang Sanyi airport expansion project belongs to the landform of the Intermountain basin, and the maximum filling thickness of the expansion area can reach 10m. The stratum of the site belongs to Quaternary alluvial lacustrine facies, with abundant groundwater, complex and messy stratification conditions of the foundation soil, and the stratum structure belongs to multi-layer type, especially with thick soft soil layer on the surface. According to the existing measured data, when the aircraft is in the ground area, the ground load on the bottom layer of the pavement structure is less than 30KPa. Therefore, according to the technical requirements of the designed compactness, the bearing capacity of the foundation soil can meet the above requirements; for settlement deformation, the post-construction settlement in the affected area of flight area pavement within the design service life (based on 30 years) shall not be greater than 0.3m, and the post-construction differential settlement shall not be greater than 2 ‰ (the measurement horizontal distance is 50m) (Civil Aviation Administration of China 2013). If the thick soft soil of the project is not treated properly, it is easy to lead to large post-construction settlement deformation and differential post-construction settlement deformation, which does not meet the requirements of civil airport construction and operation.

The key technical problem of airport construction on a thick soft foundation is the control of foundation soil strength and settlement deformation. The commonly used and effective treatment methods in engineering mainly include drainage consolidation method, mixing pile method, and combined composite foundation treatment technology. According to the engineering geological conditions, construction machinery, and construction period control of the project, four working conditions of the untreated method, plastic vertical drainage plate (PVD method), cement-soil mixing pile (CMP method), and long plastic vertical drainage plate-short cement-soil mixing pile (D-M method) are selected for analysis and numerical simulation. D-M method is a new type of foundation treatment method. Compared with the PVD method, the shallow reinforcement of cement-soil mixing pile can improve the strength of foundation soil and reduce the settlement of

*Corresponding Author: yanbincisco@qq.com

 DOI 10.1201/9781003319412-56

shallow soil at the same time; compared with the CMP method, the existence of plastic drainage plate can accelerate the consolidation settlement of deep soft soil, so as to control the post-construction settlement within a certain range. Because the D-M method can give full play to the advantages of the two treatment schemes of cement-soil mixing pile and plastic drainage plate, it has the advantages of a short construction period, obvious effect of preventing post-construction settlement and post-construction differential settlement in the treatment of deep soft soil foundation (Dong et al. 2021), and has a wide application prospect in airport construction.

Since the D-M method has not been applied in airport foundation treatment, after the optimization of the foundation treatment scheme, it is necessary to provide the best-recommended value of design parameters by analyzing the action mechanism and deformation law of the construction method, so as to meet the requirements of airport design and operation.

2 PARAMETERS AND MODELS

2.1 *Engineering geological conditions*

The soft foundation treatment area of the Lake flushing foundation in the South filling channel area of the phase III expansion project of Lijiang Sanyi airport is selected for the test. Before the test, the surface cultivated soil is excavated and replaced with 0.5m gravelly soil (stratum No. 2); under it is layered silty clay (stratum No. 3, 6, 6–5), with a total thickness of 15.0m, of which 6 is soft plastic soft soil; the bottom layer is circular gravel (stratum No. 7–2). The overall thickness of the foundation is considered 50m. The groundwater is close to the surface and can be regarded as saturated soil as a whole. According to the geological exploration data, the physical and mechanical property indexes of each soil layer are intercepted, as shown in Table 1.

Table 1. Main physical and mechanical property indexes of each soil layer.

		Thickness (H)	Gravity density (γ)	Deformation modulus (E_0)	Poisson's ratio (υ)	Cohesion (c)	Internal friction angle(φ)	Permeability coefficient (k)	Void ratio (e)
Soil layer	Stratum number	m	kN/m^3	MPa		kPa	°	m/d	
Gravelly soil	2	0.5	20.0	70.0	0.15	3	30.0	43.2	0.65
Silty clay	3	2.3	19.2	10.6	0.35	29.2	9.4	1.3×10^{-3}	0.80
Silty clay	6	7.6	18.9	8.8	0.30	11.0	5.0	3.7×10^{-3}	0.91
Silty clay	6–5	5.1	19.5	10.7	0.30	24.6	8.4	1.8×10^{-3}	0.77
Circular gravel	7–2	34.5	21.5	42.6	0.15	3.0	32.0	6.9	0.65

2.2 *Cement-soil mixing pile layout*

The cement-soil mixing pile in the composite foundation is set in the silty clay layer below the bottom of the gravel cushion. The pile length is 8.0m and the pile diameter is 0.5m. It is arranged in an equilateral triangle with a pile spacing of 1.5m. In order to facilitate the simulation, the cement-soil mixing pile and the soil between piles can be equivalent to a two-dimensional pile (Deng et al. 2005). The elastic modulus of the equivalent E_p^{cal} pile can be calculated according to formula (1), and the permeability coefficient k_p^{cal} can be calculated according to formula (2).

$$E_p^{cal} = \frac{E_p + (2d - 1)E_s}{2d} \tag{1}$$

$$k_p^{cal} = \frac{(2d - 1)k_s}{2d} \tag{2}$$

Where E_p and E_s (average value) are the elastic moduli of mixing pile and soil between piles respectively; d is the pile spacing; k_s (average value) is soil permeability coefficient. In this paper, the elastic modulus is regarded as the deformation modulus in the numerical simulation. The calculation results of the deformation modulus and permeability coefficient in this project are shown in Table 2.

Table 2. Deformation modulus and permeability coefficient.

	Deformation modulus (E)	Permeability coefficient (k)
Subitem	MPa	m/d
Cement-soil mixing pile	450.0	/
Soil (average value)	9.7	3.0×10^{-3}
Equivalent pile	156.5	2.0×10^{-3}

2.3 *Plastic vertical drainage plate layout*

The length of the plastic vertical drainage plate in the composite foundation is 15.5m and the spacing is 1.5m. It is arranged in an equilateral triangle and equally spaced with a cementsoil mixing pile. The equivalent double-layer foundation method is often used in numerical simulation, that is, the soil layer within the depth range of the drainage plate and the plastic vertical drainage plate are equivalent to a uniform two-dimensional soil layer, and the double-layer foundation is composed of the untreated soil layer below the drainage plate, which can simplify the model and improve the calculation accuracy (Deng 2013), as shown in formula (3) and (4).

$$\mu = \ln\left(\frac{n}{s}\right) + \frac{k_h}{k_s}\ln(s) - 0.75 + \pi\frac{2l^2 k_h}{3q_w} \tag{3}$$

$$k_{ve} = \left(1 + \frac{2.5l^2}{\mu d_e^{\,2}}\frac{k_h}{k_v}\right)k_v \tag{4}$$

Where, l is the drainage plate distance; d_e is the diameter of the equivalent range of a single well, for the regular triangle arrangement, d_e= 1.05S(S is the plate spacing); k_h and k_v are the horizontal and vertical permeability coefficients of foundation soil respectively; k_s is the permeability coefficient of the disturbed area; Well diameter ratio $n=d_e/d_w$; Application ratio $s=d_s/d_w$ d_s is the diameter of the disturbed area, in this paper, 5 times the equivalent diameter of the plate is taken; q_w is the annual average water flow of the slab; d_w is the equivalent diameter of plastic drainage plate, it can be calculated according to formula (5). Where b and δ are the width and thickness of the plastic drainage plate respectively.

$$d_w = \frac{2(b+\delta)}{\pi} \tag{5}$$

The permeability parameters of the plastic drainage plate are shown in Table 3. According to the comprehensive formulas (3), (4), and (5), the equivalent average permeability coefficient of each soil layer within the depth range of plastic drainage plate arranged at a spacing of 1.5m can be obtained, as shown in Table 4.

Table 3. Permeability parameters of the plastic vertical drainage plate.

l	S	d_e	b	δ	d_w	d_s	n	s	k_h/k_s	q_w
m	m	m	mm	mm	M	m				cm^3/s
15.00	1.50	1.58	100.00	4.70	0.07	0.33	23.63	5.00	5.00	50.00

Table 4. The equivalent average permeability coefficient of each soil layer within the plastic drainage plate.

Soil layer	Stratum number	Permeability coefficient (k_v) m/d	Equivalent average permeability coefficient (k_{ve}) m/d
Gravelly soil	2	43.20	45.00
Silty clay	3	1.32×10^{-3}	3.00×10^{-2}
Silty clay	6	3.69×10^{-3}	8.14×10^{-2}
Silty clay	6–5	1.77×10^{-3}	4.00×10^{-2}

2.4 *Filling and stacking steps*

The 10m thick fill adopts layered stacking, which is divided into 5 layers, and the stacking height of each layer is 2m; the stacking range is 140m, the shape of the stacking section is a trapezoid, symmetrical on both sides, the slope rate of the stacking edge on each side is 1:2, the length of the slope bottom is 20m, and the length of the stacking top surface is 100m. For the first four times of stacking, the stacking period of each layer is 1 day, the intermittent period is 89 days, the fifth stacking period is 1 day, the intermittent period is 359 days, and the total stacking period is 720 days. The stacking height duration curve is shown in Figure 1.

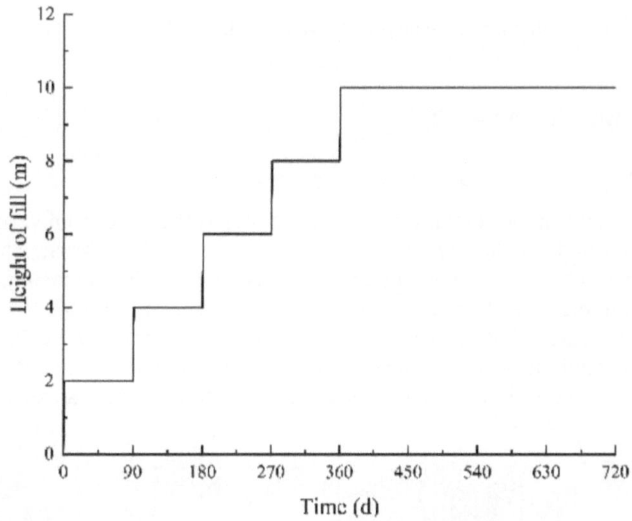

Figure 1. Stacking height duration curve.

2.5 *Establishment of numerical model*

The numerical simulation plane model based on the D-M method described above is established by using the numerical simulation software ABAQUS, as shown in Figure 2. The model is loaded after the in-situ stress balance is set. After the foundation treatment is completed, in order to detect the normal use deformation, a uniformly distributed load of 30KPa is set on the top surface of the foundation (Aircraft grounding load), and the action period is 10800 days, which is the service period. Mohr Coulomb's constitutive model is adopted for model foundation soil and surcharge soil. There are 94 equivalent piles of cement soil mixing piles, and the ideal linear elastic constitutive model is adopted. The horizontal displacement is limited by the left and right boundaries of the model, and the vertical displacement is limited by the lower boundary. The upper boundary of the foundation is completely drained, and the other boundaries are not drained. In addition, the CMP method, PVD method, and untreated method models are established for numerical analysis and comparative study.

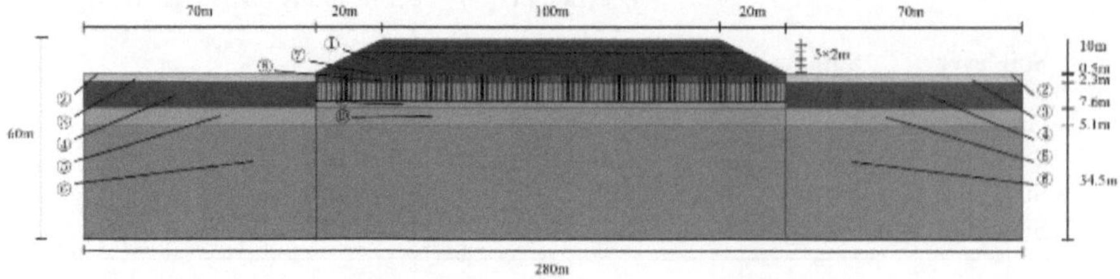

Number	Material	Color	Number	Material	Color	Material	Color
①	Fill					CMP Pile	
②	Gravelly soil		⑦	PVD-promoted gravelly soil			
③	3-Silty clay		⑧	PVD-promoted 3-silty clay			
④	6-Silty clay		⑨	PVD-promoted 6-silty clay			
⑤	6-5-Silty clay		⑩	PVD-promoted 6-5-silty clay			
⑥	7-2-Circular gravel						

Figure 2. Composite foundation model treated by D-M method.

3 SIMULATION RESULT ANALYSIS

3.1 *Settlement deformation during a surcharge period*

As shown in Figure 3, when the stacking period is completed, the maximum vertical displacement of each model is located on the central axis of the model stacking area, which is downward displacement, and the settlement to the outside of the stacking area is decreasing. Due to the action of the cement-soil mixing pile, the maximum foundation settlement of the D-M treatment method and CMP treatment method is significantly less than that of the other two construction methods, and the settlement is obvious only near the central axis of the surcharge area. Relatively speaking, the settlement of the two pile-free models is obvious in the surcharge area except for the slope.

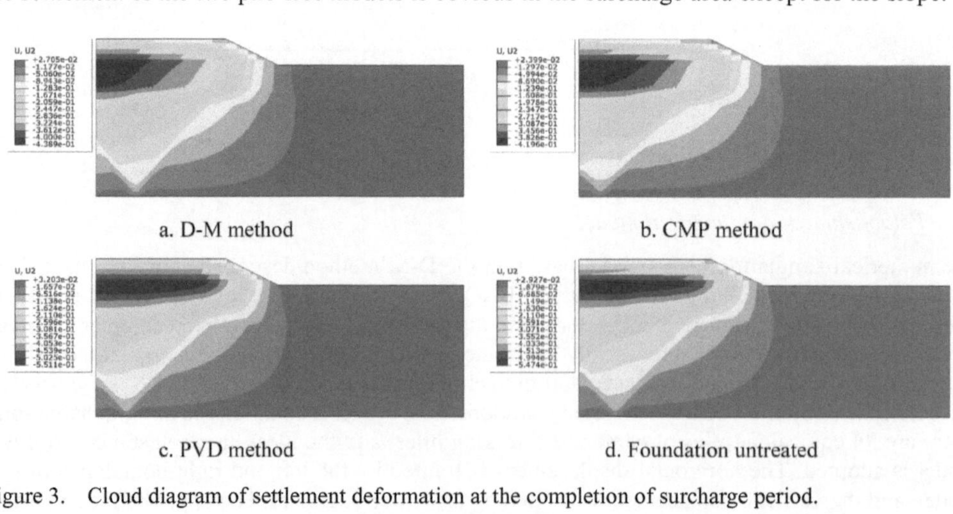

Figure 3. Cloud diagram of settlement deformation at the completion of surcharge period.

The settlement duration change curve of the surface center point of the foundation gravel cushion is shown in Figure 4; the variation law of the settlement duration curve of the center point of the foundation surface corresponds to the stacking progress. There is a sudden increase in the stacking period, and the stacking interval continues to grow slowly, and the growth rate gradually slows down with the time until it is stable. In terms of settlement degree, the maximum settlement of the D-M method and CMP method is 0.43m and 0.42m respectively, while the maximum settlement of the PVD method and untreated method is 0.54m and 0.52m respectively. The settlement of composite foundations with and without plastic drainage plates is reduced by 20.4% and 19.2% respectively; in terms of settlement change trend, the development speed of settlement in the stacking interval of the D-M method and PVD method slows down rapidly, and the settlement stability can be completed within 40 days from the beginning of the stacking interval, while the settlement speed of CMP method and untreated method slows down not significantly, and the settlement cannot reach stability within 90 days. At the end of the surcharge period, the final settlement of the composite foundation with and without piles can be increased by 2.4% and 3.8% respectively.

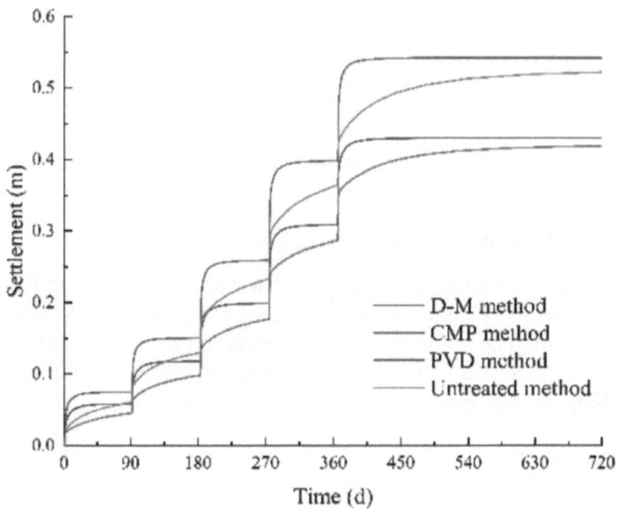

Figure 4. Settlement duration curve of center point on the foundation surface.

3.2 *Pore-water pressure during a surcharge period*

In order to explore the law of foundation consolidation in the process of graded surcharge, the lowest end of the depth of the plastic drainage board at the central axis of the foundation reinforcement area is selected as the monitoring point, and its pore water pressure duration curve is shown in Figure 5.

The variation law of pore water pressure at this place corresponds to the stacking progress. The stacking period suddenly increases to a higher level and decreases rapidly after entering the stacking interval, but the decline speed gradually slows down; the stable value of pore water pressure is about 155kPa, the maximum pore water pressure caused by surcharge is about 205kPa, and the excess pore water pressure is about 50kPa. In the D-M method and PVD method, the excess pore water pressure can dissipate completely within 40 days after the start of the surcharge interval, corresponding to the variation law of foundation settlement in Figure 4, and the completed strength of the foundation is improved; the decrease rate of pore water pressure in CMP method and Untreated method is obviously slow, which corresponds to the continuous increase of foundation settlement in Figure 4 within 90 days. The existence of excess pore water pressure means that the seepage deformation of foundation soil has not been completed, which will aggravate the post-construction settlement to a certain extent. In addition, the excess pore water pressure cannot be completely dissipated within 90 days, which will lead to the accumulation of pore water pressure to a higher value in the next stacking period. Due to the slow growth of effective stress, the new stacking carries the risk of foundation shear failure. Therefore, it will require a longer stacking period and increase the time cost.

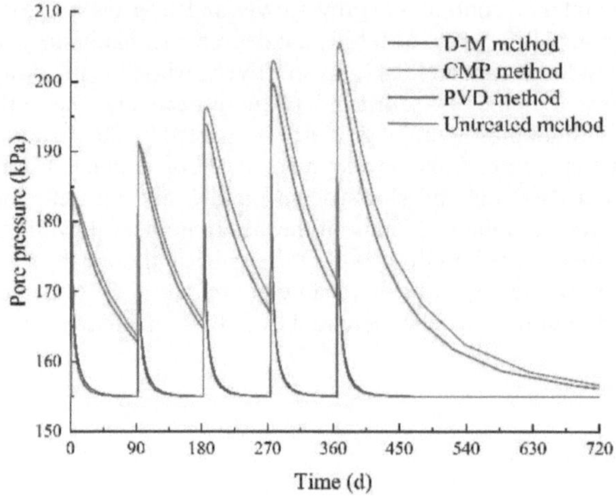

Figure 5. Pore-water pressure duration curve at the bottom depth of plastic drainage plate.

3.3 *Post construction settlement and post-construction differential settlement*

After the completion of the stacking period, the top surface of the stacking fill and the pavement will enter the operation period (the aircraft grounding load is considered as 30kPa) and the service life is considered as 30 years. The law of post-construction settlement and deformation is analyzed. The variation curve of vertical displacement and distance of filling the top surface can be drawn in the direction of filling the top surface from the center to the slope, as shown in Figure 6.

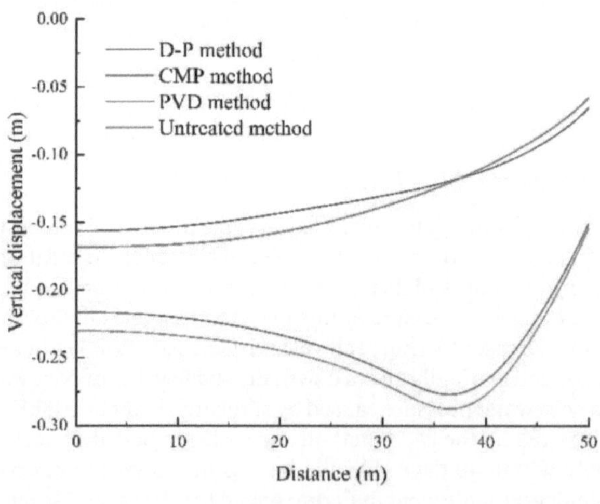

Figure 6. Variation curve of vertical displacement of filling the top surface with a horizontal distance of the center.

It can be seen from Figure 6 that for different foundation treatment methods, the vertical displacement value of the top surface of the fill presents different forms with the distribution of horizontal distance. For the D-M method and CMP method, the displacement gradually decreases from the center to both sides, which is consistent with the settlement at the end of the surcharge period; for the PVD method and untreated method, the minimum displacement is still at the edge of the top surface, but the maximum displacement is 14m away from the edge of the top surface (sliding arc cut-in position). The reason is that the filling slope has an arc sliding surface in the slope under the action of service load, and the area passing through the sliding arc shows more significant vertical displacement on the top surface of the fill. Therefore, for the foundation of the project, if there is no reinforcement measure, dangerous slope sliding may occur during use.

The maximum post-construction settlement and post-construction settlement difference of the filling top surface corresponding to different treatment methods are shown in Figure 7. In terms of the maximum post-construction settlement, the settlement deformation of the four treatment methods is less than 0.3m, but for D-M method and CMP method, the maximum post-construction settlement is 0.16m and 0.17m respectively, which is significantly less than 0.28m and 0.29m of PVD method and no foundation treatment method. As for the maximum differential settlement after construction, only the maximum settlement difference of the D-M construction method is less than 0.1m, that is, the settlement difference within 50m is not more than 2 ‰, while the corresponding maximum settlement difference of the other three construction methods does not meet this index.

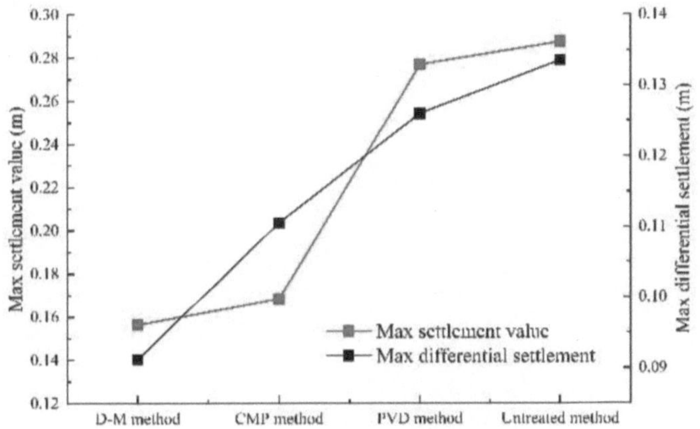

Figure 7. Maximum post-construction settlement and post-construction settlement difference of filling top surface.

4 CONCLUSIONS

As a combined construction method, the maximum settlement and maximum differential settlement of the D-M method are reduced by 5.9% and 18.2% compared with the CMP method, and 42.9% and 30.8% compared with the PVD method. This shows that the cement-soil mixing pile reinforcement in the composite foundation can effectively eliminate the post-construction settlement and differential settlement, and the plastic drainage plate to accelerate the drainage and consolidation can effectively compress the construction period and avoid the risk of foundation shear failure. In addition, the good drainage conditions can completely dissipate the excess pore water pressure of the foundation soil in the surcharge preloading, and the growth of strength is relatively uniform. Under the condition of no drainage plate, the strength of the foundation soil is difficult to increase effectively when bearing a small load near the slope. Therefore, the drainage plate can promote the elimination of differential settlement after construction.

To sum up, the PVD method and untreated method not only have large post-construction settlement and post-construction differential settlement but also have slope sliding risk in use, so they are not recommended. Both the D-M method and the CMP method can meet the index of maximum settlement. Judging by the index of maximum differential settlement, only the D-M method can meet the design requirements (the thickest fill thickness). In the actual design, the D-M method and CMP method are carried out respectively according to the filling thickness and calculation results.

REFERENCES

Civil Aviation Administration of China (2013).: Code for Geotechnical Engineering Design of Airport. China Civil Aviation Press, Beijing.

Deng, Y (2013).: Analytical theory and finite element analysis of shaft foundation consolidation. Zhejiang University.

Deng, Y., Hong, Z., Liu, S (2005).: Discussion on the simplified method of plane simulation of mixing pile composite foundation. Geotechnical mechanics (S1), 209–212.

Dong, B., Zhang, M., Zhang, P (2021).: Study on bearing capacity of dredger fill foundation treated by long plate short pile method. Journal of Wuhan University of Technology (traffic science and Engineering Edition) 45(02), 303–307.

Advances in Mineral Resources, Geotechnology and Geological Exploration – Rashid & Zhang (eds)
 ISBN 978-1-032-33377-9

Numerical analysis of pull-out bearing characteristics of torpedo anchor in deep-sea engineering

Tongkun Zhao, Mingming Liu & Gang Li*
Xijing University, Xi'an, Shaanxi, China

Jinli Zhang
Dalian University of Technology, Dalian, Liaoning, China

Jia Liu
Chang'an University, Xi'an, Shaanxi, China

ABSTRACT: Torpedo anchor is a kind of deep-sea anchorage structure with the advantages of simple installation, low cost, and broad application prospects. Its bearing capacity is closely related to the stability and anti-wind ability of the mooring structure. Based on the Abaqus software, the pull-out bearing characteristics of torpedo anchors were simulated, and the influence of pull-out angle, soil properties, buried depth, and length-diameter ratio on the bearing capacity were analyzed. The results showed that with the increase in pull-out angle, the bearing capacity increased first and then decreased. When the pull-out angle reached 45°, the bearing capacity reaches the maximum, while the bearing capacity reaches the minimum when the pull-out angle attained 90°. The uplift capacity of the torpedo anchor increased with increasing undrained strength and buried depth. When the buried depth keeps constant, the bearing capacity increased with the increase of length to diameter ratio.

1 INTRODUCTION

With the gradual depletion of land and offshore oil resources, the scope of oil exploitation has been expanded to the deep sea. The mooring system is an important part of the whole floating marine structure. With the increase in water depth and the scale of the marine structure, the complexity, cost, and installation difficulty of the mooring system are greatly increased. Therefore, the mooring system with low cost, easy installation, and high reliability has become a development trend (Yu et al. 2018). Compared with the traditional anchoring system, the torpedo anchor is a deep-sea anchoring structure with broad application prospects because of its simple and rapid installation, low cost, high pull-out ratio, and vertical pull-out force provided by the gravity penetration method (Wang et al. 2020). Therefore, it is of great significance to study the bearing characteristics of torpedo anchors for guiding engineering practice. Cheng and Qiu (2021) used the material point method to carry out a numerical simulation of the uplift capacity of torpedo anchors in normally consolidated soil and obtained the law of foundation deformation in the uplift process and the influencing factors of uplift capacity. Liu et al. (2017 a, b) studied the influence of different rate effect model parameters on the penetration depth of dynamic anchor by using a motion differential equation and concluded that the penetration depth of dynamic anchor would be significantly reduced by soil rate effect. When the rate effect parameter is large, the penetration depth of the anchor is sensitive to the change of reference strain rate. In the meantime, the drag resistance coefficient of the dynamic anchor changed from 0.93 to 1.12, and the terminal velocity of the dynamic anchor is about 28 m/s. Han et al. (2021) verified the installation performance of dynamic anchor through

*Corresponding Author: T_bag945@126.com

DOI 10.1201/9781003319412-57

field large-scale test, focused on analyzing the influence of plate anchor shape, booster weight, and release height of combined anchor in water on the final penetration depth, and established a prediction method of penetration depth based on the test results. Based on Abaqus software, Yu et al. (2019; 2020a, 2020b; 2021) conducted a systematic study on the pull-out capacity of torpedo anchors in clay soil, and discussed the influence of anchor type, soil type, pull-out load angle, the horizontal component of pull-out load and included angle of anchor wing on the pull-out capacity. Combined with the upper bound method of plastic limit analysis, a simplified model is proposed to calculate the horizontal bearing capacity of the torpedo anchor. It was found that when the rotation center is lower than the length of the anchor wing, the length or width of the anchor wing is reduced, and the rotation center is closer to the anchor tip. When the fin length of the torpedo anchor is greater than 1/2 of the length of the anchor, increasing the fin width is more effective than increasing the fin length.

The pull-out capacity of a torpedo anchor is very important for the design and construction of offshore engineering, and the factors affecting the pull-out capacity mainly included buried depth, soil strength, anchor type, anchor wing, and pull-out angle. Based on Abaqus software, the bearing characteristics of torpedo anchors were simulated, and the effect of pull-out angle (α), soil properties (S_u), buried depth ($Z_{e,tip}$), and length to diameter ratio (L/D) on the bearing capacity of torpedo anchor were analyzed. The results have an important reference value for predicting the bearing capacity of torpedo anchors in deep-sea engineering.

2 MODEL ESTABLISHMENT AND VERIFICATION

2.1 *Model building*

In order to study the effect of fins of torpedo anchors on bearing capacity, Abaqus software is used to carry out numerical simulation for two types of torpedo anchors (T-98 type wingless and T-98 type with four fins). The shape and size of the torpedo anchor are shown in Table 1. Considering the symmetry of the torpedo anchor pull-out problem, 1/2 region was selected for numerical analysis. The soil area is a cylinder with a diameter of 20 D to eliminate the influence of the boundary effect on the torpedo anchor. The height of the cylinder H is composed of torpedo anchor penetration depth $Z_{e,tip}$, and the distance H_a from the torpedo anchor tip to the bottom of the soil. The elastic modulus E of the torpedo anchor was set at 200 GPa, Poisson's ratio v was set at 0.3, and the density ρ was set at 7850 kg/m^3. The soil around the torpedo anchor is isotropic and the Tresca constitutive model is adopted. It is assumed that the undrained shear strength of the soil increases linearly with depth. In numerical calculation, the foundation soil is divided into two types: homogeneous soil and normally consolidated soil. The elastic modulus is proportional to the undrained shear strength, E=500 S_u(z), Poisson's ratio v is 0.49, the floating weight of soil is 6 kN/m^3, and the friction angle and dilatancy angle are set as 0 to simulate the undrained condition of clay. The anchor-soil interface adopts full friction contact. The boundary conditions were set as the bottom complete constraints, the side constraints in the X and Y directions, and the symmetry plane constraints in the Y direction. The loading is loaded by the displacement control method, and the corresponding

Table 1. Size of two types of torpedo anchors.

Parameters	Symbols	Wingless torpedo anchor	Four-fins torpedo anchor
Anchor length/m	L	17	17
Anchor diameter/m	D	1.07	1.07
Anchor tip length/m	L_t	2	2
Anchor wing length /m	L_f		10
Anchor wing width/m	W_f		0.9
Anchor wing thick/m	t		0.1

load when the torpedo anchorage shift reaches 0.24 *D* is used as the ultimate bearing capacity. Eight-node hexahedron reduction integral element (C3D8R) was selected as the grid division unit, and the local encryption principle was adopted to achieve the balance of calculation efficiency and accuracy. Among them, there are 4064 elements for wingless torpedo anchors and 6760 elements for four fins torpedo anchors.

2.2 *Model verification*

To verify the reliability of the numerical model, the calculated results were compared with those of the American Petroleum Institute (API) method. The APT method is only applicable to the calculation of vertical pull-out capacity, but cannot calculate the bearing capacity under inclined pull-out loading. Based on the API method, theoretical calculation of soil strength $S_u(z)=1.5z$ kPa, buried depth 1.5 times of anchor length is carried out, and the vertical bearing capacity is about 3598.95 kN, while the vertical bearing capacity obtained by numerical simulation is 3499.84 kN, and the error is about 2.75%, indicating that the theoretical and numerical calculation results are relatively close, verified the reliability of numerical calculation.

3 RESULTS AND DISCUSSION

3.1 *Influence analysis of pull-out angle*

A numerical model is established for the wingless and four fins torpedo anchors in normally consolidated clay ($S_u(z)$ = 1.5z kPa) at a depth of 1.5 L. In order to discuss the influence of pull-out angle on the bearing capacity of torpedo anchor, $\alpha = 0°$, 15°, 30°, 45°, 60°, 75°, and 90° were calculated respectively. Figure 1 shows the variation curves of bearing capacity and displacement of two anchor types under different pull-out angles. It can be observed that the loading-displacement curves of the two anchor types are similar, which can be divided into two stages: when the displacement is small, the load increases linearly with the increase of displacement, and the soil around the torpedo anchor is still in the elastic stage. With the increase of displacement, the soil around the torpedo anchor appears plastic deformation, and the load increases nonlinearly until it reaches a relatively stable stage.

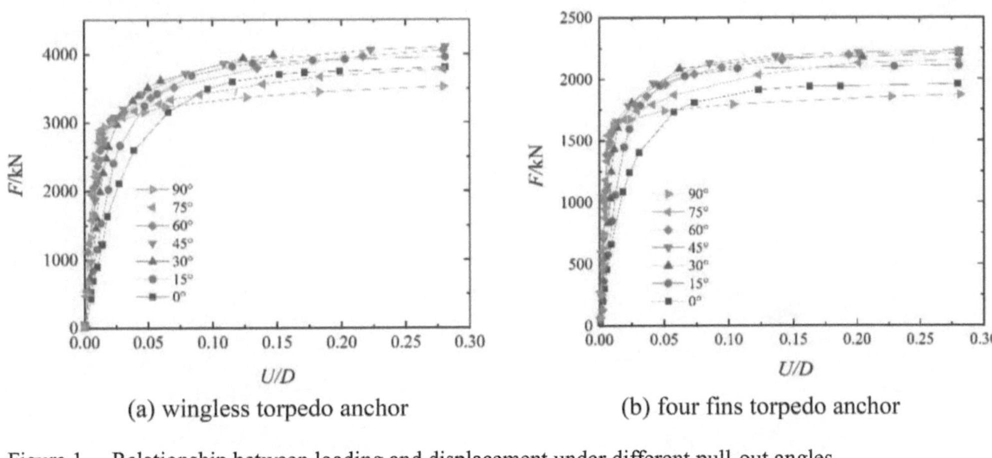

(a) wingless torpedo anchor (b) four fins torpedo anchor

Figure 1. Relationship between loading and displacement under different pull-out angles.

Figure 2 shows the relationship between the bearing capacity and pull-out angle of two anchors under different pull-out angles. As seen from the figure, with the increase in pull-out angle, the bearing capacity of the torpedo anchor increases first and then decreases. When $\alpha = 45°$, the bearing capacity reaches the maximum value; when $\alpha = 90°$ (vertical), the bearing capacity reaches the minimum value. By comparison, it can be concluded that the bearing capacity of the four fins torpedo anchor is significantly higher than that of the wingless torpedo anchor, and its bearing capacity is 88% higher than that of the wingless torpedo anchor. The main reason is that the side area of four fins increased by the torpedo anchor with an anchor wing is 70.72 m^2, while the side area of the torpedo anchor without an anchor wing is only 53.9 m^2. When the torpedo anchor is subjected to vertical pull-out loading, the side and end resistance of the four fins torpedo anchor is significantly higher than that of the wingless. When bearing horizontal pull-out loading, the horizontal bearing capacity of the four fins torpedo anchor increases with the same amplitude compared with that of the wingless torpedo anchor. The main reason is that the increase of side area makes the distribution of earth pressure increase evenly along the anchor body, which leads to the higher horizontal bearing capacity of the four fins torpedo anchor.

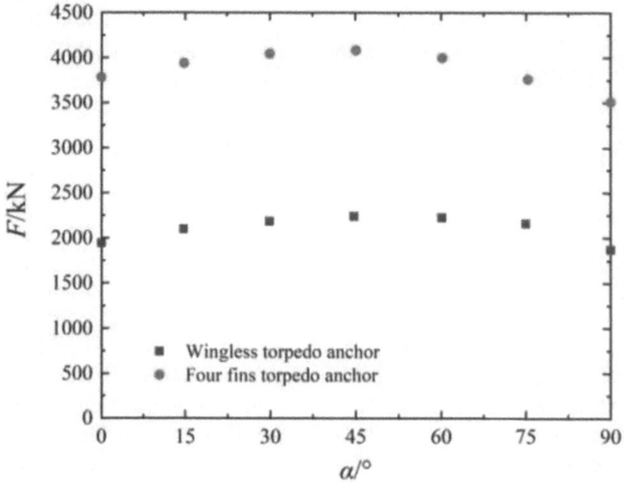

Figure 2. Relationship between loading and pull-out angle of different torpedo anchors.

3.2 *Influence analysis of soil properties*

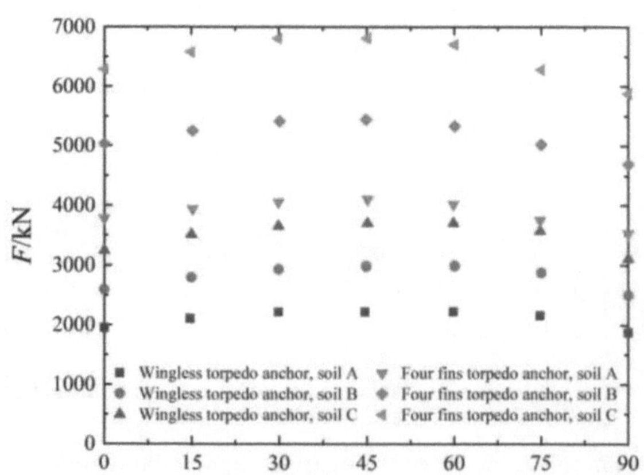

Figure 3. Relationship between loading and pull-out angle of different consolidated soils.

Soil properties are important factors affecting the uplift capacity of torpedo anchors. In order to investigate the influence of soil on the uplift capacity of torpedo anchor, three normal consolidated soils (Soil A, $S_u(z)$ = 1.5z kPa; Soil B, $S_u(z)$ = 2.0z kPa; Soil C, $S_u(z)$ = 2.5z kPa), and at the buried depth $Z_{e,tip}$=1.5 L, the numerical models of two types of torpedo anchors are established respectively, and the calculation results are shown in Figure 3. It can be seen that the uplift capacity of torpedo anchors increases gradually with the increase of soil strength. When $\alpha = 90°$ (vertical), the bearing capacity of the two torpedo anchors is the smallest. When $\alpha = 45°$, the maximum bearing capacity is obtained. When the soil strength increases from $S_u(z)$ = 1.5z kPa to $S_u(z)$ = 2.5z kPa, the vertical bearing capacity of the torpedo anchor increases from 3499.84 kN to 5834.08 kN, and the bearing capacity increases by 66.7%. The horizontal bearing capacity increased from 3770.63 kN to 6284.38 kN and also increased by 66.7%. The vertical bearing capacity of the wingless torpedo anchor increased 66.7%, while the horizontal bearing capacity increased 66.7%. It can be concluded that with the increasing undrained shear strength of soil, the vertical and horizontal uplift bearing capacity of torpedo anchor increases meanwhile.

3.3 *Influence analysis of buried depth*

In order to study the influence of buried depth on the uplift capacity of torpedo anchors, for normally consolidated soil ($S_u(z)$ = 1.5z kPa), the buried depth of the two torpedo anchors was set as $Z_{e,tip}$ = 1.5 L, 2.0 L, and 2.5 L respectively. The numerical calculation results are shown in Figure 4. It can be seen that the bearing capacity of the torpedo anchor increases with the increase of buried depth. When the pull-out angle changes from 0° to 90°, the bearing capacity increases slowly to the peak at first and then decreases slowly. When the buried depth increased from 1.5 L to 2.5 L, the vertical bearing capacity of the four fins torpedo anchor is increased from 3499.84 kN to 6689.967 kN with an increase of 91.15%, while the vertical bearing capacity of the wingless torpedo anchor increased from 1856.48 kN to 3470.85 kN with 88.96% increase. For normally consolidated soil, its undrained shear strength increases with the increase of depth. When the buried depth of the torpedo anchor increases, the undrained shear strength and bearing capacity of soil increase accordingly. In addition, the increase in buried depth leads to the corresponding increase of horizontal stress in the soil, which increases the lateral resistance and end resistance of the torpedo anchor, thereby improving its bearing capacity.

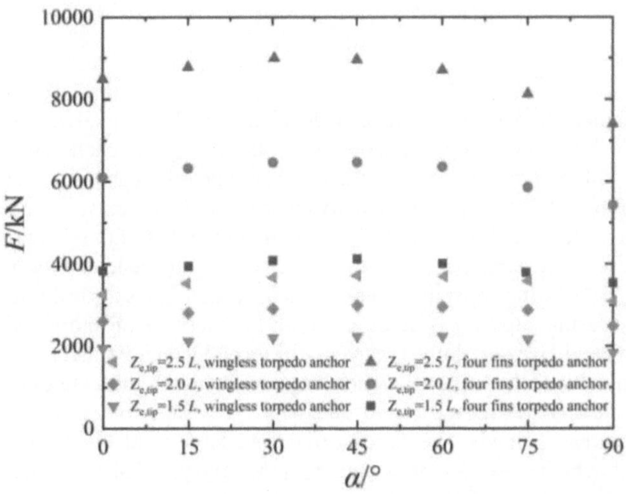

Figure 4. Relationship between loading and pull-out angle under different buried depth.

3.4 *Influence analysis of length to diameter ratio*

In order to study the influence of length to diameter ratio on the uplift capacity, numerical simulation was carried out by setting the L/D = 10, 15.89, 20 under the buried depth $Z_{e,tip}$ = 34 m for the wingless torpedo anchor in normally consolidated soil ($S_u(z)$ = 1.5z kPa). The calculation results are shown in Figure 5. It can be seen that under the same buried depth, the larger the ratio of torpedo anchor is, the higher the bearing capacity will be. When the length to diameter ratio increases from 10 to 15.89 and 20, the vertical bearing capacity increases by 32.94% and 44.22%, respectively. The main reason is that the increase of the length to diameter ratio leads to the increase of the lateral area of the torpedo anchor, which improves the lateral resistance of the torpedo anchor and then improves its vertical bearing capacity. When the ratio increases from 10 to 15.89 and 20, the horizontal bearing capacity of the torpedo anchor increases by 39.28% and 46.91%, respectively. The increase of length to diameter ratio makes the range of soil moved by torpedo anchor rotate larger and then improves the horizontal bearing capacity.

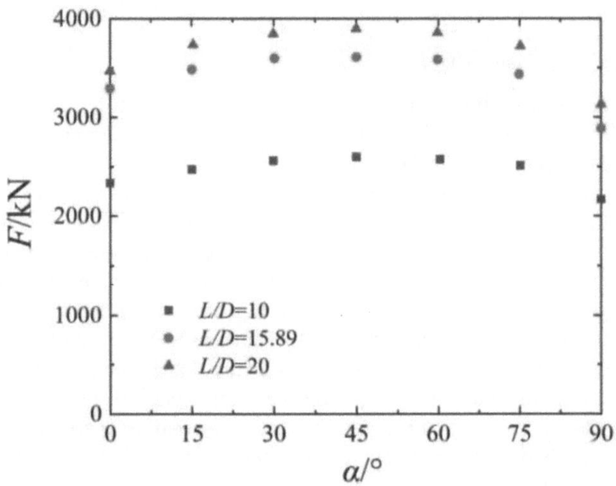

Figure 5. Relationship between loading and pull-out angle under a different length to diameter ratio.

4 CONCLUSIONS

(1) The loading-displacement curve of the torpedo anchor can be divided into the elastic stage and stable stage. With the increase in pull-out angle, the bearing capacity of the torpedo anchor increases first and then decreases. When $\alpha = 45°$, the bearing capacity reaches the maximum value, and when $\alpha = 90°$, the bearing capacity reaches the minimum value. The bearing capacity of the four fins torpedo anchor is significantly higher than that of the wingless torpedo anchor.
(2) With the increase of soil strength, the uplift capacity of the torpedo anchor increases gradually. When $\alpha = 90°$, the bearing capacity of the two torpedo anchors is the minimum. When $\alpha = 45°$, the maximum bearing capacity is obtained. With the increase of undrained shear strength of soil, the vertical and horizontal uplift capacity of torpedo anchors increases meanwhile.
(3) The bearing capacity of the torpedo anchor increased with the increase of buried depth. When the pull-out angle changes from 0° to 90°, the bearing capacity increases slowly to the peak at first and then decreases slowly. The increase of buried depth results in the increase of horizontal stress in the soil, which increases the lateral resistance and end resistance of the torpedo anchor, and then increases its bearing capacity.
(4) Under the same buried depth, the bearing capacity of the torpedo anchor increases with the increase of length to diameter ratio. The increase of length to diameter ratio results in the

increase of the lateral area of the torpedo anchor, which increases the lateral resistance and then increases its bearing capacity. When the length to diameter ratio increases from 10 to 15.89 and 20, the vertical bearing capacity increases by 32.94% and 44.22%, and the horizontal bearing capacity increases by 39.28% and 46.91%, respectively.

(5) The research results have an important reference value for predicting the bearing capacity of torpedo anchors in deep-sea engineering. Meanwhile, field model tests need to be carried out to verify the reliability of numerical results.

ACKNOWLEDGMENTS

This study was supported by the Natural Science Basic Research Program of Shaanxi Province (2021JM-535) and the Special Fund for Scientific Research by Xijing University (XJ18T01).

REFERENCES

Cheng, Y. & C. L. Qiu (2021). Numerical analysis of torpedo anchor's uplift bearing capacity by material point method. J. Water. Harbor 42(1), 114–122.

Han, C. C, Shen, K. M, Li, W, Zhao, S. W, Wang, J, Liu, J. & X. J. Kong (2021). Field tests on installation performance of a new hybrid dynamically installed anchor. Chin. J. Geotech. Eng. 43(9), 1657–1665.

Liu, J, Li, M. Z. & C. C. Han (2017a). Influence of soil strain-rate effect on embedment depth of dynamically installed anchors. J. Dalian Univ. Technol. 57(1), 68–77.

Liu, J. & X. Q. Zhang (2017b). Numerical simulation of the installation of gravity installed plane anchors in water. The Ocean Eng. 35(3), 29–36.

Wang, C, Chen, X. H. & G. L. Yu (2020). The study on the penetration depth of a torpedo anchor in a calcareous sandy bed. J. Zhejiang Univ. (Sci. Ed.). 47(2), 253–260.

Yu, G. L, Wang, W. K. & C. Wang (2018). The structure and characteristics of powered torpedo anchor. The Ocean Eng. 36(2), 143–148.

Yu, L, Yang, Q, Yang, G. & J. L. Zhang (2020a). Analysis of the resistance of elliptical tip of torpedo anchor by plastic limit analysis. Rock Soil Mech. 41(6), 1953–1962.

Yu, L, Yang, Q, Zhang, J. L. & G. Yang (2020b). A simplified model for an upper bound method to analyze the horizontal bearing capacity of torpedo anchors. Chin. J. Geotech. Eng. 42(4), 773–781.

Yu, L, Yang, Q. & J. L. Zhang (2021). Ultimate bearing capacity of the elliptical tip by upper bound analysis. Chin. J. Geotech. Eng. 43(2), 356–364.

Yu, L, Zhang, J. L, Yang, Q. & G. Yang (2019). The capacity of torpedo anchor in clay using finite element analysis. The Ocean Eng. 37(6), 122–129.

Advances in Mineral Resources, Geotechnology and Geological Exploration – Rashid & Zhang (eds)
© 2023 Copyright the Author(s), ISBN 978-1-032-33377-9

Study on landslide failure process and mechanism based on discrete element method

Liulei Bao* & Shaoping Deng*
Powerchina Sichuan Electric Power Engineering Co., Ltd, Chengdu, China

Peng Yuan*
The Nuclear Industry Geological Survey 283 Brigade, Sichuan Province, China

Liyong Wang*
Powerchina Sichuan Electric Power Engineering Co., Ltd, Chengdu, China

ABSTRACT: The discrete element model is established based on a two-dimensional particle flow program (PFC^{2D}). Its microscopic parameters were calibrated through laboratory experiments to simulate the deformation and failure process of the residual landslide, and to study its deformation failure mode and mechanism. The results show that under the action of gravity, the initial deformation and failure of the landslide are mainly caused by tension cracks at the foot of the slope and creep deformation in some parts. With the development of the landslide deformation, the cracks are pulled upward and expanded, and finally run through the whole sliding surface. The process reveals that its damage mode is typical traction progressive damage, and its damage process is divided into three stages. The research results can provide a reference for the safety problems of the same type of related slopes.

1 INTRODUCTION

China is one of the countries with the most serious slope geological disasters in the world. Especially the slopes developed in the western region have the characteristics of large scale and complex mechanisms of geological disasters, which makes the research of slope problems more important. Relevant scholars have continuously explored the numerical analysis methods of slopes and landslides (Cundall 1971; Wang 2017; Zhao 2016), mainly numerical methods such as the finite element method, discrete element method, and finite difference method, among which the finite element analysis method is mostly. Considering that the landslide body itself has the characteristics of non-uniform, non-continuous, and discrete material composition, solving the displacement by the finite element method cannot meet the requirements of its large deformation. While discrete element techniques, such as particle flow codes (PFC) (Bi 2015; Chen 2018; Liang 2013; Wang 2021; Zhang 2013; Zhao 2017; Zheng 2011; Zhou 2009; Zou 2019), have been found to be suitable not only for dealing with large deformation discontinuities, but also for studying force states and motion patterns at the microscopic scale. It has been widely used in geotechnical engineering, including the freeze-thaw spalling of the slope (Evans 2009; Zhu 2021), and the effect of rainfall on the erosion of the slope surface (Tsuji 1933); Xiong et al. (Xiong 2013) conducted a three-dimensional simulation of rainfall slope collapse through fluid-structure coupling analysis, in which rainfall

*Corresponding Authors: baoliulei-sedc@powerchina.cn, 441375451@qq.com, 1663915106@qq.com, 461779048@qq.com

 DOI 10.1201/9781003319412-58

was simulated by applying pressure on the boundary. The results show that it is feasible to study the collapse formation mechanism by PFC using the fluid-structure coupling method. The results show that it is feasible to study the formation mechanism of collapse by PFC. Song (Song 2013) and Wu (Wu 2014), and others used the above methods to analyze the evolution process of slope erosion and the distribution of slope erodibility. Ma (Ma 2012) and Wu et al. (Wu 2013) simulated the erosion damage of steep slope through a two-dimensional PFC^{2D} model, in which the effect of runoff scours was simulated by the movement of particles on the slope surface. In general, the deformation and failure process of the slope can be simulated by PFC analysis.

Therefore, taking the Jinfeng Village landslide as the research object, the motion process and deformation characteristics of the landslide were analyzed using the discrete element particle flow method (PFC^{2D}).

2 LANDSLIDE IN JINFENG VILLAGE

2.1 *Geological overview of Jinfeng Village landslide*

The Jinfeng Village landslide is located on the southeast side of the former residence of Zhang Aiping in Luojiang Town, Dazhou City, and a scarp is formed at the front edge of the landslide. The continuous heavy rainfall aggravated the landslide in Jinfeng Village, and the trailing edge fell to form cracks. Under the action of rain, some of them collapsed. The landslide is 45.0m long, and 120.0m wide. The average thickness of the soil layer is about 4.0m and the volume is 21600m^3 Small soil landslide. The slope is distributed in the range of elevation 319m–335m, the relative height difference is about 16m. The plane shape is irregular fan-shaped, the slope aspect is 44°, and the overall slope of the slope is about 25° – 35°. As shown in Figure 1, the study area is less affected by groundwater.

Figure 1. Topography.

2.2 *Stratigraphy and structure*

According to the drilling exposure data of the jinfengcun landslide, the strata in the study area are mainly quaternary Holocene eluvium and diluvium (Q_4^{el+dl}), and the Middle Jurassic Upper Shaximiao Formation (J_2S) silty mudstone, as shown in Figure 2. The rock formation is 230°∠10.

Figure 2. Borehole core.

The study area is located in the Neocathaysian structural area in eastern Sichuan. The structure in the area is dominated by a series of NNE-NE trending comb folds. The geological structure in the area is mainly affected by the Huaying Mountain Uplift fold belt. From west to east, the structures mainly include Dushi syncline, Tieshan anticline, Daxian Dazhu syncline, Tongluoxia (Zhongshan) anticline, etc., as shown in Figure 3.

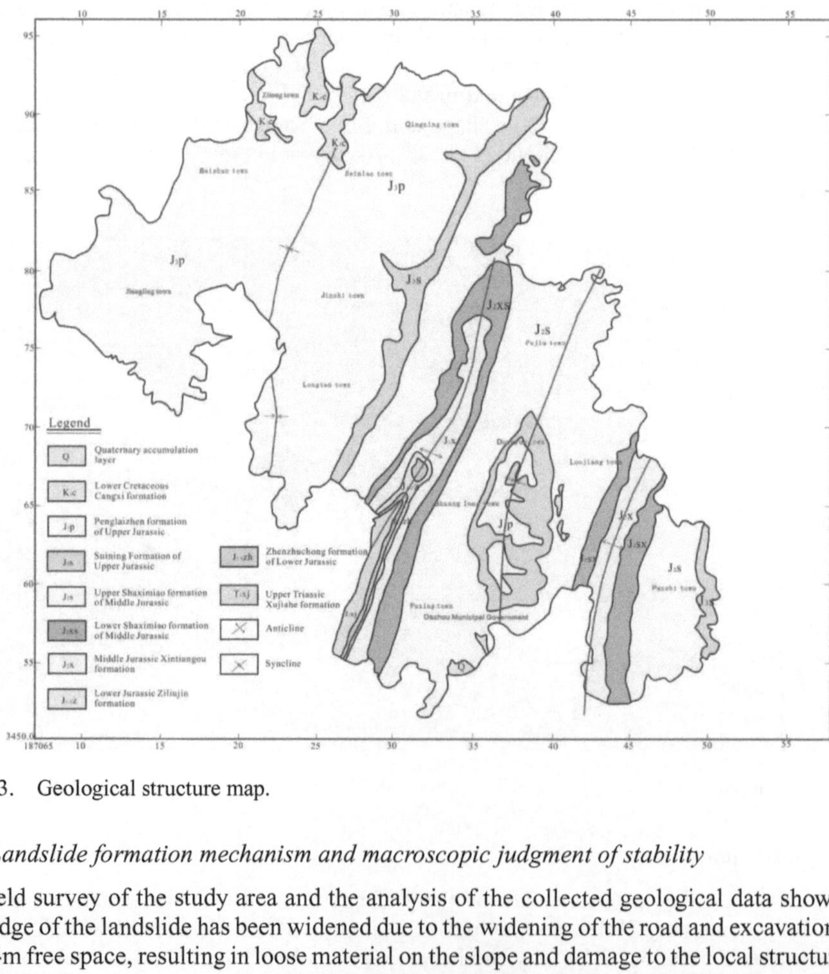

Figure 3. Geological structure map.

2.3 *Landslide formation mechanism and macroscopic judgment of stability*

The field survey of the study area and the analysis of the collected geological data show that the front edge of the landslide has been widened due to the widening of the road and excavation to form a 3m-4m free space, resulting in loose material on the slope and damage to the local structure. Since then, during the rainy season, the surface water is poured along the cracks, infiltrating and softening the slope material, reducing the shear strength of the slope and reducing the slope stability, resulting in the sliding of the upper soil along with the foundation overburden interface under the action of gravity and instability failure. To sum up, the instability failure mode of the slope is traction type.

3 MICRO PARAMETER SELECTION PROCESS

3.1 *Introduction to discrete element theory of particle flow*

PFC (particle flow code) is based on discrete element theory and simulates the actual moving process of rock and soil particles (Cundallp 1979,1992). There are two built-in bonding models in PFC^{2D}: the contact bond model and the parallel-bond model. The deformation evolution of the medium is deduced by simulating the trajectory of particles and the displacement of monitoring points, which provides a certain scientific theory for the study of practical problems. Compared with other discrete element software, PFC has more advantages in soil landslide simulation.

3.2 *Parameter selection*

The PFC^{2D} program uses granular microscopic parameters, which have no direct standard quantitative relationship with the actual macroscopic parameters of rock and soil mechanics. Therefore, it is necessary to transform the macro mechanical characteristic parameters of rock into particle meso parameters. The commonly used method is to establish a relationship through numerical experiments to obtain reasonable microscopic particle characteristic parameters (Dong 2016; Huang 2017; Wang 2015). According to the research experience on the correlation of macro and mesoscopic parameters of relevant rock and soil mass, uniaxial tests were carried out to obtain the mesoscopic parameters of the model, as shown in Figure 4 and Table 1.

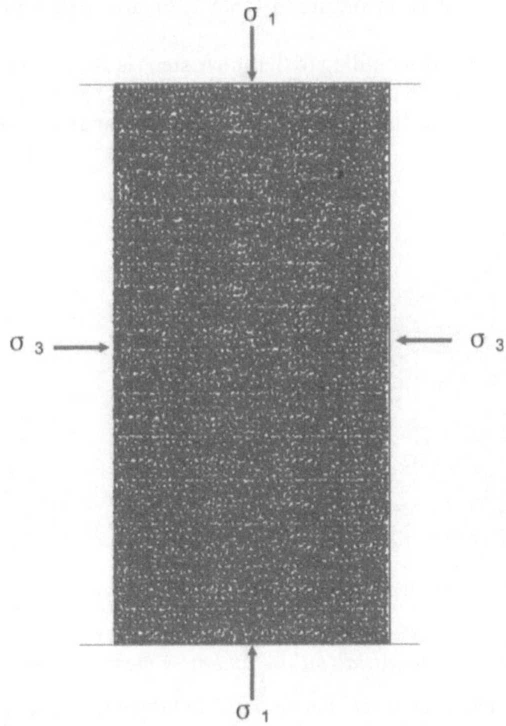

Figure 4. Uniaxial compression test model.

Table 1. Particle meso parameters.

Meso parameters	Silty clay
Particle density /(kg.m^{-3})	1960
Particle size /m	0.2–0.3
Normal contact stiffness /MPa	19.6
Tangential contact stiffness /MPa	19.6
Friction coefficient	0.5
Normal bonding stiffness /(MPa·m^{-1})	80
Tangential bonding stiffness /(MPa·m^{-1})	80
Bonding radius coefficient	0.5
Normal bond strength /Kpa	200.0
Tangential bond strength /Kpa	100.0

4 ANALYSIS OF NUMERICAL SIMULATION RESULTS

4.1 *Establishment of slope parameters*

PFC particle flow software has two built-in modeling methods, which can be modeled by the command flow of the software or by AutoCAD software. Compared with the former modeling method, the latter has higher modeling efficiency and less workload. The main modeling steps of the latter are complete the drawing of the model boundary in AutoCAD, import the DXF file into PFC through the geometry import command and the wall import geometry command, and generate the boundary wall according to the boundary line of the actual situation of the landslide, simulate the landslide mass with particles and the bedrock with walls through the ball wall method, and give the particle micro parameters according to Table 1. Finally, delete particles beyond the wall boundary.

The landslide model generated according to the above steps is shown in Figure 5, which is solved by iterative calculation. And set up monitoring points (measures) at the characteristic positions of the landslide model to monitor the displacement of the slope and analyze the damage and movement characteristics of the landslide.

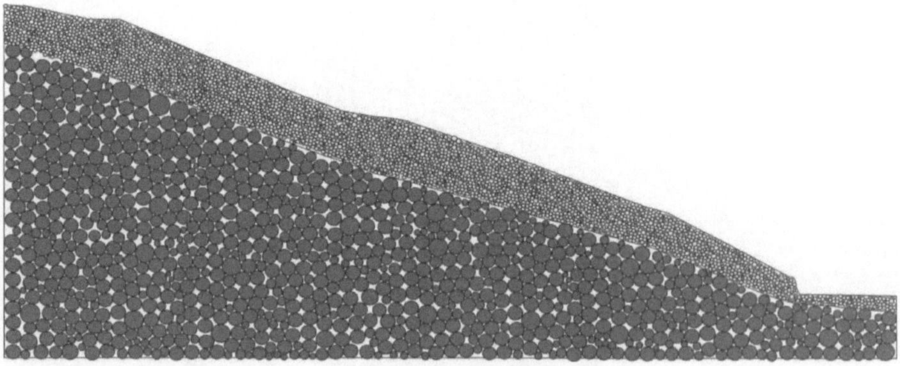

Figure 5. Initial model of landslide and distribution of monitoring points.

4.2 *Simulation and analysis of landslide failure and movement process*

The landslide model is initially balanced, the velocity and displacement are cleared, and the value is assigned. Then, delete the particles under the wall and apply gravity to make the slope particles displace under the action of gravity. Finally, the solution command is used to balance the internal stress of the landslide, monitor different parts of the landslide, and study its deformation and failure process. The slope surface contour in the figure is only for reference, and the whole failure process of the landslide is shown in Figure 6.

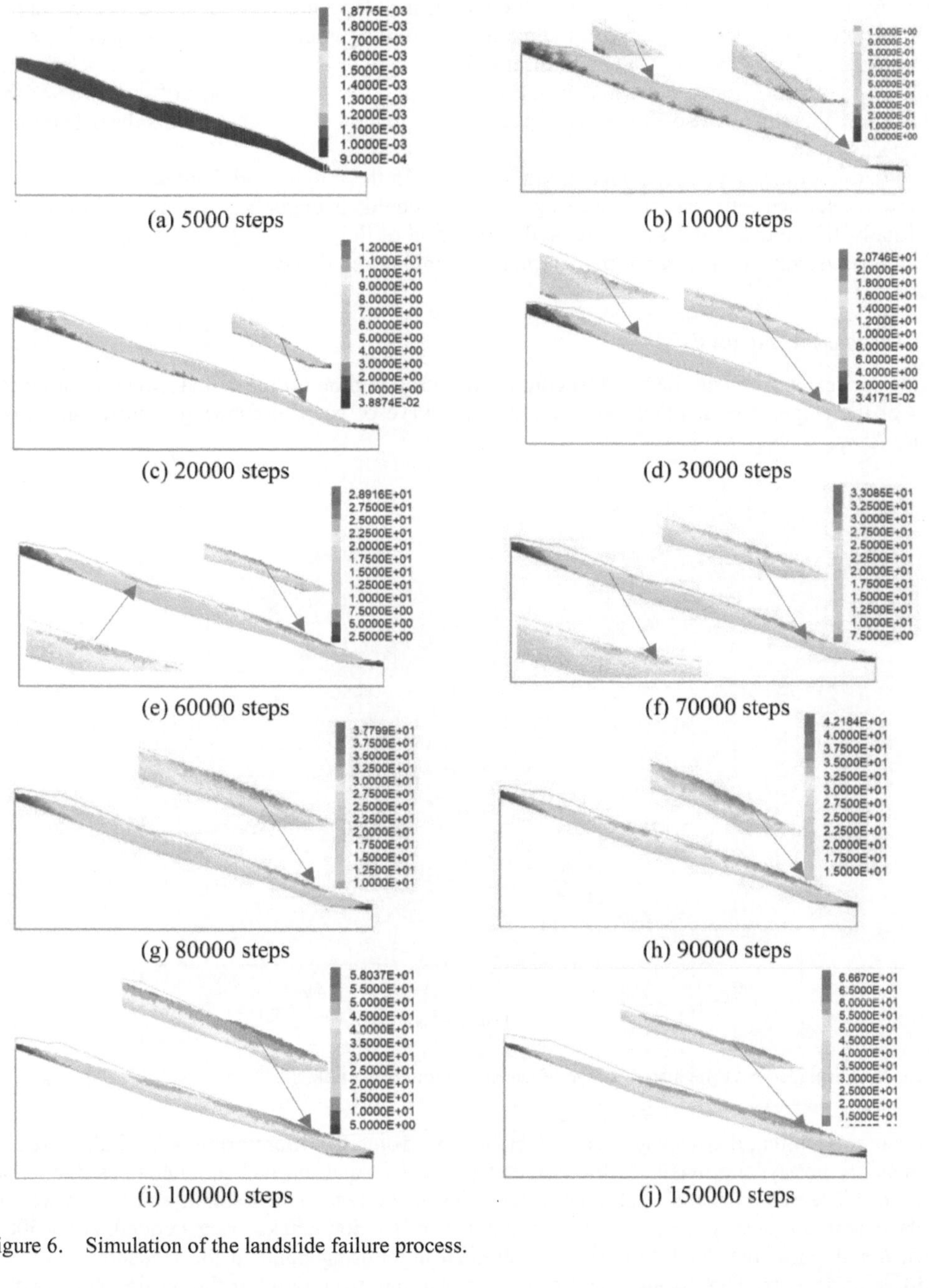

Figure 6. Simulation of the landslide failure process.

It can be seen from Figure 6(a)–6(b) that under the action of gravity, when the model runs to 5000 steps, the slope body does not have obvious deformation. At 10000 steps, the rear edge of the slope sinks significantly, and the soil body at the foot of the slope sinks. Extrusion deformation occurs, and micro-cracks and tension cracks occur locally on the surface of the slope toe and the middle and trailing edge of the trailing edge, and the deformation occurs only in the superficial region.

As the deformation increases, the front edge of the slope produces tensile crack expansion, Figure 6(c)–6(g). At 30000 steps, part of the soil at the slope toe was cut out under the upper compression, and the tensile cracks in the middle and trailing edge of the slope increased. As the cracks expand, the soil behind the cracks loses effective support and moves forward, which further leads to the compression and shrinkage of the cracks and the phenomenon of "sit down" of the soil. At the same time, it also shows that the slope deformation at this stage enters the progressive failure stage.

The deformation continued to increase, and the soil in the middle and front sections moved to create a free surface. After entering this stage, the slope could no longer bear too much deformation, and finally, the landslide mass accumulated in the front of the slope, Figure 6(h)–6(j). The whole failure form is typical traction progressive failure from bottom to top.

4.3 *Displacement analysis*

During particle movement, the ball hist command is used to monitor the displacement of different parts of the slope. The displacement time history curve of each monitoring point is shown in Figure 7.

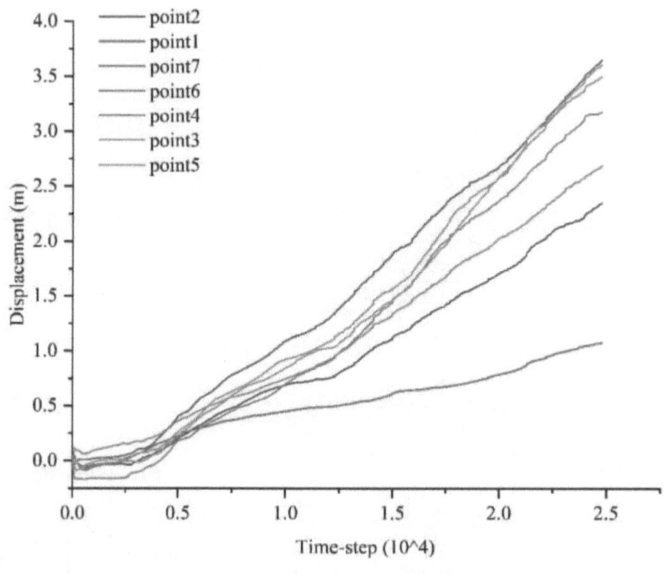

Figure 7. Displacement time history curve of landslide monitoring points.

It can be seen that the displacements of monitoring point 1 to monitoring point 7 all increase, but they are not synchronized. At the initial stage of deformation, before 5000 steps, the slope of the displacement curve of each measuring point of the landslide is small, and the growth of displacement is basically the same. A small amount of tension cracks are generated. After 5000 steps, the slope of the displacement curve at each monitoring point in the landslide increases to different degrees. The displacement of the No. 1 monitoring point at the trailing edge of the landslide is significantly lower than that of other monitoring points, because the deformation of the slope is not synchronous in the sliding process, there is a "lag" phenomenon, and the front of the landslide moves faster, while the trailing edge of the slope moves slowly due to the slow slope, so the displacement difference between the soil in the area of No. 1 measuring point and the front is getting bigger and bigger.

According to the deformation and failure process of the landslide and the displacement monitoring curve, the failure process is divided into three stages. (1) Initial deformation stage: in the early stage of the test, under the action of self-weight, the trailing edge and the leading edge of the slope produce tensile cracks, and at the same time close to the top of the slope Part of the soil at this point has the phenomenon of "sit down", the deformation is limited at this stage, and the slope is in a stable state, as shown in Figure 6(b). (2) Crack propagation and sliding stage: the slope surface first slides down locally, resulting in tension cracks; the development of tension cracks on the front edge of the slope further expands and extends upward. Sliding failure occurs, as shown in Figure 6(c). (3) Failure stage: the deformation develops further, the tensile fracture surface from the back edge of the slope to the middle of the slope expands along the original fracture surface, forming a penetrating fracture surface, and the slope loses stability, resulting in sliding failure.

5 CONCLUSIONS

Based on the results and discussions presented above, the conclusions are obtained as below:

(1) Through the numerical simulation of the landslide failure process by the particle flow program PFC^{2D}, it is known that the initial failure of the landslide is mainly tensile cracks and creep deformation at the slope toe. With the development of time, the cracks stretch upward, and the landslide slides along the foundation overburden interface and accumulates at the slope toe. Its failure mode is typical traction progressive failure, which is consistent with the actual investigation.

(2) By monitoring the displacement of the landslide movement process, it is believed that the movement speed of each part of the landslide is basically the same in the early stage of failure; after that, the movement speed of each part of the landslide increases rapidly, but the time to reach the peak speed is also from the foot of the slope to the top of the slope. It increases gradually, thus showing a time lag, which is consistent with the typical characteristics of traction landslides.

(3) The whole process of landslide movement can be divided into initial deformation stage-crack propagation and sliding stage-failure stage and accumulation stage. At the same time, PFC^{2D} simulation results well reproduce the deformation and failure process of the landslide, better simulate the interaction between particles, and provide a relevant basis for landslide treatment measures.

REFERENCES

Bi, Y.Z., He, S.M., Fu, Y.S., et al. (2015) A new protection structure for high-speed remote landslide debris flow based on discrete element method[J]. Mountain Research, 33(5): 560–570.

Cundall, P.A. (1971) A computer model for simulating progressive large scale movements in blocky rock system[C] //Symposium of International Society of Rock Mechanics. Nancy: Scientific Research Publishing, 128–132.

Chen, D., Xue, X.C., Wei, J.B. (2018) Simulation of failure process of Liujian landslide based on PFC^{2D}[J]. Coal Geology & Exploration, 46(4): 115–121.

Cundallp, P.A., Strack, O.D.L. (1979) A discrete numerical model for granular assemblies[J]. Geotechnique, 29(1): 47–65.

Cundall, P.A, Hart, R.D. (1992) Numerical modeling of discontinual[J]. Engineering Computations, 9(2): 101–113.

Dong, Q.P., Yao, H.L., Zhan, Y.X. (2016) Macro-meso-scopic mechanical behaviors of granular material based on particle flow model[J]. Yangtze River, 47(4): 68–73.

Evans, M.T., Chall, S.L., Zhao, X.L., et al. (2009) Visualization and Analysis of Microstructure in Three-Dimensional Discrete Numerical Models[J]. Journal of Computing in Civil Engineering, 23(4), 277–287.

Huang, L., Li, X.A., Cai, W.B., et al. (2017) Simulation of collapsible characteristics of Malan loess in Yan'an new area by PFC^{2D}[J]. Coal Geology&Exploration, 45(3): 119–124.
Liang, J.X., Hu, X.W., Xu, X.J. (2013) Particle Flow Simulation of Earthquake Induced Deformation Failure of Soil Slopes with Different Geological Factors under Earthquake[J]. Engineering Geology, 25(06), 1537–1546.
Ma, D.H. (2012) A Study of Water-Soil Mechanics Coupling Mechanism and Model for Loess Slope Surface Erosion on Highway[D]. Jilin: Jilin University.
Song, P.R. (2013) The Erosion Damage Characteristics and Numerical Simulation of Loess Slope[D]. Jilin: Jilin University.
Tsuji, Y., Kawaguchi, T., Tanaka, T. (1933) Discrete Particle Simulation of Two-Dimensional Fluidized Bed[J]. Powder Tech, 77(1), 79–87.
Wang, J.M, Zhang, Y.B., Zhao X.Q., et al. (2017) Using point contact the simulated sliding surface analysis method to study the stability of landslide[J]. Rock and Soil Mechanics, 38(9): 2746–2756.
Wang, X., Yin, Z.Y., Wu, X.X., et al. (2021) A Novel Approach of Random Packing Generation of Complex-Shaped 3D Particles with Controllable Sizes and Shapes[J]. Acta Geotechnica, 1–22.
Wu, Q., Wang, C.M., Song, P.R., et.al. (2014) Rainfall Erosion experiment for Steep Loess Slope and Fluid-Soil Coupling Simulation with PFC^{3D}[J]. Rock Soil Mech, 35(4), 977–985.
Wu, Q., Wang, C.M., Ma, D.H., et.al. (2013) Mechanism of Steep Loess Slope Rainfall Erosion in Western Liaoning[J]. Jilin Univ, 43(5), 1563–1571.
Wang, Y.B., Wang, L.P., Jiao, H.Z., et al. (2015) Mechanical characteristics of deformation and damage mechanism of sand-stone under different confining pressure[J]. Coal Geology & Exploration, 43(4): 63–68.
Xiong, C.X., Wang, T., Lu, X.B. (2013) Meso-mechanical Simulation of Slope Disintegration Erosion under Rainfall[J]. Mountain Sci, 31(6): 710–715.
Zhao, Y.H., Wang, H., Deng, K., et al. (2016) The finite element simulation of dynamics of the Three Gorges Shuping landslide[J]. Acta Petrologica Sinica, 32(7): 2217–2224.
Zou, H., Chen, J.G., Cai, H.H., et al. (2019) Preventive Characteristics of High-speed Long-distance Landslide-clastic Flow in Metamorphic Rock Strata in Huanggang Area. Resources Environment & Engineering, 33(S01), 43–52.
Zhou, J., Wang, J.Q., Zeng, Y., et al. (2009) Particle flow simulation for soil slope stability analysis[J]. Rock and Soil Mechanics, 30(1): 86–90.
Zhao, Z., Wei, J.B. (2017) Simulation of landslide movement process in accumulation layer based on particle flow method[J]. Coal Geology & Exploration, 45(6): 111–116.
Zheng, L.N., Xie, Q., Liu, Y.L., et al. (2011) Analysis of failure properties of Engineering Society, 28(3): 32–36.
Zhang, D., Huang, X., Zhao, Y. (2013) Algorithms for Generating Three Dimensional Aggregates and Asphalt Mixture Samples by the Discrete-Element Method[J]. Journal of Computing in Civil Engineering, 27(2), 111–117.
Zhu, T.T., Chen, J.X., Huang, D., Luo, Y.B., et al. (2021) A DEM-Based Approach for Modeling the Damage of Rock under Freeze-Thaw Cycles[J]. Rock Mechanics and Rock Engineering, 54(6), 2843–2858.

Advances in Mineral Resources, Geotechnology and Geological Exploration – Rashid & Zhang (eds)
© 2023 Copyright the Author(s), ISBN 978-1-032-33377-9

TOC content prediction of source rocks based on machine learning

Yipeng Yin*
Liaohe Oilfield Exploration and Development Research Institute, Panjin, Liaoning, China

ABSTRACT: Machine learning plays an important role in the field of geology, especially in the process of oil and gas exploration. TOC content in kerogen is an important index to determine the quality of source rock, which can affect the optimization of favorable reservoir areas. At present, the prediction accuracy of TOC content in highly heterogeneous source rocks by various seismic interpretation and logging parameter regression methods is relatively low, which seriously restricts the progress of oil and gas exploration. Based on this, this paper takes Wulanhua sag in Erlian Basin as an example, the TOC content of source rocks is predicted by ΔLogR and artificial neural network (ANN) simulation, and the differences are compared. The results show that: (1) the measured TOC values of source rocks in Wulanhua sag are distributed in 0.2%~5.3%, with an average of 2.0%, and the characteristics of source rocks are good. (2) The accuracy rates of TOC content predicted by the two methods are 79% and 90% respectively. The ANN simulation method is feasible to simulate TOC content. (3) ANN simulation method is not limited by geological conditions, and the model will take into account the effects of various parameters. In general, the TOC prediction method of lacustrine source rocks based on machine learning has a broad application prospect for source rock evaluation.

1 INTRODUCTION

The total organic carbon (TOC) content determines the effectiveness of source rocks. For highly heterogeneous source rocks, local sampling and testing can't fully reflect the characteristics of source rocks in the study area, and too much sampling and experimental testing will cause a lot of resource consumption. Therefore, most experts and scholars propose to use seismic and logging interpretation methods to predict the TOC content of source rocks. At present, the widely used methods include multiple linear regression and ΔLogR, etc., and achieved good results (Fertl 1980; Mallick & Raju 1995; Passey et al. 1990). However, with the deepening of exploration and development, the evaluation of source rocks needs to be further refined, and the usual logging parameter regression methods can't meet the needs of oil and gas exploration. Based on this, this paper proposes to use the machine learning method to predict TOC content, in order to provide help for the evaluation of source rocks in the future.

Wulanhua sag is located in the south of Erlian Basin, with an overall northeast trend and an area of about 600 km^2 (Cao et al. 2021; Yuan et al. 2016). In 2011, the successful deployment of well L-1 found about 1×10^8 t of oil reserves. According to the strata encountered in well L-1 drilling and the analysis and test results, the source rocks in Wulanhua sag are mainly located in Tenggeer Formation and Aershan Formation. Among them, the average content of TOC in the source rocks of the Tenggeer Formation is 1.93%, the average hydrocarbon generation potential is 7.45mg/g, and the kerogen is type II_2, which belongs to medium- good oil source rocks. The average content of TOC in the source rocks of the Aershan Formation is 2.08%, and the average hydrocarbon generation potential is 8.88mg/g. The kerogen is type II_2–II_1, which belongs to good oil source rocks. Due to the complex sedimentary evolution and structural development, many

*Corresponding Author: yinyp@petrochina.com.cn

types of traps are developed in Wulanhua sag, most of which are located in Tumur, Saiwusu, and Hongjing structural belts. At present, more than 10 industrial reservoirs have been found in the first member of the Tenegeer Formation, Arshan Formation, and Paleozoic. Combined with the analysis of drilled results, three reservoir trap modes are mainly developed, including ancient buried hill reservoir trap, lithologic reservoir trap, and structure-lithology reservoir trap. Oil and gas migrate along unconformities and faults horizontally and vertically and finally form reservoirs.

2 METHODS

This paper plans to use ΔLogR and ANN simulation to predict the TOC content of source rocks in Wulanhua sag, and the prediction accuracy is compared.

2.1 *ΔLogR*

The basic principle of ΔLogR technology is that in impermeable strata, the increase of organic matter content will often lead to the increase of acoustic time difference (AC) and gamma (GR), while the hydrocarbon generation of source rock will lead to the increase of corresponding depth resistivity (RT). Determine the baseline according to AC and RT curves corresponding to the non-source rock section, adjust the appropriate proportion range, and overlap AC and RT curves of the non-source rock section. At this time, if there is an interval between the two curves (recorded as ΔLogR) is the source rock section. There is a linear relationship between ΔLogR and TOC content. The higher the abundance of organic matter, the greater the ΔLogR value. The calculation formula is shown in reference (Fertl 1980).

2.2 *Artificial neural network simulation*

ANN is a network composed of many neurons with specific functions (Figure 1). It is generally composed of an input layer, hidden layer, and output layer. Its basic principle is to use error back propagation to train data on the basis of sample data. Using the multi-layer perceptron neural

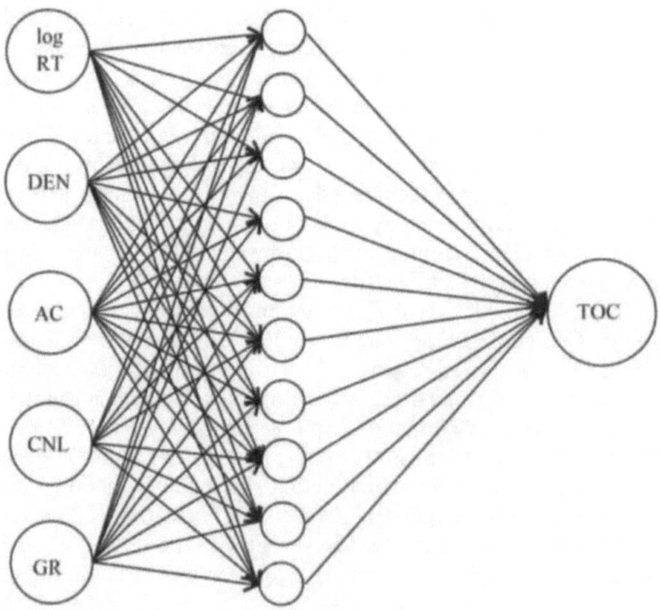

Figure 1. Model of artificial neural network simulation and prediction of TOC content.

network model in SPSS, after repeatedly adjusting the training parameters, when the network structure model is a double-layer structure, the activation function of the hidden layer is sigmoid and the activation function of the output layer is softmax, the training test ratio is 7:3, using the batch training type and the adjusted conjugate gradient optimization algorithm, and other parameters are the default values, good results can be achieved (Shi et al. 2016).

In this paper, gamma (GR), neutron (CNL), the acoustic time difference (AC), resistivity (logrt), density (DEN), and other logging parameters are used as the input layer, and the output layer is set as the predicted value of TOC content, as shown in Figure 1.

3 RESULTS AND DISCUSSION

As mentioned above, the measured TOC values of source rocks of Tenggeer Formation and Aershan Formation in Wulanhua sag are distributed in the range of 0.2%~5.3%, with an average of 2.0%. The distribution range of TOC calculated by the ΔLogR method is 0.5%~4.5%, with an average of 1.7%. The distribution range of TOC calculated by the ANN simulation method is 0.18%~5.2%, with an average of 1.9%. Comparing the calculated results of the two methods with the measured values, it is found that the average accuracy of the ΔLogR method is about 79%, and the average accuracy of ANN simulation is about 90%, as shown in Figure 2. The results show that the ANN simulation algorithm is more suitable for the prediction of the TOC value of source rocks.

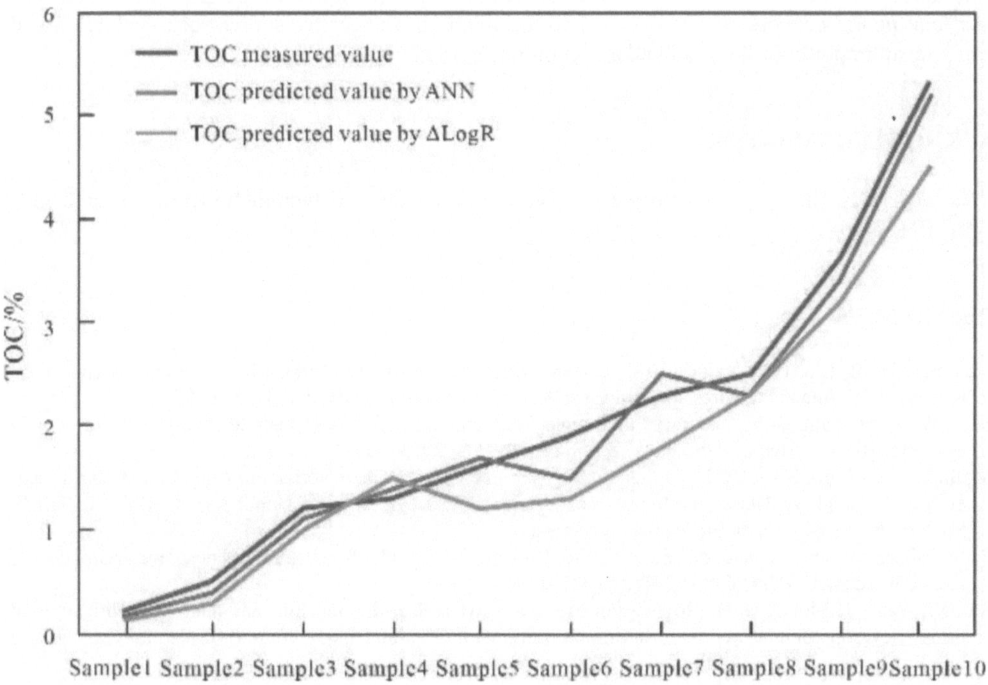

Figure 2. Comparison between ANN and ΔLogR on the prediction of TOC content.

Comparison of ANN simulation and ΔLogR method, ANN simulation method has a good advantage in predicting the TOC content of source rocks. Its main characteristics are: firstly, the ANN simulation method is not limited to geological conditions, and a large number of geological parameters can be added to modeling. Secondly, the ANN model is a complex optimization algorithm different from multiple linear regression. Probability-weighted matching will be carried out for each parameter in the calculation. However, there are still some deficiencies in this method. So far, there is no unified and complete theoretical guidance for the selection of neural network structure, which can only be selected by experience. If the selection of network structure is too large, the training efficiency is not high, and overfitting may occur, resulting in low network performance and poor fault tolerance. If the network structure is too small, the network may not converge. Therefore, in practical application, we should select the appropriate network structure adjust the operation function in time and optimize the calculation parameters according to the geological conditions.

4 CONCLUSION

According to the sedimentary and structural characteristics of Wulanhua sag, the geochemical characteristics of its source rocks are summarized in this paper. On this basis, the organic carbon content of its source rock is predicted by using the ΔLogR method and the Artificial neural network simulation method, and their advantages and disadvantages are compared. The source rocks in Wulanhua sag are of type II_2-II_1, which have a great hydrocarbon generation potential and belong to medium-good source rocks. The measured TOC ranges from 0.2% to 5.3%. The distribution range of TOC predicted by the ΔLogR method is 0.5%~4.5%, with an average of 1.7%. The distribution range of TOC predicted by the ANN simulation method is 0.18%~5.2%, with an average of 1.9%. The prediction accuracy of the TOC value by the ΔLogR method and artificial neural network simulation method is about 90%, which is about 11% higher than that of the ΔLogR method. The artificial neural network simulation method has great advantages for source rock evaluation and can play an important role in oil and gas exploration in the future.

ACKNOWLEDGMENTS

This work was financially supported by the Major science and technology projects of CNPC (2017E-1504).

REFERENCES

Cao, S.J., Li, P., Liu, D.M., et al. (2021) Characteristics of the structural transition zone and its control on reservoirs in Wulanhua depression. Oil Geophysical Prospecting. 56(02): 380–388+155.

Fertl, W. (1980) Gamma-ray spectral evaluation techniques identify fractured shale reservoirs and source-rock characteristics. Journal of Petroleum Technology. 32(11): 2053–2062.

Mallick, R.K., Raju, S.V. (1995) Application Of Wireline Logs In Characterization And Evaluation Of Generation Potential Of Palaeocene-lower Eocene Source Rocks In Parts Of Upper Assam Basin, India[M]. Society of Petrophysicists and Well-Log Analysts.

Passey, Q.R., Moretti, F.J., Kulla, J.B., et al. (1990) A practical model for organic richness from porosity and resistivity logs. AAPG Bulletin. 74(12):1777–1794.

Shi, X., Wang, J., Liu, G., et al. (2016) Application of extreme learning machine and neural networks in total organic carbon content prediction in organic shale with wireline logs[J]. Journal of Natural Gas Science and Engineering, 33:687–702.

Yuan, H.Q., Liu, C.L., Xiao, Y., et al. (2016) Sedimentary characteristics of Wulanhua sag in Erlian basin. Progress in Geophysics. 31(5): 2239–2245.

Advances in Mineral Resources, Geotechnology and Geological Exploration – Rashid & Zhang (eds)
© 2023 Copyright the Author(s), ISBN 978-1-032-33377-9

Analysis of the multi-functional spatiotemporal pattern of cultivated land and its influencing factors in the area of Zuoyoujiang in Guangxi Province

Feng Wei* & Yan Fei Wei*
School of Natural Resources and Geomatics, Nanning Normal University, Nanning, China

Xin Hua Tong*
School of Geographical Sciences and Planning, Nanning Normal University, Nanning, China

ABSTRACT: In order to reveal the spatial distribution characteristics of agricultural production, social security, and ecological security functions of cultivated land at the county-level of the Zuoyoujiang area of Guangxi, this paper analyzes the multi-functional spatial status of cultivated land in the Zuoyoujiang area of Guangxi. By constructing a multi-functional evaluation system of cultivated land, this paper analyzes the relationship of multi-functional cultivated land in 32 counties of the Zuoyoujiang area of Guangxi and uses the grey correlation analysis method and R/S analysis method to reveal the influence of each index on the multifunctional system of cultivated land and the future trend of the multifunctional index of cultivated land. The results show that from 2005 to 2020, the overall function of the cultivated land in the Zuoyoujiang area of Guangxi has been enhanced; from 2005 to 2020, the average value of the multi-functional comprehensive index of the cultivated land in the Zuoyoujiang area of Guangxi is in a state of "high at the edge and low in the middle"; the impact of each indicator on the overall function of cultivated land in the Zuojiang River in Guangxi is mainly due to the transition from the population, cultivated land resource endowment and factor input in 2005 to the level of economic development and urbanization in 2020; by calculating the Hurst value with the R/S analysis method, it can be seen that most counties have persistent characteristics, and only a few regions have anti-persistence characteristics.

1 INTRODUCTION

Cultivated land is the basic resource and condition for human survival, an important basic condition for the formation and continuation of human civilization, and a complex ecosystem with the highest degree of human dependence (Song 2014). The report of the 19th National Congress of the Communist Party of China put forward the strategy of rural revitalization, and proper handling of agricultural and rural peasant problems has become a new social hotspot. As an important carrier to solve the "three rural" problems, it is more important to re-understand the function and status of cultivated land. The traditional functions of cultivated land mainly focus on the supply of crops and other food, which cannot meet the needs of today's multi-objective orientation. With the development of economy and society, in order to alleviate the pressure on resources and the environment and promote the construction of ecological civilization, cultivated land is no longer limited to single-purpose use, and the importance of multiple functions such as life and ecology has become more and more prominent. At the same time, social and economic development promotes the expansion of multi-functional areas of arable land, which also leads to problems such as marginalization of arable land resources, non-agricultural utilization of arable land, and fragmentation of arable land (Song 2012). Therefore, to have a deep understanding of the meaning of multifunctional cultivated land, and to balance the relationship between cultivated land function development and cultivated land protection while continuously enriching cultivated land functionality, there is not only the need

*Corresponding Authors: 525608447@qq.com, 65358105@qq.com, tongxh001@163.com

DOI 10.1201/9781003319412-60

of the times but also a scientific exploration to maximize the use-value of cultivated land. Therefore, to have a deep understanding of the meaning of multi-functional cultivated land, and to balance the relationship between cultivated land function development and cultivated land protection, it needs a scientific exploration not only to maximize the use-value of cultivated land, but also to promote the transformation of my country's cultivated land protection from a single quantitative management or protection of both quantity and quality to multi-functional comprehensive management, which can provide a theoretical basis and policy support for deepening the understanding of cultivated land protection and development and maintaining social harmony and stability.

2 RESEARCH STATUS

The United Nations Conference on Environment and Development put forward the concept of "agricultural versatility" in the 1990s (Peng 2014), and scholars at home and abroad have carried out multi-level exploration based on the functional perspective. Foreign scholars' research on cultivated land function mainly focuses on the concept of cultivated land function (Ferrari 2009); (Aubry 2012), multi-functional management of cultivated land (Coyle 2016; Jongeneel 2008), multi-functional evaluation and pattern recognition of cultivated land (Van der Ploeg 2009), and the influencing factors and quantification of cultivated land value (Bastian 2002). Domestic scholars mainly focus on the construction of a multi-functional index system of cultivated land (Liu 2021; Xiang 2019; Yang 2014, 2022), the relationship between time and space evolution (Wang, 2018; Yang & Tan, 2014; Zhang, 2018), the coupling and synergy relationship (Fu, 2021; Lu, 2021; Zhang, 2020), and the analysis of functions (Ke 2016; Tan 2012; Wang 2020). In the choice of research space, scholars at home and abroad not only focus on macro-scales such as countries (Liu 2021; Song & Ou 2012; Xiang 2019), provinces (Luo 2016), cities (Shi 2015; Tang 2021), and counties (Lu 2021); (Jin 2019), but also include typical areas such as special landform areas (Du 2015; Fan 2018) and major economic areas (Chen 2018; Dong 2019); in the subdivision of research units, scholars mainly divide different types of cultivated land functional divisions from two aspects: grid (Qian 2020) and administrative unit. The above studies have promoted the in-depth study of multi-functional cultivated land, but there are generally more studies on the quantitative relationship between the various functions of cultivated land, while relatively ignoring the changes in the spatial pattern.

3 OVERVIEW OF THE STUDY AREA AND DATA SOURCES

3.1 *Overview of the study area*

The Zuoyoujiang area is located in the southwest of Guangxi, between 104°28′ ~ 109°09′E and 21°36′ ~ 25°37′N. In 2015, the State Council approved the implementation of the "Plan for the Revitalization of the Old Revolutionary Areas of Zuoyoujiang (2015–2025)", which includes the whole territory of Baise, Hechi, and Chongzuo in Guangxi, as well as Mashan and Long'an counties in Nanning, covering an area of 91,700 square kilometers. The cultivated land area is about 1.4378 million hectares, accounting for 15.68% of the total area of the study area. The basin is rich in forest and wetland resources, with diverse ecosystems, and karst landforms account for more than 66% of the basin area. While creating rich tourism resources, it has strong constraints on local agricultural development and is relatively backward in the economy.

Relevant statistics show that in 2020, the total sown area of crops in the area of Zuoyoujiang in Guangxi will decrease by 94,747 hectares compared with 2005, and the grain output in 2020 will decrease by 157,269 tons compared with 2005. Under this background, it is of great significance to study the multifunctional changes and spatial differences of cultivated land in the Zuoyoujiang area of Guangxi for optimizing the utilization mode of cultivated land and rationally allocating cultivated land resources in the research area.

3.2 *Data source and preprocessing*

The land use data of the old revolutionary base area of Zuoyoujiang in Guangxi comes from Landsat8/TM data (spatial resolution of 15m*15m) in 2005, 2010, 2015, and 2020 and the Resource and

Environmental Science Data Center of the Chinese Academy of Sciences. This paper uses counties and districts as the basic unit to carry out data statistics. The data on the total cultivated land area, population, agricultural production, economy, and ecological environment in the evaluation index system are all from the "Guangxi Statistical Yearbook" and "China County Statistical Yearbook" on the official website of the Guangxi Bureau of Statistics, the original values of the required indicators can be obtained directly or indirectly by consulting the yearbook, and some missing data are filled by linear interpolation.

3.3 *Research method*

Construction of multi-function evaluation index system of cultivated land

Cultivated land is a complex spatial system with multiple interlaced functions. The movement of its internal energy produces various service functions (Jiang 2011), covering production activities, living space, and ecosystems. The interpenetration between functions is an important manifestation of the complexity of the cultivated land system. Drawing on the basis of relevant scholars for the selection of evaluation indicators and combining the actual characteristics of cultivated land in the Zuoyoujiang area of Guangxi, this paper defines the function of cultivated land into three categories: basic life security function, agricultural economic contribution function, and ecological security maintenance function, and 12 indicators are selected to construct evaluation index system (Table 1).

The basic living security function is an indicator used to measure the ability of cultivated land to solve the problem of people's food and clothing. It is the core function of cultivated land, and it is reflected in the ability to meet basic survival needs. The contribution function of the agricultural economy is reflected in the contribution of cultivated land to the overall economy while satisfying the problem of food and clothing, as the most basic means of production for human beings. The function of maintaining ecological security is reflected in the protection of cultivated land, which is an important form of ecological civilization construction, the basis of cultivated land productivity, and plays an important role in landscape beautification and optimization of the overall ecological environment of the region.

Table 1. The evaluation index system of the multifunctional value of cultivated land in the old revolutionary base area of Zuojiang River in Guangxi.

Criterion layer	Indicator layer	Indicator description	Attribute
Agricultural production function	Grain production	Grain production/total arable land	+
	Vegetable production	Vegetable production/total arable land	+
	Fruit production	Fruit production/total arable land	+
	Multiple cropping index	The total sown area of crops/total arable land	+
Social security function	Income ratio of urban and rural residents	Per capita disposable income of rural residents/per capita disposable income of urban residents	+
	Grain self-sufficiency rate	Grain output/(regional population × 400kg)	+
	Contribution of the primary industry to the economy	Value-added to the primary industry/GDP	+
	Farmers' income level	Farmers' per capita net income	+
Ecological security maintenance function	An environmental load of arable land use	Fertilizer application converted pure quantity/total arable land area	-
	Cropland type ecological dominance degree	paddy field area/total cultivated land area	+
	Population per unit area of arable land	Permanent population/total area of arable land	-
	Land reclamation rate	Cultivated land area/total land area	+

Calculation of functional value of cultivated land

Entropy weight assignment method

The entropy weight method weights the indicators by measuring the discrete degree of the difference information. By using the entropy value to judge the degree of dispersion of an index, the smaller the information entropy value, the greater the degree of dispersion of the index, and the greater the influence (i.e., weight) of the index on the comprehensive evaluation. If the information entropy value of an indicator is larger, it means that the discrete degree of the indicator is lower, and the weight of the indicator is correspondingly smaller. Therefore, the information entropy tool can be used to calculate the weight of each index, and then the weighted sum method can be used to calculate the functional values of cultivated land[12].

(1) Calculate the entropy value of the j index. The calculation formula is:

$$e_j = -a\sum_{i=1}^{n} q_{ij} \ln \quad (q_{ij}) \tag{1}$$

When ej ≥ 0; $a > 0$, $a = \frac{1}{\ln(n)}$; $q_{ij} = \frac{A_{ij}}{\sum_{j=1}^{m} A_{ij}}$, When qij=0, $q_{ij} \ln(q_{ij}) = 0$.

(2) Calculate the weight of the j index, the calculation formula is:

$$w_j = \frac{f_j}{\sum_{j=1}^{m} f_j} \tag{2}$$

When $1 \leq j \leq m$; $\sum_{i=1}^{m} f_j = 1$; $E_e = \sum_{i=1}^{m} e_j$; $f_j = \frac{1-e_j}{m-E_e}$

R/S analysis method

R/S analysis method, also known as the scalar range analysis method, R/S analysis method can distinguish a random sequence from a non-random sequence, and through R/S analysis method can also carry out the long-term memory process of nonlinear systems quest. According to the H value, it is judged whether the future change trend of the research phenomenon is persistent or anti-persistent. If H=0.5, it means that the autocorrelation coefficient of the time series difference is 0, that is, the changes before and after the time series are irrelevant; if $0 < H < 0.5$, it means that the autocorrelation coefficient of the time series difference is less than 0, that is, the changes before and after the time series are negatively correlated; If $0.5 < H < 1$, it means that the autocorrelation coefficient of the time series difference is greater than 0, that is, the changes before and after the time series are positively correlated, and the larger the H, the stronger the continuity.

Grey relational analysis

In this study, the grey relational method was used to explore the correlation degree between each system index and the multi-function of cultivated land. The grey relational method refers to measuring the degree of correlation between factors according to the degree of similarity or dissimilarity of the development trends between the factors and judges the degree of influence of the influencing factors on the research object by comparing the degree of each correlation. The calculation formula is as follows:

$$\xi_i(k) = \frac{\min_i \min_k |y(k) - x_i(k)| + \rho \max_i \max_x |y(k) - x_i(k)|}{|y(k) - x_i(k)| + \rho \max_i \max_x |y(k) - x_i(k)|} \tag{3}$$

$$r_i = \frac{1}{n}\sum_{k=1}^{n} \xi_i(k) \tag{4}$$

In the formula, is the correlation coefficient; is the correlation degree; is the normalized parameter value; is the normalized comparison value, i is the number of comparison series (i=1,2...,n k is the number of indicators for each comparison object (k=1,2,…,m); A is the resolution coefficient, usually 0.5; the degree of correlation is divided according to the interval of the correlation degree value, where [1, 0.8] is the strongest, (0.8, 0.6] is strong, (0.6, 0.4] is medium, (0.4, 0.2] is weak, and [0.2, 0) is the weakest.

4 RESULTS AND ANALYSIS

4.1 *The spatial and temporal evolution characteristics of multifunctional cultivated land in the Zuoyoujiang area of Guangxi*

The evolution of the multifunctional comprehensive index of cultivated land

(1) Analysis of multifunctional comprehensive index of cultivated land based on time dimension
Judging from Figure 1, the average value of the multi-functional value of cropland in the Zuoyoujiang area of Guangxi is 0.3479 at the four research points, showing a fluctuating upward trend. The lowest value was 0.3057 in 2010, and the peak value was 0.3872 in 2020. It shows that with the scientific and systematic improvement of cultivated land utilization, cultivated land as a composite space system no longer only has the production function of providing food, while the value of social security function and ecological maintenance function continues to increase, and the overall function is constantly enriched. Judging from the coefficient of variation of the multi-functional comprehensive index of cultivated land in the Zuoyoujiang area of Guangxi, compared with the change of the time series mean, it shows obvious time imbalance, and generally shows a fluctuating growth trend, with a minimum value of 0.1874 in 2015 and a maximum of 0.2354 value in 2020. On the whole, the fluctuation range of the coefficient of variation from 2005 to 2010 was small, which means that the inter-annual variation of the multi-functional comprehensive index of cultivated land in the Zuoyoujiang area of Guangxi from 2005 to 2010 was small and the development was stable. During the period from 2010 to 2020, the difference in the coefficient of variation between the research time points increased, and the development was unstable. Therefore, when participating in the management of cultivated land, more attention should be paid to the synergistic utilization of cultivated land for multi-functions, so as to avoid the negative phenomenon of mutual exclusion among the various functions of cultivated land, and to improve the comprehensive function benefit of cultivated land.

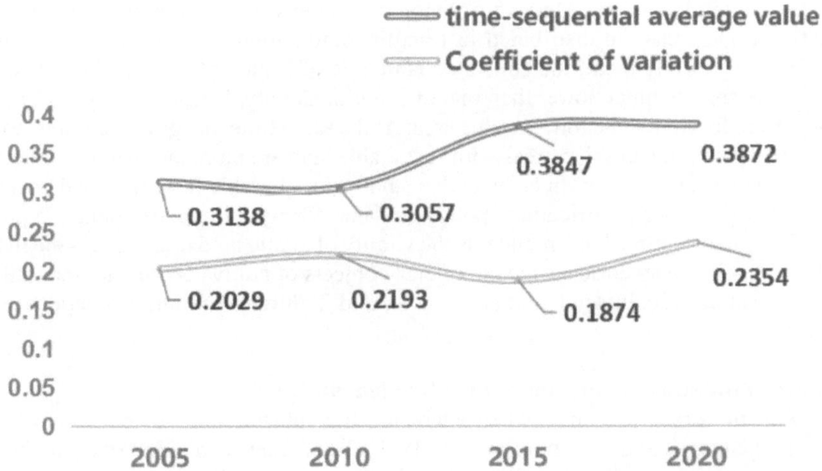

Figure 1. Time evolution trend of the multi-functional comprehensive index of arable land in the area of Zuoyoujiang in Guangxi in 2005, 2010, 2015 and 2020.

(2) Analysis of multifunctional comprehensive index of cultivated land based on the spatial dimension from Figure 2, in order to more intuitively reflect the spatial trend of the multi-functional comprehensive index of Guangxi's left and right rivers, based on the natural fracture method in ArcGIS 10.2 software, this paper divides the average value of the 2005-2020 annual comprehensive index into five grades: highest, high, medium, low, and lowest.

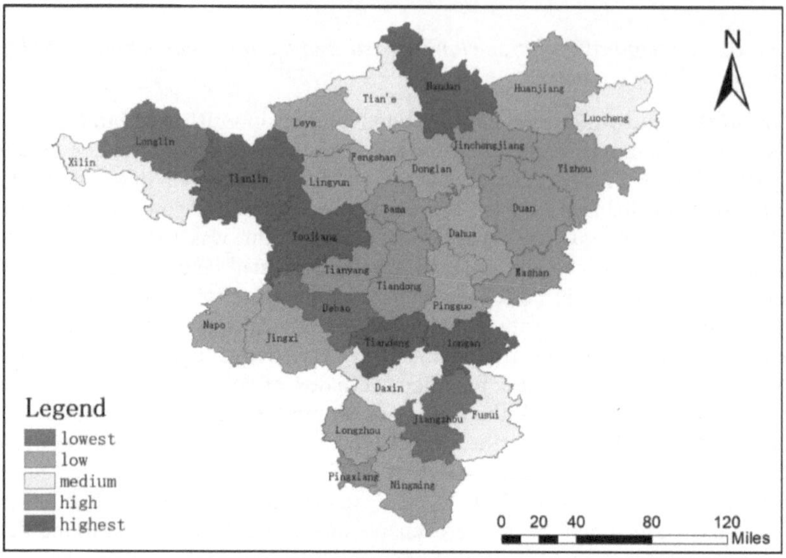

Figure 2. Spatial distribution map of the average value of the multifunctional comprehensive index of arable land in the Zuoyoujiang area of Guangxi in 2005, 2010, 2015 and 2020.

From Figure 2, it can be seen that the multi-functional highest-value areas of Zuoyoujiang arable land are distributed in Tianlin County, Youjiang District, Tiandeng County, Nandan County, and Long'an County, and the high-value areas are distributed in Jinchengjiang District and Du'an Yao Autonomous County and 8 counties, the medium area is distributed in 5 counties including Xilin County, Tian'e County, Luocheng Mulao Autonomous County, the low-value area is distributed in 11 counties including Huanjiang Maonan Autonomous County, Leye County, Ningming County, and the lowest-value areas are distributed in Longlin Autonomous County, Debao County, and Jiangzhou District. Among them, the cultivated land area of Nandan County, which is located in the highest-value area, is much lower than that of Longlin County, Debao County, and Jiangzhou District, which are located in the lowest-value area. At the same time, the grain output is leading by a large margin, which indicates that areas with vast arable land are more prone to inefficient use of arable land or even abandonment. In areas with a small area of arable land, the local government, through scientific training and agricultural policy guidance for agricultural personnel, improves the utilization efficiency of arable land, maintains the vitality of arable land use, reduces the illegal use of arable land, continuously enriches the beneficiary objects of cultivated land system utilization, improves the function of cultivated land as a whole, and highlights the multi-functional value of cultivated land.

The evolution of the sub-function index of cultivated land

Consistent with the previous paper, when analyzing the sub-functions of cultivated land, this paper uses the natural fracture method in ArcGIS 10.2 software to divide the agricultural production function, social security function, and ecological security maintenance function into five grades, in order to reflect the evolution trend of the sub-function level of cultivated land in each county.

(1) Agricultural production function

As can be seen from Figure 3, the highest-value areas of agricultural production functions in the Zuoyoujiang region of Guangxi have decreased from 8 in 2005 to 4 in 2020, and the lowest-value areas have decreased from 8 in 2005 to 4 in 2020. The 4 high-value areas in 2005 increased to 8 in 2020, and the average functional value increased from 0.0883 in 2005 to 0.1324 in 2020, which indicated that the overall agricultural production function of the area of Zuoyoujiang in Guangxi in 2020 was better than that in 2005, the gap in agricultural production function between regions gradually weakened. From a local point of view, under the background of the reduction of arable land, the grade intensity of Xilin County has changed significantly, and the grain output in 2020 has increased by 183.87% compared with 2005, indicating that the crop production capacity in this region plays an important role in the improvement of the overall agricultural production function.

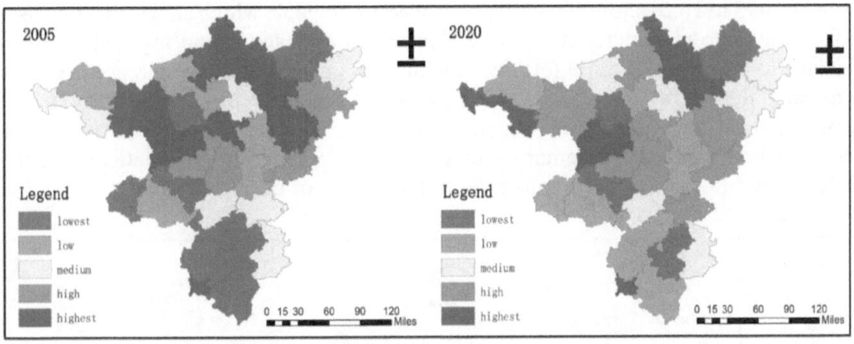

Figure 3. Evolution of agricultural production functions in the Zuoyoujiang.

(2) Social security function

From Figure 4, it can be seen that the highest-value areas of the Zuoyoujiang region of Guangxi social security function increased from 4 in 2005 to 8 in 2020, the lowest-value areas increased from 4 in 2005 to 7, and the average value decreased from 0.1667 in 2005 to 0.1548 in 2020, which indicated that the social security function of arable land in the areas of Zuoyoujiang region of Guangxi weakened from 2005 to 2020 and the gap between regions tended to widen, indicating that local governments should strengthen regional agricultural cooperation, and should strengthen the role of cultivated land in ensuring the living standards of agricultural personnel and prevent the loss of agricultural personnel. From a local point of view, Youjiang District has developed from the lowest-value area in 2005 to the highest-value area in 2020, and the income of the primary industry has increased from 2,323.1 yuan in 2005 to 4,020 yuan in 2020, which is higher than the growth rate of other counties. It is conducive to stimulating farmers' enthusiasm for farming, mainly reflected

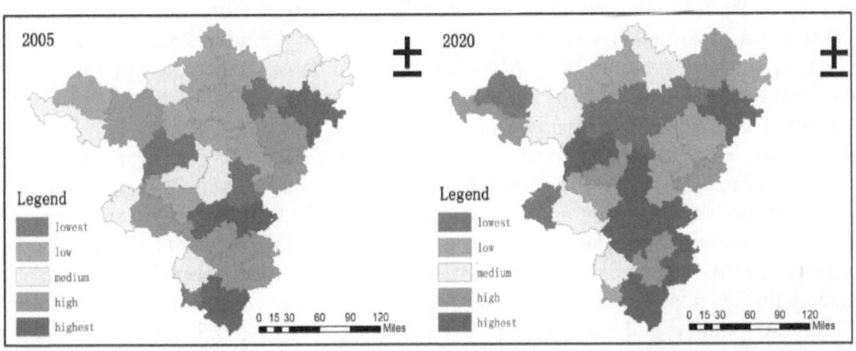

Figure 4. Evolution of social security functions in the Zuoyoujiang region of Guangxi in 2005 and 2020.

in the substantial increase in vegetable and fruit production in Youjiang District, and the per capita net income of farmers in this area has increased by 383.70%. From the perspective of spatial distribution, the social security function level of the counties in the southeast of the Zuoyoujiang region of Guangxi is higher, and the function level in the northwest is lower.

(3) Ecological security maintenance function

As can be seen from Figure 5, the highest-value areas of the ecological security maintenance function of the area of Zuoyoujiang in Guangxi have decreased from 4 in 2005 to 3 in 2020, the lowest-value areas increased from 5 in 2005 to 6, and the average increased from 0.0588 in 2005 to 0.0999 in 2020. It can be seen that the overall performance of the ecological security maintenance function of the area of Zuoyoujiang in Guangxi is poor. Taking Huanjiang Maonan Autonomous County as an example, the amount of chemical fertilizer application in this area has increased from 10,561 tons in 2005 to 16,574 tons in 2020, an increase of 56.94%, which shows that the district and counties focus on improving regional agricultural production capacity, ignoring the limited ability of cultivated land to digest agricultural fertilizers, which not only limited the increase in grain output, but also affected the quality of cultivated land and weakened the ecological security maintenance function of cultivated land. Therefore, local governments should realize that the various functions of cultivated land should be complementary and synergistic, and only by systematically participating in cultivated land governance can promote the improvement of overall benefits.

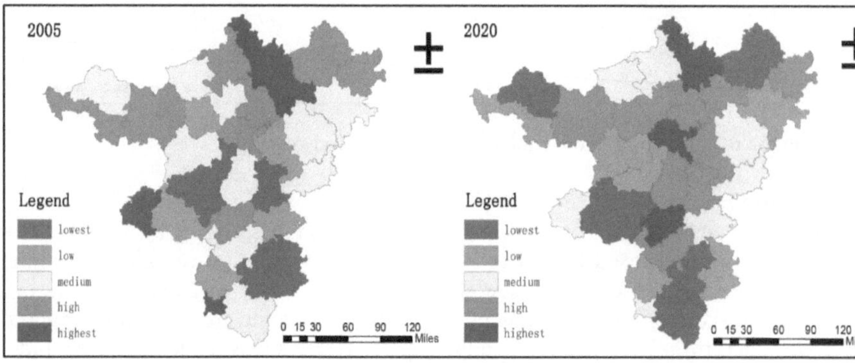

Figure 5. Evolution of the ecological security maintenance function of the area of Zuoyoujiang in Guangxi in 2005 and 2020.

4.2 *R/S Analysis*

Based on the R/S analysis method, this paper calculates the Hurst value to predict the sustainable characteristics of the multi-functional time series development of cultivated land in each county. Based on the multifunctional comprehensive index of cultivated land in 23 counties of the area of Zuoyoujiang in Guangxi from 2005 to 2020, the Hurst value was calculated by univariate linear regression on the ln(R/S) and ln(n) of the multifunctional comprehensive index of cultivated land in each county. The results are shown in Figure 6.

In general, the Hurst values of most counties are in the range of (0.5-1], that is, the multi-functional comprehensive index of cultivated land in these counties generally continues the trend of past development, showing continuous characteristics; while Pingxiang City and Jingxi City are between (0, 0.5], indicating that the multi-functional comprehensive index of cultivated land in these two regions is negatively correlated with the past development trend, showing the characteristics of anti-sustainability; in addition, Bama Yao Autonomous County, Luocheng Mulao Autonomous County, Long'an County, and Napo County are slightly larger than 0.5, indicating that although these three areas also have certain continuity, these three areas are less likely to continue past trends than counties with larger Hurst values.

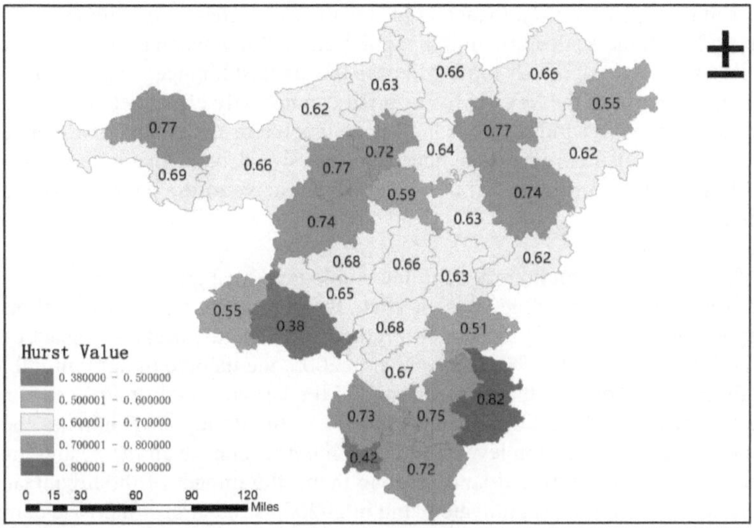

Figure 6. Distribution map of Hurst value in the area of Zuoyoujiang in Guangxi.

4.3 *Grey relational analysis*

Taking the total function index and sub-function index of the arable land in the area of Zuoyoujiang in Guangxi in 2005 and 2020 as the reference series, 9 impact index values are selected as the comparison sequence to calculate the gray correlation degree of the comparison sequence relative to the reference sequence (Table 2).

Table 2. Grey correlation degree between multi-functional level and influencing factors of arable land of the area of Zuoyoujiang in Guangxi in 2005 and 2020.

	General function		Agricultural production function		Social security function		Ecological security maintenance function	
Influencing factors	2005	2020	2005	2020	2005	2020	2005	2020
Cultivated area	0.7535	0.7298	0.6323	0.4785	0.8524	0.5582	0.4950	0.7846
Land Reclamation rate	0.7528	0.7302	0.6317	0.4786	0.8532	0.8379	0.4947	0.7649
GDP	0.3580	0.7517	0.3357	0.3728	0.3388	0.3457	0.5220	0.3576
GDP per capita	0.3763	0.3357	0.3553	0.3641	0.3523	0.3405	0.5600	0.3513
Per capita net income of rural residents	0.3787	0.7305	0.3579	0.5464	0.3541	0.9031	0.5650	0.4714
Total population	0.9970	0.5561	0.8399	0.4920	0.6929	0.4632	0.5846	0.6609
Fertilizer usage	0.8143	0.6772	0.9170	0.5027	0.6071	0.4645	0.6771	0.8142
Total Mechanical power	0.5418	0.6343	0.5448	0.9486	0.4626	0.5658	0.9894	0.6636
Urbanization rate	0.3422	0.7426	0.3006	0.6377	0.3256	0.8566	0.4632	0.7921

(1) Agricultural production function

In 2005, the use of chemical fertilizers and the total population had the greatest impact on the agricultural productivity of arable land. In 2020, the greatest impact on the productivity of arable land was the total mechanical power, which shows that in the early stage of cultivated land utilization, the agricultural production function is the main manifestation of cultivated land function, and the function of cultivated land is mainly reflected by the high input of chemical fertilizers and the large-scale agricultural population. In 2020, when agricultural technology is mature, the amount of machinery input will have the greatest impact on agricultural production capacity. The increase in the level of factor input can not only effectively transfer agricultural personnel to higher-level industries, which is conducive to the extension of the agricultural industry chain, but also eliminate the need for excessive dependence on large-scale input of chemical fertilizers for farming output, which is conducive to the construction of ecological civilization.

(2) Social security function

In 2005, the land reclamation rate and the arable land area had the greatest impact on the social security capacity of arable land. In 2020, the per capita income of rural residents, the urbanization rate, and the land reclamation rate would have the greatest impact on the social security ability of arable land. This shows that in 2005, the income of agricultural households was mainly derived from the farming output, and the key to social security capacity is "land available for farming". In 2020, with the changes in the policy environment, the economic development level, urbanization level, and factor input level have an increasing impact on the social security level of cultivated land. Among them, the impact of the urbanization rate on the social security function of cultivated land in 2020 has the largest increase compared with 2005, which also shows that the level of urbanization has an impact on the direction of regional cultivated land use.

(3) Ecological security maintenance function

In 2005, the total mechanical power and the use of chemical fertilizers had the greatest impact on the ability to maintain the ecological security of farmland. In 2020, the amount of chemical fertilizer, the urbanization rate, and the area of farmland had the greatest impact on the ability to maintain the ecological security of farmland. This shows that the level of agricultural science and technology in 2005 was relatively low. While using machinery and chemical fertilizers to increase production, the negative impact of water, soil, and air pollution on the ecological environment was ignored.; In 2020, when the country emphasizes ecological protection, the improvement of the quality of green fertilizers and the key protection of arable land elements are conducive to the improvement of the ecological safety maintenance function of arable land. However, the continuous improvement of the level of urbanization and the construction of roads and other infrastructure facilities have fragmented some arable land resources, destroying the integrity of arable land to a certain extent, and the fragmentation of arable land will have a negative impact on the ability to maintain ecological security.

5 CONCLUSION

Based on the panel data of 32 counties in the Zuoyoujiang area of Guangxi from 2005 to 2020, this paper reveals the temporal and spatial evolution characteristics of the multi-functional cultivated land in the Zuoyoujiang area of Guangxi from 2005 to 2020, the R/S analysis method was used to analyze the future trend of multi-functional indicators of cultivated land, and the grey relational analysis method was used to identify the impact of each index on the agricultural production function, social security function and ecological security maintenance function in each year and the following conclusions were drawn:

(1) From 2005 to 2020, the overall multi-functional value of arable land in the Zuoyoujiang area of Guangxi showed a fluctuating upward trend. While the agricultural production capacity was continuously improved, the social security capacity and ecological security maintenance capacity of the arable land also continued to improve, and the overall function continued to increase. However, during the period from 2010 to 2020, the difference in the coefficient of variation of the multifunctional index of cultivated land increased, and attention should be paid to the coordinated development of multifunctional cultivated land.

(2) From 2005 to 2020, the average value of the multi-functional comprehensive index of cultivated land in the counties of the Zuoyoujiang area of Guangxi showed differences. The high-value areas are generally distributed on the fringes of the area of Zuoyoujiang in Guangxi, and the regional multifunctional index of cultivated land does not only depend on the area of cultivated land. For example, in Nandan County, where the cultivated land area is not dominant, the grain output is outstanding, and the average multi-functional comprehensive index of cultivated land is high.
(3) On the whole, the impact of each indicator on the overall function of cultivated land in the Zuoyoujiang area of Guangxi is mainly due to the transition from the population, cultivated land resource endowment, and factor input in 2005 to the level of economic development and urbanization in 2020.
(4) Through the R/S analysis method to calculate the Hurst value, it can be seen that most counties have persistent characteristics, and only a few areas have anti-sustained characteristics.

ACKNOWLEDGMENTS

This article is financially supported by the National Natural Science Foundation of China "Analysis and Time-Space Simulation of the Mechanism of Mutual Feedback between Land Use and Poverty in the Old Revolutionary Areas of the Left and Right Rivers of Guangxi" (Approval Number: 41861039), 2017 Guangxi Philosophy and Social Science Planning Research Project "Research on the Formation and Spatial Pattern of Poverty Characteristics in Guangxi under the Constraints of Resources and Environment" (Approval Number: 17FGL014) and 2017 Guangxi Philosophy and Social Science Planning Research Project "Research on the Innovation of Land Poverty Alleviation Policy under the Constraints of Resources and Environment" (Approval Number: 17FGL016).

REFERENCES

Aubry, C., Ramamonjisoa, J., Dabat, M.H., et al. Urban agriculture and land use in cities: An approach with the multi-functionality and sustainability concepts in the case of Antananarivo (Madagascar)[J]. Land use policy, 29(2): 429–439 (2012).

Bastian, C.T., McLeod, D.M., Germino M J, et al. Environmental amenities and agricultural land values: a hedonic model using geographic information systems data[J]. Ecological Economics, 40(3): 337–349 (2002).

Coyle, C., Creamer, R.E., Schulte, R.P.O., et al. A functional land management conceptual framework under soil drainage and land-use scenarios[J]. Environmental Science & Policy, 56: 39–48 (2016).

Chen, L., Liu, J., Hao, J.M., et al. Comprehensive evaluation of multi-functional operation effects of cultivated land systems in metropolitan areas: Taking Beijing as an example [J]. Journal of Beijing Normal University (Natural Science Edition), 54(03): 284–291 (2018).

Du, J.F., Yuan, Z.Y. Discussion on the protection threshold of mega-urban based on multi-functional requirements of arable land: Taking the Pearl River Delta as an example [J]. Journal of Natural Resources, 30(08): 1255–1266 (2015).

Dong, P.Y., Zhao, H.F. Research on the multi-functional trade-off and synergy of cultivated land: Taking Qingpu District of Shanghai as an example [J]. Resources and Environment of the Yangtze River Basin, 28(02): 368–375 (2019).

Ferrari, S., Rambonilaza, M. Agricultural Activities, Rural Areas and Natural Environment: Drawing Up the Frontiers of the Multifunctionality Concept[M]//Rural Landscapes and Agricultural Policies in Europe. Springer, Berlin, Heidelberg, 25: 21–34 (2009).

Fu H.Z., Guo, X., Zhou, B.J., et al. Research on the multi-functional spatiotemporal evolution, coordination and trade-off of cultivated land resources in southern hilly and mountainous areas: Taking Anfu County as an example [J]. Anhui Agricultural Science, 49(06): 85–91+153 (2021).

Fan, Y.T., Jin, X.B., Xiang, X.M, et al. Multifunctional evaluation and spatial characteristic analysis of cultivated land in southern Jiangsu [J]. Resources Science,40(05): 980–992 (2018).

Jongeneel, R.A., Polman, N.B.P., Slangen, L.H.G.Why are Dutch farmers going multifunctional? [J]. Land use policy, 25(1): 81–94 (2008).

Jin, W.J., Bian, Z.X., Wei, Z.Y, et al. Multifunctional evaluation of cultivated land around Shenyang city [J]. China Agricultural Resources and Zoning, 40(05): 166–173 (2019).
Jiang, Guang.H., Zhang, F.R., Kong, X.B., et al. (2011) Hierarchy of multi-function of cultivated land and its protection [J]. China Land Science, 25(08): 42–47 (2011).
Ke, X.L., Li, H.Y., Liu, R.X. Spatial Matching Pattern of Recreational Function and Accessibility of Cultivated Land Landscape in Wuhan City [J]. Yangtze River Basin Resources and Environment,25(05): 751–760 (2016).
Liu, B.T., Lu, P., Liu, S., et al. Multifunctional spatiotemporal pattern of cultivated land in major grain-producing areas in China and its optimization [J]. East China Economic Management, 35(12): 92–99 (2021).
Lu, C., Geng, H., Liu, R., et al. Analysis on the spatiotemporal pattern of multi-functional coupling coordination degree of cultivated land in Shandong Province [J]. Journal of Agricultural Machinery,52(03): 265–274 (2021).
Luo, C., Cai, Y.Y. The temporal and spatial evolution of the function of cultivated land resources in the main agricultural producing areas of Hubei Province [J]. Economic Geography, 36(03): 153–161 (2016).
Peng, J., Liu, Z.C., Liu, Y.X. Research progress on the evaluation of agricultural multifunctionality [J]. China Agricultural Resources and Zoning, 35(06): 1–8 (2014).
Qian, F, Chi Y., Lal R. Spatiotemporal characteristics analysis of multifunctional cultivated land: A case-study in Shenyang, Northeast China[J]. Land Degradation & Development, 31(14): 1812–1822 (2020).
Song, X.Q., Wu, Z.F. Changes of cultivated land function in China since 1949 [J]. Acta Geographica Sinica, 69(4): 435–447 (2014).
Song, X.Q., O, Y.Z. Multifunctional connotation of arable land and its enlightenment to arable land protection [J]. Advances in Geographical Sciences, 31(07): 859–868 (2012).
Song, Xi.Q., Ou, Y.Z. Discussion on the practice path of multi-functional management of cultivated land in China [J]. Journal of Natural Resources, 27(04): 540–551 (2012).
Shi, Y.Y., Zhao, H.F., Yun, W.J., et al. Interpretation of multi-functional spatial differentiation of cultivated land in Beijing and its social and economic coordination model [J]. Resources Science, 37(02): 247–257 (2015).
Tan, Y.Z., Wang, Q.R., Chen, J., et al. Research progress and review of non-market value evaluation methods for cultivated land resources [J]. Journal of Natural Resources, 27(05): 883–892 (2012).
Tang, C.Y., Zang, J.M. Multi-function evaluation and spatial evolution analysis of cultivated land in Guangzhou in the new era [J]. Journal of Agricultural Resources and Environment, 38(02): 332–343. DOI: 10.13254/j.jare.2020.0648(2021).
Van der Ploeg, J.D., Laurent, C., Blondeau, F., et al.Farm diversity, classification schemes and multifunctionality[J]. Journal of environmental management, 90: S124–S131 (2009).
Wang, C., Peng, Q., Tang, N., et al. Research on the temporal and spatial evolution of multifunctional cultivated land and its coordination and trade-off from 2005 to 2015: Taking Shapingba District of Chongqing as an example [J]. Geographical Sciences, 38(04) :590–599 (2018).
Wang, Y.H., Li, X.B., Xin, L.J., et al. Research on the Spatial Differentiation of the Social Security Function of Cultivated Land Assets: A Comparison of Different Agricultural Type Areas [J]. Advances in Geography, 39(09): 1473–1484 (2020).
Xiang, J.W., Liao, X.L., Song, X.Q., et al. Regional convergence of multifunctional cultivated land in China [J]. Resources Science, 41(11): 1959–1971 (2019).
Yang, F.Y.Z., Hu, W.Y., Liu, T., et al. The scale effect of the multifunctional trade-off and synergy of arable land: Taking Wuhan City Circle as an example [J/OL]. Zhejiang Agricultural Journal: 1–12 (2022).
Yang, X., Tan, M.H. Multifunctional evolution of cultivated land in Beijing in recent years and its correlation [J]. Journal of Natural Resources, 29(05): 733–743 (2014).
Yang, X., Tan, M.H. Spatial differences and evolution of cultivated land functions in Beijing [J]. Geographical Research, 33(06): 1106–1118 (2014).
Zhang, Y.N., Long, H.L., Ge Dajun, et al.The spatiotemporal characteristics and driving mechanism of the evolution of cultivated land functions in the Huanghuaihai Plain [J]. Acta Geographica Sinica, 73(03): 518–534 (2018).
Zhang, Y.D., Liu, X.L., Ren, J., et al. Research on the transformation of cultivated land use based on the multifunctional trade-off and synergistic analysis of cultivated land—taking Beijing as an example [J]. China Agricultural Resources and Zoning, 41(06): 25–33 (2020).

Advances in Mineral Resources, Geotechnology and Geological Exploration – Rashid & Zhang (eds)

Model test study on thermodynamic characteristics and bearing deformation behavior of energy pile in loess foundation

Qing Li & Weiping Cao*
School of Civil Engineering, Xi'an University of Architecture and Technology, Xi'an, Shanxi, China

ABSTRACT: Energy pile is a new shallow geothermal energy mining technology that involves the exchange of heat with surrounding soil through the built-in circulating pipe. To analyze the variation law of the thermodynamic characteristics of the pile body during the cold and heat cycle of the energy pile in the loess foundation, based on the indoor model test, the variation processes of the pile body temperature, the pile top settlement, the pile internal force of the energy pile under the two working conditions of temperature load and thermal-mechanical coupling are measured. The test results show that the temperature of the energy pile in the loess foundation decreases with the increase in effective depth of the pile; at the end of the cold and thermal cycles, the pile top of the energy pile produces cumulative settlement; the strain of pile body in sandy soil layer is greater than that in loess layer.

1 GENERAL INTRODUCTION

Energy pile technology combines traditional pile foundation with ground source heat pump technology, and the heat exchange with the surrounding soil is carried out through the pipe built in the pile to exploit the shallow geothermal energy. This has many advantages, such as high space utilization, economy, and green.

Loria et al. (2015) analyzed the thermodynamic characteristics of energy piles in saturated sand through numerical simulation and found that the displacement of the pile top will accumulate continuously and that the heat of the pile body will transfer significantly when the energy pile is subjected to thermal load. Melis et al. (2019) compared and analyzed the influence of pile top constraint based on a field test, and found that the thermal behavior of energy pile highly depends on the constraint level of pile top and pile end. Guo et al. (2019) found through in situ test analysis that the greater the strength of the soil around the pile, the stronger the restraint on the pile body, and the greater the additional stress on the pile body. Yang et al. (2021) found through the indoor model test that the thermal load will cause the temperature of pile and soil to rise, resulting in the gradual accumulation of pile internal force and pile top displacement.

Most relevant studies focus on the thermodynamic characteristics of energy piles in a single temperature cycle, and there are still few studies on energy piles in loess foundations with small heat capacities. Based on the model test, this paper studies the thermodynamic characteristics and bearing deformation behavior of energy pile in loess foundation under multiple cold and heat cycles, and obtains some regular understanding, which can be used as useful reference for the application of energy pile technology in practical engineering.

*Corresponding Author: 2428840921@qq.com

DOI 10.1201/9781003319412-61

2 MODEL TEST

2.1 *Test device*

The test device is mainly composed of a model tank, loading and measuring device, and temperature cycling device, as shown in Figure 1.

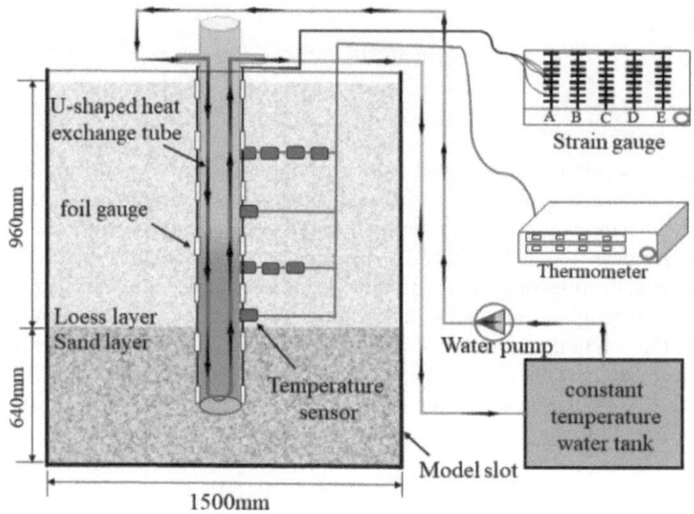

Figure 1. Schematic diagram of model test.

The size of the model box is 1500 mm (length) × 800 mm (width) × 1800 mm (high); the loading and measuring devices are RSC-5 hydraulic cylinder and DYLY-104 axial force meter, respectively; the temperature measuring device adopts PT100-A rolling sensor with an accuracy of 0.15%, and the sensor is connected with EWP eight loop digital display instrument to display the temperature; MP-10RZ micro water pump with stable flow rate is adopted for circulating power equipment.

During the thermal cycle, a constant heating rod with power 2 kW is used to heat the circulating liquid in the water tank; cold circulation adopts the method of ice water mixing.

Table 1. Model pile parameters.

Pile length /mm	Pile diameter /mm	Compressive strength /MPa	Modulus of elasticity /GPa	Coefficient of thermal expansion/°C	Thermal conductivity λ kJ/(m· h·°C)
1400	81.6	30.6	30	10^{-5}	10.6

BX120-20AA resistance strain gauge is pasted at the symmetrical position of the pile body surface every 200 mm along the pile length and protected with epoxy resin. The base size of the strain gauge is 20 mm × 3 mm, the sensitivity coefficient is 2.08±1%, and the resistance value is 119.8±0.1 Ω. CML-1H-32 electronic strain gauge is used to measure and read the strain of the pile body in the way of a 1/4 bridge, and the average value of the readings of two strain gauges at the symmetrical position of the same section is taken as the strain of the pile body at this position.

In the model test, sand is used as pile tip soil and loess is used as pile side soil. The measured maximum and minimum dry density of sand are 1800 kg/m^3 and 1690 kg/m^3, respectively, and the relative density is 55.6%.

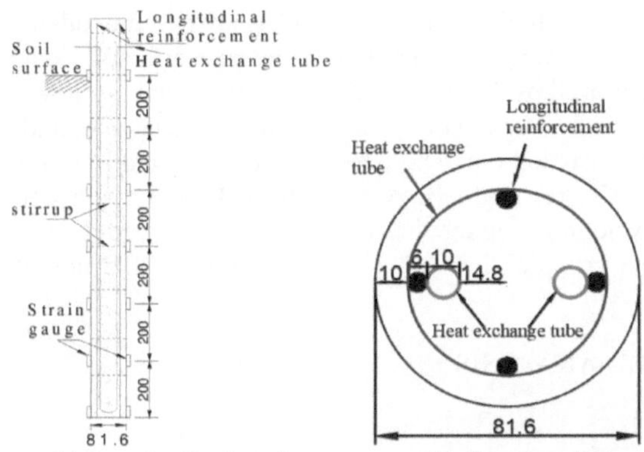

Figure 2. Reinforcement layout of model pile.

Undisturbed loess is taken from Lantian, Shanxi, and its basic physical properties are shown in Table 2. The moisture content of remolded loess after drying and crushing is about 5% and the remolded loess C φ. The values are 53.6 kPa and 23.5°, respectively.

Table 2. Physical property parameters of undisturbed loess.

Soil sample	Moisture content/%	Natural density g/cm^3	The specific gravity of soil/Gs	Particles liquid limit/%	Plastic limit/%
Undisturbed loess	17.41	1.65	2.69	31.6	19.5

2.2 *Test process*

Fill the model trench in layers according to the thickness of each layer of 200 mm, and use the vibration trowel to control the density of sand and loess to be 1750 kg/m^3 and 1450 kg/m^3, respectively. First, fill the sand bearing layer with a thickness of 5D (D is the pile diameter) at the bottom of the model groove, then continue to fill the sand at the pile end with a thickness of 3D, and finally, the remolded loess is filled with a thickness of 960 mm. The fixed position of the model pile is shown in Figure 3.

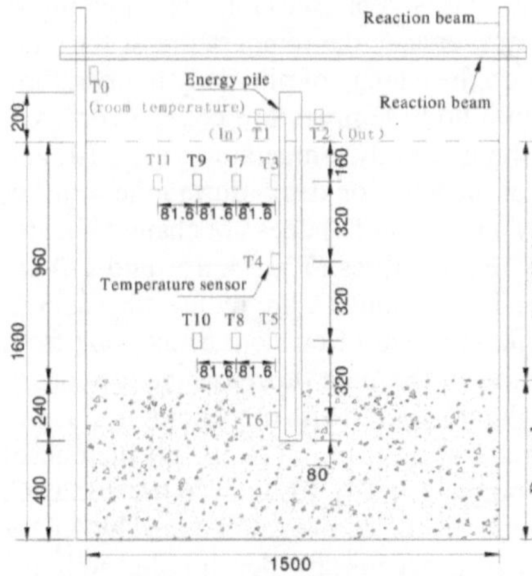

Figure 3. The layout of model pile.

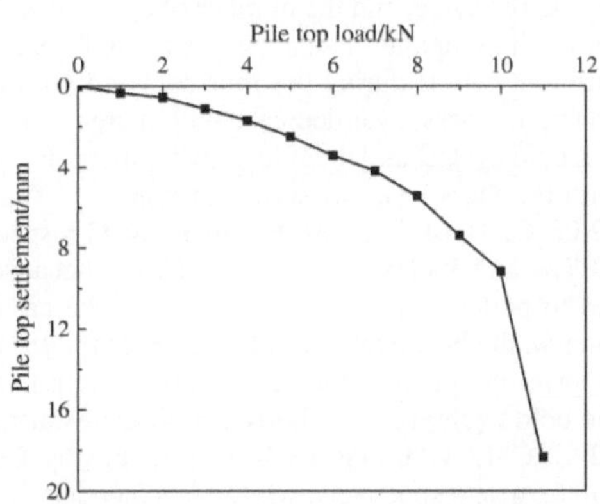

Figure 4. Q-S curve.

According to the above method, the pile and soil shall be embedded into the model groove and stand still. After the self-weight settlement of the pile and soil is stable, the settlement of the pile top shall be monitored by 100%, and the vertical downward load shall be applied on the pile top by stages until it is 5 kN and continues to remain unchanged, and then the cold and heat cycle of the energy pile is carried out. The temperature of circulating liquid during hot and cold circulation is 35°C and 4°C respectively. A thermal cycle and a subsequent cold cycle form a complete temperature cycle in which the thermal cycle is 5 h, the cold cycle is 4 h after an interval of 10 h, and then the next group of thermal cycles is carried out after an interval of 8 h.

3 TEST RESULTS AND ANALYSIS

3.1 *Pile temperature*

Figure 5 shows the change process of pile temperature at different positions with time when the energy pile is only subjected to a cold and heat cycle. During the test, the PVC hose connected between the water pump and the inlet of the heat exchange pipe of the energy pile will exchange heat with the surrounding environment during the cold and hot cycles. The heat loss makes the temperature at the inlet of the heat exchange pipe of the energy pile unable to reach the water source temperature. Finally, after the heat and cold cycles are stable, the inlet temperature is about 33°C and 12°C respectively.

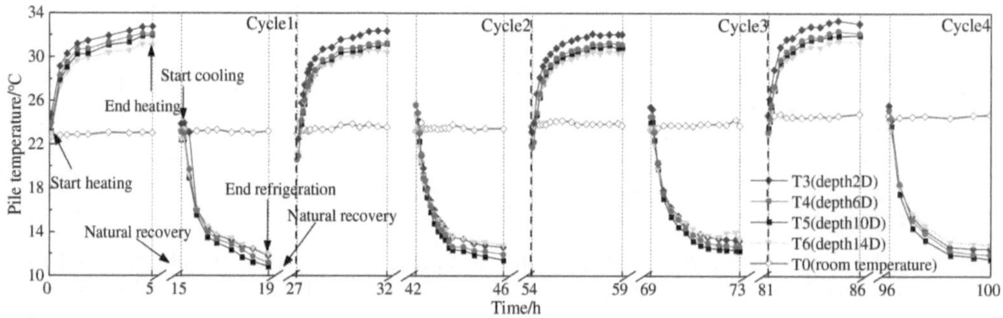

Figure 5. Temperature variation curve of the pile.

It can be seen from the time variation law of pile body temperature at different soil depths in four temperature cycles in Figure 5 that when the temperature load is kept constant, the temperature at different measuring points of the pile body gradually approaches the temperature at the inlet during the cycle, and the number of cycles has little effect on the change of pile body temperature. Through comparative analysis, it can be seen from the pile body temperature change curves at different depths during the four cold and hot cycles that the pile body temperatures (T3, T4, and T5) in the loess layer decrease with increase in effective soil penetration depth during the four hot and cold cycles, and the pile body temperature in the sandy soil layer (T6) does not change linearly with the loess layer, as shown in Figure 7. The pile body temperatures (T3, T4, T5, and T6) are 32.08°C, 31.14°C, 30.84°C and 30.44°C respectively. T3, T4, T5, and T6 are reduced by 2.93%, 3.87%, and 5.11%, respectively. This is because the circulating liquid flows out of the pipe from top to bottom, and the temperature of the circulating liquid in the pipe decreases continuously. Finally, the heat exchange efficiency of the lower part of the pile body with a deeper burial depth is lower than that of the upper part of the pile body with a shallower burial depth. At the end of the cold cycle, the pile body temperatures at measuring points T3, T4, T5, and T6 are 13.20°C, 12.67°C, 12.31°C, and 13.34°C, respectively. Compared with T3, T4 and T5 are reduced by 4.02% and 6.74%, respectively, while T6 is increased by 1.06%. This is because the density of sand in the model tank is higher than that of loess, and the mineral composition of foundation soil also has certain impact on the thermal conductivity. Because the mineral content in the sand is higher than that of loess. As a result, the specific heat capacity of sandy soil is greater than that of loess (Dong et al., 2017), so the temperature change rate at the pile end of the sandy soil layer is less than that of pile body in loess layer, and finally, the temperature of sandy soil layer is slightly higher than that of loess layer during the cold cycle.

3.2 *Pile top settlement*

Figure 6 shows the pile top settlement accumulation curve of the energy pile in the process of the cold and heat cycles. It is specified that the pile top displacement is positive upward and negative downward.

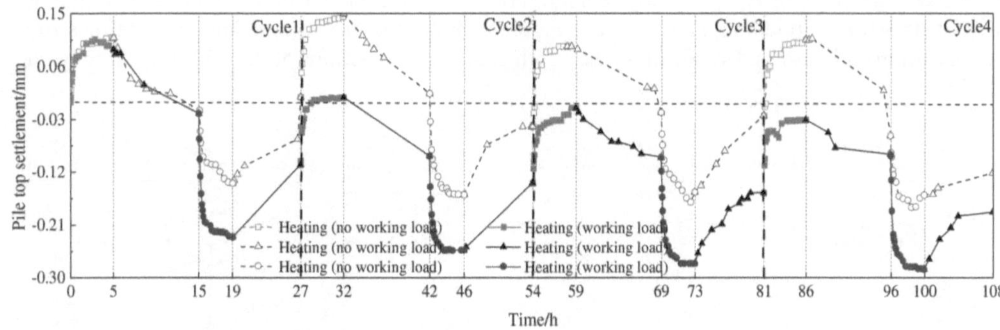

Figure 6. Pile top settlement.

In Figure 6, when the energy pile is only subjected to temperature load, the pile top will produce –0.1225 mm accumulated settlement after the end of four cycles (108 h). Some recovery deformation will occur in the natural recovery period after the end of each cycle, but they cannot be restored to the initial state. At the end of 1–4 thermal cycles, the accumulated settlement of pile top is 0.109, 0.150, 0.094, and 0.105 mm, respectively; At the end of the 1st–4th cold cycle, the accumulated settlement of pile top is –0.140, –0.1595, –0.172, and –0.181 mm, respectively. At the end of the 2nd–4th cold cycle, the accumulated settlement of pile top increases by 13.93%, 22.86%, and 29.29%, respectively, compared with the first cold cycle. When the energy pile is only subjected to temperature load, the cumulative settlement of the pile top after each cold and the thermal cycle is found to be in direct proportion to the number of cold cycles. The influence of the 4°C cold cycles on the cumulative settlement of the pile top is significantly greater than that of the 35°C thermal cycles. Therefore, the influence of the cold cycle on the cumulative settlement of energy piles needs to be considered in engineering applications.

Through the analysis of pile top settlement in the process of a single cold and thermal cycle, it was found that when only subjected to temperature load, the pile top displacement of energy pile after the end of one to four thermal cycles is 0.109, 0.134, 0.1365, and 0.140 mm, respectively. Compared with the first thermal cycle, the pile top displacement of the second, third, and fourth thermal cycles increases by 22.94%, 25.23%, and 28.44%, respectively; at the end of the first to fourth cold cycles, the pile top displacement increments of each cycle are –0.177, –0.158, –0.1385, and –0.1225 mm, respectively. Compared with the first cold cycle, the pile top displacement of the second, third, and fourth cycles is reduced by 10.73%, 21.75%, and 30.79%, respectively.

3.3 *Variation of internal force of pile body*

The pile will expand and contract during the cold and heat cycle, and the measured strain value is less than the free thermal strain due to constraints of the mechanical load on the pile side soil and the pile top. Therefore, additional heating strain will occur during the cold and heat cycle, and the additional strain will lead to the additional heating stress in the pile. Based on this, Amatya et al. (2012) proposed a calculation method for additional thermal strain and additional thermal stress inside the energy pile during the cold and heat cycle, as shown in Formulas (1)–(3):

$$\varepsilon_{R-Str} = \varepsilon_{T-Free} - \varepsilon_{T-Obs} \tag{1}$$

$$\varepsilon_{T,Free} = \alpha_c \Delta T \tag{2}$$

$$\sigma_T = -E\,(\varepsilon_{T-Free} - \varepsilon_{T-Obs}) \tag{3}$$

Among these, there are observed strain, free thermal strain, and thermal strain (constrained strain) in the axial direction of the pile body, the thermal expansion coefficient of the model pile, and the change value of ΔT pile body temperature, which are the elastic modulus and additional thermal stress of the pile body, respectively. Figure 7 shows the distribution law of four times free thermal strain and observed strain along the depth at different positions of the pile body after cyclic stability.

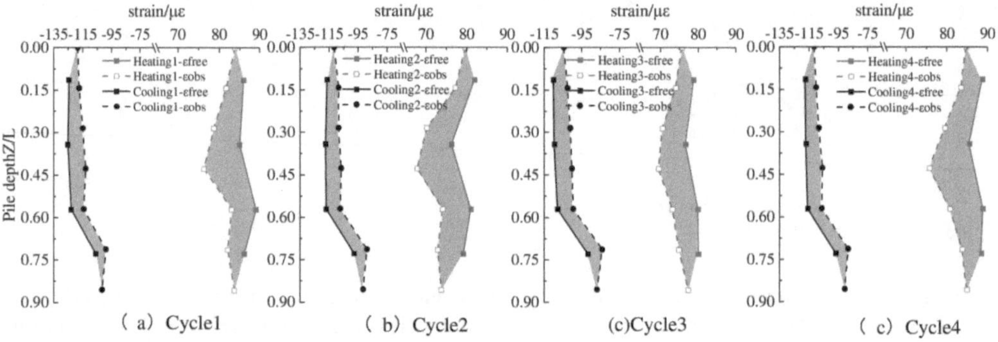

Figure 7. The distribution of strain in a pile.

The shaded part in the figure is the difference between and, that is, the attached heating strain of the pile body. Compared with Figure 7, it is found that the free thermal strain of the pile body is positively correlated with the observed strain, that is, the greater the free thermal strain, the higher the observed strain value. During the thermal cycle, the observed strain value of the pile body increases first and then decreases along the depth, and the maximum value is taken at the middle of the pile body (Z/L = 0.57). During the cold cycle, the observed strain values of the upper part of the pile body (Z/L < 0.57) show little difference, while the lower value is found to be significantly less than the upper position. Comparing the observed strain values at the pile end of four cycles, it was found that the observed strain values at the pile end (Z/L = 0.86) in the cold and hot cycle stage are slightly larger than the adjacent measuring points above (Z/L = 0.71), which is because the density of the pile end is sand in loess, and the greater the constraint, the higher is the observed strain.

4 CONCLUSIONS

Based on the indoor model test, the thermodynamic characteristics of the energy pile in the loess foundation are analyzed, and the following conclusions are obtained:

(1) After the stabilization of cold and thermal cycles, the temperature of the pile body in the loess layer decreases with increase in effective soil depth, and the temperature change rate in the sandy soil layer at the pile end is less than that in the loess layer.

(2) After the cold and heat cycle, the pile top will have cumulative settlement and cannot be restored to its initial state. When the energy pile is subjected to thermal-mechanical coupling, the cumulative settlement of the pile top is greater than that under the condition of cold and heat cycles only.
(3) After the cold and thermal cycles are stable, the pile strain at the sandy soil layer is greater than that at the relevant measuring points in the loess layer.

REFERENCES

Amatya B L, Soga K, Bourne-webb P J, et al., (2012). Thermo-mechanical behavior of energy piles. J. Géotechnique, 62(6): 503–519.

Dong X H, Ye W J, Yang G S, et al., (2017). Experimental study of the influence of temperature on thermal properties of loess. J. Rock and Soil Mechanics, 38(10): 2888–2894+2900.

Guo Y M, Zhou X, Liu S Y, et al., (2019). Prototype experimental investigation on the thermo-mechanical behaviors of free constrained full-scale PHC energy piles in multi-layer strata. J. Chinese Journal of Rock Mechanics and Engineering, 38(03): 582–590.

JGJ106-2014. Technical code for testing of building foundation piles. Beijing: China Architecture and Building Press: 4.

Kong G Q, Wang C L, Liu H L, et al., (2017). Analysis of pile head displacement of energy pile under repeated temperature cycling. J. Rock and Soil Mechanics, 38(4): 958–964.

Loria A, Gunawan A, Chao S, et al., (2015). Numerical modelling of energy piles in saturated sand subjected to thermo-mechanical loads. J. Geomechanics for Energy and the Environment, 1(1): 1–15.

Melis A, Tb B, Cgo C. (2019). Full-scale in-situ tests on energy piles: Head and base-restraining effects on the structural behavior of three energy piles. J. Geomechanics for Energy and the Environment, 18: 56–68.

Yang W B, Sun T F, Yang B B, et al., (2021). Laboratory study on the thermomechanical behavior of a phase change concrete energy pile in summer mode. J. Journal of Energy Storage, 41: 1–12.

Advances in Mineral Resources, Geotechnology and Geological Exploration – Rashid & Zhang (eds)

Study on the application of the relationship between rock mass basic quality and rock firmness coefficient

Huxiang Hou
China Road and Bridge Corporation, Beijing, China

Yirui Zhao
School of Civil Engineering, Xi'an University of Architecture and Technology, Xi'an, China

Jie Li
China Road and Bridge Corporation, Beijing, China

Zhanping Song*
School of Civil Engineering, Xi'an University of Architecture and Technology, Xi'an, China
Shaanxi Provincial Key Laboratory of Geotechnical and Underground Space Engineering/Xi'an University of Architecture and Technology, Xi'an, China

Xianze Tao
CCCC Fourth Highway Engineering Bureau Co., LTD., Beijing, China

ABSTRACT: With the development of the Belt and Road construction project, the contradictory situation of different specifications put forward a serious examination of the construction. This paper is based on the principle of volumetric joint equivalence, the empirical formula of volumetric joint coefficient expressed by integrity coefficient (K_V) and rock quality index (*RQD*) was used, and the relationship between K_V and *RQD* was deduced. Based on the principle of equivalent uniaxial compressive strength, rock firmness coefficient (f) and rock mass basic quality index (*BQ*) respectively are used to represent the rock uniaxial compressive strength of empirical formula, the relationship between *BQ* and f suggested in this paper is obtained, and the corresponding table of the relationship between F Methods and BQ Methods is established. The research results have been successfully applied in the scheme modification and optimization design of the south-north trans-ridge tunnel in Kyrgyzstan, which provides a reference for tunnel construction in countries along the "Belt and Road".

1 INTRODUCTION

There is no unified international standard for engineering rock mass classification method and selected index. For example, rock quality index classification (RQD Methods) (Deere 1964), Platts rock firmness coefficient (F Methods) (Mining Department of Northeast Institute of Technology 1959), Barton rock mass quality classification (Q System) (Barton 1974), RMR Methods (Bieniawski 1989) and engineering rock mass classification (BQ Methods) (GB/T 50218-2014). Among them, BQ methods have been widely applied and studied in China. Fan Xinran (Zhu et al. 2017) and Zhu Xunguo (FAN 2016) made a comparative analysis of the relevant contents of the *Standard for Engineering Classification of Rock mass* (GB/T 50218–2014) and obtained a simplified calculation method of basic rock mass quality index (*BQ*). Wu Aiqing (Wu & Wang 2014) proposed a

*Corresponding Author: songzhpyt@xauat.edu.cn

DOI 10.1201/9781003319412-62

rock mass classification method for rock slope engineering based on *BQ* on the basis of in-depth study and systematic digestion of rock mass classification results of slope engineering.

With the deepening of research and the emergence of a large number of international projects, many scholars try to establish the relationship between rock mass classification in various countries. Barton (Barton 2002) studied the relationship between *RMR* and *Q*. CAI Bin (Cai et al, 2001) analyzed the relationship between BQ Methods, Q System, and RMR Methods through the statistical analysis of hundreds of field measured data. Jiang Ping and Meng Ping (Jang & Weng 2004) analyzed the measured data of slope rock mass in the middle section of Yixing Road of Ning-Hang Expressway and obtained the correlation between *RMR* and *BQ*. Guo Chenye (Guo et al. 2008) regressed the test results of firmness coefficient (*f*) and uniaxial compressive strength (*UCS*) of different rocks and obtained a linear correlation between them. The above studies on the relationship between various rock classification systems focus on RMR Methods, Q systems, and BQ Methods, and do not involve the studies on F Methods and BQ Methods.

In this paper, the relationship between BQ methods and F methods was studied based on the south-north trans-ridge tunnel in Kyrgyzstan, and the gap in this research field was also made up. It provides a theoretical reference for reconciling the differences of different surrounding rock classifications, and provides a reference for tunnel engineering construction in countries along the “Belt and Road”.

2 PROJECT

The second north-south Road construction project in Kyrgyzstan (phase I) is composed of a road tunnel and a service guide tunnel, excluding composite buildings at the entrance. The total length of the tunnel is 3750m. Tunneling is carried out over Paleozoic strata with sedimentary rocks of moderate strength and fragmentation. The tunnel passes through several tectonic faults and rock fragmentation zones are widely distributed near faults. The engineering-geological cross-section is shown in Figure 1.

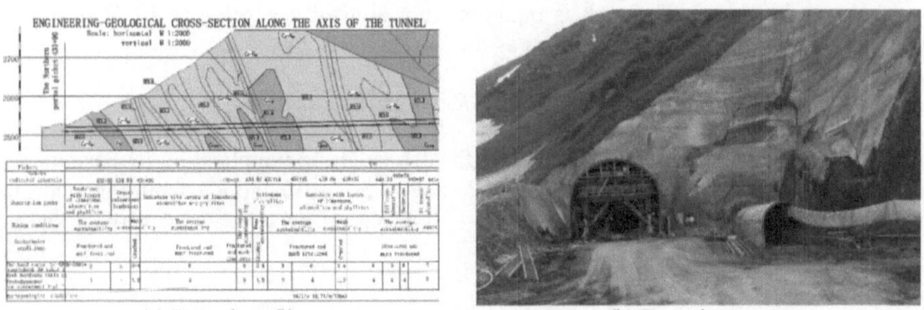

(a) Tunnel profile (b) Tunnel entrance

Figure 1. Engineering-geological cross-section along the axis of the tunnel.

Kyrgyzstan’s north-south trans-ridge tunnel was designed by the Russian standard, according to China’s tunnel construction concept, and Kyrgyzstan serves as the state supervision of the road tunnel. The tunnel was originally designed to use a cantilever tunnelling machine for construction. Due to hard sandstone and limestone are not suitable for mechanical excavation and the area within 20km near the entrance of the cave is a no-man’s land, it is not conducive to large-scale mechanical construction. Therefore, it is necessary to change the construction method of the original design to drilling and blasting.

3 ANALYSIS OF THE RELATIONSHIP BETWEEN *BQ* AND *F*

Although there are great differences between BQ Methods and F Methods, the *UCS* is used in both of them, indicating that there is a certain correlation between the two classification methods. This paper establishes the relationship between BQ Methods and F Methods based on the above content.

3.1 *Rock firmness*

Rock firmness reflects the ability of the rock to resist damage under the action of any external force. The f is determined by various methods, such as compressive strength method, crushing method, etc. Prototyakonov (Mining Department of Northeast Institute of Technology, 1959) revised the previous calculation method of f based on years of accumulated experience and proposed to use *UCS* to calculate f, as shown in Formula (1).

$$f = \frac{UCS}{10} \tag{1}$$

Guo Chenye (Guo et al., 2008) obtained the linear relationship between the f and *UCS* experimental results of different rocks by regression.

$$UCS = 19.078f + 18.361 \tag{2}$$

The striking feature of rocks compared with other materials is that the data vary greatly when the firmness is repeatedly measured. Therefore, Formulas (1) and (2) are quite different. It is difficult to avoid the inhomogeneity of the objective existence of rocks and the discreteness of the measured data, thus this paper considered the difference between them, and took the two expressions as the basis.

3.2 *Engineering rock mass classification*

According to the *Standard for Engineering Classification of Rock mass* (GB/T 50218-2014), the rock is qualitatively divided according to its hardness, integrity, weathering, and combination degree of structural planes. The rock is quantitatively divided according to *UCS* and integrity coefficient (K_V). The calculation formula is as follows.

$$BQ = 90 + 3R_C + 250K_V \tag{3}$$

Where R_C is the uniaxial compressive strength of rock. In the application of Formula (3), it should be noted that when $R_C > 90 \bullet K_V + 30$, $R_C = 90 \bullet K_V + 30$ and K_V should be put into the formula to calculate BQ; When $K_V > 0.04 \bullet R_C + 0.4$, $K_V = 0.04 \bullet R_C + 0.4$ and R_C should be substituted into the formula to calculate BQ.

3.3 *The relationship between BQ and f*

Both classification methods use the index of the uniaxial compressive strength of rock, so the relationship between BQ and f can be established according to the principle of equivalent *UCS*. Substituting Formulas (1) and (2) into Formula (3), the following formula is obtained.

$$BQ = 90 + 30f + 250K_V \tag{4}$$

$$BQ = 145.1 + 57.2f + 250K_V \tag{5}$$

Formulas (4) and (5) obtained the relation between BQ and f through *UCS* equivalent substitution, but BQ still cannot be obtained. In combination with the study of Jiang Zhaohui (Jiang et al. 2013), Merritt A H (Merritt, 1968) and WANG Shi-chun (Wang et al. 2007), as well as Formulas

(4) and (5) obtained above and based on the principle of equivalent substitution, the following eight formulas are finally obtained.

$$BQ = 182.3 + 30f + 1.205RQD \tag{6}$$

$$BQ = 79.51 + 30f + 2.335RQD \tag{7}$$

$$BQ = 237.40 + 57.2f + 1.205RQD \tag{8}$$

$$BQ = 137.61 + 57.2f + 2.335RQD \tag{9}$$

$$BQ = 80.36 + 30f + 2.283RQD \tag{10}$$

$$BQ = 88.36 + 30f + 2.335RQD \tag{11}$$

$$BQ = 135.5 + 57.2f + 2.283RQD \tag{12}$$

$$BQ = 143.5 + 57.2f + 2.335RQD \tag{13}$$

3.4 *Discussion*

In this paper, according to the BQ limit value of each surrounding rock grade, the stability of rock mass is considered to be poor when the f value is too large *and* RQD is too small, f is too small and the RQD value is too large. Upper and lower limit conditions are set for rock mass classification.

In Figure 2, the mapping method of limiting conditions is adopted to draw the contour map of BQ with RQD as abscissa and f as ordinate, so as to facilitate the selection of an appropriate rock mass classification basis. Generally speaking, when $RQD \leq 25$, the integrity of the surrounding rock is very poor, which is not conducive to tunnel excavation. Even if the rock is harder, it cannot raise the grade of the surrounding rock. When $f \leq 1$, the stability of such soft texture layers as tight clay, gravel, and sand is poor even though the integrity is very good, Therefore, the text will be used as the upper and lower limits of $RQD \leq 25$ and $f \leq 1$.

According to Figure 2, Figure (f) corresponding to Formula (17) can best reflect the classification characteristics of rock mass, so Formula (17) is selected in this paper as the basis for qualitative classification. In the application, it should be noted that when $f > 0.08 \bullet RQD + 2.94$, $f = 0.08 \bullet RQD + 2.94RQD$ should be substituted into the formula to calculate BQ; when the $RQD > 42.46f + 41.77$, $RQD > 42.46f + 41.77$ and f should be substituted into the formula to calculate BQ.

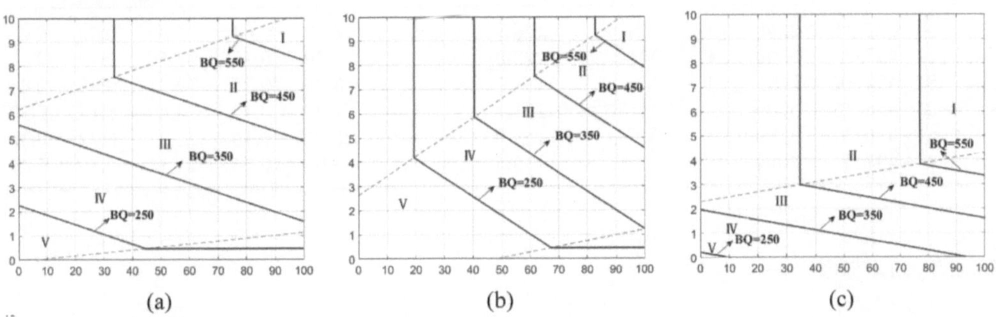

Figure 2. Classification diagram of rock mass quality.

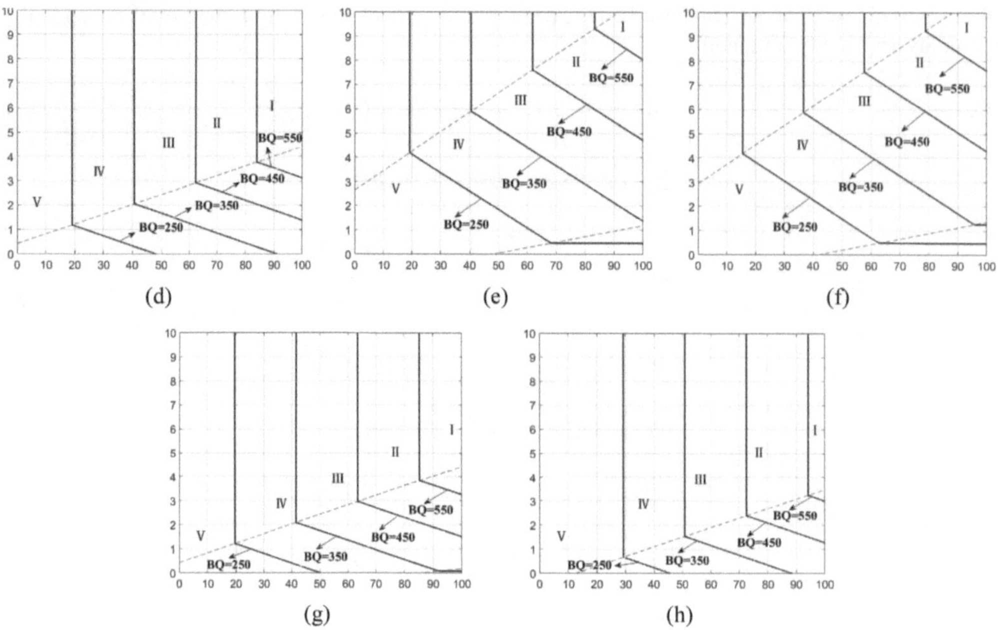

Figure 2. Continued.

4 APPLICATION

Based on the Kyrgyzstan north-south trans-ridge tunnel, the F method and BQ method are used to classify surrounding rock and estimate the physical and mechanical parameters. Table 1 gives the classification of tunnel surrounding rock of typical tunnel sections.

It can be seen from Table 1 that there is a clear correspondence between F Methods and BQ Methods. The main difference between the two methods is that the quantitative classification index of rock firmness only represents rock strength, while BQ Methods uses both *UCS* and K_V.

According to the classification results of surrounding rock and the corresponding mechanical parameters of rock mass, the tunnel entrance section is determined as the construction risk area, and the corresponding construction scheme is formulated.

Table 1. Classification of tunnel surrounding rock and estimation of rock parameters.

Pile no.	*RQD*/ %	*f*	F Methods	*BQ*	BQ Methods	γ/kN •m^{-3}	*c*/ MPa	φ/ °	*E*/ GPa	μ
K431+90~433+59	2.74	1	VII	124.7	V	18.59	0.04	15.96	0.32	0.41
K433+59~433+84	5.84	4	V	222.0	V	21.29	0.15	24.58	1.12	0.37
K433+84~434+06	2.11	1.5	VIa	138.2	V	18.99	0.05	17.01	0.38	0.41
K434+06~436+51	5.46	4	V	221.1	V	21.27	0.15	24.49	1.11	0.37
K436+51~436+92	3.74	3	Va	187.0	V	20.36	0.10	21.21	0.72	0.38
K436+92~437+14	2.64	1.5	VIa	139.5	V	19.02	0.05	17.11	0.39	0.41

5 CONCLUSION

In order to meet the needs of engineering change, the relationship between BQ Methods and F Methods is studied in this paper.

(1) Based on the principle of equivalent volume joint coefficient and uniaxial compressive strength, eight relation formulas between *BQ* and *f* were established. The proposed expression between *BQ* and *f* is obtained.
(2) According to the results of this study, it is found that the F Methods' grade IV corresponds to the BQ Methods' grade IV, and the F Methods' grade V–VII corresponds to the BQ Methods' grade V. The F Methods is not corresponding to the BQ Methods but corresponds to different engineering rock classification according to different *RQD*.
(3) According to the classification results of surrounding rock and the corresponding mechanical parameters of rock mass, the tunnel entrance section is determined as the construction risk area, and the corresponding construction scheme is formulated.

ACKNOWLEDGMENTS

This work was financially supported by the Innovation Capability Support Plan of Shaanxi Province-Innovation team (2020TD-005); Shaanxi Province Housing urban and rural Construction Science and Technology project (No2019-K39).

REFERENCES

Barton N. Some new-value correlations to assist in site characterization and tunnel design[J]. International Journal of Rock Mechanics & Mining Sciences, 2002, 39(2): 185–216.

Barton. A review of the shear strength of filled discontinuities in rock[J]. Saertrykk, 1974, 19.

Bieniawski Z T. Engineering rock mass classifications[M]. New York: The Wiley-Interscience Publication, 1989.

Cai Bin, Yu-Yong, Wu Xiao-ming. Relationship among national code, Q system, and RMR in rock mass classification and evaluation of deformation parameter. Engineering, Chinese Journal Rock Mechanics and 2001 (S1): 1677–1679.

Deere D C. Technical description of rock cores for engineering purposes[J]. Rock Mech,1964, 1(4): 17–22.

Fan Xin-ran. Simplified graphical method of basic quality index BQ in "Standard for Engineering Classification of Rock Mass" [J]. Water Resources and Hydropower Engineering, 2016, 47(5): 154–156 +170.

GB/T 50218-2014, Standard for engineering classification of rock masses[S].

Guo Chen-ye, Xian Xue-fu, Wu Xuan-hong, et al. The Relationship among Rock Gushing Energy, the Protodyakonov Coefficient and Rock Strength[J]. Chongqing: Journal of Chongqing Jianzhu University, 2008, 30(06): 28–31.

Jang Ping, Weng Wei. Research on rock mechanics parameters of rock quality grade-oriented[J]. Journal of China Three Gorges University (Natural Sciences), 2004, 26(5): 424–427.

Jiang Zhaohui, Meng Yiping, Zhang Lianfu. Study of the correlation of rock mass classification's quantitative indices in a coal mine [J]. Journal of Hefei University Technology (Natural Science), 2013, 36(06): 722–726.

Merritt A H. Engineering Classification of In-Situ rock[D]. Urbana Univ Illinois, 1968.

Mining Department of Northeast Institute of Technology, Discussion on Rock classification in China[M], Beijing, China Coal Industry Publishing House, 1959.

Wang Shi-chun, HE Fa-liang, LI Cang-song. Rock mass classification of Tunnel Engineering [M]. Chengdu, Southwest Jiaotong University Press, 2007.

Wu Ai-qing, Wang Bin. Engineering rock mass classification method based on rock mass quality index BQ for rock slope[J]. Chinese Journal of Rock Mechanics and Engineering, 2014, 33(4): 699–706.

Zhu Xun-guo, Xia Hong-chun, Wang Zhong-zhang. Application and analysis of the standard of classification of engineering rocks in the classification of surrounding rock of deep roadway[J]. Coal Geology & Exploration, 2017, 45(2): 118–125.

Advances in Mineral Resources, Geotechnology and Geological Exploration – Rashid & Zhang (eds)
© 2023 Copyright the Author(s), ISBN 978-1-032-33377-9

Application of early identification of water invasion and prediction of water breakthrough time in M gas reservoir

Kuncheng Li* & Jingxin Ruan*
Southwest Petroleum University, Chengdu, China

Jifu Ruan*
Chuanzhong Oil and Gas Mine of PetroChina Southwest Oil and Gas Field Company, Suining, China

Yuhang Liu*
Engineering Technology Research Institute of PetroChina Southwest Oil and Gas Field Company, Chengdu, China

Hong Luo*
Chuanzhong Oil and Gas Mine of PetroChina Southwest Oil and Gas Field Company, Suining, China

ABSTRACT: The early identification of water invasion risk and the prediction of water breakthrough time of gas wells are helpful to better formulate the gas reservoir development plan and reasonably arrange the daily management of gas wells. It is of great significance to adjust the technical countermeasures of gas reservoir development. This paper illustrates the water production characteristics and water invasion mode of M gas reservoir, identifies the water invasion risk wells by using the early water invasion identification method, verifies the applicability of the water breakthrough time prediction model of edge water gas reservoir in M gas reservoir by using its water breakthrough time, and predicts the water breakthrough time of water invasion risk wells. The research shows that the water production characteristics of M gas reservoir mainly include three types: the rapid rise of water production and water-gas ratio, the slow rise and gradual stability of water-gas ratio, and the low value of water production and water-gas ratio. There are four principal water invasion modes: edge water tongue (as the main mode), edge water finger, bottom water cone and sealed water tongue. The unstable well test method, water invasion volume coefficient method, and apparent geological reserve method can be used for the early identification of water invasion. The prediction model of water breakthrough time of gas wells in edge water gas reservoir is applicable to the prediction of water breakthrough time of M gas reservoir, and its prediction results provide a reference value for further investigation.

1 INTRODUCTION

Natural gas is an important low-carbon clean energy. Improving the development and utilization level of natural gas in China has strategic significance for the realization of China's double carbon goal (Chen 2021). The main purpose of gas field development is to maximize natural gas exploitation in the gas reservoir. For the edge and bottom water gas reservoir, with the progressing production of natural gas, the formation pressure decreases, and the edge and bottom water invades the gas reservoir to supplement the formation pressure until the water cone breaks through the bottom of the well, and the gas well produces a large amount of water and then stops production. Therefore, if the water invasion risk of gas wells can be identified early and the water breakthrough

*Corresponding Authors: 399606911@qq.com, 506077509@qq.com, ruanjifu@petrochina.com.cn, 1529517263@qq.com, luohong123@petrochina.com

 DOI 10.1201/9781003319412-63

time of gas wells can be predicted, the gas reservoir development plan can be formulated more reasonably, and the daily management of gas wells can be arranged relatively properly, which is of great significance to the adjustment of technical countermeasures for gas reservoir development (Feng 2015).

2 OVERVIEW OF GAS RESERVOIR

2.1 *Geological characteristics*

The structure of the M gas reservoir extends from northeast to East and is dominated by a carbonate platform sedimentary system. The lithology is mainly grain dolomite, sandy dolomite, oolitic dolomite, argillaceous limestone, and argillaceous dolomite mixed with a small amount of sandstone. The thickness remains unchanged, mainly fracture pore (hole) type and pore type. Through sampling and analysis, the M gas reservoir contains medium to low CO_2 and H_2S, which is characterized by high chloride ion content and high salinity. The middle pressure is 75.74~76.09 MPa, the pressure coefficient is 1.64~1.69, and the formation temperature is 138.5~145.8°C. It is a high-temperature and high-pressure gas reservoir.

2.2 *Water production characteristics*

There is a certain volume of edge water in the north and south of the M gas reservoir. The microfracture system of the reservoir is developed. The water production types include fractured water channeling and edge water advancing mode. Based on the analysis of the water-gas ratio change curve of the water production well (Figure 1), the water production characteristics of the water production well of the M gas reservoir fall principally into three categories as follows.

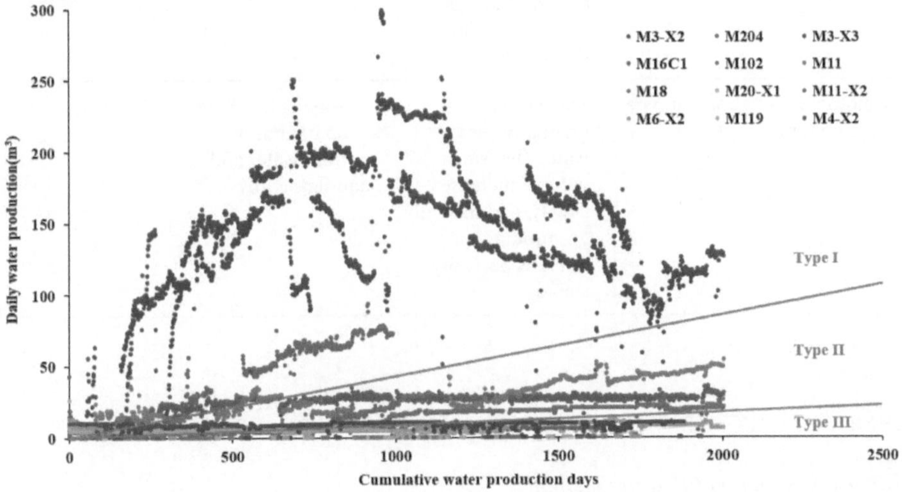

Figure 1. Water-gas ratio curve of a production well in M gas reservoir.

Type I: High yield water well, which is close to the water body, and the fracture directly connects the water area. After the gas well touches water, the water yield and water-gas ratio increase sharply to a high level.

Type II: After the water breakthrough in the gas well, the increase of water production is slower than that of the fractured water channeling well group, and the production period of a stable water zone is relatively long, with an obvious aggravation period of water invasion.

Type III: The gas well is relatively far away from the active water body. After water invasion, the daily water production and water-gas ratio rise slowly and remain at a low value all the way.

2.3 *Water invasion model*

Considering the water production source of the gas well (Ruan 2020), the dynamic change in water production of the gas well is observed. Four water invasion modes of M gas reservoir are derived: edge water finger type, edge water tongue type (as the main mode), bottom water coning type, and primary movable water tongue type. See Table 1 for the classification and statistics of well water invasion modes.

Table 1. Classification and statistics of water invasion modes of water producing gas wells.

Formation water	Mode	Features	Descriptions	Well number
Edge water	Finger type	The gas well suddenly sees water, the water volume increases rapidly, and the daily gas production shows a downward trend	Fractures are developed, and formation water rapidly flows along the fractures	M3–X2, M3–X3
	Tongue type		It is located at the edge of the well area and adjacent to the edge water or gas water transition area	M8, M10, M11, M16C1, M18, M102, M202, M205, M15–H1, M3–X1, M7–H1, M12–X1
Bottom water	Conical type	The gas well suddenly sees water, the water volume increases rapidly, and the daily gas production shows a downward trend	The well logging water absorption height shows that there is a water layer below	M204
Primary movable water in the bottom water layer	Tongue type	The gas well gradually sees water, the water volume increases slowly, and the daily gas production remains basically stable	Geologically, there is no boundary bottom water and aquifer nearby	M9-H9

3 IDENTIFICATION OF WATER INVASION IN THE GAS RESERVOIR

3.1 *Water invasion identification method*

Following the literature research to date, water invasion identification methods fall primarily into five types: production fluid identification, material balance identification, production data identification, well test analysis identification, and numerical simulation identification. Each water invasion identification method employs different principles, observation parameters, and data requirements, and their applicable objects and stages are also different. By comparing and analyzing the characteristics and limitations of ten water invasion identification methods (Table 2), it is found that the methods that can predict the early identification of water invasion (water invasion is identified before it reaches the wellbore), including H_2S content detection, water invasion volume coefficient, apparent geological reserves, unstable well test, and numerical simulation. At the same time, these early identification methods also have some limitations. For example, the H2S content

Table 2. Identification methods, characteristics, and limitations of water invasion.

Type	Method	Features	Limitations
Fluid production	Hydrochemical characteristic analysis	The total salinity, chloride ion content, and trace element content of produced water (formation water) increased significantly (Ruan 2020).	Unpredictability (Kang 2004). Regular sampling and determination are required.
	H_2S content detection	The H_2S content in the produced gas increases significantly, with certain predictability.	It has poor applicability to gas wells that produce H_2S in the early stage (Liu 2011).
Material balance	Apparent formation pressure	There is a nonlinear relationship (upturning) between apparent formation pressure and cumulative gas production (Xing 2015).	Unpredictability (Li 2016). It is not suitable for abnormally high-pressure gas reservoirs.
	Water invasion volume coefficient	The actual formation curve relative to apparent pressure and recovery degree is above the diagonal (Chen 1978), with certain predictability.	A regular static pressure test is required to affect the well opening rate (Deng 2020). Dynamic reserves are difficult to obtain in the early stage (Yuan 2020).
	Apparent geological reserves	There is a nonlinear relationship (upturning) between the apparent geological reserves and cumulative gas production (Liu 1999), with certain predictability.	A regular static pressure test is required to affect the well opening rate of gas wells (Deng 2020).
Production data	Water-gas ratio	The water-gas ratio will rise rapidly after water invasion.	Unpredictability.
	Production decline curve	The gas production decreased significantly, and the production water-gas ratio and water production increased (Zhang 2017).	Unpredictability. The data on gas well production and pressure should be accurate.
	Bottom hole flowing pressure	The bottom hole flow pressure has a curve relationship with time (Liu 2011).	Unpredictability. According to the inlet pressure, there is a large error in calculating the bottom hole pressure (Feng 2015).
Well test	Unstable well test	The pressure derivative curve is warped in the later stage (Li 2009; Sun 2011; Tao 2003), with certain predictability.	Multiple well tests are required in different production periods (Song 2019).
Numerical simulation	Numerical simulation	Simulate reality through the numerical model of Geology and production (Peng 2018), with certain predictability.	The model establishment and numerical solution are complex (Feng 2015). It is not convenient for field applications.

detection has poor applicability to gas wells that produce H2S in the early stage. The numerical simulation has heavy modeling and calculation workload and is not convenient for field application. When identifying the early water invasion risk of gas wells, the identification method will be comprehensively optimized according to the relevant static and dynamic data of gas wells, taking into account factors such as production economic benefits, and one or more methods can be selected for identification.

3.2 *Water invasion volume coefficient method*

For a constant volume closed gas reservoir, when there is no water invasion, the relationship curve between formation relative apparent pressure and recovery degree is a straight line with an inclination of 45°. When there is water invasion, the relationship curve between formation relative apparent pressure and recovery degree has an inclination of more than 45°. The water invasion volume coefficient method was used to identify the water invasion risk in the early production stage of well M9. The curve began to deviate from the diagonal and rise in September 2015 (Figure 2). It was determined that it began to be affected by water invasion. By June 2020, the cumulative water production had been $64.65\times10^4 m^3$.

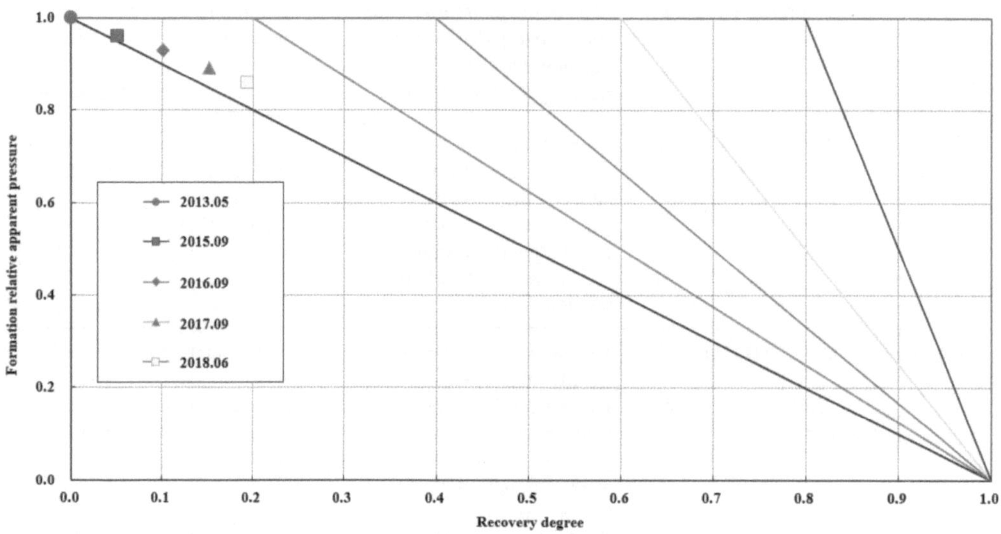

Figure 2. Analysis curve of water invasion volume coefficient method in well M9.

3.3 *Apparent geological reserve method*

For a constant volume closed gas reservoir, when there is no water invasion, the apparent geological reserve is equal to the original geological reserve; when there is water invasion, the apparent geological reserve is greater than the original geological reserve. For example, since September 2015, the apparent geological reserves of well M9 are above the original geological reserves (Figure 3), which are determined to be affected by water invasion.

3.4 *Unstable well test method*

The dynamic characteristic response of the water invasion process on the well test curve of a gas well is the change of boundary distance. With the advance of edge water, the radius of the gas reservoir decreases continuously. By arranging appropriate times of pressure recovery well tests and analyzing the time when the boundary reflection of the pressure derivative of the gas well in

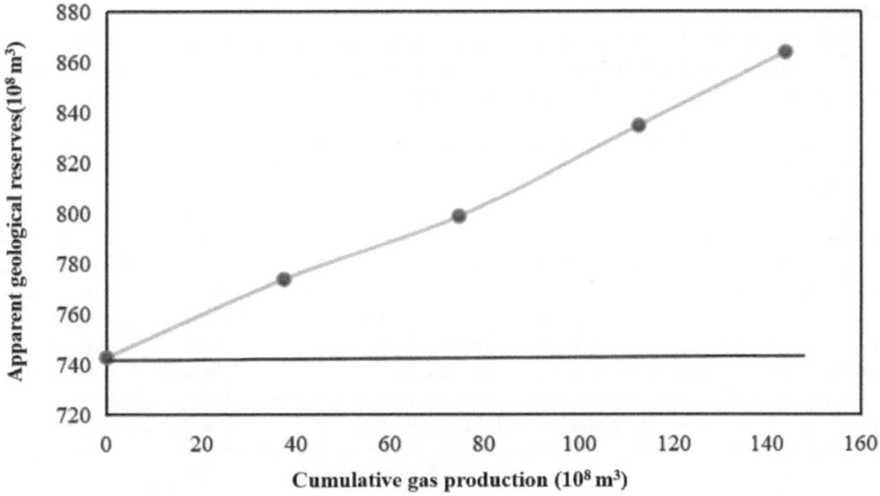

Figure 3. Apparent geological reserve analysis curve of well M9.

different periods, the advancing distance of the edge water in different periods can be calculated, so as to identify the water invasion of the gas reservoir (Li 2009; Song 2019). See Figure 4 for the interpretation and comparison curve of two pressure recovery well tests of well M8–H8. The effective horizontal section length, i.e., the radius of the gas reservoir, is reduced from 276 m to 269 m. It is analyzed that it is affected by water invasion, so it is judged to be a water invasion risk well.

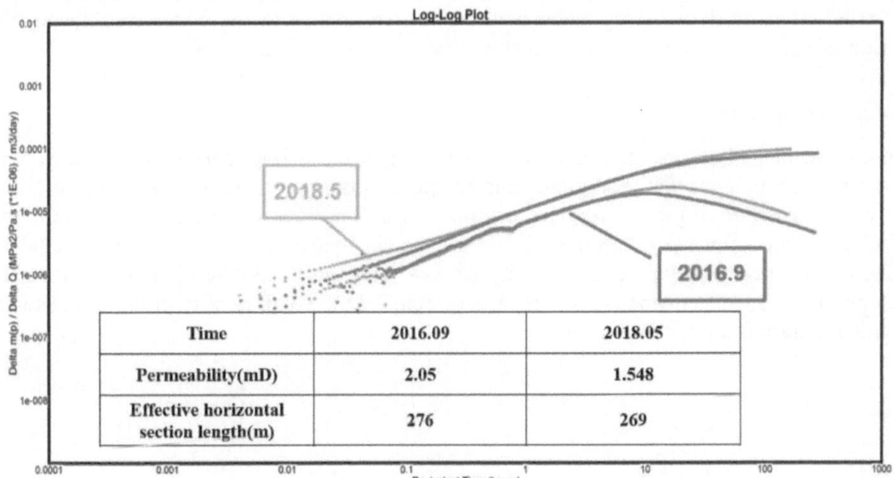

Figure 4. Analysis curve of the unstable well test method for well M8–H8.

3.5 *Identification of water invasion risk wells*

Using the above early identification methods of water invasion, combined with the geological characteristics, engineering characteristics, production, and inter-well connectivity of the M gas reservoir, the gas wells with a high possibility of water invasion are selected for water invasion risk identification based on the following factors: ① the distance between the perforation bottom boundary and – 4385 m, with the longitudinal distance less than 30; ② the distance between the plane and the gas-water interface, less than 1500 m; ③ the distance from the produced water well target, less than 3 km; ④ the fracture water invasion channel; ⑤ the well test interpretation well area with high permeability). The results (Table 3) show that well M7–X2, M18–X1, M8–X1, M17–X1, M11–X2, M116, and M8–H8 are water invasion risk wells.

Table 3. Identification results of water invasion risk wells in M gas reservoir.

Well number	Early identification method of water intrusion	Recognition results
M7–X2	Unstable well test, water invasion volume coefficient	There is a risk of water invasion
M18–X1	Water invasion volume coefficient	There is a risk of water invasion
M8–X1	Unstable well test, water invasion volume coefficient	There is a risk of water invasion
M17–X1	Unstable well test, water invasion volume coefficient	There is a risk of water invasion
M11–X2	Unstable well test, water invasion volume coefficient	There is a risk of water invasion
M116	Water invasion volume coefficient, apparent geological reserve	There is a risk of water invasion
M8–H8	Unstable well test, water invasion volume coefficient	There is a risk of water invasion

4 PREDICTION OF WATER BREAKTHROUGH TIME

There are many studies on calculating the water breakthrough time of gas wells. Ming Ruiqing et al. compared and analyzed the applicable conditions and applicability of different model methods for predicting the water breakthrough time (Ming 2018; Wang 2008; Wu 2011). According to the non-condensate gas reservoir and other geological characteristics of M gas reservoir, based on the theoretical method of water breakthrough time prediction of gas wells in edge water gas reservoir proposed by Wang Huiqiang (Wang 2008), combined with the reservoir physical parameters of well test interpretation, the prediction formula of water breakthrough time of edge water is fitted and verified by using the existing water wells, and then applied to the water breakthrough time prediction of risk wells. The calculation results have a certain reference value.

4.1 *Model characterization*

Assuming that there is a production well B near the gas water boundary of one side water gas reservoir, the irregular gas water boundary can be approximately considered a straight line, so as to simplify the derivation of the calculation formula of water breakthrough time of gas well. Taking the initial gas-water interface position as the r-axis, the distance between the gas well and the r-axis is a. There is a water quality point at the original gas-water interface r for plane radial flow, and it reaches point A (x, y) after time t. The seepage velocity of the water quality point is set as v_g. The schematic diagram is shown in Figure 5.

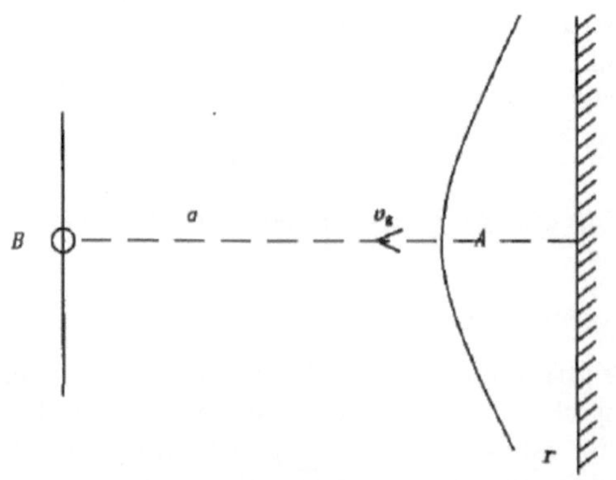

Figure 5. Schematic diagram of single production well at the edge of linear supply of edge water gas reservoir (Wang 2008).

Assumption: the inclined reservoir is equal in thickness and homogeneous. In the process of gas-water two-phase seepage, it is displaced in the form of a piston, and residual gas is formed in the water invasion area. Ignoring the influence of capillary force and gravity, viscous force controls the flow of gas and water in the formation. The fluid is slightly compressible, and the density and viscosity of the gas phase and liquid phase are fixed values. Both air and water flow obey Darcy's law.

According to the derivation and calculation in reference (Wang 2008), the water breakthrough time of gas well in edge water gas reservoir is

$$t_{bt} = M_{gw} \frac{\pi h \phi (1 - S_{wi} - S_{gr}) a^2}{q_g B_g} \tag{1}$$

Therein

$$M_{gw} = \frac{K_{gwi}/\mu_g}{K_{wgr}/\mu_w} \tag{2}$$

Where, t_{bt} is the water breakthrough time (water cone breakthrough) of a gas well in an edge water gas reservoir, a; M_{gw} is the gas water mobility ratio, decimal; h is the effective thickness of the production layer, m; ϕ is the reservoir porosity, decimal; S_{wi} and S_{gr} are the original irreducible water saturation and the residual gas saturation, percentage; a is the distance from the gas well to the edge water, m; q_g is the gas well production, m^3/d; B_g is the volume factor of natural gas, decimal; K_{gwi} is the gas phase permeability under S_{wi} and K_{wgr} is the water phase permeability under S_{gr}, μm^2; μ_g and μ_w are the viscosity of gas phase and the water phase, mPa•s.

4.2 *Model validation*

Well M8 is selected as an example. The physical parameters of the well are shown in Table 4, and the edge water breakthrough time is calculated.

Table 4. Physical parameters of well M8.

Well number	Average porosity (%)	Average permeability (mD)	Gas phase viscosity (mPa•s)	Aqueous viscosity (mPa•s)	Effective thickness of reservoir (m)	Original gas volume coefficient
M8	5	530	0.02875 (Check chart)	0.2113 (Petrel model)	49.6	0.0026

According to the relative permeability curve (Figure 6), the irreducible water saturation is 0.418, and the relative permeability of the gas phase is 0.356. The residual gas saturation is 0.054, and the relative permeability of the water phase is 0.854. The plane distance between M8 well and the gas water boundary is about 900 m, and the average daily output before the water breakthrough is $50\times10^4 m^3$. According to the prediction formula of the water breakthrough time of gas wells in edge water gas reservoir, the water breakthrough time is calculated to be four years, while the actual water breakthrough time of an M8 well is 4.6 years. The gas production curve of the M8 well is shown in Figure 7.

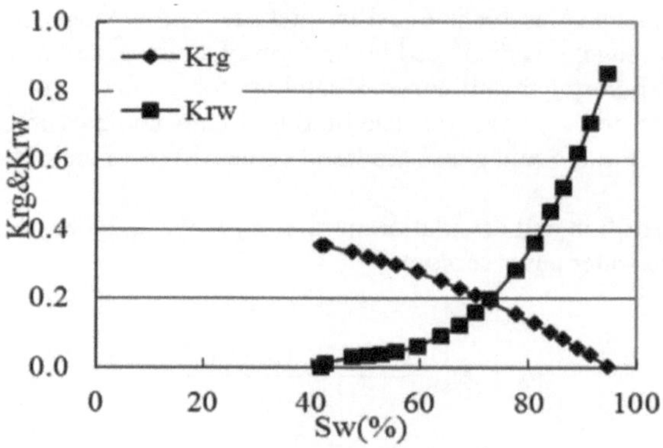

Figure 6. Gas water phase permeability curve of a rock sample of M gas reservoir (taken from well M13).

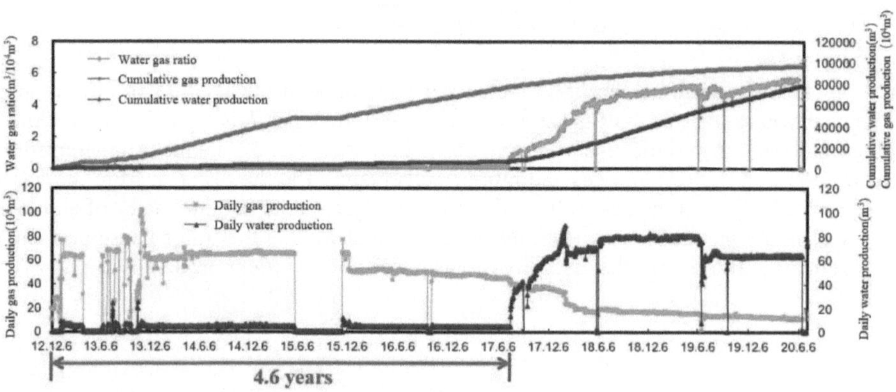

Figure 7. Gas production curve of well M8.

This method predicts the water breakthrough time of ten wells with water breakthrough. The difference between the predicted results and the actual water breakthrough time is less than one year (Table 5), indicating that this calculation method is applicable in M gas reservoir and the predicted results have a certain reference value.

Table 5. Comparison between predicted water breakthrough time and actual water breakthrough time of some production wells.

Well number	Porosity (%)	Effective thickness (m)	Edge water distance (m)	Gas production (10^4m^3)	Production time	Water breakthrough time	Actual years (a)	Predicted years (a)	Gap (a)
M12-X1	3.8	46.5	500	67	2016.04.21	2020.01.23	3.75	3.9	+0.15
M15-H1	3.5	29	550	68	2015.11.13	2018.11.10	3	2.67	−0.33
M7-H1	4.1	38	500	73.1	2014.11.12	2019.01.01	4.05	3.16	−0.89
M8-H3	3.1	14.43	250	25	2016.05.27	2016.10.18	0.45	0.66	+0.21
M8-X2	3	24.2	450	49.7	2016.10.21	2017.09.02	0.9	1.75	+0.85
M10	4.5	42	300	74.4	2013.10.31	2014.11.24	1.1	1.35	+0.25
M102	4.8	41.5	850	47	2018.05.09	2020.06.16	2.1	3.07	+0.93
M16C1	5.5	52.5	100	65	2014.12.25	2015.04.21	0.33	0.26	−0.07
M18	4	14.5	235	24.2	2014.09.27	2015.10.11	1	0.79	−0.21
M8	5	49.6	350	60	2012.12.06	2017.07.21	4.6	4	−0.6

4.3 *Prediction of water breakthrough time of risk wells*

This method is used to predict the water breakthrough time of water invasion risk wells in the M gas reservoir. The prediction results are shown in Table 6.

Table 6. Prediction of water breakthrough time of water production risk wells.

Well number	Porosity (%)	Effective thickness of reservoir (m)	Edge water distance (m)	Gas production (10^4m^3)	Production time	Predicted water breakthrough time
M7-X2	3.5	37.4	1000	60	2015.12.13	2027.11
M18-X1	4.1	30.4	1000	72	2015.12.18	2026.02
M8-X1	3.8	45.5	700	81	2016.03.01	2022.05
M17-X1	4.9	56.5	1100	117	2015.08.04	2032.08
M11-X2	3.2	21.7	980	76	2016.11.19	2022.02
M116	5.8	33.5	735	19	2018.11.30	2022.03
M8-H8	3.7	36.1	1060	41	2014.10.30	2031.06

5 CONCLUSION

(1) The water production characteristics of M gas reservoir mainly include three types: the rapid rise of water production and water-gas ratio, the slow rise and gradual stability of water-gas ratio, and the low value of water production and water-gas ratio.
(2) There are four principal water invasion modes: edge water tongue(as the main mode), edge water finger, bottom water cone, and sealed water tongue.
(3) There are mainly five predictive early identification methods of water invasion: H_2S content detection, water invasion volume coefficient, apparent geological reserves, unstable well test, and numerical simulation, each of which has its applicability and limitations. Unstable well tests, water invasion volume coefficients, and apparent geological reserves can be used for the early identification of water invasion in M gas reservoirs.
(4) The applicability of the prediction model of water breakthrough time of gas wells in edge water gas reservoir in M gas reservoir is verified by using the actual water breakthrough time of produced wells. It is identified that M7-X2, M18-X1, M8-X1, M17-X1, M11-X2, M116, and M8-H8 wells are water invasion risk wells, and their water breakthrough time is predicted.

REFERENCES

Chen Junhua, Li Qiaochu, He Jing Regional differences of low carbon efficiency in Sichuan Province under the goal of carbon neutralization [J] Natural gas industry, 2021,41 (06): 162–170.
Chen Yuanqian Judgment method of gas field natural water invasion [J] Petroleum exploration and development, 1978 (03): 51–57.
Deng Chenggang, Li Jiangtao, Chai Xiaoying, et al. Early identification method of water invasion in weak water drive gas reservoir in Sebei gas field [J] Lithologic reservoir, 2020, 32 (01): 128–134.
Feng Xi, Zhong Bing, Yang Xuefeng, et al. Problems and understanding of effectively controlling the impact of water invasion in the process of gas reservoir development [J] Natural gas industry, 2015, 35 (02): 35–40.
Kang Xiaodong, Li Xiangfang, Zhang Guosong Identification method of early water invasion in gas reservoir [J] Geoscience, 2004.7–6306.
Liu Yun Research on early identification method and calculation of water influx in gas reservoir [D] Chengdu University of Technology, 2011.
Li Xinhong, Gong Ning, song Linke, et al. Application of early water invasion identification method in Moxi Longwangmiao gas reservoir [C]. Proceedings of 2016 National Natural Gas academic annual conference. 2016: 1033–1040

Liu Shuzhi, Huang Bingguang, Li Daoxuan. Comparison and Discussion on identification methods of water drive gas reservoir [J] Natural gas industry, 1999 (04): 50–53
Li Xiaoping, Zhang Liehui, Li Yun. Application of unstable seepage theory in water drive gas reservoir water invasion identification [J] Journal of applied basic and Engineering Sciences, 2009,17 (3): 10.
Ming Ruiqing, He Huiqun, Hu Qiangfa, et al. A new model for predicting water breakthrough time of high-yield wells in edge water condensate gas reservoir [J] Special oil and gas reservoir, 2018, 25 (2): 4.
Peng Han, Zhao Xian Qian, et al. Research on fine numerical simulation of super large gas reservoir based on fine geological modeling — Taking Moxi Longwangmiao formation gas reservoir as an example [C]. 2018 National Natural Gas academic annual meeting.
Ruan Jifu, Ou Jiaqiang, Zhou Lianlian, et al. Water invasion mode, and key water control technology of Longwangmiao formation gas reservoir in Moxi [C] / / Proceedings of the 32nd National Natural Gas Academic Annual Conference (2020). 2020: 777–785 DOI:10.26914/c.cnkihy. 2020.064945.
Sun Hedong Well test curve characteristics and typical case analysis of gas wells in water drive gas reservoirs [C] National Conference on seepage mechanics. 2011.
Song Gangxiang Application of water invasion identification technology in water drive gas field [J] Special oil and gas reservoir, 2019,26 (6): 4.
Tao Shiping, Feng Xi, Xiao Shihong. Application of unstable well test analysis method to identify early water invasion of gas reservoir [J] Natural gas industry, 2003, 23 (4): 3.
Wang Huiqiang, Li Xiaoping, Wu Feng, et al. Prediction method of water breakthrough time of gas well in edge water gas reservoir [J] Special oil and gas reservoir, 2008,15 (4): 3.
Wu Keliu, Li Xiangfang, Zhang Gongtao, et al. A new method for predicting water breakthrough time of edge water condensate gas reservoir considering reverse condensation [J] Science, technology and engineering, 2011,11 (19): 4.
Xing Dongping Study on early water invasion identification method and technology of Heba Feisan water-bearing gas reservoir [J] China chemical trade, 2015, (25): 13–14.
Yuan Lin, Chen Xi, Jiang Linxi, et al. Optimization and application of early water invasion identification method in reef beach gas reservoir [J] Sino foreign energy, 2020,25 (07): 36–43
Zhang Mingdi, Zhao Yong, Wang Bencheng, et al. Early identification and Adjustment Countermeasures of water invasion in reef gas reservoir of Changxing formation [J] Fault block oil and gas field, 2017,24 (06): 800–804

Advances in Mineral Resources, Geotechnology and Geological Exploration – Rashid & Zhang (eds)
© 2023 Copyright the Editor(s), ISBN 978-1-032-33377-9

Author index